Lecture Notes in Computer Science

Lecture Notes in Artificial Intelligence 14649

Founding Editor

Jörg Siekmann

Series Editors

Randy Goebel, *University of Alberta, Edmonton, Canada*
Wolfgang Wahlster, *DFKI, Berlin, Germany*
Zhi-Hua Zhou, *Nanjing University, Nanjing, China*

The series Lecture Notes in Artificial Intelligence (LNAI) was established in 1988 as a topical subseries of LNCS devoted to artificial intelligence.

The series publishes state-of-the-art research results at a high level. As with the LNCS mother series, the mission of the series is to serve the international R & D community by providing an invaluable service, mainly focused on the publication of conference and workshop proceedings and postproceedings.

De-Nian Yang · Xing Xie · Vincent S. Tseng ·
Jian Pei · Jen-Wei Huang · Jerry Chun-Wei Lin
Editors

Advances in Knowledge Discovery and Data Mining

28th Pacific-Asia Conference on Knowledge Discovery and Data Mining, PAKDD 2024
Taipei, Taiwan, May 7–10, 2024
Proceedings, Part V

Editors
De-Nian Yang
Academia Sinica
Taipei, Taiwan

Xing Xie
Microsoft Research Asia
Beijing, China

Vincent S. Tseng
National Yang Ming Chiao Tung University
Hsinchu, Taiwan

Jian Pei
Duke University
Durham, NC, USA

Jen-Wei Huang
National Cheng Kung University
Tainan, Taiwan

Jerry Chun-Wei Lin
Silesian University of Technology
Gliwice, Poland

ISSN 0302-9743 ISSN 1611-3349 (electronic)
Lecture Notes in Artificial Intelligence
ISBN 978-981-97-2264-8 ISBN 978-981-97-2262-4 (eBook)
https://doi.org/10.1007/978-981-97-2262-4

LNCS Sublibrary: SL7 – Artificial Intelligence

This Springer imprint is published by the registered company Springer Nature Singapore Pte Ltd.
The registered company address is: 152 Beach Road, #21-01/04 Gateway East, Singapore 189721, Singapore

Paper in this product is recyclable.

General Chairs' Preface

On behalf of the Organizing Committee, we were delighted to welcome attendees to the 28th Pacific-Asia Conference on Knowledge Discovery and Data Mining (PAKDD 2024). Since its inception in 1997, PAKDD has long established itself as one of the leading international conferences on data mining and knowledge discovery. PAKDD provides an international forum for researchers and industry practitioners to share their new ideas, original research results, and practical development experiences across all areas of Knowledge Discovery and Data Mining (KDD). This year, after its two previous editions in Taipei (2002) and Tainan (2014), PAKDD was held in Taiwan for the third time in the fascinating city of Taipei, during May 7–10, 2024. Moreover, PAKDD 2024 was held as a fully physical conference since the COVID-19 pandemic was contained.

We extend our sincere gratitude to the researchers who submitted their work to the PAKDD 2024 main conference, high-quality tutorials, and workshops on cutting-edge topics. The conference program was further enriched with seven high-quality tutorials and five workshops on cutting-edge topics. We would like to deliver our sincere thanks for their efforts in research, as well as in preparing high-quality presentations. We also express our appreciation to all the collaborators and sponsors for their trust and cooperation. We were honored to have three distinguished keynote speakers joining the conference: Ed H. Chi (Google DeepMind), Vipin Kumar (University of Minnesota), and Huan Liu (Arizona State University), each with high reputations in their respective areas. We enjoyed their participation and talks, which made the conference one of the best academic platforms for knowledge discovery and data mining. We would like to express our sincere gratitude for the contributions of the Steering Committee members, Organizing Committee members, Program Committee members, and anonymous reviewers, led by Program Committee Chairs De-Nian Yang and Xing Xie. It is through their untiring efforts that the conference had an excellent technical program. We are also thankful to the other Organizing Committee members: Workshop Chairs, Chuan-Kang Ting and Xiaoli Li; Tutorial Chairs, Jiun-Long Huang and Philippe Fournier-Viger; Publicity Chairs, Mi-Yen Yeh and Rage Uday Kiran; Industrial Chairs, Kun-Ta Chuang, Wei-Chao Chen and Richie Tsai; Proceedings Chairs, Jen-Wei Huang and Jerry Chun-Wei Lin; Registration Chairs, Chih-Ya Shen and Hong-Han Shuai; Web and Content Chairs, Cheng-Te Li and Shan-Hung Wu; Local Arrangement Chairs, Yi-Ling Chen, Kuan-Ting Lai, Yi-Ting Chen, and Ya-Wen Teng. We feel indebted to the PAKDD Steering Committee for their constant guidance and sponsorship of manuscripts. We are also grateful to the hosting organizations, National Yang Ming Chiao Tung University and Academia Sinica, and all our sponsors for continuously providing institutional and financial support to PAKDD 2024.

May 2024

Vincent S. Tseng
Jian Pei

PC Chairs' Preface

It is our great pleasure to present the 28th Pacific-Asia Conference on Knowledge Discovery and Data Mining (PAKDD 2024) as Program Committee Chairs. PAKDD is one of the longest-established and leading international conferences in the areas of data mining and knowledge discovery. It provides an international forum for researchers and industry practitioners to share their new ideas, original research results, and practical development experiences in all KDD-related areas, including data mining, data warehousing, machine learning, artificial intelligence, databases, statistics, knowledge engineering, big data technologies, and foundations.

This year, PAKDD received a record number of 720 submissions, among which 86 submissions were rejected at a preliminary stage due to policy violations. There were 595 Program Committee members and 101 Senior Program Committee members involved in the double-blind reviewing process. For submissions entering the double-blind review process, each one received at least three quality reviews from PC members. Furthermore, each valid submission received one meta-review from the assigned SPC member, who also led the discussion with the PC members. The PC Co-chairs then considered the recommendations and meta-reviews from SPC members and looked into each submission as well as its reviews and PC discussions to make the final decision.

As a result of the highly competitive selection process, 175 submissions were accepted and recommended to be published, with 133 oral-presentation papers and 42 poster-presentation papers. We would like to thank all SPC and PC members whose diligence produced a high-quality program for PAKDD 2024. The conference program also featured three keynote speeches from distinguished data mining researchers, eight invited industrial talks, five cutting-edge workshops, and seven comprehensive tutorials.

We wish to sincerely thank all SPC members, PC members, and external reviewers for their invaluable efforts in ensuring a timely, fair, and highly effective paper review and selection procedure. We hope that readers of the proceedings will find the PAKDD 2024 technical program both interesting and rewarding.

May 2024

De-Nian Yang
Xing Xie

Organization

Organizing Committee

Honorary Chairs

Philip S. Yu	University of Illinois at Chicago, USA
Ming-Syan Chen	National Taiwan University, Taiwan

General Chairs

Vincent S. Tseng	National Yang Ming Chiao Tung University, Taiwan
Jian Pei	Duke University, USA

Program Committee Chairs

De-Nian Yang	Academia Sinica, Taiwan
Xing Xie	Microsoft Research Asia, China

Workshop Chairs

Chuan-Kang Ting	National Tsing Hua University, Taiwan
Xiaoli Li	A*STAR, Singapore

Tutorial Chairs

Jiun-Long Huang	National Yang Ming Chiao Tung University, Taiwan
Philippe Fournier-Viger	Shenzhen University, China

Publicity Chairs

Mi-Yen Yeh	Academia Sinica, Taiwan
Rage Uday Kiran	University of Aizu, Japan

Industrial Chairs

Kun-Ta Chuang	National Cheng Kung University, Taiwan
Wei-Chao Chen	Inventec Corp./Skywatch Innovation, Taiwan
Richie Tsai	Taiwan AI Academy, Taiwan

Proceedings Chairs

Jen-Wei Huang	National Cheng Kung University, Taiwan
Jerry Chun-Wei Lin	Silesian University of Technology, Poland

Registration Chairs

Chih-Ya Shen	National Tsing Hua University, Taiwan
Hong-Han Shuai	National Yang Ming Chiao Tung University, Taiwan

Web and Content Chairs

Shan-Hung Wu	National Tsing Hua University, Taiwan
Cheng-Te Li	National Cheng Kung University, Taiwan

Local Arrangement Chairs

Yi-Ling Chen	National Taiwan University of Science and Technology, Taiwan
Kuan-Ting Lai	National Taipei University of Technology, Taiwan
Yi-Ting Chen	National Yang Ming Chiao Tung University, Taiwan
Ya-Wen Teng	Academia Sinica, Taiwan

Steering Committee

Chair

Longbing Cao	Macquarie University, Australia

Vice Chair

Gill Dobbie	University of Auckland, New Zealand

Treasurer

Longbing Cao	Macquarie University, Australia

Members

Ramesh Agrawal	Jawaharlal Nehru University, India
Gill Dobbie	University of Auckland, New Zealand
João Gama	University of Porto, Portugal
Zhiguo Gong	University of Macau, Macau SAR
Hisashi Kashima	Kyoto University, Japan
Hady W. Lauw	Singapore Management University, Singapore
Jae-Gil Lee	KAIST, Korea
Dinh Phung	Monash University, Australia
Kyuseok Shim	Seoul National University, Korea
Geoff Webb	Monash University, Australia
Raymond Chi-Wing Wong	Hong Kong University of Science and Technology, Hong Kong SAR
Min-Ling Zhang	Southeast University, China

Life Members

Longbing Cao	Macquarie University, Australia
Ming-Syan Chen	National Taiwan University, Taiwan
David Cheung	University of Hong Kong, China
Joshua Z. Huang	Chinese Academy of Sciences, China
Masaru Kitsuregawa	Tokyo University, Japan
Rao Kotagiri	University of Melbourne, Australia
Ee-Peng Lim	Singapore Management University, Singapore
Huan Liu	Arizona State University, USA
Hiroshi Motoda	AFOSR/AOARD and Osaka University, Japan
Jian Pei	Duke University, USA
P. Krishna Reddy	IIIT Hyderabad, India
Jaideep Srivastava	University of Minnesota, USA
Thanaruk Theeramunkong	Thammasat University, Thailand
Tu-Bao Ho	JAIST, Japan
Vincent S. Tseng	National Yang Ming Chiao Tung University, Taiwan
Takashi Washio	Osaka University, Japan
Kyu-Young Whang	KAIST, Korea
Graham Williams	Australian National University, Australia
Chengqi Zhang	University of Technology Sydney, Australia

Ning Zhong	Maebashi Institute of Technology, Japan
Zhi-Hua Zhou	Nanjing University, China

Past Members

Arbee L. P. Chen	Asia University, Taiwan
Hongjun Lu	Hong Kong University of Science and Technology, Hong Kong SAR
Takao Terano	Tokyo Institute of Technology, Japan

Senior Program Committee

Aijun An	York University, Canada
Aris Anagnostopoulos	Sapienza Università di Roma, Italy
Ting Bai	Beijing University of Posts and Telecommunications, China
Elisa Bertino	Purdue University, USA
Arnab Bhattacharya	IIT Kanpur, India
Albert Bifet	Université Paris-Saclay, France
Ludovico Boratto	Università degli Studi di Cagliari, Italy
Ricardo Campello	University of Southern Denmark, Denmark
Longbing Cao	University of Technology Sydney, Australia
Tru Cao	UTHealth, USA
Tanmoy Chakraborty	IIT Delhi, India
Jeffrey Chan	RMIT University, Australia
Pin-Yu Chen	IBM T. J. Watson Research Center, USA
Bin Cui	Peking University, China
Anirban Dasgupta	IIT Gandhinagar, India
Wei Ding	University of Massachusetts Boston, USA
Eibe Frank	University of Waikato, New Zealand
Chen Gong	Nanjing University of Science and Technology, China
Jingrui He	UIUC, USA
Tzung-Pei Hong	National University of Kaohsiung, Taiwan
Qinghua Hu	Tianjin University, China
Hong Huang	Huazhong University of Science and Technology, China
Jen-Wei Huang	National Cheng Kung University, Taiwan
Tsuyoshi Ide	IBM T. J. Watson Research Center, USA
Xiaowei Jia	University of Pittsburgh, USA
Zhe Jiang	University of Florida, USA

Toshihiro Kamishima	National Institute of Advanced Industrial Science and Technology, Japan
Murat Kantarcioglu	University of Texas at Dallas, USA
Hung-Yu Kao	National Cheng Kung University, Taiwan
Kamalakar Karlapalem	IIIT Hyderabad, India
Anuj Karpatne	Virginia Tech, USA
Hisashi Kashima	Kyoto University, Japan
Sang-Wook Kim	Hanyang University, Korea
Yun Sing Koh	University of Auckland, New Zealand
Hady Lauw	Singapore Management University, Singapore
Byung Suk Lee	University of Vermont, USA
Jae-Gil Lee	KAIST, Korea
Wang-Chien Lee	Pennsylvania State University, USA
Chaozhuo Li	Microsoft Research Asia, China
Gang Li	Deakin University, Australia
Jiuyong Li	University of South Australia, Australia
Jundong Li	University of Virginia, USA
Ming Li	Nanjing University, China
Sheng Li	University of Virginia, USA
Ying Li	AwanTunai, Singapore
Yu-Feng Li	Nanjing University, China
Hao Liao	Shenzhen University, China
Ee-peng Lim	Singapore Management University, Singapore
Jerry Chun-Wei Lin	Silesian University of Technology, Poland
Shou-De Lin	National Taiwan University, Taiwan
Hongyan Liu	Tsinghua University, China
Wei Liu	University of Technology Sydney, Australia
Chang-Tien Lu	Virginia Tech, USA
Yuan Luo	Northwestern University, USA
Wagner Meira Jr.	UFMG, Brazil
Alexandros Ntoulas	University of Athens, Greece
Satoshi Oyama	Nagoya City University, Japan
Guansong Pang	Singapore Management University, Singapore
Panagiotis Papapetrou	Stockholm University, Sweden
Wen-Chih Peng	National Yang Ming Chiao Tung University, Taiwan
Dzung Phan	IBM T. J. Watson Research Center, USA
Uday Rage	University of Aizu, Japan
Rajeev Raman	University of Leicester, UK
P. Krishna Reddy	IIIT Hyderabad, India
Thomas Seidl	LMU München, Germany
Neil Shah	Snap Inc., USA

Yingxia Shao	Beijing University of Posts and Telecommunications, China
Victor S. Sheng	Texas Tech University, USA
Kyuseok Shim	Seoul National University, Korea
Arlei Silva	Rice University, USA
Jaideep Srivastava	University of Minnesota, USA
Masashi Sugiyama	RIKEN/University of Tokyo, Japan
Ju Sun	University of Minnesota, USA
Jiliang Tang	Michigan State University, USA
Hanghang Tong	UIUC, USA
Ranga Raju Vatsavai	North Carolina State University, USA
Hao Wang	Nanyang Technological University, Singapore
Hao Wang	Xidian University, China
Jianyong Wang	Tsinghua University, China
Tim Weninger	University of Notre Dame, USA
Raymond Chi-Wing Wong	Hong Kong University of Science and Technology, Hong Kong SAR
Jia Wu	Macquarie University, Australia
Xindong Wu	Hefei University of Technology, China
Xintao Wu	University of Arkansas, USA
Yiqun Xie	University of Maryland, USA
Yue Xu	Queensland University of Technology, Australia
Lina Yao	University of New South Wales, Australia
Han-Jia Ye	Nanjing University, China
Mi-Yen Yeh	Academia Sinica, Taiwan
Hongzhi Yin	University of Queensland, Australia
Min-Ling Zhang	Southeast University, China
Ping Zhang	Ohio State University, USA
Zhao Zhang	Hefei University of Technology, China
Zhongfei Zhang	Binghamton University, USA
Xiangyu Zhao	City University of Hong Kong, Hong Kong SAR
Yanchang Zhao	CSIRO, Australia
Jiayu Zhou	Michigan State University, USA
Xiao Zhou	Renmin University of China, China
Xiaofang Zhou	Hong Kong University of Science and Technology, Hong Kong SAR
Feida Zhu	Singapore Management University, Singapore
Fuzhen Zhuang	Beihang University, China

Program Committee

Zubin Abraham	Robert Bosch, USA
Pedro Henriques Abreu	CISUC, Portugal
Muhammad Abulaish	South Asian University, India
Bijaya Adhikari	University of Iowa, USA
Karan Aggarwal	Amazon, USA
Chowdhury Farhan Ahmed	University of Dhaka, Bangladesh
Ulrich Aïvodji	ÉTS Montréal, Canada
Esra Akbas	Georgia State University, USA
Shafiq Alam	Massey University Auckland, New Zealand
Giuseppe Albi	Università degli Studi di Pavia, Italy
David Anastasiu	Santa Clara University, USA
Xiang Ao	Chinese Academy of Sciences, China
Elena-Simona Apostol	Uppsala University, Sweden
Sunil Aryal	Deakin University, Australia
Jees Augustine	Microsoft, USA
Konstantin Avrachenkov	Inria, France
Goonmeet Bajaj	Ohio State University, USA
Jean Paul Barddal	PUCPR, Brazil
Srikanta Bedathur	IIT Delhi, India
Sadok Ben Yahia	University of Southern Denmark, Denmark
Alessandro Berti	Università di Pisa, Italy
Siddhartha Bhattacharyya	University of Illinois at Chicago, USA
Ranran Bian	University of Sydney, Australia
Song Bian	Chinese University of Hong Kong, Hong Kong SAR
Giovanni Maria Biancofiore	Politecnico di Bari, Italy
Fernando Bobillo	University of Zaragoza, Spain
Adrian M. P. Brasoveanu	Modul Technology GmbH, Austria
Krisztian Buza	Budapest University of Technology and Economics, Hungary
Luca Cagliero	Politecnico di Torino, Italy
Jean-Paul Calbimonte	University of Applied Sciences and Arts Western Switzerland, Switzerland
K. Selçuk Candan	Arizona State University, USA
Fuyuan Cao	Shanxi University, China
Huiping Cao	New Mexico State University, USA
Jian Cao	Shanghai Jiao Tong University, China
Yan Cao	University of Texas at Dallas, USA
Yang Cao	Hokkaido University, Japan
Yuanjiang Cao	Macquarie University, Australia

Sharma Chakravarthy	University of Texas at Arlington, USA
Harry Kai-Ho Chan	University of Sheffield, UK
Zhangming Chan	Alibaba Group, China
Snigdhansu Chatterjee	University of Minnesota, USA
Mandar Chaudhary	eBay, USA
Chen Chen	University of Virginia, USA
Chun-Hao Chen	National Kaohsiung University of Science and Technology, Taiwan
Enhong Chen	University of Science and Technology of China, China
Fanglan Chen	Virginia Tech, USA
Feng Chen	University of Texas at Dallas, USA
Hongyang Chen	Zhejiang Lab, China
Jia Chen	University of California Riverside, USA
Jinjun Chen	Swinburne University of Technology, Australia
Lingwei Chen	Wright State University, USA
Ping Chen	University of Massachusetts Boston, USA
Shang-Tse Chen	National Taiwan University, Taiwan
Shengyu Chen	University of Pittsburgh, USA
Songcan Chen	Nanjing University of Aeronautics and Astronautics, China
Tao Chen	China University of Geosciences, China
Tianwen Chen	Hong Kong University of Science and Technology, Hong Kong SAR
Tong Chen	University of Queensland, Australia
Weitong Chen	University of Adelaide, Australia
Yi-Hui Chen	Chang Gung University, Taiwan
Yile Chen	Nanyang Technological University, Singapore
Yi-Ling Chen	National Taiwan University of Science and Technology, Taiwan
Yi-Shin Chen	National Tsing Hua University, Taiwan
Yi-Ting Chen	National Yang Ming Chiao Tung University, Taiwan
Zheng Chen	Osaka University, Japan
Zhengzhang Chen	NEC Laboratories America, USA
Zhiyuan Chen	UMBC, USA
Zhong Chen	Southern Illinois University, USA
Peng Cheng	East China Normal University, China
Abdelghani Chibani	Université Paris-Est Créteil, France
Jingyuan Chou	University of Virginia, USA
Lingyang Chu	McMaster University, Canada
Kun-Ta Chuang	National Cheng Kung University, Taiwan

Robert Churchill	Georgetown University, USA
Chaoran Cui	Shandong University of Finance and Economics, China
Alfredo Cuzzocrea	Università della Calabria, Italy
Bi-Ru Dai	National Taiwan University of Science and Technology, Taiwan
Honghua Dai	Zhengzhou University, China
Claudia d'Amato	University of Bari, Italy
Chuangyin Dang	City University of Hong Kong, China
Mrinal Das	IIT Palakkad, India
Debanjan Datta	Virginia Tech, USA
Cyril de Runz	Université de Tours, France
Jeremiah Deng	University of Otago, New Zealand
Ke Deng	RMIT University, Australia
Zhaohong Deng	Jiangnan University, China
Anne Denton	North Dakota State University, USA
Shridhar Devamane	KLE Institute of Technology, India
Djellel Difallah	New York University, USA
Ling Ding	Tianjin University, China
Shifei Ding	China University of Mining and Technology, China
Yao-Xiang Ding	Zhejiang University, China
Yifan Ding	University of Notre Dame, USA
Ying Ding	University of Texas at Austin, USA
Lamine Diop	EPITA, France
Nemanja Djuric	Aurora Innovation, USA
Gillian Dobbie	University of Auckland, New Zealand
Josep Domingo-Ferrer	Universitat Rovira i Virgili, Spain
Bo Dong	Amazon, USA
Yushun Dong	University of Virginia, USA
Bo Du	Wuhan University, China
Silin Du	Tsinghua University, China
Jiuding Duan	Allianz Global Investors, Japan
Lei Duan	Sichuan University, China
Walid Durani	LMU München, Germany
Sourav Dutta	Huawei Research Centre, Ireland
Mohamad El-Hajj	MacEwan University, Canada
Ya Ju Fan	Lawrence Livermore National Laboratory, USA
Zipei Fan	Jilin University, China
Majid Farhadloo	University of Minnesota, USA
Fabio Fassetti	Università della Calabria, Italy
Zhiquan Feng	National Cheng Kung University, Taiwan

Len Feremans	Universiteit Antwerpen, Belgium
Edouard Fouché	Karlsruher Institut für Technologie, Germany
Dongqi Fu	UIUC, USA
Yanjie Fu	University of Central Florida, USA
Ken-ichi Fukui	Osaka University, Japan
Matjaž Gams	Jožef Stefan Institute, Slovenia
Amir Gandomi	University of Technology Sydney, Australia
Aryya Gangopadhyay	UMBC, USA
Dashan Gao	Hong Kong University of Science and Technology, China
Wei Gao	Nanjing University, China
Yifeng Gao	University of Texas Rio Grande Valley, USA
Yunjun Gao	Zhejiang University, China
Paolo Garza	Politecnico di Torino, Italy
Chang Ge	University of Minnesota, USA
Xin Geng	Southeast University, China
Flavio Giobergia	Politecnico di Torino, Italy
Rosalba Giugno	Università degli Studi di Verona, Italy
Aris Gkoulalas-Divanis	Merative, USA
Djordje Gligorijevic	Temple University, USA
Daniela Godoy	UNICEN, Argentina
Heitor Gomes	Victoria University of Wellington, New Zealand
Maciej Grzenda	Warsaw University of Technology, Poland
Lei Gu	Nanjing University of Posts and Telecommunications, China
Yong Guan	Iowa State University, USA
Riccardo Guidotti	Università di Pisa, Italy
Ekta Gujral	University of California Riverside, USA
Guimu Guo	Rowan University, USA
Ting Guo	University of Technology Sydney, Australia
Xingzhi Guo	Stony Brook University, USA
Ch. Md. Rakin Haider	Purdue University, USA
Benjamin Halstead	University of Auckland, New Zealand
Jinkun Han	Georgia State University, USA
Lu Han	Nanjing University, China
Yufei Han	Inria, France
Daisuke Hatano	RIKEN, Japan
Kohei Hatano	Kyushu University/RIKEN AIP, Japan
Shogo Hayashi	BizReach, Japan
Erhu He	University of Pittsburgh, USA
Guoliang He	Wuhan University, China
Pengfei He	Michigan State University, USA

Yi He	Old Dominion University, USA
Shen-Shyang Ho	Rowan University, USA
William Hsu	Kansas State University, USA
Haoji Hu	University of Minnesota, USA
Hongsheng Hu	CSIRO, Australia
Liang Hu	Tongji University, China
Shizhe Hu	Zhengzhou University, China
Wei Hu	Nanjing University, China
Mengdi Huai	Iowa State University, USA
Chao Huang	University of Hong Kong, Hong Kong SAR
Congrui Huang	Microsoft, China
Guangyan Huang	Deakin University, Australia
Jimmy Huang	York University, Canada
Jinbin Huang	Hong Kong Baptist University, Hong Kong SAR
Kai Huang	Hong Kong University of Science and Technology, China
Ling Huang	South China Agricultural University, China
Ting-Ji Huang	Nanjing University, China
Xin Huang	Hong Kong Baptist University, Hong Kong SAR
Zhenya Huang	University of Science and Technology of China, China
Chih-Chieh Hung	National Chung Hsing University, Taiwan
Hui-Ju Hung	Pennsylvania State University, USA
Nam Huynh	JAIST, Japan
Akihiro Inokuchi	Kwansei Gakuin University, Japan
Atsushi Inoue	Eastern Washington University, USA
Nevo Itzhak	Ben-Gurion University, Israel
Tomoya Iwakura	Fujitsu Laboratories Ltd., Japan
Divyesh Jadav	IBM T. J. Watson Research Center, USA
Shubham Jain	Visa Research, USA
Bijay Prasad Jaysawal	National Cheng Kung University, Taiwan
Kishlay Jha	University of Iowa, USA
Taoran Ji	Texas A&M University - Corpus Christi, USA
Songlei Jian	NUDT, China
Gaoxia Jiang	Shanxi University, China
Hansi Jiang	SAS Institute Inc., USA
Jiaxin Jiang	National University of Singapore, Singapore
Min Jiang	Xiamen University, China
Renhe Jiang	University of Tokyo, Japan
Yuli Jiang	Chinese University of Hong Kong, Hong Kong SAR
Bo Jin	Dalian University of Technology, China

Ming Jin	Monash University, Australia
Ruoming Jin	Kent State University, USA
Wei Jin	University of North Texas, USA
Mingxuan Ju	University of Notre Dame, USA
Wei Ju	Peking University, China
Vana Kalogeraki	Athens University of Economics and Business, Greece
Bo Kang	Ghent University, Belgium
Jian Kang	University of Rochester, USA
Ashwin Viswanathan Kannan	Amazon, USA
Tomi Kauppinen	Aalto University School of Science, Finland
Jungeun Kim	Kongju National University, Korea
Kyoung-Sook Kim	National Institute of Advanced Industrial Science and Technology, Japan
Primož Kocbek	University of Maribor, Slovenia
Aritra Konar	Katholieke Universiteit Leuven, Belgium
Youyong Kong	Southeast University, China
Olivera Kotevska	Oak Ridge National Laboratory, USA
P. Radha Krishna	NIT Warangal, India
Adit Krishnan	UIUC, USA
Gokul Krishnan	IIT Madras, India
Peer Kröger	CAU, Germany
Marzena Kryszkiewicz	Warsaw University of Technology, Poland
Chuan-Wei Kuo	National Yang Ming Chiao Tung University, Taiwan
Kuan-Ting Lai	National Taipei University of Technology, Taiwan
Long Lan	NUDT, China
Duc-Trong Le	Vietnam National University, Vietnam
Tuan Le	New Mexico State University, USA
Chul-Ho Lee	Texas State University, USA
Ickjai Lee	James Cook University, Australia
Ki Yong Lee	Sookmyung Women's University, Korea
Ki-Hoon Lee	Kwangwoon University, Korea
Roy Ka-Wei Lee	Singapore University of Technology and Design, Singapore
Yue-Shi Lee	Ming Chuan University, Taiwan
Dino Lenco	INRAE, France
Carson Leung	University of Manitoba, Canada
Boyu Li	University of Technology Sydney, Australia
Chaojie Li	University of New South Wales, Australia
Cheng-Te Li	National Cheng Kung University, Taiwan
Chongshou Li	Southwest Jiaotong University, China

Fengxin Li	Renmin University of China, China
Guozhong Li	King Abdullah University of Science and Technology, Saudi Arabia
Huaxiong Li	Nanjing University, China
Jianxin Li	Beihang University, China
Lei Li	Hong Kong University of Science and Technology (Guangzhou), China
Peipei Li	Hefei University of Technology, China
Qian Li	Curtin University, Australia
Rong-Hua Li	Beijing Institute of Technology, China
Shao-Yuan Li	Nanjing University of Aeronautics and Astronautics, China
Shuai Li	Cambridge University, UK
Shuang Li	Beijing Institute of Technology, China
Tianrui Li	Southwest Jiaotong University, China
Wengen Li	Tongji University, China
Wentao Li	Hong Kong University of Science and Technology (Guangzhou), China
Xin-Ye Li	Bytedance, China
Xiucheng Li	Harbin Institute of Technology, China
Xuelong Li	Northwestern Polytechnical University, China
Yidong Li	Beijing Jiaotong University, China
Yinxiao Li	Meta Platforms, USA
Yuefeng Li	Queensland University of Technology, Australia
Yun Li	Nanjing University of Posts and Telecommunications, China
Panagiotis Liakos	University of Athens, Greece
Xiang Lian	Kent State University, USA
Shen Liang	Université Paris Cité, France
Qing Liao	Harbin Institute of Technology (Shenzhen), China
Sungsu Lim	Chungnam National University, Korea
Dandan Lin	Shenzhen Institute of Computing Sciences, China
Yijun Lin	University of Minnesota, USA
Ying-Jia Lin	National Cheng Kung University, Taiwan
Baodi Liu	China University of Petroleum (East China), China
Chien-Liang Liu	National Yang Ming Chiao Tung University, Taiwan
Guiquan Liu	University of Science and Technology of China, China
Jin Liu	Shanghai Maritime University, China
Jinfei Liu	Emory University, USA
Kunpeng Liu	Portland State University, USA

Ning Liu	Shandong University, China
Qi Liu	University of Science and Technology of China, China
Qing Liu	Zhejiang University, China
Qun Liu	Louisiana State University, USA
Shenghua Liu	Chinese Academy of Sciences, China
Weifeng Liu	China University of Petroleum (East China), China
Yang Liu	Wilfrid Laurier University, Canada
Yao Liu	University of New South Wales, Australia
Yixin Liu	Monash University, Australia
Zheng Liu	Nanjing University of Posts and Telecommunications, China
Cheng Long	Nanyang Technological University, Singapore
Haibing Lu	Santa Clara University, USA
Wenpeng Lu	Qilu University of Technology, China
Simone Ludwig	North Dakota State University, USA
Dongsheng Luo	Florida International University, USA
Ping Luo	Chinese Academy of Sciences, China
Wei Luo	Deakin University, Australia
Xiao Luo	UCLA, USA
Xin Luo	Shandong University, China
Yong Luo	Wuhan University, China
Fenglong Ma	Pennsylvania State University, USA
Huifang Ma	Northwest Normal University, China
Jing Ma	Hong Kong Baptist University, Hong Kong SAR
Qianli Ma	South China University of Technology, China
Yi-Fan Ma	Nanjing University, China
Rich Maclin	University of Minnesota, USA
Son Mai	Queen's University Belfast, UK
Arun Maiya	Institute for Defense Analyses, USA
Bradley Malin	Vanderbilt University Medical Center, USA
Giuseppe Manco	Consiglio Nazionale delle Ricerche, Italy
Naresh Manwani	IIIT Hyderabad, India
Francesco Marcelloni	Università di Pisa, Italy
Leandro Marinho	UFCG, Brazil
Koji Maruhashi	Fujitsu Laboratories Ltd., Japan
Florent Masseglia	Inria, France
Mohammad Masud	United Arab Emirates University, United Arab Emirates
Sarah Masud	IIIT Delhi, India
Costas Mavromatis	University of Minnesota, USA

Mario Prado-Romero	Gran Sasso Science Institute, Italy
Bardh Prenkaj	Sapienza Università di Roma, Italy
Jianzhong Qi	University of Melbourne, Australia
Buyue Qian	Xi'an Jiaotong University, China
Huajie Qian	Columbia University, USA
Hezhe Qiao	Singapore Management University, Singapore
Biao Qin	Renmin University of China, China
Zengchang Qin	Beihang University, China
Tho Quan	Ho Chi Minh City University of Technology, Vietnam
Miloš Radovanović	University of Novi Sad, Serbia
Thilina Ranbaduge	Australian National University, Australia
Chotirat Ratanamahatana	Chulalongkorn University, Thailand
Chandra Reddy	IBM T. J. Watson Research Center, USA
Ryan Rossi	Adobe Research, USA
Morteza Saberi	University of Technology Sydney, Australia
Akira Sakai	Fujitsu Laboratories Ltd., Japan
David Sánchez	Universitat Rovira i Virgili, Spain
Maria Luisa Sapino	Università degli Studi di Torino, Italy
Hernan Sarmiento	UChile & IMFD, Chile
Badrul Sarwar	CloudAEye, USA
Nader Shakibay Senobari	University of California Riverside, USA
Nasrin Shabani	Macquarie University, Australia
Ankit Sharma	University of Minnesota, USA
Chandra N. Shekar	RGUKT RK Valley, India
Chih-Ya Shen	National Tsing Hua University, Taiwan
Wei Shen	Nankai University, China
Yu Shen	Peking University, China
Zhi-Yu Shen	Nanjing University, China
Chuan Shi	Beijing University of Posts and Telecommunications, China
Yue Shi	Meta Platforms, USA
Zhenwei Shi	Beihang University, China
Motoki Shiga	Tohoku University, Japan
Kijung Shin	KAIST, Korea
Kai Shu	Illinois Institute of Technology, USA
Hong-Han Shuai	National Yang Ming Chiao Tung University, Taiwan
Zeren Shui	University of Minnesota, USA
Satyaki Sikdar	Indiana University, USA
Dan Simovici	University of Massachusetts Boston, USA
Apoorva Singh	IIT Patna, India

Bikash Chandra Singh	Islamic University, Bangladesh
Stavros Sintos	University of Illinois at Chicago, USA
Krishnamoorthy Sivakumar	Washington State University, USA
Andrzej Skowron	University of Warsaw, Poland
Andy Song	RMIT University, Australia
Dongjin Song	University of Connecticut, USA
Arnaud Soulet	Université de Tours, France
Ja-Hwung Su	National University of Kaohsiung, Taiwan
Victor Suciu	University of Wisconsin, USA
Liang Sun	Alibaba Group, USA
Xin Sun	Technische Universität München, Germany
Yuqing Sun	Shandong University, China
Hirofumi Suzuki	Fujitsu Laboratories Ltd., Japan
Anika Tabassum	Oak Ridge National Laboratory, USA
Yasuo Tabei	RIKEN, Japan
Chih-Hua Tai	National Taipei University, Taiwan
Hiroshi Takahashi	NTT, Japan
Atsuhiro Takasu	National Institute of Informatics, Japan
Yanchao Tan	Fuzhou University, China
Chang Tang	China University of Geosciences, China
Lu-An Tang	NEC Laboratories America, USA
Qiang Tang	Luxembourg Institute of Science and Technology, Luxembourg
Yiming Tang	Hefei University of Technology, China
Ying-Peng Tang	Nanjing University of Aeronautics and Astronautics, China
Xiaohui (Daniel) Tao	University of Southern Queensland, Australia
Vahid Taslimitehrani	PhysioSigns Inc., USA
Maguelonne Teisseire	INRAE, France
Ya-Wen Teng	Academia Sinica, Taiwan
Masahiro Terabe	Chugai Pharmaceutical Co. Ltd., Japan
Kia Teymourian	University of Texas at Austin, USA
Qing Tian	Nanjing University of Information Science and Technology, China
Yijun Tian	University of Notre Dame, USA
Maksim Tkachenko	Singapore Management University, Singapore
Yongxin Tong	Beihang University, China
Vicenç Torra	University of Umeå, Sweden
Nhu-Thuat Tran	Singapore Management University, Singapore
Yash Travadi	University of Minnesota, USA
Quoc-Tuan Truong	Amazon, USA

Yi-Ju Tseng	National Yang Ming Chiao Tung University, Taiwan
Turki Turki	King Abdulaziz University, Saudi Arabia
Ruo-Chun Tzeng	KTH Royal Institute of Technology, Sweden
Leong Hou U	University of Macau, Macau SAR
Jeffrey Ullman	Stanford University, USA
Rohini Uppuluri	Glassdoor, USA
Satya Valluri	Databricks, USA
Dinusha Vatsalan	Macquarie University, Australia
Bruno Veloso	FEP - University of Porto and INESC TEC, Portugal
Anushka Vidanage	Australian National University, Australia
Herna Viktor	University of Ottawa, Canada
Michalis Vlachos	University of Lausanne, Switzerland
Sheng Wan	Nanjing University of Science and Technology, China
Beilun Wang	Southeast University, China
Changdong Wang	Sun Yat-sen University, China
Chih-Hang Wang	Academia Sinica, Taiwan
Chuan-Ju Wang	Academia Sinica, Taiwan
Guoyin Wang	Chongqing University of Posts and Telecommunications, China
Hongjun Wang	Southwest Jiaotong University, China
Hongtao Wang	North China Electric Power University, China
Jianwu Wang	UMBC, USA
Jie Wang	Southwest Jiaotong University, China
Jin Wang	Megagon Labs, USA
Jingyuan Wang	Beihang University, China
Jun Wang	Shandong University, China
Lizhen Wang	Yunnan University, China
Peng Wang	Southeast University, China
Pengyang Wang	University of Macau, Macau SAR
Sen Wang	University of Queensland, Australia
Senzhang Wang	Central South University, China
Shoujin Wang	Macquarie University, Australia
Sibo Wang	Chinese University of Hong Kong, Hong Kong SAR
Suhang Wang	Pennsylvania State University, USA
Wei Wang	Fudan University, China
Wei Wang	Hong Kong University of Science and Technology (Guangzhou), China
Weicheng Wang	Hong Kong University of Science and Technology, Hong Kong SAR

Wei-Yao Wang	National Yang Ming Chiao Tung University, Taiwan
Wendy Hui Wang	Stevens Institute of Technology, USA
Xiao Wang	Beihang University, China
Xiaoyang Wang	University of New South Wales, Australia
Xin Wang	University of Calgary, Canada
Xinyuan Wang	George Mason University, USA
Yanhao Wang	East China Normal University, China
Yuanlong Wang	Ohio State University, USA
Yuping Wang	Xidian University, China
Yuxiang Wang	Hangzhou Dianzi University, China
Hua Wei	Arizona State University, USA
Zhewei Wei	Renmin University of China, China
Yimin Wen	Guilin University of Electronic Technology, China
Brendon Woodford	University of Otago, New Zealand
Cheng-Wei Wu	National Ilan University, Taiwan
Fan Wu	Central South University, China
Fangzhao Wu	Microsoft Research Asia, China
Jiansheng Wu	Nanjing University of Posts and Telecommunications, China
Jin-Hui Wu	Nanjing University, China
Jun Wu	UIUC, USA
Ou Wu	Tianjin University, China
Shan-Hung Wu	National Tsing Hua University, Taiwan
Shu Wu	Chinese Academy of Sciences, China
Wensheng Wu	University of Southern California, USA
Yun-Ang Wu	National Taiwan University, Taiwan
Wenjie Xi	George Mason University, USA
Lingyun Xiang	Changsha University of Science and Technology, China
Ruliang Xiao	Fujian Normal University, China
Yanghua Xiao	Fudan University, China
Sihong Xie	Lehigh University, USA
Zheng Xie	Nanjing University, China
Bo Xiong	Universität Stuttgart, Germany
Haoyi Xiong	Baidu, Inc., China
Bo Xu	Donghua University, China
Bo Xu	Dalian University of Technology, China
Guandong Xu	University of Technology Sydney, Australia
Hongzuo Xu	NUDT, China
Ji Xu	Guizhou University, China

Tong Xu	University of Science and Technology of China, China
Yuanbo Xu	Jilin University, China
Hui Xue	Southeast University, China
Qiao Xue	Nanjing University of Aeronautics and Astronautics, China
Akihiro Yamaguchi	Toshiba Corporation, Japan
Bo Yang	Jilin University, China
Liangwei Yang	University of Illinois at Chicago, USA
Liu Yang	Tianjin University, China
Shaofu Yang	Southeast University, China
Shiyu Yang	Guangzhou University, China
Wanqi Yang	Nanjing Normal University, China
Xiaoling Yang	Southwest Jiaotong University, China
Xiaowei Yang	South China University of Technology, China
Yan Yang	Southwest Jiaotong University, China
Yiyang Yang	Guangdong University of Technology, China
Yu Yang	City University of Hong Kong, Hong Kong SAR
Yu-Bin Yang	Nanjing University, China
Junjie Yao	East China Normal University, China
Wei Ye	Tongji University, China
Yanfang Ye	University of Notre Dame, USA
Kalidas Yeturu	IIT Tirupati, India
Ilkay Yildiz Potter	BioSensics LLC, USA
Minghao Yin	Northeast Normal University, China
Ziqi Yin	Nanyang Technological University, Singapore
Jia-Ching Ying	National Chung Hsing University, Taiwan
Tetsuya Yoshida	Nara Women's University, Japan
Hang Yu	Shanghai University, China
Jifan Yu	Tsinghua University, China
Yanwei Yu	Ocean University of China, China
Yongsheng Yu	Macquarie University, Australia
Long Yuan	Nanjing University of Science and Technology, China
Lin Yue	University of Newcastle, Australia
Xiaodong Yue	Shanghai University, China
Nayyar Zaidi	Monash University, Australia
Chengxi Zang	Cornell University, USA
Alexey Zaytsev	Skoltech, Russia
Yifeng Zeng	Northumbria University, UK
Petros Zerfos	IBM T. J. Watson Research Center, USA
De-Chuan Zhan	Nanjing University, China

Huixin Zhan	Texas Tech University, USA
Daokun Zhang	Monash University, Australia
Dongxiang Zhang	Zhejiang University, China
Guoxi Zhang	Beijing Institute of General Artificial Intelligence, China
Hao Zhang	Chinese University of Hong Kong, Hong Kong SAR
Huaxiang Zhang	Shandong Normal University, China
Ji Zhang	University of Southern Queensland, Australia
Jianfei Zhang	Université de Sherbrooke, Canada
Lei Zhang	Anhui University, China
Li Zhang	University of Texas Rio Grande Valley, USA
Lin Zhang	IDEA Education, China
Mengjie Zhang	Victoria University of Wellington, New Zealand
Nan Zhang	Wenzhou University, China
Quangui Zhang	Liaoning Technical University, China
Shichao Zhang	Central South University, China
Tianlin Zhang	University of Manchester, UK
Wei Emma Zhang	University of Adelaide, Australia
Wenbin Zhang	Florida International University, USA
Wentao Zhang	Mila, Canada
Xiaobo Zhang	Southwest Jiaotong University, China
Xuyun Zhang	Macquarie University, Australia
Yaqian Zhang	University of Waikato, New Zealand
Yikai Zhang	Guangzhou University, China
Yiqun Zhang	Guangdong University of Technology, China
Yudong Zhang	Nanjing Normal University, China
Zhiwei Zhang	Beijing Institute of Technology, China
Zike Zhang	Hangzhou Normal University, China
Zili Zhang	Southwest University, China
Chen Zhao	Baylor University, USA
Jiaqi Zhao	China University of Mining and Technology, China
Kaiqi Zhao	University of Auckland, New Zealand
Pengfei Zhao	BNU-HKBU United International College, China
Pengpeng Zhao	Soochow University, China
Ying Zhao	Tsinghua University, China
Zhongying Zhao	Shandong University of Science and Technology, China
Guanjie Zheng	Shanghai Jiao Tong University, China
Lecheng Zheng	UIUC, USA
Weiguo Zheng	Fudan University, China

Aoying Zhou	East China Normal University, China
Bing Zhou	Sam Houston State University, USA
Nianjun Zhou	IBM T. J. Watson Research Center, USA
Qinghai Zhou	UIUC, USA
Xiangmin Zhou	RMIT University, Australia
Xiaoping Zhou	Beijing University of Civil Engineering and Architecture, China
Xun Zhou	University of Iowa, USA
Jonathan Zhu	Wheaton College, USA
Ronghang Zhu	University of Georgia, China
Xingquan Zhu	Florida Atlantic University, USA
Ye Zhu	Deakin University, Australia
Yihang Zhu	University of Leicester, UK
Yuanyuan Zhu	Wuhan University, China
Ziwei Zhu	George Mason University, USA

External Reviewers

Zihan Li	University of Massachusetts Boston, USA
Ting Yu	Zhejiang Lab, China

Sponsoring Organizations

Accton

ACSI

Appier

Chunghwa Telecom Co., Ltd

DOIT, Taipei

ISCOM

Metaage

NSTC

Pegatron

Quanta Computer

TWS

Wavenet Co., Ltd

Contents – Part V

Multimedia and Multimodal Data

Recommender Systems

Spatio-temporal Data

Multimedia and Multimodal Data

Re-thinking Human Activity Recognition with Hierarchy-Aware Label Relationship Modeling

Jingwei Zuo(✉) and Hakim Hacid

Technology Innovation Institute, Abu Dhabi, UAE
{jingwei.zuo,hakim.hacid}@tii.ae

Abstract. Human Activity Recognition (HAR) has been studied for decades, from data collection, learning models, to post-processing and result interpretations. However, the inherent hierarchy in the activities remains relatively under-explored, despite its significant impact on model performance and interpretation. In this paper, we propose H-HAR, by rethinking the HAR tasks from a fresh perspective by delving into their intricate global label relationships. Rather than building multiple classifiers separately for multi-layered activities, we explore the efficacy of a flat model enhanced with graph-based label relationship modeling. Being hierarchy-aware, the graph-based label modeling enhances the fundamental HAR model, by incorporating intricate label relationships into the model. We validate the proposal with a multi-label classifier on complex human activity data. The results highlight the advantages of the proposal, which can be vertically integrated into advanced HAR models to further enhance their performances.

Keywords: Human Activity Recognition · Hierarchical Label Modeling · Graph Neural Networks · Hierarchical Human Activity

1 Introduction

Human activity recognition (HAR) has gained, in recent years, a great interest from both the research community and industry players. The activity data can be collected from multiple data sources [19], such as GPS trajectories, web browsing records, smart sensors, etc. Among which, the human physical activity with wearable sensors are widely studied [22], which are represented by multivariate time series (MTS). When studying the HAR tasks, the focus is typically on the complexity of the activity data itself, such as dealing with complex MTS formats, processing noisy data, or identifying multiple sequential actions within a single activity. Existing research often explores a flat classifier approach that learns complex activity data. However, the complexity in the label relationships between activity class labels is usually overlooked.

As shown in Fig. 1, an inner label structure always exists in physical activities, providing rich information for building a reliable HAR classifier. Recent

D.-N. Yang et al. (Eds.): PAKDD 2024, LNAI 14649, pp. 3–14, 2024.
https://doi.org/10.1007/978-981-97-2262-4_1

work [4,8,15,17] consider the hierarchy features between human activities, and have proved that a hierarchy-aware model shows better performance in terms of model's reliability and efficiency. However, these methods employ a straightforward top-down approach, constructing individual classifiers at each hierarchical level. Classifiers in lower layers are based on predictions from upper-layer classifiers, a local-based process that introduces several limitations: i) Multiple classifiers need to be built at each level, leading to escalating complexity within the hierarchical structure; ii) Classifiers at each level focus on relationships within the same layer, under a common parent node, ignoring the broader, global relationships between the cross-layer activities; iii) Classifiers at lower levels rely on the predictions from upper layers as their training annotations, leading to larger accumulative errors in a deeper hierarchical structure. Moreover, as depicted in Fig. 1, the relationships between class labels depend not only on the pre-defined hierarchical structure, but also on the implicit, hidden links between certain activities. Therefore, relying on a pre-defined label structure risks overlooking these vital relationships between activities.

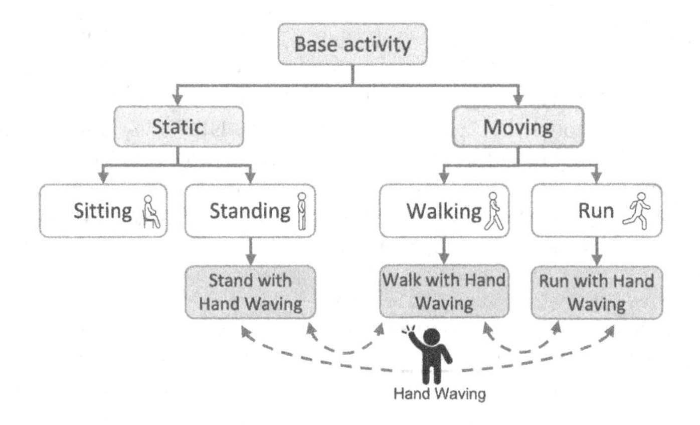

Fig. 1. The label structure in physical activities, including predefined relationships (solid lines) and implicit relationships (dashed lines).

To tackle the above-mentioned challenges, we propose H-HAR, a Hierarchy-aware Model for HAR tasks. Instead of building multiple local HAR classifiers, we learn a flat model to process the activities in a global manner. Precisely, we embed and project the label hierarchy into the representation space of the data. The label and data hierarchy will be carefully aligned, allowing the model to benefit the rich information from both data and the hierarchy features. To build the representation space, a graph-based label encoder and an activity data encoder are proposed: the label encoder learns complex label embeddings by combining a predefined label hierarchy and a learnable graph structure; whereas the data encoder builds data embeddings of input activities. The aligned label-data embeddings are learned via a supervised contrastive loss [6], considering inter-class and intra-class embeddings. Concretely, the nearby neighbors in the hierarchy will stay close to each other in the representation space, while distant nodes will keep far away from each other. A multi-label classifier is jointly built over the representation space.

To summarize, our contributions in this paper are as follows:

- **Label relationship modeling**: we propose a label encoder that automatically learns the label relationships without a predefined label structure.
- **Embeddable label encoder with scalability**: the label encoder can be seamlessly integrated into other HAR models to learn better representations.
- **Label-data semantic alignment**: we align the label and data semantics in the representation space, allowing building class-separable data embeddings.
- **Joint embedding & multi-label classifier optimization**: we jointly optimize the embedding space and classifier, providing reliable performance.

2 Related Work

In this section, we describe the most related work of our proposal in Human Activity Recognition (HAR) tasks and Hierarchical Label Modeling.

2.1 Human Activity Recognition (HAR)

Human Activity Recognition (HAR) is a largely investigated domain. In our context, we consider human physical activities, with data acquired easily from smart sensors. The sensor-based activity data is generally represented by Multivariate Time Series (MTS) [22]. As a classification task, the HAR can be based on various feature extractors (i.e., feature representations) and classifiers. For instance, one can use handcrafted statistical features [19,20] to feed any classifiers, which is easy-to-deploy and requires linear processing time. Other work in MTS Classification domain, where researchers aim to build general ML models covering various applications [22], including HAR tasks. For instance, Shapelet features [21] with a kNN classifier, or end-to-end neural network models [22].

However, these approaches usually focus on handling the complex activity data, overlooking the complex label relationships, that provides rich information for building reliable feature representations.

2.2 Hierarchical Label Modeling

Inherently, there exists a hierarchical label structure in human activities. The label hierarchy, as shown in Fig. 1, can be considered as a tree-based structure. It allows enriching the data embeddings and forge a robust representation space to learn class-separable embeddings. Rarely investigated in HAR tasks, the label modeling is usually studied in Natural Language Processing (NLP) applications, where the hierarchy features widely exist in the semantic labels [18]. A typical example is Hierarchical Text Classification (HTC) [16], for which a sentence can be tagged with different labels. A multi-label classifier can be built over the text representations, which can be improved by considering the label relationships.

The hierarchical label modeling in previous studies [18] can be either local or global approaches. Local approaches [1,5,10] build multiple classifiers at each hierarchical level. However, they basically ignore the rich structural interactions between nodes at a global scale, i.e., the activities can share common patterns even though they do not share the same parent nodes. For instance, in Fig. 1, *still with hand waving* and *walking with hand waving* are two activities under *still* and *walking.* Though having different parent nodes, they share the same action of *hand waving.* In consequence, the local modeling approaches only capture limited interactions between neighboring activity nodes in the hierarchy structure. Previous HAR models [4,8,15,17] usually model the hierarchy in this manner, i.e., training multiple classifiers at different hierarchical levels.

As for global approaches [3,10,13,18], they build a flat-label classifier for all classes. Therefore, how to integrate the hierarchy information into the model becomes the research focus of the recent studies, i.e., building a hierarchy-aware flat-label classifier. Various work has studied the joint modeling of label and data embeddings in HTC tasks. For instance, authors in [12,16] designed a generalized triplet loss with hierarchy-aware margin, which allows differentiating fine and coarse-label classes. With more considerations on the hierarchical information, the work in [18] introduced Prior Hierarchy Information from the training set, which serves to encode the label structures. The label structure can be either encoded by a Bidirectional Tree-LSTM, or a Graph Convolutional Network (GCN). Consequently, the hierarchy-aware label embedding can be combined with text embeddings to feed a multi-label classifier. HiMatch [3] further aligns the text semantics and label semantics, and adopt a similar Triplet loss with a hierarchy-aware margin to accelerate the computation process.

However, the above-mentioned work, both local and global approaches, heavily relies on prior knowledge of the label hierarchy information. In consequence, the implicit, hidden relationships between label nodes are usually ignored, leading to a less optimal modeling of the label relationships.

3 Problem Formulation

In this section, we formulate our research problems on HAR with learnable label relationship modeling. Table 1 summarizes the notations used in the paper.

Definition 1. *(Hierarchical Human Activity). We denote the Hierarchical Human Activity data as* $\mathcal{D} = \{X, L\}$ *with a sequence of activity sets* $X = \{X_1, ..., X_N\}$ *and a sequence of label sets* $L = \{l_1, ..., l_N\}$*. Each label set* l_i *contains a set of classes, belong to either one or more sub-paths in the hierarchy.*

As shown in Fig. 1, the hierarchical class labels can be formulated as a graph structure. Therefore, each label set l_i *represents the labels passed through the root node to a terminal node, that can be a leaf or a non-leaf node.*

Definition 2. *(Hierarchical Human Activity Recognition). Given a data set* $\mathcal{D} = \{X, L\}$*, we aim to learn a multi-label classifier* f *from* $\mathcal{D}$*. For an unseen activity* x_i*, the classifier* f *can accurately predict its label set* $\hat{l}_i = \{\hat{y}_i\}^m$*, where* m *is the number of labels.*

Table 1. Notation

Notation	Description
$\mathcal{D} = \{X, L\}$	Activity and label sets
$X = \{X_1, ..., X_N\}$ *or* $\{x_1, ..., x_n\}$	A sequence of activity sets $X_1, ..., X_N$, sample $x_1, ..., x_n$
$L = \{l_1, ..., l_N\}$ *or* $\{l_1, ..., l_n\}$	A sequence of label sets $l_1, ..., l_N$. (Note: *multi-label for* x_i)
N, n	Number of label sets, number of samples
$\mathbf{E}_L = \{e_1, ..., e_N\}$	Label embeddings
$\mathbf{E}_X = \{e'_1, ..., e'_N\}$	Data embeddings
$\mathcal{G} =< \mathcal{V}, \mathcal{E} >$	A graph including the vertex and edge sets
$\varphi : \mathcal{X} \rightarrow \mathcal{R}^d$	Feature map function, e.g., a linear layer
Θ	Model parameters

Definition 3. *(Hierarchical Label Embedding). Given a sequence of label sets* $L = \{l_1, ..., l_N\}$*, we aim to learn a set of hierarchy-aware label embeddings* $\mathbf{E}_L = \{e_1, ..., e_N\}$*, integrating hierarchical features from L for each target instance.*

Learning hierarchical human activities requires considering not only the features of activity data, i.e., data embeddings, but also relationships (*explicit* and *implicit*) between activities, i.e., label embeddings. We aim to learn a representation space $\mathcal{H}$ where the raw activity data are embedded and aligned with learnable label relationships. The learning objective is to minimize the classification loss $\theta = min_{\theta \in \Theta} \mathrm{L}(f_\theta(X, \mathbf{E}_L), Y)$.

4 Our Proposals

To handle the aforementioned challenges, we propose H-HAR, a Hierarchy-aware model for HAR tasks. As shown in Fig. 2, H-HAR relies on a graph-based label encoder and an activity data encoder. The graph-based label encoder extracts the complex hierarchical relationships between labels, with a predefined label hierarchy and a learnable graph structure. The data encoder simply builds data embeddings of input activities. By aligning the hierarchical label and data embeddings, H-HAR is able to learn data representations with hierarchy semantics.

4.1 Hierarchy-Aware Label Encoding

In the label hierarchy, the nodes under the same parent node share similar patterns. Unlike previous studies [4,8,15,17] considering only child-parent and child-child relationships, we consider global node relationships, coming with more discriminative features. Due to the intricate global relationships among the label nodes, it is natural for us to represent the label hierarchy as a graph.

Definition 4. *(Hierarchy as Graph). We define a label graph* $\mathcal{G} =< \mathcal{V}, \mathcal{E} >$ *to represent the hierarchical structure among the labels, where* $\mathcal{V} = \{v_1, v_2, ..., v_N\}$ *denotes the label set with N nodes,* $\mathcal{E} = \{(v_i, v_j) | v_i \in \mathcal{V}, v_j \in link(i)\}$ *indicates the directed edge connections between* v_i *and it's linked nodes.*

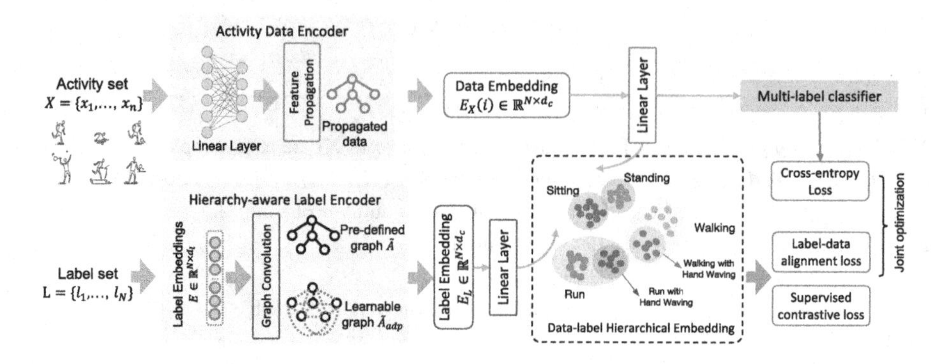

Fig. 2. Global system architecture of H-HAR

In the label graph $\mathcal{G}$, each activity is regarded as a node, and can be connected or disconnected from others. The graph edges represent the node relationships. We should note that the relationships are not fully decided by a pre-defined label hierarchy, i.e., edge connections. As aforementioned in Fig. 1, the implicit hidden relationships exist for unconnected nodes in the pre-defined hierarchy. Therefore, we propose to learn the implicit node relationships via a learnable hidden graph. The pre-defined and learnable graphs are jointly considered in a Graph Convolution Network (GCN) [23], to build hierarchy-aware label embeddings.

Predefined Graph Structure. In GCNs [7], the adjacency matrix represents the graph connections or relationships between the nodes. We define $A_{i,j} = \frac{|H(v_i) \cap H(v_j)|}{|H(v_i)|}$ as the connection weight between node v_i and v_j, where $H(v)$ denotes a set of higher level nodes (i.e., all the parent nodes of v). Intuitively, $|H(v_i) \cap H(v_j)|$ represents the number of shared parent nodes between v_i and v_j, $A_{i,j}$ shows the proportion of common ancestors over the node v_i. A larger value of $A_{i,j}$ represents a closer relationship between v_i and v_j.

Let $\tilde{\mathbf{A}} = I + D^{-\frac{1}{2}} A D^{-\frac{1}{2}} \in \mathcal{R}^{N \times N}$ denote the normalized adjacency matrix with self-loops, where D is the degree matrix representing the degree of each vertex in the graph. Given a sequence of label set $L = \{l_1, ..., l_N\}$, we define the intermediate label embeddings $E = e(L) \in \mathcal{R}^{N \times d_l}$ as the input signals, where e is the embedding function. Then the label embeddings $\mathbf{E}_L$ integrating the graph structural features is defined as the output of a graph convolution layer [7]:

$$\mathbf{E}_L = \sigma(\tilde{\mathbf{A}} E \mathbf{W_p}) \in \mathcal{R}^{N \times d_c} \tag{1}$$

where σ is ReLU activation, $\mathbf{W_p} \in \mathcal{R}^{d_l \times d_c}$ denotes GCN's weight matrix.

Learnable Graph Structure. The hidden interactions allow the model to enrich the label embeddings from a global view (i.e., interacting nodes from different layers and branches). To capture the implicit connections, as a complement of the predefined graph, we learn a self-adaptive graph, that does not require any prior knowledge and is learned end-to-end through stochastic gradient descent. We initialize two random matrices E_1, $E_2 \in \mathcal{R}^{N \times d_f}$, representing source and target node embeddings [14]. We define the self-adaptive adjacency matrix as:

$$\tilde{\mathbf{A}}_{adp} = SoftMax(ReLU(E_1 E_2^T)) \quad (2)$$

$E_1 E_2^T$ shows the dependency weights between source/target nodes. *ReLU* serves to filter weak connections. *SoftMax* is used to normalize the adjacency matrix.

With the predefined graph in Eq. 1, we re-define the graph layer as:

$$\mathbf{E}_L = \sigma(\tilde{\mathbf{A}} E \mathbf{W}_\mathbf{p} + \tilde{\mathbf{A}}_{adp} E \mathbf{W}_{adp}) \in \mathcal{R}^{N \times d_c} \quad (3)$$

4.2 Activity Data Encoding

Raw physical activity data is usually represented as Multivariate Time Series, i.e., $X = \{x_1, ..., x_n\} \in \mathcal{R}^{n \times m \times t}$, where n, m, t represent number of instances, sensors and timestamps. In this paper, we focus on the model's label encoding behavior. Therefore, aligned with pre-processed data of multiple HAR datasets (e.g., DaliAc [8], mHealth [2]), we consider $X = \{x_1, ..., x_n\} \in \mathcal{R}^{n \times d}$, where d is the input feature dimension. More advanced feature extractors on raw data can be explored and integrated into our framework. This is orthogonal to our work.

Given $X \in \mathcal{R}^{n \times d}$, we define the intermediate data embedding $E_d = e(X) \in \mathrm{R}^{n \times d_x}$. Following previous work [18], we further introduce a graph-based feature propagation module to encode label hierarchy information. The propagation module first reshapes activity features E_d to align with the graph node input:

$$V = E_d W_{res} \in \mathrm{R}^{n \times N \times d_c} \quad (4)$$

where $W_{res} \in \mathrm{R}^{d_x \times N \times d_c}$. Then the GCNs built in Eq. 3 can be employed to integrate label hierarchical information:

$$\mathbf{E}_X = \sigma(\tilde{\mathbf{A}} V \mathbf{W}_\mathbf{p}' + \tilde{\mathbf{A}}_{adp} V \mathbf{W}'_{adp}) \in \mathcal{R}^{n \times N \times d_c} \quad (5)$$

Note that $\tilde{\mathbf{A}}$ and $\tilde{\mathbf{A}}_{adp}$ are shared graphs between the label and data encoding.

4.3 Label-Data Joint Embedding Learning

Label-Data Alignment. Even though the data embeddings are reshaped to align with the graph node input, there is no explicit matching between data embeddings and label embeddings, that contains rich label relationships. To this end, we jointly built label-data embeddings in the representation space to align data and label semantics. Concretely, we apply the L2 loss between data and label embeddings:

$$\mathrm{L}_{\mathrm{align}} = \sum_{i=1}^{n} \|\varphi_x(\mathbf{E}_X(i)) - \varphi_l(\mathbf{E}_L)\|^2 \quad (6)$$

where φ_x and φ_l are linear layers to project $\mathbf{E}_X, \mathbf{E}_L$ to a common latent space.

Class-Separable Embedding Building. The label-data alignment loss only captures the correlations between activity data and labels, while the label embeddings are not clearly separable. To learn class-separable embeddings, we employ a supervised contrastive loss [6] to the representation space:

$$\begin{aligned} \mathrm{L}_{con}\left(\mathbf{E}_X(i), \mathbf{E}_X(j), Y\right) = Y * \left\|\varphi_X\left(\mathbf{E}_X(i)\right) - \varphi_X\left(\mathbf{E}_X(j)\right)\right\|^2 \\ + (1-Y) * \left\{\max\left(0, \mathrm{m}^2 - \left\|\varphi_X\left(\mathbf{E}_X(i)\right) - \varphi_X\left(\mathbf{E}_X(j)\right)\right\|^2\right)\right\} \end{aligned} \tag{7}$$

where $\mathrm{m} > 0$ is the margin parameter, $Y = 1$ if $l_i = l_j$, otherwise $Y = 0$.

Classification and Joint Optimization. As shown in Fig. 1, the hierarchy can be flattened for multi-label classification. The data embedding $\mathbf{E}_X$ is followed by a linear layer and a sigmoid function to output the probability on label j:

$$p_{ij} = \mathrm{sigmoid}\left(\varphi_X\left(\mathbf{E}_X(i)\right)\right)_j \tag{8}$$

Therefore, a binary cross-entropy loss is applied:

$$\mathrm{L}_{ce} = \sum_{i=1}^{n}\sum_{j=1}^{N} -y_{ij}\log\left(p_{ij}\right) - \left(1 - y_{ij}\right)\log\left(1 - p_{ij}\right) \tag{9}$$

where y_{ij} is the ground truth: $y_{ij} = 1$ if x_i contains a label j, otherwise 0.

We jointly optimize the model by combining the label-data alignment loss, contrastive loss and cross-entropy loss:

$$\mathrm{L} = \mathrm{L}_{align} + \lambda_1 \mathrm{L}_{con} + \lambda_2 \mathrm{L}_{ce} \tag{10}$$

where λ_1, λ_2 are hyperparameters controlling the weight of the related loss. During inference, we only use the Activity Data Encoder for classification.

5 Experiments

In this section, we validate H-HAR with real-life human activity datasets. The experiments were designed to answer the following Research Questions (RQs):

RQ 1 *H-HAR Performance*: How does H-HAR compare to other (hierarchical) models in HAR tasks?

RQ 2 *Label Encoding Efficiency*: How effective is our graph-based label encoding compared to other label modeling methods in HAR?

RQ 3 *Impact of Joint Optimization*: What are the benefits of using multiple objective functions together in improving HAR model performance?

5.1 Experimental Settings

Dataset Descriptions. We choose DaliAc [8] and UCI HAPT [9] as testing datasets because of their rich label relationships. As shown in Fig. 3, the DaliAc dataset contains 13 activities collected by 19 participants. The UCI HAPT dataset was collected from 30 volunteers, with 6 basic activities and 6 postural transitions. We follow [4,11] as for the data preprocessing and training/testing split. As both datasets are relatively class-balanced, for simplicity, we report the average accuracy of all classes in each dataset.

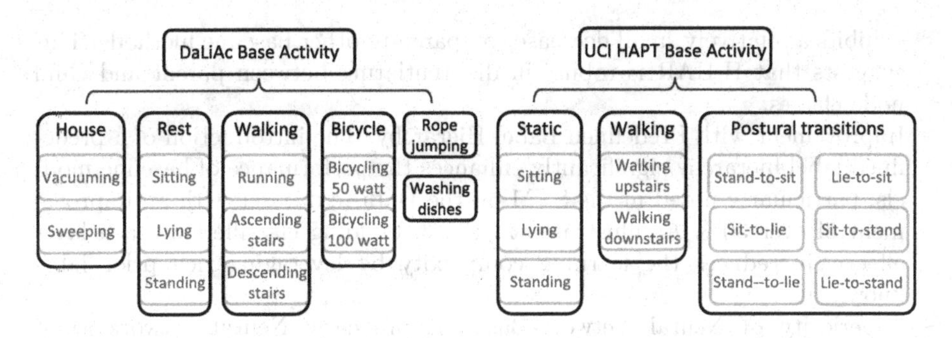

Fig. 3. Predefined label hierarchy in DaLiAc and UCI HAPT.

Execution and Parameter Settings. The proposed model is implemented in PyTorch 1.6.0 and is trained using the Adam optimizer in one single Nvidia A100 (40G). We set an adaptive learning rate regarding training epochs, i.e., the learning rate starts from 0.01 and decreases by half every training epoch. We set the balancing weight $\lambda_1 = \lambda_2 = 1$.

Baselines. For HAR tasks, much of the existing research adopts local-based approaches. These typically involve constructing multiple classifiers in a top-down manner. They can be essentially simplified to a flat classifier model when not considering the predefined label hierarchy. Therefore, we selected popular conventional ML models for evaluation, including AdaBoost, kNNs (k=7), SVM, and Multi-layer Perceptron (MLP) with a Softmax activation function.

It's important to note that while there are numerous advanced models that could potentially yield superior HAR performance, our focus is primarily on examining the impact of label relationship modeling within HAR tasks, rather than identifying the most advanced model architectures.

Additionally, we assessed the performance of these models both with and without considering label hierarchy. For the baseline models, we did not incorporate any predefined hierarchy. In contrast, for H-HAR, we substituted the graph-based label encoding layer with a linear layer. We also extended our evaluation to both single-label and multi-label classification tasks to comprehensively understand the models' behavior.

5.2 Experimental Results

Table 2 presents a comparison of the accuracy of various HAR models, both with and without considering label hierarchy (denoted as w/o H. and w/ H. respectively). With advanced label-data embedding learning and joint classifier building, it is not surprising that H-HAR shows superior performances of others. However, the results offer several key insights:

- Robustness in Multi-label Classification (**RQ 1**): While there is a general decline in model performance for multi-label classification tasks, H-HAR exhibits a relatively small decrease compared to other baseline methods. This suggests that H-HAR is robust in differentiating between parent and child node classes.

- Improvement with Predefined Label Hierarchy: The introduction of a predefined label hierarchy significantly enhances the performance of baseline models, particularly noted in the SVM on the DaLiAc dataset with an improvement of over 30%. As illustrated in Fig. 3, building classifiers at each layer effectively reduces the learning complexity by leveraging rich prior label knowledge.
- Superiority of Neural Network-Based Approaches: Neural network-based models generally outperform traditional ML models in this context, where the data is straightforward, and the feature space is limited. Exploring more advanced network architectures could further augment the model's performance, which is orthogonal to this work.

However, due to a larger parameter space, H-HAR performs less efficient than MLP, taking 39 s for one training epoch, compared to 12 s for MLP. Conventional ML models are not compared on efficiency as they are running on CPU.

Table 2. Accuracy (%) comparison between models w/o or w/ label hierarchy

		AdaBoost		kNN		SVM		MLP		H-HAR	
Dataset	Classifier	w/o H	w/ H	w/o H	w/ H	w/o H	w/ H	w/o H	w/ H	w/o H	w H
DaLiAc	single-label	80.0	86.64	68.71	85.48	54.13	87.12	88.92	94.62	91.64	**97.43**
	multi-label	76.28	83.34	64.53	76.32	52.34	82.34	88.32	92.43	90.98	**97.23**
UCI HAPT	single-label	88.96	92.39	75.62	88.92	89.26	94.25	90.54	96.77	95.45	**97.98**
	multi-label	84.23	89.23	72.43	84.34	87.23	92.34	90.23	95.88	94.32	**97.82**

5.3 Ablation Study

To understand why our model performs effectively, we conduct ablation studies on various parameters that might impact or enhance the model's performance. Specifically, as detailed in Table 3, we examine several H-HAR variants:

- Label Hierarchy
  - None: replace the graph modeling layer in Eq. 3 with a linear layer;
  - $\hat{A}$: only use the predefined label hierarchy for label modeling;
  - $\hat{A}_{adp}$: only employ a learnable graph-based label modeling.
- Feature Propagation (None): replace Feature Propagation by a linear layer
- Objective Function
  - $L_{align}+L_{ce}$: label-data alignment loss with cross-entropy loss;
  - $L_{con}+L_{ce}$: contrastive loss with cross-entropy loss;
  - L_{ce}: only cross-entropy loss.

Table 3. Ablation study: model accuracy (%) w.r.t. various parameters

Dataset	Classifier	Label Hierarchy			Feat. Propag.	Objective Function			H-HAR
		None	$\hat{A}$	$\hat{A}_{adp}$	None	$L_{align}+L_{ce}$	$L_{con}+L_{ce}$	L_{ce}	
DaLiAc	single-label	91.64	94.23	**97.69**	96.23	94.32	97.33	94.42	97.43
	multi-label	90.98	94.12	97.21	95.67	93.23	96.59	92.38	**97.23**
UCI HAPT	single-label	95.45	97.32	**98.20**	97.45	97.28	97.89	96.73	97.98
	multi-label	94.32	97.24	97.12	97.12	96.52	97.65	95.72	**97.82**

From the results, we observe that i) The model performs better in single-label classification with just the learnable graph than when combined with a predefined label hierarchy. This suggests that learning relationships directly from data can be more effective than using pre-set connections (**RQ 2**); ii) Adding feature propagation improves the model's performance. This likely happens because it helps align data better with the graph's structure; iii) The biggest boost in performance comes from supervised contrastive learning, which helps build class-separable embeddings. Joint optimization of these techniques also helps enhance the model's overall effectiveness (**RQ 3**).

6 Discussions and Conclusion

Modeling and integrating label relationships into HAR models allows regularizing the representation space, thus building better feature embeddings. The proposed H-HAR brings multiple research opportunities, which are not fully addressed in the paper: i) the hierarchy-aware label modeling allows us to handle data with heterogeneous-granular labels, leading to less effort and better flexibility in practice for data annotations; ii) the contrastive learning can be further explored in the context of label relationship modeling. For instance, a hierarchy-aware margin parameter can be investigated [3]; etc.

Conclusion. In this work, we propose H-HAR and rethink Human Activity Recognition (HAR) tasks from a perspective of graph-based label modeling. The proposed hierarchy-ware label encoding can be seamlessly integrated into other HAR models to improve further models' performance. For future work, one can be exploring more complex data with a deeper hierarchy and intricate label relationships. Human activities with multi-modality will also be one of the research directions in the future.

References

1. Banerjee, S., Akkaya, C., Perez-Sorrosal, F., Tsioutsiouliklis, K.: Hierarchical transfer learning for multi-label text classification. In: ACL, pp. 6295–6300 (2019)
2. Banos, O., Garcia, R., Saez, A.: MHEALTH Dataset. UCI Machine Learning Repository (2014). https://doi.org/10.24432/C5TW22
3. Chen, H., Ma, Q., Lin, Z., Yan, J.: Hierarchy-aware label semantics matching network for hierarchical text classification. In: ACL, pp. 4370–4379 (2021)

4. Debache, I., Jeantet, L., Chevallier, D., Bergouignan, A., Sueur, C.: A lean and performant hierarchical model for human activity recognition using body-mounted sensors. Sensors **20**(11), 3090 (2020)
5. Dumais, S., Chen, H.: Hierarchical classification of web content. In: SIGIR, pp. 256–263 (2000)
6. Khosla, P., Teterwak, P., Wang, C., Sarna, A., Tian, Y., Isola, P., Maschinot, A., Liu, C., Krishnan, D.: Supervised contrastive learning. NeurIPS **33**, 18661–18673 (2020)
7. Kipf, T.N., Welling, M.: Semi-supervised classification with graph convolutional networks. In: International Conference on Learning Representations (2017)
8. Leutheuser, H., Schuldhaus, D., Eskofier, B.M.: Hierarchical, multi-sensor based classification of daily life activities: comparison with state-of-the-art algorithms using a benchmark dataset. PLoS ONE **8**(10), e75196 (2013)
9. Reyes-Ortiz, J.L., Oneto, L., Samà, A., Parra, X., Anguita, D.: Transition-aware human activity recognition using smartphones. Neurocomputing **171**, 754–767 (2016)
10. Shimura, K., Li, J., Fukumoto, F.: HFT-CNN: learning hierarchical category structure for multi-label short text categorization. In: ACL, pp. 811–816 (2018)
11. Thu, N.T.H., Han, D.S.: HiHAR: a hierarchical hybrid deep learning architecture for wearable sensor-based human activity recognition. IEEE Access **9**, 145271–145281 (2021)
12. Tonioni, A., Di Stefano, L.: Domain invariant hierarchical embedding for grocery products recognition. CVIU **182**, 81–92 (2019)
13. Wang, Z., Wang, P., Huang, L., Sun, X., Wang, H.: Incorporating hierarchy into text encoder: a contrastive learning approach for hierarchical text classification. In: ACL, pp. 7109–7119 (2022)
14. Wu, Z., Pan, S., Long, G., Jiang, J., Zhang, C.: Graph wavenet for deep spatial-temporal graph modeling. In: IJCAI, pp. 1907–1913 (2019)
15. Zhang, S., McCullagh, P., Nugent, C., Zheng, H.: Activity monitoring using a smart phone's accelerometer with hierarchical classification. In: 2010 Sixth International Conference on Intelligent Environments, pp. 158–163. IEEE (2010)
16. Zhang, X., Zhou, F., Lin, Y., Zhang, S.: Embedding label structures for fine-grained feature representation. In: CVPR, pp. 1114–1123 (2016)
17. Zheng, Y.: Human activity recognition based on the hierarchical feature selection and classification framework. J. Elect. Comput. Eng. **2015**, 34–34 (2015)
18. Zhou, J., et al.: Hierarchy-aware global model for hierarchical text classification. In: ACL, pp. 1106–1117 (2020)
19. Zuo, J., Arvanitakis, G., Hacid, H.: On handling catastrophic forgetting for incremental learning of human physical activity on the edge. In: EDBT (2023)
20. Zuo, J., Arvanitakis, G., Ndhlovu, M., Hacid, H.: Magneto: edge AI for human activity recognition - privacy and personalization. In: EDBT (2024)
21. Zuo, J., Zeitouni, K., Taher, Y.: Exploring interpretable features for large time series with se4tec. In: EDBT (2019)
22. Zuo, J., Zeitouni, K., Taher, Y.: SMATE: semi-supervised spatio-temporal representation learning on multivariate time series. In: ICDM, pp. 1565–1570 (2021)
23. Zuo, J., Zeitouni, K., Taher, Y., Garcia-Rodriguez, S.: Graph convolutional networks for traffic forecasting with missing values. DMKD **37**(2), 913–947 (2023)

Geometrically-Aware Dual Transformer Encoding Visual and Textual Features for Image Captioning

Yu-Ling Chang[1], Hao-Shang Ma[2], Shiou-Chi Li[1], and Jen-Wei Huang[1](✉)

[1] Department of Electrical Engineering, National Cheng Kung University, Tainan, Taiwan
jwhuang@mail.ncku.edu.tw

[2] Department of Computer Science and Information Engineering, National Taichung University of Science and Technology, Taichung, Taiwan
hsma@nutc.edu.tw

Abstract. When describing pictures from the point of view of human observers, the tendency is to prioritize eye-catching objects, link them to corresponding labels, and then integrate the results with background information (i.e., nearby objects or locations) to provide context. Most caption generation schemes consider the visual information of objects, while ignoring the corresponding labels, the setting, and/or the spatial relationship between the object and setting. This fails to exploit most of the useful information that the image might otherwise provide. In the current study, we developed a model that adds the object's tags to supplement the insufficient information in visual object features, and established relationship between objects and background features based on relative and absolute coordinate information. We also proposed an attention architecture to account for all of the features in generating an image description. The effectiveness of the proposed Geometrically-Aware Dual Transformer Encoding Visual and Textual Features (GDVT) is demonstrated in experiment settings with and without pre-training.

Keywords: image captioning · computer vision · visual-language

1 Introduction

Image captioning involves generating a textual description of an image. The two challenges in this process are the collection of meaningful features from image data and the conversion of those features into natural language. Most automated image caption systems use an encoder-decoder architecture to deal with these problems. In [5,7], the authors used a strong baseline, Transformer, with an extra attention gate or connection structure to enhance feature combination. In [6], the authors further considered relative geometric information, whereas in [8], the focus was on capturing textual descriptions of grid features. Note however that the above-mentioned methods perform discrete analyses of objects and grid features without considering the complementarity of object tags and object regions or the dependency between grid features and regional features. In

D.-N. Yang et al. (Eds.): PAKDD 2024, LNAI 14649, pp. 15–27, 2024.
https://doi.org/10.1007/978-981-97-2262-4_2

Table 1. The example of human intuition doing image caption considers three features.

Step 1. Find important objects	Step 2. Associate object to words	Step 3. Consider background information
	car, bus, truck, person, ...	on the road/street, line up, ... ⇒ Sentence: Several cars and bus wait in line on the road.

the current study, we developed an image-captioning model based on an encoder-decoder architecture to address these shortcomings.

In developing the proposed encoder, we referred to the human habit of generating stories from pictures. As shown in Table 1, this process involves identifying important objects in the picture, finding the corresponding vocabulary to describe it, and then considering the background information in order to establish a setting or relationship between multiple objects. We developed a dual-path encoder architecture to encode the relationship among the grid feature, visual features of object areas, and the textual features of object tags. The encoder also includes a modified self-attention mechanism tasked with adding information related to relative and absolute positions. The decoder is based on a cross-on-cross attention, wherein the two features output by the encoder are used in two cross attention operations with the previous sequence, after which the corresponding outputs mutually perform an additional cross attention operation to elucidate all possible relationships. We also adjust the position embedding layer in the decoder to extract additional contextual features of the previous sequence, rather than relying on the order of tokens in the statement.

In experiments involving online testing and the MS COCO dataset, the proposed Geometrically-Aware Dual Transformer Encoding Visual and Textual Features (GDVT) outperformed state-of-the-art methods in multiple evaluation matrices. The main contributions of this work are summarized as follows:

- We embed object labels for use by a dual-path-encoder to extract supplementary contextual information, while paying particular attention to objects that are emphasized repeatedly.
- Grid features are dealt with by integrating relative and absolute coordinates to elucidate dependencies between the grid and regional features.
- A proprietary cross-on-cross attention module maximizes the utility of image features.

2 Related Works

Image captioning is a popular research field aimed at facilitating human-robot interactions. Most previous studies can be divided into two categories: templates and deep neural networks. There are also retrieval-based methods that retrieving pre-existing captions from repositories to describe images. Due to the lack of caption diversity in retrieval-based methods, we focus on generation-based methods, template-based [4] and neural-network-based methods, rather than retrieval-based methods.

Captioning based on a neural network usually involves an encoder-decoder architecture with a specific network (e.g., graph convolutional networks (GCNs) with scene graphs). In [16] , the authors used spatial GCNs to embed scene graphs into visual representations. Another common approach involves capturing image features using a convolutional neural network (CNN) and then performing sequence generation using the seq2seq-based architecture. Transformer [14] is used in the latest seq2seq models in conjunction with a CNN-based model for the encoding of visual information and generation of captions. In [6], the authors implemented Transformer using relative geometric information to obtain additional attention weights. In [7], the authors extended the attention mechanisms originally developed for Transformer to determine the relevance of attention results to corresponding queries. In [5], the authors introduced two novel attention operators with a special connection between the encoder and decoder to enable the inclusion of prior knowledge and elucidate the relationship between the various layers in the encoder and decoder. In [12], the authors proposed an X-Linear Attention Networks to conduct bilinear pooling operations. In [8], the authors generated textual descriptions of grid feature to strengthen the interaction between the object and the background in which it appears.

However, in most of previous works, they often separately consider the relationship or interaction between region features or grid features without considering the complementary between object tags and object regions and the dependency between grid features and region features. Also, the positional embedding in traditional Transformer considers only word order without feature of previous content.

3 Proposed Approach

In the following, we outline the model into a features extractor and a caption generator both of which comprise stacks of attentive layers. Note that these two modules are respectively responsible for extracting features related to objects or backgrounds from image and combining the extracted features with predicted sequence to predict the very next word until the whole sentence has been generated. In accordance with previous work [7], we replaced the attention block in Transformer [14] to filter out unrelated attentive results. In the following subsections, we detail the features extractor, caption generator, attention block and training strategy. The model architecture is illustrated in Fig. 1.

3.1 Features Extractor

In order to consider the detail of the objects in the image and the interaction between objects or the scene information of image, we extract both object features and background features. Object feature extraction is performed using a faster R-CNN to detect objects, which then outputs visual features, textual features with corresponding object tags, and bounding boxes indicating the region of the image in which the object of interest is located. Background features are obtained using the Swin Transformer through the extraction of grid feature.

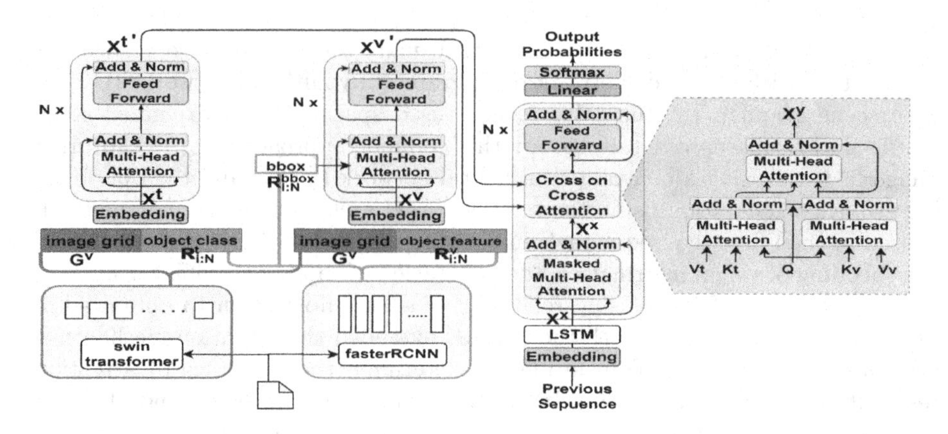

Fig. 1. Overview of proposed model architecture

$$R^v_{i:N}, R^t_{i:N}, R^{bbox}_{i:N} = fastRCNN(I) \tag{1}$$

$$G^v = SwinTransform(I) \tag{2}$$

where $R^v_{i:N}, R^t_{i:N}, R^{bbox}_{i:N}$ indicates region-based features, including visual features, textual features and bounding boxes for N object regions detected in image I. G^v is a grid-based feature of image I. After capturing $R^v_{i:N}, R^t_{i:N}, R^{bbox}_{i:N}$ and G^v, two Transformer encoders are used to encode visual and textual relationship between grid-based and region-based feature.

Textual Relationship. We directly concatenate the grid-based feature G^v with the word embedding of textual region-based feature $\hat{R}^t_{i:N}$ as a new input for the encoder. The operation of the textual relationship encoder is defined as follows, where the dimensionalities of G^v and $\hat{R}^t_{i:N}$ are projected to the same by trainable linear projection since we assume that it can learn the relationship between these two features by itself during the model training and makes an efficient mapping; Π^t is the word embedding matrix; and W^Q_i, W^K_i, W^V_i and W^O are trainable projection weights. This calculation is used to derive textual relationship, $X^{t'}$.

$$X^t = Concat(G^v, \hat{R}^t_{i:N})\ ,\ \hat{R}^t_{i:N} = R^t_{i:N} \cdot \Pi^t \tag{3}$$

$$head_i = SelfAttention(X^t W^Q_i, X^t W^K_i, X^t W^V_i) = softmax(\frac{QK^T}{\sqrt{d_k}})V \tag{4}$$

$$X^{t'} = MultiHead(X^t) = Concat(head_1, ..., head_h)W^O \tag{5}$$

Visual Relationship. We also concatenate grid-based feature G^v with the region-based feature. Here we use visual region-based feature $\hat{R}^v_{i:N}$ as a region-based feature. This operation is similar to the method used to obtain textual relationship $X^{t'}$. The attention operates on three vectors, as follows: queries Q, keys K, and values V, projected from concatenation matrix X^v and takes a weighted sum of V in accordance with the similarity distribution between Q and K. The operation is as follows, where the dimensionalities of G^v and $\hat{R}^v_{i:N}$ are the same. Π^v is the projection matrix; and W_i^Q, W_i^K, W_i^V and W^O are trainable projection weights. This calculation is used to derive visual relationship, $X^{v\prime}$.

$$X^v = Concat(G^v, \hat{R}^v_{i:N}) \ , \ \hat{R}^v_{i:N} = R^v_{i:N} \cdot \Pi^v \tag{6}$$

$$head_i = SelfAttention(X^v W_i^Q, X^v W_i^K, X^v W_i^V) = softmax(\frac{QK^T}{\sqrt{d_k}})V \tag{7}$$

$$X^{v\prime} = MultiHead(X^v) = Concat(head_1, ..., head_h)W^O \tag{8}$$

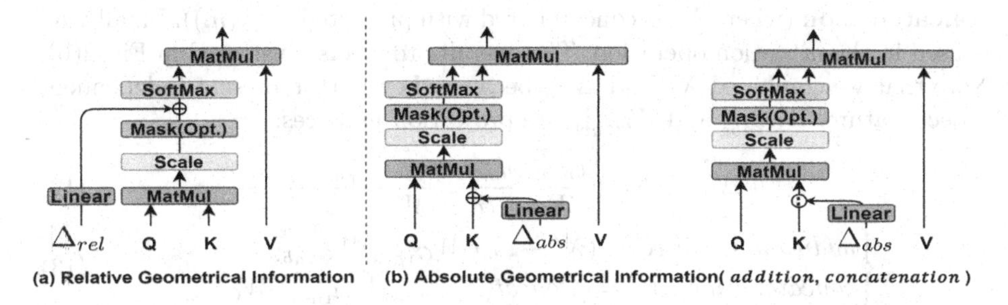

Fig. 2. Model structure involving the addition of relative and absolute geometric information where $\oplus$ is addition and projection; and $\odot$ is concatenation and projection.

However, concatenating two features directly would make it difficult for the model to differentiate between G^v and $\hat{R}^v_{i:N}$, considering that they are both visual features. We therefore differentiate between them using geometric features which also provide the interaction information among objects, as outlined below.

Relative Geometric Information. Each object possesses its own bounding box $(x_{min}, y_{min}, x_{max}, y_{max})$. We define $(0, 0, W, H)$ as the bounding box of the grid feature, where W and H respectively indicate the width and height of the image. We respectively compute (x_m, y_m, w_m, h_m) and (x_n, y_n, w_n, h_n) from the bounding boxes of objects m and n, where x_m and y_m are the center coordinates along the x and y-axes, and w_m and h_m respectively indicate the width and height of bounding box m. We then calculate relative geometric relationship between each pair of objects. Take as an example objects m and n, wherein the relative center coordinate and the ratio of the box size are dealt with as a 4D vector $\Delta_{rel}(m, n)$. We then project $\Delta_{rel}(m, n)$ using two linear layers $W_{G_1 rel}$ and $W_{G_2 rel}$ as geometric wight X^G_{mn} to be added to the self-attention weight as weight bias. The two linear projections are used to learn the relationship or importance between each object. A log operation is used to prevent the geometric weight from dominating the attention weight. The operation is defined below and illustrated in

Fig. 2(a). Note that we added X^G_{mn} to $\frac{QK^T}{\sqrt{d_k}}$ because they both represent the relationship among objects.

$$\Delta_{rel}(m,n) = (\frac{|x_m - x_n|}{w_m}, \frac{|y_m - y_n|}{h_m}, \frac{w_n}{w_m}, \frac{h_n}{h_m}) \tag{9}$$

$$X^G_{mn} = log(ReLU(\Delta_{rel}\, W_{G_1 rel}\, W_{G_2 rel})) \tag{10}$$

$$SelfAttention(Q,K,V) = softmax(\frac{QK^T}{\sqrt{d_k}} + X^G_{mn})V \tag{11}$$

Absolute Geometric Information. As with relative geometric information, we have a bounding box for each object; however, in this case, we rescale it according to the width and height of the image as $\Delta_{abs}(m)$. We then use $\Delta_{abs}(m)$ to obtain a new key K' via **addition** (where $\Delta_{abs}(m)$ is added to X^v after mapped) and **concatenation** (where X^v is concatenated with projected $\Delta_{abs}(m)$). Finally, K' is used in the attention operation. The overall process is illustrated in Fig. 2(b). Note that we combined X^v and Δ_{abs} because they both represent independent object feature. $W_{G_1 abs}$ and $W_{G_2 abs}$ are projection matrices.

$$\Delta_{abs}(m) = (\frac{x_{min}}{W}, \frac{y_{min}}{H}, \frac{x_{max}}{W}, \frac{y_{max}}{H}) \tag{12}$$

$$\begin{cases} addition & : K' = (X^v + \Delta_{abs} W_{G_1 abs}) W_{G_2 abs} \\ concatenation & : K' = (Concat(X^v, \Delta_{abs} W_{G_1 abs})) W_{G_2 abs} \end{cases} \tag{13}$$

3.2 Caption Generator

Our caption generator is based on the Transformer decoder which combines features derived using the features extractor with the previous sentence $y_{1:t-1}$ to predict the next word y_t. We replaced the position embedding layer in Transformer with LSTM to encode the contextual semantics of $y_{1:t-1}$. Π^t is a word embedding matrix. A cross-on-cross attention block is used to combine features extracted using the features extractor with those of the previous sequence.

Cross-on-Cross Attention. Two features (visual relationship $X^{v'}$ and textual relationship $X^{t'}$) are used for the attention operation with sentence representation X^x; therefore, we employed two cross-attention blocks to perform discrete attention operations for visual and textual relationships with the previous sentence. Finally, we combine the output of the two cross-attentions by performing another cross-attention operation, as detailed on the right side of Fig. 1.

$$X^x = LSTM(y_{1:t-1} \cdot \Pi^t) \tag{14}$$

$$\begin{cases} head^t_i = CrossAttention(X^x W^Q_i, X^{t'} W^K_i, X^{t'} W^V_i) \\ head^v_i = CrossAttention(X^x W^Q_i, X^{v'} W^K_i, X^{v'} W^V_i) \end{cases} \tag{15}$$

$$head^y_i = CrossAttention(head^v_i\, W^Q_i, head^t_i\, W^K_i, head^t_i\, W^V_i) \tag{16}$$

$$X^y = MultiHead(X^t) = Concat(head^y_1, ..., head^y_h) W^O \tag{17}$$

3.3 Attention Block

Attention block is used to replace the self-attention and cross-attention in the proposed model. In accordance with AoANet [7], we used a sigmoid gate to filter out irrelevant attention results and implemented a residual connection to avoid the problem of gradient descent associated with the sigmoid function. The process is detailed below and illustrated in Figs. 3(a) and 3(b), where Q and A are the query and the attention result of multi-head attention; σ and $\otimes$ are sigmoid function and element-wise product; $h(\hat{A})$ is the operation of "attention on attention" which multiplies the output of the sigmoid function with the output of the linear projection; and $g(\hat{A})$ is the alternate calculation with extra residual connection adding $\gamma\hat{A}$ to $h(\hat{A})$.

$$h(\hat{A}) = (\hat{A} \cdot W_1) \otimes \sigma(\hat{A} \cdot W_2), \quad \hat{A} = Concat(Q, A) \tag{18}$$

$$g(\hat{A}) = h(x) + \gamma\hat{A} = (\hat{A} \cdot W_1) \otimes \sigma(\hat{A} \cdot W_2) + \gamma\hat{A} \tag{19}$$

The maximum value of the derivative of the sigmoid function is $\frac{1}{4}$, which could lead to the problem of gradient descent due to a mathematical concept that values of less than 1 become increasingly smaller (approaching zero) when multiplied together. However, the derivative of $g(\hat{A})$ in Eq. 19 will be at least γ and never reach zero , thereby preventing gradient descent, as shown on the right side of Fig. 3.

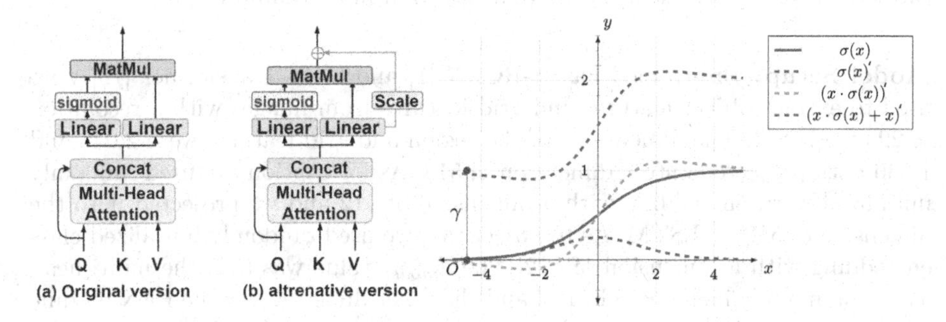

Fig. 3. (Left side) Original and alternative versions of attention model; (right side) derivative function of σ, $x \cdot \sigma(x)$ and $x \cdot \sigma(x) + x$ plotted on the graph.

3.4 Training and Objectives

The proposed model underwent two-stage training, based on cross entropy loss followed by fine-tuning via self-critical sequence training based on the CIDEr-D score [13]. The entire model including the feature extractor and caption generator underwent end-to-end learning process on training set during both of the two training stages.

Training via Cross Entropy Loss. Here, the model is optimized via cross entropy loss L_{CE}. The previously predicted sequence $y_{1:t-1}$ is used to predict the next word y_t and obtain the word probability distribution p_θ based on model parameter θ.

$$L_{CE}(\theta) = -\sum_{t=1}^{T}(log(p_\theta(y_t|y_{1:t-1}))) \tag{20}$$

Fine-Tuning via CIDEr-D. After each batch training session, we generate a sample sentence y^s using the probability distribution and baseline sentence $\hat{y}$, which is the result of greedy decoding. We then evaluate them using the CIDEr-D evaluator $r(\cdot)$.

$$\nabla_\theta L_{RL}(\theta) \approx -(r(y^s_{1:T}) - r(\hat{y}_{1:T}))\, \nabla_\theta\, log(p_\theta(y^s_{1:T})) \tag{21}$$

4 Experiments

4.1 Experiments Setup

Dataset. The MS COCO dataset was used to evaluate the performance of the proposed system. We adopted the 'Karpathy' splits widely used in previous works to split the data into 113,287 images for training (5 captions each) and two additional 5k images for validation and testing. Note that the generation of captions involved collecting words that occurred more than 5 times.

Model Setup. Pre-trained Faster-RCNN [1] and Swin Transformer [10] were used to extract object features and grid features from images with a resolution of $224 \times 224 \times 3$. The dimensions of the region and grid features were 2,048 and 1,536 both projected onto a dimension of 512. As for caption, we used randomly initialized word embedding with a dimension of 512 and we projected it to the dimension of 512 by LSTM. As to object tag, we used randomly initialized class embedding with a dimension of 512. The d_{model} value was 512, the multi-head attention module included 8 heads, and the inner dimension of the FFN module was 2,048. For ablation studies, those without encoder and decoder layers marked **L**(=3) are applicable layers. For testing, we applied a beam size of 3.

Evaluation Matrices. This study adopted matrices [3] used in previous works including BLEU-1/2/3/4, METEOR, ROUGE-L, CIDEr, and SPICE, which are marked as B1/2/3/4, M, R, C, and S.

Comparison Methods. SCST [13] and Up-Down [2] use attention machines. GCN-LSTM [17] uses GCN to encode semantic and spatial relationships. AoANet [7] alerts attention to filter out useless results. M^2 Transformer [5] comprises memory-augmented encoding layers with the encoder and decoder

Table 2. Comparison of proposed network and SOTA in offline evaluation: the quality of the captions was assessed using standard evaluation matrices, where **GDVT** refers to "SB w/ grid, textual,bbox (rel+abs$_C$)" and **L** indicates the number of encoder and decoder layers.

(A) Offline Evaluation

model	Cross-Entropy Loss								CIDEr Score Optimization							
	B1	B2	B3	B4	M	R	C	S	B1	B2	B3	B4	M	R	C	S
SCST[13]	-	-	-	30.0	25.9	53.4	99.4	-	-	-	-	34.2	26.7	55.7	114.0	-
Up-Down[2]	77.2	-	-	36.2	27.0	56.4	113.5	20.3	79.8	-	-	36.3	27.7	56.9	120.1	21.4
GCN-LSTM[17]	77.3	-	-	36.8	27.9	57.0	116.3	20.9	80.5	-	-	38.2	28.5	58.3	127.6	22.0
AoANet[7]	77.4	-	-	37.2	28.4	57.5	119.8	21.3	80.2	-	-	38.9	29.2	58.8	129.8	22.4
M^2 Transformer[5]	-	-	-	-	-	-	-	-	80.8	-	-	39.1	29.2	58.6	131.2	22.6
X-LAN[12]	78.0	62.3	48.9	38.2	28.8	58.0	122.0	21.9	80.8	65.6	51.4	39.5	29.5	59.2	132.0	23.4
NG-SAN[6]	-	-	-	-	-	-	-	-	-	-	-	39.9	29.3	59.2	132.1	23.3
CLIP[8]	-	-	-	-	-	-	-	-	81.5	-	-	39.7	**30.0**	59.5	**135.9**	23.7
GDVT [L=3]	77.5	61.7	48.2	37.4	28.9	57.7	121.6	21.8	82.6	67.4	53.0	40.7	29.7	60.0	134.0	23.8
GDVT [L=6]	79.4	63.9	49.5	37.7	28.3	58.4	123.4	21.9	**82.8**	**67.6**	**53.2**	**40.8**	29.9	**60.1**	135.1	**23.9**

(B) Online Evaluation

model	B1		B2		B3		B4		M		R		C	
	C5	C40	C5	C40	C5	C40	C5	C40	C5	C40	C5	C40	C5	C40
SCST[13]	78.1	93.7	61.9	86	47	75.9	35.2	64.5	27	35.5	56.3	70.7	114.7	116.7
Up-Down[2]	80.2	95.2	64.1	88.8	49.1	79.4	36.9	68.5	27.6	36.7	57.1	72.4	117.9	120.5
GCN-LSTM[17]	80.8	95.2	65.5	89.3	50.8	80.3	38.7	69.7	28.5	37.6	58.5	73.4	125.3	126.5
AoANet[7]	81	95	65.8	89.6	51.4	81.3	39.4	71.2	29.1	38.5	58.9	74.5	126.9	129.6
M^2 Transformer[5]	81.6	96	66.4	90.8	51.8	82.7	39.7	72.8	29.4	39	59.2	74.8	129.3	132.1
X-LAN[12]	81.1	95.3	66	89.8	51.5	81.5	39.5	71.4	29.4	38.9	59.2	74.7	128	130.3
NG-SAN[6]	80.8	95.0	65.4	89.3	50.8	80.6	38.8	70.2	29.0	38.4	58.7	74.0	126.3	128.6
GDVT [L=6]	**83.0**	**96.4**	**67.9**	**91.6**	**53.2**	**83.7**	**40.7**	**73.7**	**30.0**	**39.8**	**60.3**	**76.1**	**132.8**	**136.5**

connected by a mesh structure. XLAN [12] uses X-Linear attention to conduct bilinear pooling for high-order features. NG-SAN [6] uses instance normalization to solve internal covariate shifts and applies geometric weights. CLIP [8] retrieves text descriptions of grid features to strengthen the relationship between objects and images.

In experiment settings involving huge pretrained datasets, [9,11,15,18,19] were pre-trained using a multiple image-text corpus other than the COCO dataset to avoid overfitting. GRIT [11] is trained using an end-to-end scheme to encode both grid and region features. SimVLM [15] is pretrained using extremely large datasets.

4.2 Experiment Result

Comparison with SOTA. The proposed scheme was compared with SOTA in terms of captioning performance. Note that an excessive number of trainable parameters for the image caption model training can lead to overfitting. Thus, we either trained the model using only the COCO dataset until overfitting became an issue [2,5–8,12–14,17] , or we pre-trained the model using multiple corpus in addition to COCO to avoid overfitting, [9,11,15,18,19]. Thus, we treat the experiment as two settings mentioned above. In the following two paragraphs, we denote "SB w/ grid, textual,bbox (rel+abs$_C$)" as **GDVT**. **SB** is "Transformer with LSTM, AOA". This is described in detail in ablation studies. All the experiment results of previous works are excerpted from the original papers.

Normal Setting: Training with COCO based on "Karpathy" Split Only.
Offline Evaluation. Table 2(A) lists the offline evaluation results, in which "Karpathy" split test data is used as an evaluation target. When applied to BLEU-1/2/3/4, the proposed network with six layers outperformed previous models. When applied to METEOR and SPICE, the results were comparable. This was also the case for the model with three layers. The 3-layer model with 59.6M model parameters, less than current SOTA model with 96.1M model parameters, outperformed previous models when applied to most of the evaluation matrices, such as BLEU-1/2/3/4, ROUGE, and SPICE. That is, the 3-layer model can achieve comparable performance based on low computational cost.

Table 3. Comparison of proposed network and SOTA models pretrained using a huge dataset, where **L** indicates the number of layers; **VL** and **OD** respectively indicate the types of pretrained data, visual-language corpus and object detection datasets; and **4DS** contains 4 datasets: COCO, Visual Genome, Open Images, and Object365.

model	# pretrain data	Cross-Entropy Loss							
		B1	B2	B3	B4	M	R	C	S
UVLP[19]	3.0M (VL)	-	-	-	39.5	29.3	-	129.3	23.2
Oscar[9]	6.5M (VL)	-	-	-	40.5	29.7	-	137.6	22.8
VinVL[18]	8.9M (VL)	-	-	-	41.0	31.1	-	140.9	25.2
SimVLM[15]	1.8B (VL)	-	-	-	40.6	33.7	-	143.3	25.4
GRIT[11]	4DS (OD)	84.2	-	-	42.4	30.6	60.7	144.2	24.3
GDVT+Oscar [L=6]	-	83.8	68.9	54.5	42.1	30.4	61.1	139.2	24.4
GDVT+GRIT [L=6]	-	84.0	69.3	54.9	42.6	30.6	61.2	140.2	24.5

Online Evaluation. Table 2(B), lists the online evaluation results obtained using two test datasets: C5 (5 ground-truth captions) and C40 (40 ground-truth captions). The proposed network outperformed previous works in all evaluation metrics. Note that **CLIP** [8] did not present online evaluation results; therefore, a comparison could not be performed.

Huge Dataset Setting: Preraining with Huge Corpus. Table 3 lists comparisons with works using large pre-training datasets or advanced pre-training methods, where **GDVT+Oscar** indicates that the object features obtained by Faster R-CNN are replaced with the object features obtained after the pre-training of Oscar. Similarly, **GDVT+GRIT** indicates object features have been replaced with GRIT's. The performance of the proposed network was superior to those of the original Oscar and GRIT in most evaluation matrices, which indicates that the proposed architecture can supply the information beyond visual features of objects.

Table 4. Ablation study aimed at investigating the impact of each component where **V**, **G**, **T**, **B** refer to visual object features, image grids, object tags, and object bboxes. (note that all of the models included three layers).

model	V	G	T	B	Cross-Entropy Loss (above row) CIDEr Score Optimization (below row)							
					B1	B2	B3	B4	M	R	C	S
(A) Ablation Study: impact of each component in our model on caption generation												
B	✓				76.1	59.9	45.2	34.0	27.6	56.2	113.3	21.0
					80.0	64.6	50.2	38.3	28.6	58.4	128.0	22.4
B w/ LSTM	✓				76.1	60.1	45.8	34.7	27.9	56.6	115.2	21.4
					80.2	65.0	50.6	38.8	28.9	58.6	128.4	22.5
B w/ LSTM, AOA	✓				76.3	60.3	46.9	36.4	28.2	56.7	116.6	21.5
= SB					81.6	65.8	51.1	38.9	29.0	59.0	129.0	22.8
SB w/ grid	✓	✓			76.9	61.1	47.0	36.9	28.3	57.2	120.2	21.6
					81.5	66.3	52.0	39.9	29.3	59.4	132.0	23.2
SB w/ grid, textual	✓	✓	✓		77.4	61.7	48.2	37.4	28.8	57.7	121.1	21.6
					82.0	66.6	52.2	39.9	29.4	59.4	132.0	23.4
SB w/ grid, textual,bbox	✓	✓	✓	✓	**77.5**	**61.7**	**48.2**	**37.4**	**28.9**	**57.7**	**121.6**	**21.8**
(rel+abs$_C$)					**82.6**	**67.4**	**53.0**	**40.7**	**29.7**	**60.0**	**134.0**	**23.8**
(B) Ablation Study: impact adding geometric feature using three methods												
SB w/ grid, textual,bbox	✓	✓	✓	✓	77.4	61.7	48.2	37.4	28.8	57.7	121.1	21.7
(rel)					82.4	67.2	52.7	40.4	29.7	59.7	133.8	23.8
SB w/ grid, textual,bbox	✓	✓	✓	✓	77.3	61.5	47.2	35.8	28.6	57.4	120.4	21.7
(rel+abs$_A$)					82.4	67.1	52.5	40.3	29.6	59.7	133.6	23.6
SB w/ grid, textual,bbox	✓	✓	✓	✓	**77.5**	**61.7**	**48.2**	**37.4**	**28.9**	**57.7**	**121.6**	**21.8**
(rel+abs$_C$)					**82.6**	**67.4**	**53.0**	**40.7**	**29.7**	**60.0**	**134.0**	**23.8**

Ablation Study. This analysis was meant to elucidate the impact of each model component (see Table 4(A)). We first used Transformer as a baseline (denoted as **B**) to receive object features from Faster-RCNN and decode sentences. We then replaced the position embedding layer in B with LSTM (denoted as **B w/ LSTM**), which benefits from additional contextual information. Note that we also altered the attention module to filter attention output as **B w/ LSTM, AOA**. Based on Transformer with LSTM and AOA using only the visual features of object areas as a stronger baseline (denoted as **SB**), we concatenated visual region features with grid features (denoted as **SB w/ grid**). This change made a significant improvement. These results indicate that grid features contain background information and the relationships among objects. We also extended the model to a dual-path encoder (denoted as **SB w/ grid, textual**), which encodes the relationship between grid features and visual object features as well as between grid features and textual object features. The fact that this further improved performance indicates that visual and textual features both provide valuable information. Finally, we added positional features to augment the available geometric information. As shown in Table 4(B), we tesed three methods:

SB w/ grid, textual, bbox(rel) (adding only relative position information), **SB w/ grid, textual, bbox(rel+abs$_A$)** (adding relative and absolute position via addition) and **SB w/ grid, textual, bbox(rel+abs$_C$)**(adding relative as well as absolute position via concatenation). **SB w/ grid, textual, bbox(rel+abs$_C$)** made the most pronounced improvement due to its ability to integrate absolute and relative position information without interfering with the original visual features.

5 Conclusions

We propose a Transformer-based duel-path architecture, GDVT, for caption generation. For region-based features, we incorporate complementary features, object tag embedding, to reinforce repeatedly emphasized information. For grid-based features, we add absolute and relative position information to enhance the relationship between grid and region features. Finally, we design cross-on-cross attention to synthesize features for prediction. We improved performance on both experiments with and without pretraining. The results of the case study confirm that object tags, geometric information and grid features indeed improve image captioning.

References

1. Anderson, P., et al.: Bottom-up and top-down attention for image captioning and visual question answering. In: CVPR (2018)
2. Anderson, P., et al.: Bottom-up and top-down attention for image captioning and visual question answering. In: 2018 IEEE/CVF Conference on Computer Vision and Pattern Recognition, pp. 6077–6086 (2018)
3. Chen, X., et al.: Microsoft coco captions: data collection and evaluation server. CoRR (2015)
4. Cornia, M., Baraldi, L., Cucchiara, R.: Show, control and tell: a framework for generating controllable and grounded captions. In: 2019 IEEE/CVF Conference on Computer Vision and Pattern Recognition (CVPR), pp. 8299–8308 (2019)
5. Cornia, M., Stefanini, M., Baraldi, L., Cucchiara, R.: Meshed-memory transformer for image captioning. In: 2020 IEEE/CVF Conference on Computer Vision and Pattern Recognition (CVPR), pp. 10575–10584
6. Guo, L., Liu, J., Zhu, X., Yao, P., Lu, S., Lu, H.: Normalized and geometry-aware self-attention network for image captioning. In: 2020 IEEE/CVF Conference on Computer Vision and Pattern Recognition (CVPR), pp. 10324–10333 (2020)
7. Huang, L., Wang, W., Chen, J., Wei, X.Y.: Attention on attention for image captioning. In: 2019 IEEE/CVF International Conference on Computer Vision (ICCV), pp. 4633–4642 (2019)
8. Kuo, C., Kira, Z.: Beyond a pre-trained object detector: cross-modal textual and visual context for image captioning. In: 2022 IEEE/CVF Conference on Computer Vision and Pattern Recognition (CVPR), pp. 17948–17958 (2022)
9. Li, X., et al.: OSCAR: object-semantics aligned pre-training for vision-language tasks. In: Vedaldi, A., Bischof, H., Brox, T., Frahm, J.-M. (eds.) ECCV 2020. LNCS, vol. 12375, pp. 121–137. Springer, Cham (2020). https://doi.org/10.1007/978-3-030-58577-8_8

10. Liu, Z., et al.: Swin transformer: Hierarchical vision transformer using shifted windows. In: Proceedings of the IEEE/CVF International Conference on Computer Vision (ICCV) (2021)
11. Nguyen, V.Q., Suganuma, M., Okatani, T.: Grit: faster and better image captioning transformer using dual visual features, pp. 167–184 (2022)
12. Pan, Y., Yao, T., Li, Y., Mei, T.: X-linear attention networks for image captioning. In: 2020 IEEE/CVF Conference on Computer Vision and Pattern Recognition (CVPR), pp. 10968–10977 (2020)
13. Rennie, S.J., Marcheret, E., Mroueh, Y., Ross, J., Goel, V.: Self-critical sequence training for image captioning. In: 2017 IEEE Conference on Computer Vision and Pattern Recognition (CVPR), pp. 1179–1195 (2017)
14. Vaswani, A., et al.: Attention is all you need. In: Guyon, I., et al. (eds.) Advances in Neural Information Processing Systems (2017)
15. Wang, Z., Yu, J., Yu, A.W., Dai, Z., Tsvetkov, Y., Cao, Y.: SimVLM: Simple visual language model pretraining with weak supervision. In: International Conference on Learning Representations (2022)
16. Yang, X., Tang, K., Zhang, H., Cai, J.: Auto-encoding scene graphs for image captioning. In: 2019 IEEE/CVF Conference on Computer Vision and Pattern Recognition (CVPR), pp. 10677–10686 (2019)
17. Yao, Ting, Pan, Yingwei, Li, Yehao, Mei, Tao: Exploring visual relationship for image captioning. In: Ferrari, Vittorio, Hebert, Martial, Sminchisescu, Cristian, Weiss, Yair (eds.) Computer Vision – ECCV 2018. LNCS, vol. 11218, pp. 711–727. Springer, Cham (2018). https://doi.org/10.1007/978-3-030-01264-9_42
18. Zhang, P., et al.: VinVL: Revisiting visual representations in vision-language models. In: Proceedings of the IEEE/CVF Conference on Computer Vision and Pattern Recognition (CVPR), pp. 5579–5588 (2021)
19. Zhou, L., Palangi, H., Zhang, L., Hu, H., Corso, J.J., Gao, J.: Unified vision-language pre-training for image captioning and VQA. ArXiv (2020)

MHDF: Multi-source Heterogeneous Data Progressive Fusion for Fake News Detection

Yongxin Yu[1], Ke Ji[1](✉), Yuan Gao[1], Zhenxiang Chen[1], Kun Ma[1], and Jun Wu[2]

[1] School of Information Science and Engineering, University of Jinan, Jinan, China
{ise_jik,ise_czx,ise_mak}@ujn.edu.cn
[2] School of Computer and Information Technology, University of Beijing Jiaotong, Beijing, China
wuj@bjtu.edu.cn

Abstract. Social media platforms are inundated with an extensive volume of unverified information, most of which originates from heterogeneous data from a variety of diverse sources, spreading rapidly and widely, thereby posing a significant threat to both individuals and society. An existing challenge in multimodal fake news detection is its limitation to acquiring textual and visual data exclusively from a single source, which leads to a high level of subjectivity in news reporting, incomplete data coverage, and difficulties in adapting to the various forms and sources of fake news. In this paper, we propose a fake news detection model (MHDF) for multi-source heterogeneous data progressive fusion. Our approach begins with gathering, filtering, and cleaning data from multiple sources to create a reliable multi-source multimodal dataset, which involved obtaining reports from diverse perspectives on each event. Subsequently, progressive fusion is achieved by combining features from diverse sources. This is achieved by inputting the features obtained from the textual feature extractor and visual feature extractor into the news textual and visual feature fusion module. We also integrated sentiment features from the text into the model, allowing for multi-level feature extraction. Experimental results and analysis indicate that our approach outperforms other methods.

Keywords: Fake News detection · Multi-source heterogeneous data · Feature fusion

1 Introduction

The spread of fake news is characterized by its rapid dissemination, extensive reach, emotional content, and click-driven nature. While manual identification of fake news may have a high accuracy rate, the associated high costs and inefficiencies make it challenging to address the vast scale of fake news. Therefore, the investigation of methods for detecting fake news is crucial for upholding the

D.-N. Yang et al. (Eds.): PAKDD 2024, LNAI 14649, pp. 28–39, 2024.
https://doi.org/10.1007/978-981-97-2262-4_3

credibility of news, protecting the public from misinformation, and preserving social stability.

With the boom in social media, news content has evolved into a multimodal form, incorporating text, images, and video. Therefore, effectively extracting and fusing multimodal features has become a hot research topic. Convolutional neural networks, recurrent neural networks, and pre-trained models like BERT [1] and GPT [2] have gained widespread adoption in the domain of fake news detection. Powerful feature extractors such as TextCNN [3], VGG19 [4], and Swin Transform [5], etc. can learn higher-order feature representations. In addition, researchers have focused on investigating methods for fusing multimodal data. Some studies also employ data amplification techniques, such as text reverse translation [10], generating image descriptions, and extracting text from images, to enhance the quantity and quality of available data.

Despite significant advancements in fake news detection technology, several issues with existing methods still persist. Firstly, the current methodology lacks focus on the connections between diverse reports of the same event. There is a single source of information, which hinders a comprehensive understanding and objective conclusions about fake news. Secondly, existing methods are typically limited to amplifying a single data modality, which hinders achieving a comprehensive data augmentation effect and depends on the implementation and impact of another model. Furthermore, current methods overlook the deep sentiment features of the text, and extracting such features could be significantly helpful in identifying highly inflammatory fake news.

To address the aforementioned issues, this paper initially amplified the original dataset by incorporating multi-source heterogeneous data obtained through web crawling techniques from three data-rich websites. Subsequently, we adopt the SimBERT [14] to carefully clean and filter the data for constructing a multi-source, multimodal dataset. The model extracts and represents the textual features and visual features from the dataset separately and designs a progressive feature fusion approach to fuse diverse reports of the same event. Additionally, the inclusion of sentiment features helps capture the emotional polarity in the text. The contributions of this paper are three folds:

(1) We propose a fake news detection model (MHDF) for multi-source heterogeneous data progressive fusion. By using web crawler technology, our model expands the dataset to obtain reports about the same event from diverse perspectives. The method of progressive fusion of features is used to build the textual feature fusion module and visual feature fusion module, and simultaneously introduce sentiment features. The experimental results confirm the effectiveness of our proposed method.
(2) A progressive feature fusion approach is introduced, which proficiently integrates information from multiple sources by initially modeling them separately and eventually combining them gradually.
(3) We created a multi-source, multimodal dataset. This dataset not only includes textual reports about the same event from multiple perspectives

but also encompasses image data of the event, enabling comprehensive data collection.

2 Related Work

Contemporary research emphasis on extracting textual and visual features using powerful feature extractors. Subsequently, information from different modalities is spliced and fused, or inter-modal comparison learning and inter-modal information enhancement methods are employed to improve rumour detection.

Wang et al. [6] constructed event adversarial neural networks to acquire textual and visual feature representations in news using event discriminators. Khattar et al. [7] proposed a multimodal variational autoencoder to overcome the inability of existing fake news detection methods to effectively learn a shared representation of multimodal information. Jin et al. [8] enhance the interaction between multimodal information by introducing an attention mechanism and improve fake news detection by incorporating external information. Wu et al. [9] draw on individuals' habit of observing text before images when reading news, and propose a multimodal co-attention network for fake news detection. Hua et al. [10] employed data amplification and contrastive learning techniques to further analyze the impact of diverse image formats on fake news detection. Guo et al. [11] addresses the multimodal fusion of large amounts of information on social networks through the mutual attention module.

To enhance the information available to the model, previous approaches have attempted to employ various methods aimed at increasing both the quantity and quality of the available data. Reverse translation can be used for news texts to obtain text data that conveys the same meaning despite differences in structure. We can enrich textual data by extracting relevant information from images, such as embedded text, or by generating image descriptions using a specialized text generation model.

3 MHDF Model

3.1 Model Overview

We define a news instance $I = \{Ori_{text}, Ori_{image}, S_{text}, S_{image}\}$ as a tuple representing four different modalities of contents: the original textual content Ori_{text}, the original visual content Ori_{image}, the multi-source textual content S_{text}, and the multi-source visual content S_{image}. The overall structure of our proposed MHDF model is shown in Fig. 1. The model aims to extract features from diverse modalities and multiple data sources, learning a unified representation that is used as input for the classifier.

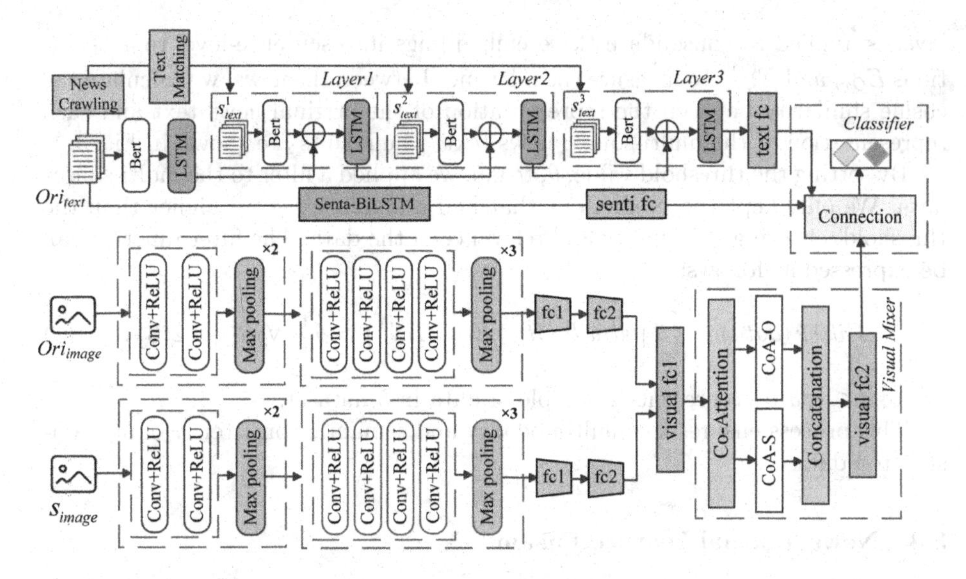

Fig. 1. The network structure of the proposed MHDF model consists of five main parts: multi-source heterogeneous data amplification at the top branch, news textual feature fusion and sentiment feature extractor in the middle branch, news visual feature fusion at the bottom branch, and feature integration in the classifier.

3.2 Multi-source Heterogeneous Data Amplification

This section focuses on collecting diverse descriptions of news reports for each news sample, addressing issues in fake news detection caused by relying on a single data source. This approach mitigates subjective reporting, incomplete information, and challenges in capturing diversity.

We extracted a minimum of five keywords, each consisting of at least two characters, from each news text using TextRank4ZH [12] and utilized web crawling to acquire reports related to these keywords. The reports were collected from three data-rich, large websites, providing diverse perspectives on each news event, including textual and visual information. In cases where we were unable to collect relevant reports, we implemented a zero-padding strategy to construct an initial multi-source, multimodal dataset.

To ensure the accuracy of dataset construction, this section employed the SimBERT model from the field of text matching for data screening. The model consists of three main parts: Embedding, Encoder, and Prediction. The prediction result p ($p \in [0, 1]$) is obtained by calculating the similarity between the original news text and each multi-source news text.

Specifically, the input news text sentences $Ori_{text} = [o_1, o_2, ..., o_n]$ and $S_{text}^{(j)} = [s_1, s_2, ..., s_n]$ are first encoded. Here n is the number of words in the text, and j ($j \in \{1, 2, 3\}$) is the number of sources. And add the $[CLS]$ tag at the beginning of the sentence to generate token lists T_{Ori} and $T_S^{(j)}$. The token lists are inputted into BERT to generate word-level embeddings. Then, a pooling layer is applied to consolidate these embeddings into sentence-level representations E_{Ori} and $E_S^{(j)}$. To measure the relevance between the news, we calculate the cosine similarity between the representation of the original news text and each representation of the multi-source news text. The results are shown in Fig. 3.

By setting the threshold value, θ, to 0.5, we applied a filter to the multi-source data. We only kept the predictions that had a similarity score higher than the threshold, ensuring the quality and relevance of the data. The filter function can be expressed as follows:

$$Filter\left(data\right)=\left\{d\mid similarity\left(E_{Ori},E_S^{(j)}\right)>\theta\right\},\forall j\in\{1,2,3\}, \tag{1}$$

where $d \in \{data\}$ represents a sample of data in a dataset.

This process ensures the multimodality and semantic consistency of our constructed dataset.

3.3 News Textual Feature Fusion

To achieve the combining of news text from multiple sources, we propose a progressive fusion strategy in this section. The strategy combines a multilayer network with the conventional LSTM [15] to systematically integrate news textual features from multiple sources. The fusion method is shown in Fig. 2.

We input the original news text Ori_{text} and the multi-source news text $S_{text}^{(j)}$ into the pre-training model BERT, which maps words to representations in a high-dimensional vector space, respectively: $f_{text}^{original}$, $f_{text}^{multi-source^{(j)}}$.

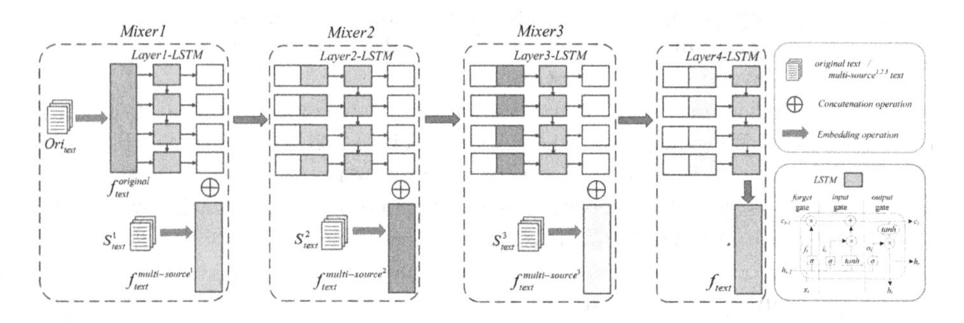

Fig. 2. Achieving progressive fusion of heterogeneous data from multiple sources by constructing three mixers.

We fused the news text vectors from each source using three mixers in a progressive manner. Add multiple source texts to each $Mixer$ and mix them sequentially, one after another. The fusion process consists of four layers, where each layer combines the current input news text vectors with the hidden states from the previous layer. Firstly, we input the original news text vectors $f_{text}^{original}$ into the $Layer1-LSTM$ to obtain its hidden states f_t^1. Subsequently, the first multi-source news text vectors $f_{text}^{multi-source^1}$ are concatenated with the hidden states f_t^1 of $Layer1-LSTM$. The combined input is then passed to the $Layer2-LSTM$ to generate the new hidden states f_t^2. The process continues until the third mixer, which merges news text vectors from two additional sources. At each step, we combine the hidden states of the previous layer with the current multi-source news text vectors and input them into the next LSTM layer. After the $Layer4-LSTM$, a fully-connected layer $(text-fc)$ is introduced to adjust the dimensionality of the textual features and generate a joint representation of the textual semantic features f_{text}.

$$f_t^1 = Layer1 - LSTM\left(f_{text}^{original}\right), \tag{2}$$

$$f_t^2 = Layer2 - LSTM\left(concat\left(f_t^1, f_{text}^{multi-source^1}\right)\right), \tag{3}$$

$$f_t^3 = Layer3 - LSTM\left(concat\left(f_t^2, f_{text}^{multi-source^2}\right)\right), \tag{4}$$

$$f_t^4 = Layer4 - LSTM\left(concat\left(f_t^3, f_{text}^{multi-source^3}\right)\right), \tag{5}$$

$$f_{text} = \varphi\left(W_{tf} \cdot f_t^4\right), \tag{6}$$

where f_t^* are the hidden states of each LSTM processed layer, W_{tf} is the weight matrix of the fully connected layer, and $\varphi(\cdot)$ is the tanh activation.

3.4 News Visual Feature Fusion

The visual network (the bottom branch in Fig. 1) takes original news images Ori_{image} and multi-source news images S_{image} as inputs and used a VGG19 network to extract semantic features of the images. A fully connected layer $(visual-fc1)$ was added after the fc2 layer to adjust the dimensionality of visual features.

$$f_{image}^{original} = \varphi\left(W_{vf1} \cdot R_{V_{vgg}}^{original}\right), \tag{7}$$

$$f_{image}^{multi-source} = \varphi\left(W_{vf1} \cdot R_{V_{vgg}}^{multi-source}\right), \tag{8}$$

where $R_{V_{vgg}}^{original}$ and $R_{V_{vgg}}^{multi-source}$ are the original news visual features and multi-source news visual features, extracted by the VGG19 network, W_{vf1} is the weight matrix of the fully connected layer, and $\varphi(\cdot)$ is the tanh activation.

To effectively combine visual features from original and multi-source news, we utilize the Co-Attention [9] mechanism to capture relevant components. We implement two Co-Attention layers to learn the most relevant components of both types of visual features. The structure is illustrated in Fig. 4, where the left Co-Attention layer learns information related to multi-source news visual features in original news visual features.

$$CoA-O = Softmax\left(Q_O K_S^T / \sqrt{d_k}\right) V_S, d_k = d/h_{num}, \tag{9}$$

where Q stands for query, K stands for key, and V stands for value. The variable d represents the dimension of the vectors in each attention head, while h_{num} is the number of heads in the multi-head attention mechanism.

The multi-head attention mechanism projects the original 512-dimensional Q_O, K_S, and V_S to multiple sets of low-dimensional representations through multiple different linear projections. The specific calculation method is as follows:

$$MultiHead\left(Q_O, K_S, V_S\right) = Concat\left(head_1, ..., head_{num}\right) W^O, \tag{10}$$

$$head_i = Attention\left(Q_O W_i^{Q_O}, K_S W_i^{K_S}, V_S W_i^{V_S}\right), \tag{11}$$

where W^O represents the ultimate linear transformation matrix utilized in each iteration of the computational process of the multi-head attention mechanism.

The Co-Attention layer on the right side of Fig. 4 is responsible for learning the information related to the original news visual features within the multi-source news visual features, which has a similar computational process. Ultimately, we combine the outputs of both Co-Attention layers to obtain the corresponding features M_V.

$$M_V = CoA\left(Q_O, K_S, V_S\right) \oplus CoA\left(Q_S, K_O, V_O\right), \tag{12}$$

where CoA represents computation using the Co-Attention layer and $\oplus$ represents the concatenation operation.

Afterwards, a fully connected layer ($visual - fc2$) is added to adjust the dimensionality of M_V and generate a joint representation of the visual semantic features f_{image}.

$$f_{image} = \varphi\left(W_{vf2} \cdot M_V\right), \tag{13}$$

where W_{vf2} is the weight matrix of the fully connected layer, and $\varphi\left(\cdot\right)$ is the tanh activation.

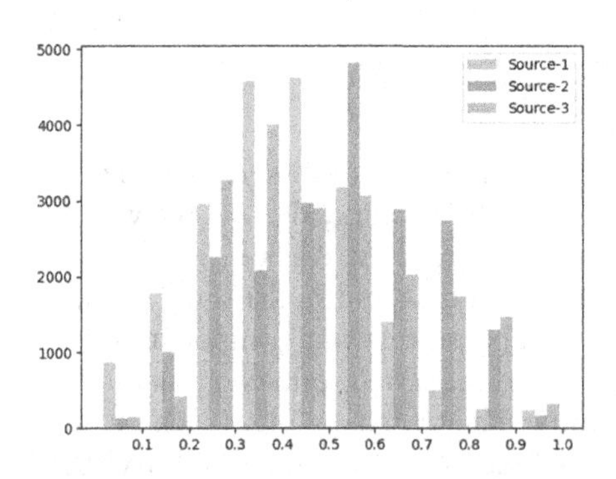

Fig. 3. The graph illustrates the distribution of similarity between each original news text in the dataset and news texts from various sources.

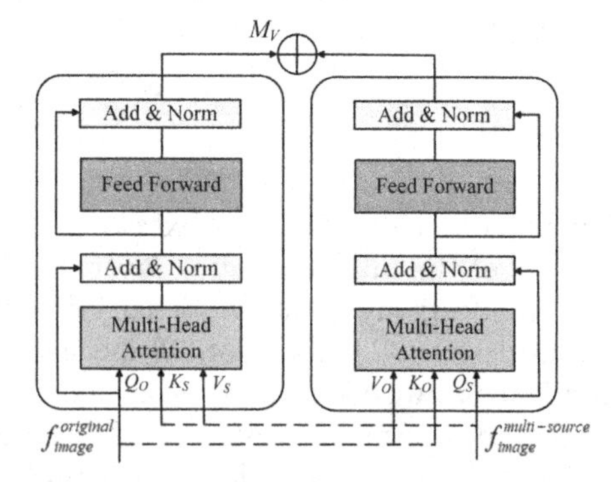

Fig. 4. By interacting original visual features with multi-source visual features, the correlation information between them is obtained.

3.5 Sentiment Feature Extractor

News texts typically contain abundant sentiment information. Through the use of sentiment analysis techniques, we can identify the sentiment polarity in the text, thereby revealing potential sentiment manipulation in fake news. We use the sentiment analysis model Senta-BiLSTM to predict the sentiment classification of the original news text and obtain the textual sentiment feature e.

$$e = Senta - BiLSTM\left(Ori_{text}\right). \tag{14}$$

Following this, we introduce a fully-connected layer ($senti - fc$) to adjust the dimensionality of the sentiment feature e in order to generate the final textual sentiment features f_{senti}.

$$f_{senti} = \varphi\left(W_{sf} \cdot e\right), \tag{15}$$

where W_{sf} is the weight matrix of the fully connected layer, and $\varphi\left(\cdot\right)$ is the tanh activation.

3.6 Feature Integration Classifier

We combine the textual semantic features f_{text}, the visual semantic features f_{image}, and textual sentiment features f_{senti} to generate a joint representation r. Its formal representation is as follows:

$$r = \left[f_{text}, f_{image}, f_{senti}\right], \tag{16}$$

where [ ] represents the connection operation.

To utilize the joint representation r, we apply a fully connected layer with a sigmoid activation function to obtain the probability of the news being true or false. The specific form is as follows:

$$\hat{y_k} = sigmoid\left(W \times r + b\right), \tag{17}$$

where W is the weight matrix of the fully connected layer, b is the bias term. y_k denotes the true label of the sample, while $\hat{y_k}$ denotes the predicted label of the model. The model uses cross-entropy as the loss function for news classification.

$$L = -\sum_{k=1}^{K}\left[y_k log\left(\hat{y_k}\right) + \left(1 - y_k\right) log\left(1 - \hat{y_k}\right)\right], \tag{18}$$

where K is the number of samples.

4 Experiments

4.1 Dataset

The dataset[1] is obtained from a data science competitions platform Biendata and jointly published by the Institute of Beijing Academy of Artificial Intelligence

[1] https://www.biendata.xyz/competition/falsenews/.

and Computer Technology of the Chinese Academy of Sciences. The original dataset only covered news until 2019. To enhance the comprehensiveness and timeliness of the dataset, we used web crawling techniques to gather the latest data and its corresponding reports for each event. The dataset was divided into training and testing sets with an 8:2 ratio. Detailed information about the dataset can be found in Table 1.

Table 1. Statistics of dataset.

Divide	Training Set	Testing Set
Fake News	8039	1961
Real News	8230	2106
Multi-source News	19527	5509

4.2 Experimental Settings

We employ the TensorFlow framework for our model and evaluate it using metrics including *Accuracy*, *Precision*, *Recall*, and $F1 - measure$. The LSTM network employs a cell size of 32 for text processing. For visual content, we resized all images to $224 \times 224 \times 3$ pixels and extracted 4096-dimensional feature vectors from the fc2 layer of the VGG19 network. After obtaining each feature vector, we introduced a fully-connected layer to adjust the feature dimension to 32, followed by another fully-connected layer with sigmoid activation for classification. The learning rate was set to 10^{-4}, and the optimization method used was Adam.

4.3 Performance Comparison

Baselines. We chose two types of baseline models for comparison with the proposed MHDF: the single modality models and the multimodal models.

Single Modality Models. (1) Textual, which only uses the textual content from the original news for classification. After converting the news text into word vectors using Word2Vec [13], an LSTM network is employed to extract textual features and generate prediction results; (2) Visual, which only uses the visual content from the original news for classification. The 4096-dimension visual features extracted from a pre-trained VGG19 are used to train the model.

Multimodal Models. (1) Simple Fusion, which adjusts the feature dimensions through a fully connected layer and then combines the features from the two single modality models. Next, the fused textual-visual features are inputted into a fake news classifier to generate the final prediction; (2) Original and Multi-source News (O-M), the textual and visual features from the original and multi-source news were transformed into the same dimensions and connected through fully connected layers. The resulting joint features were then inputted into the fake news classifier for the final prediction. (3) AttRNN [8], which integrates textual, visual, and social contextual features by incorporating an attention mechanism. In our experiments, we excluded the section that addressed social contextual information to ensure a fair comparison. (4) EANN [6], the Event Adversarial Neural Network, comprises three primary components: a multimodal feature extractor, a fake news detector, and an event discriminator. In our experiments, we utilized a modified version of EANN that omits the event discriminator. (5) MVAE [7], the Multimodal Variational Auto-Encoder, consists of three essential elements: an encoder to convert text and images into vector representations, a decoder to reconstruct the original image and textual content, and a fake news detector that utilizes shared representations learned for detecting fake news.

Table 2. Comparison of multiple methods.

Method	Accuracy	Fake News			Real News		
		Precision	Recall	F1	Precision	Recall	F1
Textual	0.864	0.881	0.831	0.855	0.850	0.896	0.872
Visual	0.634	0.593	0.772	0.670	0.704	0.506	0.589
Simple Fusion	0.883	0.915	0.834	0.873	0.857	0.928	0.891
O-M	0.896	0.889	0.896	0.892	0.902	0.896	0.899
att-RNN	0.903	0.933	0.861	0.896	0.879	0.943	0.910
EANN	0.911	0.909	0.907	0.908	0.913	0.916	0.915
MVAE	0.895	0.919	0.857	0.887	0.875	0.929	0.901
MHDF	**0.967**	**0.962**	**0.971**	**0.966**	**0.973**	**0.964**	**0.968**

Results Analysis. The experimental results are presented in Table 2. The accuracy rate of the visual model using a single modality is lower than that of the textual model using a single modality. This observation suggests that the semantic information conveyed by text is more substantial than that provided by images in detecting fake news. It emphasizes that text remains the dominant medium for expressing news language. By carefully cleaning and incorporating data from diverse perspectives of the same event, the diversity of the dataset can be enhanced, resulting in improved model training, with an increase in accuracy from 88.3% to 89.6%. The accuracy of the MVAE is 0.1% lower compared to the O-M. This is due to the fact that only 55.2% of the dataset contains news text with images, making it challenging for MVAE to effectively learn shared representations of both images and text. This imbalance can potentially hinder the ability of MVAE to learn effective shared representations across modalities, thereby impacting its overall performance. In comparison to all baselines, our MHDF demonstrates superior performance. These results further confirm the superiority of our approach in terms of improved classification performance and reliability.

4.4 Ablation Experiments and Validity Verification

Baselines. To further validate the effectiveness of our proposed method, we compare MHDF with the following ablation experiments: (1) MHDF(-BERT), we use Word2Vec instead of BERT; (2) MHDF(-1), a new model structure, is obtained by removing the multi-source data fusion module ($Layer3$, the top branch in Fig. 1) and the multi-source visual content (S_{image}, the bottom branch in Fig. 1), aimed at validating the use of multiple data sources in the model; (3) MHDF(-2), which is based on MHDF(-1), removes the multi-source data fusion module ($Layer2$, the top branch in Fig. 1); (4) MHDF(-3), which is based on MHDF(-2), removes the multi-source data fusion module ($Layer1$, the top branch in Fig. 1); (5) MHDF(-S), we remove the sentiment feature extractor (Table 3).

Table 3. The results of component analysis.

Method	Accuracy	Fake News			Real News		
		Precision	Recall	F1	Precision	Recall	F1
MHDF	**0.967**	**0.962**	**0.971**	**0.966**	**0.973**	**0.964**	**0.968**
MHDF(-BERT)	0.922	0.933	0.902	0.917	0.911	0.940	0.925
MHDF(-1)	0.958	0.957	0.956	0.957	0.959	0.960	0.960
MHDF(-2)	0.954	0.945	0.960	0.952	0.962	0.948	0.955
MHDF(-3)	0.949	0.943	0.952	0.947	0.955	0.946	0.951
MHDF(-S)	0.960	0.951	0.967	0.959	0.969	0.946	0.951

Component Analysis. Comparing MHDF(-BERT) to MHDF, we found that using BERT improved accuracy by about 4.5% due to its powerful feature extraction ability. Enabling a better understanding of backward and forward associations in news texts compared to Word2Vec. Additionally, when comparing the MHDF(-1) and MHDF(-2), it was found that removing specific modules during the fusion improved the performance of the MHDF. The findings provide evidence that including diverse perspectives of data on the same event enhances the effectiveness of fake news detection. Furthermore, MHDF(-3), trained only on the original dataset without multi-source data fusion, performs better than the baseline models in Table 2, further confirming the effectiveness of our proposed approach. Comparing with MHDF(-S), incorporating text deep sentiment features in MHDF improves the detection accuracy by 0.7%, confirming the effectiveness of their combination.

4.5 Conclusions

The research in this paper focuses on detecting fake news in social media by considering reports from multiple diverse perspectives of the same news event. It adopts a progressive fusion approach to gradually integrate multi-source data, by constructing a textual feature fusion module and a visual feature fusion module to combine features from diverse data sources. At the same time, we integrated sentiment information from news text to achieve multi-level feature extraction. Through a series of comparative and ablation experiments conducted on the dataset, the results demonstrate the feasibility and effectiveness of the model presented in this paper.

References

1. Devlin, J., Chang, M.W., Lee, K., Toutanova, K.: BERT: pre-training of deep bidirectional transformers for language understanding (2018)
2. Radford, A., Narasimhan, K.: Improving language understanding by generative pre-training (2018)
3. Kim, Y.: Convolutional neural networks for sentence classification. Eprint Arxiv (2014)
4. Simonyan, K., Zisserman, A.: Very deep convolutional networks for large-scale image recognition. Comput. Sci. (2014)
5. Liu, Z., et al.: Swin transformer: hierarchical vision transformer using shifted windows (2021)
6. Wang, Y., Ma, F., Jin, Z., Yuan, Y., Jha, K.: EANN: event adversarial neural networks for multi-modal fake news detection. In: ACM SIGKDD International Conference (2018)
7. Khattar, D., Goud, J.S., Gupta, M., Varma, V.: MVAE: multimodal variational autoencoder for fake news detection. In: The World Wide Web Conference (2019)
8. Jin, Z., Cao, J., Guo, H., Zhang, Y., Luo, J.: Multimodal fusion with recurrent neural networks for rumor detection on microblogs. In: The 2017 ACM (2017)
9. Wu, Y., Zhan, P., Zhang, Y., Wang, L., Xu, Z.: Multimodal fusion with co-attention networks for fake news detection. In: Findings of ACL 2021 (2021)
10. Hua, J., Cui, X., Li, X., Tang, K., Zhu, P.: Multimodal fake news detection through data augmentation-based contrastive learning. Appl. Soft Comput. **136**, 110125 (2023)
11. Guo, Y.: A mutual attention based multimodal fusion for fake news detection on social network. Appl. Intell. **53**(12), 15 311–15 320 (2023)
12. Mihalcea, R., Tarau, P.: TextRank: bringing order into texts (2004)
13. Mikolov, T., Chen, K., Corrado, G., Dean, J.: Efficient estimation of word representations in vector space. Comput. Sci. (2013)
14. Su, J.: SimBERT: integrating retrieval and generation into BERT. Technical report (2020)
15. Hochreiter, S., Schmidhuber, J.: Long short-term memory. Neural Comput. **9**(8), 1735–1780 (1997)

Accurate Semi-supervised Automatic Speech Recognition via Multi-hypotheses-Based Curriculum Learning

Junghun Kim, Ka Hyun Park, and U Kang(✉)

Seoul National University, Seoul, South Korea
{bandalg97,kahyun.park,ukang}@snu.ac.kr

Abstract. How can we accurately transcribe speech signals into texts when only a portion of them are annotated? ASR (Automatic Speech Recognition) systems are extensively utilized in many real-world applications including automatic translation systems and transcription services. Due to the exponential growth of available speech data without annotations and the significant costs of manual labeling, semi-supervised ASR approaches have garnered attention. Such scenarios include transcribing videos in streaming platforms, where a vast amount of content is uploaded daily but only a fraction of them are transcribed manually. Previous approaches for semi-supervised ASR use a pseudo labeling scheme to incorporate unlabeled examples during training. Nevertheless, their effectiveness is restricted as they do not take into account the uncertainty linked to the pseudo labels when using them as labels for unlabeled cases. In this paper, we propose MOCA (Multi-hypotheses-based Curriculum learning for semi-supervised ASR), an accurate framework for semi-supervised ASR. MOCA generates multiple hypotheses for each speech instance to consider the uncertainty of the pseudo label. Furthermore, MOCA considers the various degrees of uncertainty in pseudo labels across speech instances, enabling a robust training on the uncertain dataset. Extensive experiments on real-world speech datasets show that MOCA successfully improves the transcription performance of previous ASR models.

1 Introduction

Given speech signals, how can we train an accurate transcription model in a semi-supervised setting? Speech is one of the most fundamental and prevalent forms of human communication. To efficiently utilize the extensive amount of speech data, the development of automatic speech recognition (ASR) methods is essential. ASR models are applied to various fields, including voice search [20],

J. Kim and K. H. Park—These authors contributed equally to this work.

D.-N. Yang et al. (Eds.): PAKDD 2024, LNAI 14649, pp. 40–52, 2024.
https://doi.org/10.1007/978-981-97-2262-4_4

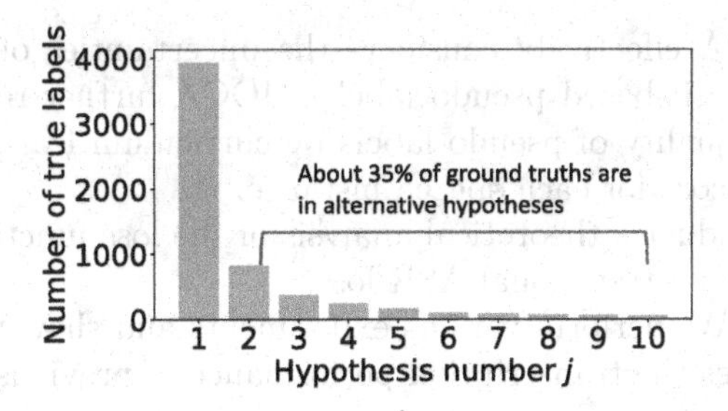

Fig. 1. Case study on LJSpeech dataset showing the distribution of true labels in the 10-best hypotheses. Note that the ground-truth labels appear frequently in alternative hypotheses, rather than being limited to the 1-best hypothesis.

speech command recognition [4], automatic transcription of spoken content [12], information extraction [24], and automatic translation [6].

Recently, semi-supervised ASR is actively studied due to the excessive costs associated with manual labeling and the exploding amount of unlabeled speeches [22]. Such scenarios often arise in customer service calls, where numerous interactions occur daily, yet only specific calls are manually transcribed for quality assurance. Semi-supervised ASR aims to leverage unlabeled data in addition to a few labeled data for training. Semi-supervised ASR improves existing ASR approaches by removing the requirement for fully observed labels, thus expanding the range of applications to more challenging scenarios.

Previous approaches for semi-supervised ASR generate pseudo labels for the unlabeled instances with an initial ASR model that has been trained using a few labeled speech instances [15]. The pseudo labels are then used to further retrain the ASR model [23]. The main limitation of those approaches is that they consider only the 1-best ASR hypothesis, ignoring the power of alternative hypotheses with potentially fewer errors. To further elaborate this idea, we show the distribution of true labels in Fig. 1 where the x-axis represents the jth-best hypothesis of 10-best hypotheses, and y-axis represents the number of true labels found in each hypothesis. Note that about 35% of ground-truths are overlooked with 1-best-based approaches in LJSpeech. To prevent such information loss, there arises the need for approaches incorporating multiple hypotheses.

In this work, we propose MOCA (<u>M</u>ULTI-HYP<u>O</u>THESES-BASED <u>C</u>URRICULUM LEARNING FOR SEMI-SUPERVISED <u>A</u>SR), an accurate method for semi-supervised ASR. MOCA incorporates multi-hypotheses-based pseudo labels for each unlabeled instance to reflect uncertainties inherently present in the 1-best pseudo label. MOCA reduces the heavy reliance on the quality of the generated pseudo labels by curriculum learning with our novel difficulty score for each speech instance. Since easier instances have lower uncertainty, MOCA initially enhances the accuracy of the ASR model with relatively certain examples and then addresses uncertain ones. This enables the ASR model to acquire more information before learning uncertain and difficult examples, allowing robust training. That is, the model is less affected by the quality of pseudo labels for difficult instances. Our contributions are summarized as follows:

- **Method.** MOCA effectively considers the uncertainties of the seed model by multi-hypothesis-based pseudo labels. MOCA further reduces the heavy reliance on the quality of pseudo labels by curriculum learning, proposing a novel hardness score for each speech instance.
- **Theory.** We conduct a theoretical analysis of the loss function of MOCA by comparing it with a traditional ASR loss.
- **Experiments.** We perform various experiments and show that MOCA successfully improves the transcription performance of previous ASR models.

The code and datasets are available at https://github.com/asrbest2023/MOCA.

2 Related Works

2.1 Automatic Speech Recognition Methods

Advancements in speech recognition have been invigorated by the emergence of unsupervised pre-training methods that extract general features from speeches [1,9,17]. The models are trained using large amounts of unlabeled audio data, and the representations are subsequently used in an end-to-end manner to enhance the performance of many downstream tasks including speech emotion recognition [11], disease detection [8], voice conversion [14], etc. The representations based on the inherent properties of the audio are then used to yield probable transcriptions, which are called hypotheses in ASR [10].

Previous methods for semi-supervised ASR employ the 1-best hypothesis approach. Higuchi et al. [7] utilize a pair of online and offline models such that they learn better representations for ASR from interactions. Park et al. [16] adopt Noisy Student Training for ASR with various levels of augmentation on input features. However, relying on a single model prediction does not capture the potential diversity of choices that the model can generate through predictions.

In ASR, wav2vec [19] and wav2vec 2.0 [3] are widely used as pre-trained general-feature extraction models due to their high performance [2,18]. The model is trained in a contrastive learning manner, distinguishing similar and dissimilar audio pairs. The wav2vec and wav2vec 2.0 models take a speech signal $x_i = [x_{i;1}, ..., x_{i;T}]$ where $x_{i;j} \in \mathbb{R}$ is a sampled value of the speech signal at time $j \in \{1, 2, ..., T\}$ as their input. The encoder of them produces feature representation $h_i = [h_{i;1}, ..., h_{i;T}]$ for each x_i, which is then mapped into a list of states $y_i = [y_{i;1}, ..., y_{i;T}]$ by a decoder. Any decoder such as MLP can be used. The j-th element $y_{i;j} \in \mathbb{R}^C$ of y_i is the prediction vector at timestamp j of a speech x_i where C is the number of alphabets and other characters such as a blank, question mark, etc.

2.2 Connectionist Temporal Classification (CTC) Loss

To measure the difference between the predicted list of states $y_i = [y_{i;t} \mid t = 1, ..., T]$ and the ground-truth text label l_i for each speech instance x_i, the connectionist temporal classification (CTC) loss have been widely used in ASR

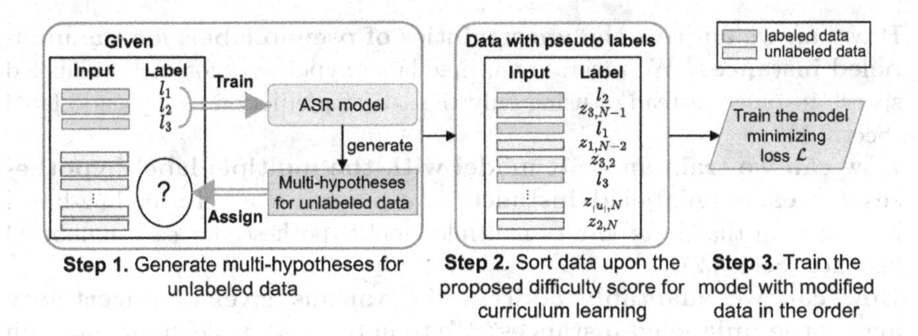

Fig. 2. Overall structure of MOCA. MOCA first trains an ASR model with a few speech examples with labels. Then MOCA utilizes the trained ASR model to generate multiple hypotheses for each unlabeled instance. MOCA retrains the ASR model utilizing the labeled and unlabeled instances with multiple hypotheses minimizing the loss function $\mathcal{L}$. Curriculum learning strategy with our new difficulty score is used during training to relieve the heavy reliance on the pseudo labels.

tasks [21]. The loss enables the model to learn the alignment between input audio frames and output characters without pre-segmented data.

The CTC algorithm aims to maximize the likelihood of generating a ground-truth transcription l_i when presented with a speech instance x_i as follows:

$$p(l_i \mid x_i) = \sum_{\pi_i \in \mathcal{B}^{-1}(l_i)} p(\pi_i \mid x_i) \tag{1}$$

where $\mathcal{B}$ is a function that merges y_i into a transcription by eliminating blanks and duplicated labels. Thus, $\mathcal{B}^{-1}(l_i)$ represents a set of all possible sequences of prediction vectors that merges to the label l_i. Then the CTC loss, which is the negative log likelihood of $p(l \mid x)$, is expressed as follows:

$$\mathcal{L}_{\text{CTC}} = -\sum_{i=1}^{M} \log p(l_i \mid x_i) \tag{2}$$

where M is the number of data instances.

3 Proposed Method

We propose MOCA (MULTI-HYPOTHESES-BASED CURRICULUM LEARNING FOR SEMI-SUPERVISED ASR), an accurate method for semi-supervised ASR. We illustrate the overall process of MOCA in Fig. 2. MOCA first trains an ASR model with the labeled speech instances, which is used for generating pseudo labels for unlabeled ones. Then MOCA retrains the ASR model with both labeled and unlabeled instances, upon the defined order considering the difficulty and uncertainty of each example. The main challenges and our approaches are:

- **How can we address the uncertainties of pseudo labels for the unlabeled instances?** We generate multiple label hypotheses for each unlabeled speech instance instead of using only the 1-best prediction as a pseudo label (Sect. 3.1).
- **How can we train an ASR model with the multiple label hypotheses for each unlabeled instance?** We propose a novel sampling-based loss function that incorporates multiple label hypotheses for each unlabeled instance (Sect. 3.2).
- **How can we adaptably address the various level of uncertainty across the unlabeled instances?** We train the ASR model beginning with easy examples, and progressing to harder ones. This provides the model with more information before encountering unlabeled instances with highly uncertain pseudo labels, enabling robust training on these instances with inaccurate pseudo labels (Sect. 3.3).

3.1 Multiple Hypotheses for Unlabeled Instances

MOCA utilizes an ASR model f, which is trained only with a few labeled instances, to generate pseudo labels for unlabeled instances. A straightforward approach would be to use the predicted label $\hat{l}_i$ from the trained model f as a pseudo label for an unlabeled instance x_i. However, there are inevitable uncertainties in $\hat{l}_i$, primarily due to the limited number of labeled examples, which in turn degrades the ASR performance. Another approach would be to use soft labels instead of the hard label $\hat{l}_i$ as in other deep learning area [13]. This prevents f from overfitting on the potentially incorrect predictions, therefore expected to be more robust than the previous 1-best based approach. Nevertheless, utilizing a soft labeling strategy in ASR tasks is impractical due to the intractably vast number of potential target labels. For example, the number of K-letter English words that can be formed using the alphabets is 26^K.

MOCA addresses the uncertainties associated with the pseudo labels by generating a set $Z_i = \{z_{i;1}, z_{i;2}, ..., z_{i;N}\}$ of N label hypotheses for each unlabeled speech signal x_i. With multiple candidates for each speech instance, MOCA prevents the possibility of fully trusting the inaccurate 1-best pseudo labels. MOCA takes into account the confidence level of multiple hypotheses by evaluating the likelihoods associated with each hypothesis of a given speech instance. A higher likelihood indicates a greater probability of observing a hypothesis, making it a more reliable candidate. This can be understood in a context similar to the soft labeling strategy employed in ASR tasks with a limited set of possible target labels. This approach facilitates an efficient approximation of the computationally challenging task inherent in the naive soft labeling approach.

We define the set Z_i of label hypotheses for each x_i as the top-N candidates of labels. The top-N candidates are derived through beam search using f_θ, which is the pre-trained ASR model trained on a few labeled instances, and parameterized by θ. We define the probability $p(z_{i;j} \mid x_i, \theta, Z_i)$ of sampling each j-th hypothesis $z_{i;j}$ from the hypotheses set Z_i given a speech signal x_i as the normalized form of beam search scores which represent the log-likelihoods of candidates. The sampling probability of $z_{i;j}$ from Z_i is as follows:

$$p(z_{i;j} \mid x_i, \theta, Z_i) = \frac{e^{s_{i;j}}}{\sum_{k=1}^{|Z_i|} e^{s_{i;k}}} \tag{3}$$

where $s_{i;j}$ is the beam search score for $z_{i;j}$. Exponentiating the log-likelihood $s_{i;j} = \log p(z_{i;j} \mid x_i, \theta)$ transforms it into the likelihood $p(z_{i;j} \mid x_i, \theta)$. Thus, $p(z_{i;j} \mid x_i, \theta, Z_i)$ in Eq. (3) represents the normalized likelihood $p(z_{i;j} \mid x_i, \theta)$ within the hypotheses set Z_i.

3.2 Training ASR Model with Multiple Hypotheses

Previous approaches for ASR uses the CTC loss, which is the sum of negative log likelihoods for all speech instances, to train an ASR model. For a labeled speech instance x_i with label l_i, the likelihood $p(l_i \mid x_i)$ is computed as in Eq. (1). However, it is not easy to define the likelihoods for unlabeled instances since ground-truth labels are not available; we have multiple label hypotheses instead.

We define the likelihood for each unlabeled instance as the probability of observing pseudo labels sampled from the set of hypotheses Z_i. The key idea is that each element $z_{i;j}$ is selected with corresponding probabilities $p(z_{i;j} \mid x_i, \theta, Z_i)$, where θ is the parameter of the pre-trained ASR model f_θ with labeled instances. This enables MOCA to adaptably generate pseudo labels for the unlabeled instances considering their uncertainty levels.

Furthermore, sampling N-best pseudo labels from the hypotheses set enhances the robustness of the model. For instance, sampling 3 times from a set with 10 hypotheses includes diverse pseudo labels compared to that with 3 hypotheses. Considering the substantial presence of true labels within the alternative hypotheses (see Fig. 1), incorporating a broader set of hypotheses as candidates for pseudo labels leads to improved pseudo label selection and enhanced robustness of the model. The distribution of certainty level among hypotheses is reflected in generated pseudo labels, and results in a more robust model.

Given a set Z_i of hypotheses for each unlabeled example x_i, the likelihood $p(S(Z_i) \mid x_i, \theta)$ of observing the sampled hypotheses $S(Z_i)$ is expressed as:

$$p(S(Z_i) \mid x_i, \theta) = \prod_{r_{i;k} \in S(Z_i)} p(r_{i;k} \mid x_i, \theta) \tag{4}$$

where $S(Z_i)$ is a list of sampled hypotheses from Z_i.

Using $p(S(Z_i) \mid x_i, \theta)$ as a likelihood for an unlabeled instance x_i, MOCA minimizes the negative log likelihood for both labeled and unlabeled instances when retraining the ASR model $f_{\theta^{\text{new}}}$ with a new parameter θ^{new}. The loss function $\mathcal{L}$ of MOCA for retraining $f_{\theta^{\text{new}}}$ is expressed as follows:

$$\mathcal{L} = -\sum_{x_i \in X_L} \log p(l_i \mid x_i, \theta^{\text{new}}) - \sum_{x_i \in X_U} \sum_{r_{i;k} \in S(Z_i)} \log p(r_{i;k} \mid x_i, \theta^{\text{new}}) \tag{5}$$

where X_L and X_U are sets of labeled and unlabeled speeches, respectively.

3.3 Curriculum Learning

If every pseudo label in the sampled hypothesis set $S(Z_i)$ for an unlabeled instance x_i is inaccurate, considering multiple pseudo labels also suffers from the uncertainty of the seed ASR model. MOCA reduces the heavy reliance on the quality of pseudo labels by implementing a curriculum learning strategy, which involves proposing a novel difficulty score for each speech instance. The main idea is to train the ASR model with easier examples first, which are more certain ones, and train with harder and more uncertain ones later. This enables the model to easily learn the complicated decision boundary.

We propose two difficulty scores for curriculum learning. The first approach treats a speech as more challenging and uncertain when the speech is uttered rapidly or it contains a longer sentence. The first difficulty score is expressed as

$$\text{difficulty\_score}(x_i) = \frac{\text{len}(l_i)}{\text{len}(x_i)} * \text{len}(l_i) \tag{6}$$

where l_i is a (pseudo) label for a speech instance x_i, $\text{len}(\cdot)$ is a function that returns the length of an input. The score is computed by multiplying the speed of a speech, which is $\text{len}(l_i)/\text{len}(x_i)$, and the length $\text{len}(l_i)$ of the uttered sentence.

The second difficulty score assesses a speech as more challenging if the spoken sentences are longer, similar to the first approach. The difference is that instances of the same length are considered easier if they have higher prediction confidence. The score is expressed as follows:

$$\text{difficulty\_score}(x_i) = \text{len}(l_i) + (1 - p(l_i \mid x_i)). \tag{7}$$

For every instance including the unlabeled ones, MOCA computes the difficulty scores and trains the ASR model starting from examples with lower scores.

3.4 Theoretical Analysis

MOCA minimizes the loss $\mathcal{L}$ in Eq. (5) to optimize the parameters θ^{new} of the ASR model $f_{\theta^{\text{new}}}$. We aim to theoretically show that the loss of MOCA is the expectation of the CTC loss in Eq. (2).

Let z_i be a latent variable representing the pseudo label of x_i; $z_i \in Z_i$ where Z_i is the set of label hypothesis for x_i. Recall that the distribution of z_i is defined with f_θ that is pre-trained with a few labeled instances (see Eq. (3)). Then, the expectation of the CTC loss in terms of z_i for retraining $f_{\theta^{\text{new}}}$ on an instance x_i is expressed as follows:

$$\mathbb{E}_{z_i \sim p(z_i|x_i,\theta,Z_i)}[-\log p(z_i|x_i,\theta^{\text{new}})] = -\sum_{z_i} p(z_i|x_i,\theta,Z_i) \log p(z_i|x_i,\theta^{\text{new}}) \tag{8}$$

where $p(z_i \mid x_i, \theta^{\text{new}})$ is the likelihood of observing z_i using the new model $f_{\theta^{\text{new}}}$.

For each unlabeled instance x_i, MOCA selects multiple pseudo labels by sampling from the hypothesis set Z_i. Then MOCA computes the loss for x_i by summing up the negative log likelihoods of all the sampled labels (see Eq. (5)). We approximate the likelihood $p(z_i|x_i, \theta, Z_i)$ in Eq. (8) as the ratio of the number of times z_i is sampled to the total number sampled labels: $p(z_i|x_i, \theta, Z_i) \approx \frac{\#z_i}{|S(Z_i)|}$ where $S(Z_i)$ is a list of sampled pseudo labels from Z_i. As a result, we have the following equation:

$$\begin{aligned} \mathbb{E}_{z_i \sim p(z_i|x_i,\theta,Z_i)}[-\log p(z_i \mid x_i, \theta^{\text{new}})] &\approx -\sum_{z_i} \frac{\#z_i}{|S(Z_i)|} \log p(z_i \mid x_i, \theta^{\text{new}}) \\ &= \frac{1}{|S(Z_i)|} \sum_{r_{i;k} \in S(Z_i)} -\log p(r_{i;k} \mid x_i, \theta^{\text{new}}). \end{aligned} \tag{9}$$

Comparing Eqs. (9) and (5) shows that MOCA minimizes the approximated expectation of CTC loss in terms of the unknown pseudo labels for unlabeled instances, with a balancing parameter $|S(Z_i)|$.

Table 1. Summary of datasets.

Dataset	Hours	Number of speakers	Language
LJSpeech[a]	24.0	1	English
LibriSpeech-dev-clean[b]	5.4	40	English
LibriSpeech-test-clean[b]	5.4	40	English

[a]https://keithito.com/LJ-Speech-Dataset/.
[b]http://www.openslr.org/12.

4 Experiments

We conduct experiments on real-world datasets with diverse settings to provide answers to the following questions.

Q1 Transcription performance (Sect. 4.2**).** How accurately does MOCA transcribe speech signals into texts compared to the baselines in a semi-supervised setting?

Q2 Speed of Convergence (Sect. 4.3**).** How does the number of hypotheses for each unlabeled instance affect the convergence speed?

Q3 Ablation Study (Sect. 4.4**).** Does each module of MOCA contribute to the transcription performance?

4.1 Experimental Settings

We introduce the experimental settings including datasets, baselines, and evaluation metrics. All the experiments are performed with a single GPU machine with GTX 3080.

Dataset. We use four speech datasets summarized in Table 1. LJSpeech dataset consists of 13,100 audio clips, totaling around 24 h of clear English speech recorded by a single female speaker. The dataset contains passages from seven non-fiction books, and a transcription is provided for each clip as a label. LibriSpeech dataset is sourced from the LibriVox project audiobooks, and categorized into dev-clean and test-clean.

Baselines. We compare MOCA with other general previous approaches for each data and difficulty score function. Supervised [19] is a simple ASR model that loads pre-trained wav2vec 2.0, and finetunes it using labeled speech data. 1-best is an ASR model that uses the trained Supervised model to generate 1-best pseudo labels for the unlabeled speech instances. Self-train [5] is a self-training-based method for semi-supervised ASR, which generates the pseudo labels dynamically during training.

Evaluation and Settings. To evaluate the ASR performance, we use WER (word error rate) as the main evaluation metric and CER (character error rate) as an additional metric. We use wav2vec 2.0 as our ASR model, where the initial parameters are set following [17]. We split LJSpeech and LibriSpeech datasets into training-labeled, training-unlabeled, and test sets with ratio 0.1 : 0.4 : 0.5 and 0.5 : 0.4 : 0.1, respectively, considering the small number of speech instances in LibriSpeech. The training-labeled set comprises speech instances with transcriptions, whereas the training-unlabeled set consists of those without transcriptions. We set the number of candidates in the hypotheses pool to 10, and vary the number of sampled pseudo labels in $\{3, 5, 10\}$.

Table 2. ASR performance of MOCA and baselines in terms of WER and CER. Bold and underlined numbers indicate the best and the second-best performance, respectively. Note that MOCA with various settings outperforms the competitors.

Model	# of sampled hypotheses K	Difficulty score	LJSpeech		dev-clean		test-clean	
			WER	CER	WER	CER	WER	CER
Supervised	N/A	N/A	10.08	2.23	21.60	6.75	19.68	6.19
1-best	N/A	N/A	8.21	1.74	16.22	5.28	15.20	4.30
Self-train	N/A	N/A	8.99	2.01	16.16	5.34	17.16	5.07
MOCA	3	confidence	6.85	1.44	15.23	4.80	15.11	4.19
	3	speed	6.88	1.48	15.32	4.94	**15.01**	4.22
	5	confidence	**6.76**	**1.43**	15.69	4.87	15.05	**4.10**
	5	speed	6.98	1.51	15.21	**4.72**	15.32	4.48
	10	confidence	6.85	1.47	15.86	**4.72**	15.24	4.23
	10	speed	8.87	2.14	**14.83**	4.76	15.75	4.63

4.2 Transcription Performance (Q1)

We compare the transcription performance of MOCA and the baselines in Table 2. Note that MOCA shows significant improvement in terms of both WER and CER than the 1-best model. Specifically, MOCA with the number $K = 5$ of sampled pseudo labels presents higher transcription accuracy than MOCA with $K = 3$. The result indicates that considering more numbers of pseudo labels from the pool of hypotheses improves the prediction performance while too large K confuses model with many possible answers, degrading the performance.

4.3 Speed of Convergence (Q2)

To study the effect of the number K of pseudo labels for each unlabeled instance in MOCA, we show the change of transcription performance as the epoch proceeds in Fig. 3. MOCA-K-conf and MOCA-K-speed represent MOCA with confidence-based and speed-based difficulty scores for curriculum learning, respectively, while having K sampled pseudo labels in common. Figure 3 shows that all cases of MOCA converge in earlier epochs than 1-BEST, which uses the 1-best hypothesis as a pseudo label for each unlabeled instance. This indicates that considering multiple hypotheses enhances the speed of convergence and transcription performance. Also note that a larger K in MOCA results in faster convergence. This is due to the significant presence of alternative hypotheses containing true labels; incorporating more hypotheses as candidates leads to improved selection of pseudo labels and enhances the robustness of the model.

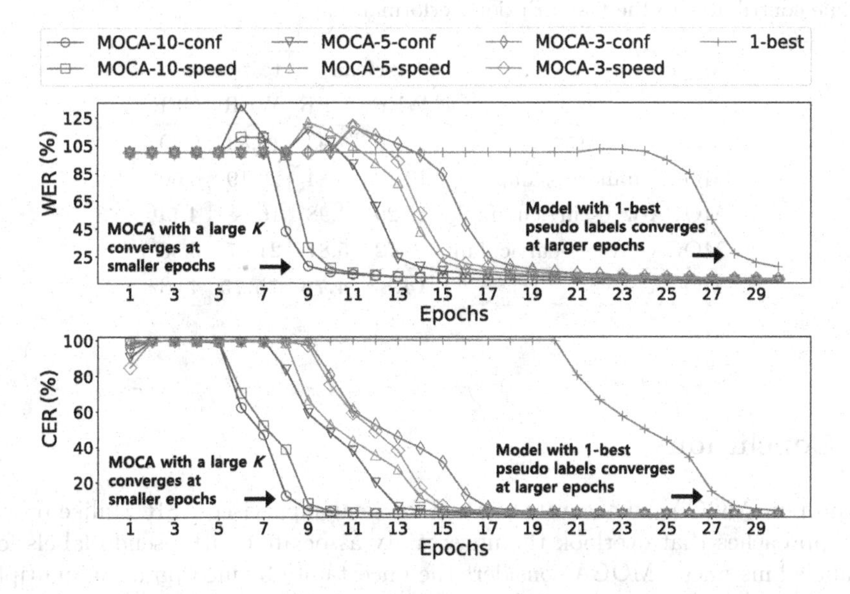

Fig. 3. Transcription performance of MOCA as the epoch proceeds in LJSpeech dataset. MOCA always converges at smaller epochs than the 1-best model, which uses a single pseudo label for each unlabeled one. Also note that more number of pseudo labels for each unlabeled instance results in quicker convergence.

4.4 Ablation Study (Q3)

We study the contribution of each module in MOCA for transcription performance in Table 3. MOCA-1-best and MOCA-uniform-sampling refer to MOCA without considering multiple pseudo labels, and MOCA that uniformly samples pseudo labels from the hypotheses set, respectively. MOCA-w/o-curriculum and MOCA-inverse-curriculum indicate MOCA without curriculum learning, and MOCA with the reversed learning order of instances, respectively. We set the number K of sampled hypotheses as 10, and use the speed-based difficulty score for curriculum learning through the ablation study.

Table 3 shows that each module of MOCA effectively contributes to the transcription performance. MOCA consistently outperforms MOCA-1-best and MOCA-uniform-sampling in terms of error rates, highlighting the significance of sampling pseudo labels based on sampling probabilities from the hypotheses set. MOCA demonstrates better performance compared to MOCA-w.o.curriculum and MOCA-w.inverse-curriculum, with MOCA-w.inverse-curriculum showing the worst performance. This indicates that the proposed curriculum score effectively improves the transcription performance.

Table 3. Ablation study. Bold numbers indicate the best performance. Note that each module contributes to the transcription performance.

Model	dev-clean		test-clean	
	WER	CER	WER	CER
MOCA-1-best	15.76	4.93	16.83	5.19
MOCA-uniform-sampling	16.20	4.84	23.19	6.60
MOCA-w/o-curriculum	15.29	4.98	16.08	**4.59**
MOCA-inverse-curriculum	18.72	5.81	21.67	6.36
MOCA	**14.83**	**4.76**	**15.75**	4.63

5 Conclusions

We propose MOCA, an accurate method for semi-supervised ASR. Unlike previous approaches that overlook the uncertainty associated with pseudo labels for unlabeled instances, MOCA considers the uncertainty by incorporating multiple hypotheses through sampling from a pool of probable hypotheses. Furthermore, MOCA relieves the problem of the dependence on the quality of the multiple pseudo labels by implementing a curriculum learning strategy with our novel difficulty scores. Experiments on real-world datasets show that MOCA outperforms other baselines in terms of transcription accuracy by addressing uncertainties in pseudo labels.

Acknowledgement. This work was supported by Institute of Information & communications Technology Planning & Evaluation(IITP) grant funded by the Korea government(MSIT) [No. 2022-0-00641, XVoice: Multi-Modal Voice Meta Learning], [No. 2021-0-01343, Artificial Intelligence Graduate School Program (Seoul National University)], and [NO. 2021-0-02068, Artificial Intelligence Innovation Hub (Artificial Intelligence Institute, Seoul National University)].

References

1. Al-Zakarya, M.A., Al-Irhaim, Y.F.: Unsupervised and semi-supervised speech recognition system: a review. AL-Rafidain J. Comput. Sci. Math. (2023)
2. Baevski, A., Mohamed, A.: Effectiveness of self-supervised pre-training for ASR. In: ICASSP (2020)
3. Baevski, A., Zhou, Y., Mohamed, A., Auli, M.: Wav2Vec 2.0: a framework for self-supervised learning of speech representations. In: NeurIPS (2020)
4. Cantiabela, Z.: Deep learning for robust speech command recognition using convolutional neural networks (CNN). In: IC3INA (2022)
5. Chen, Y., Wang, W., Wang, C.: Semi-supervised ASR by end-to-end self-training. arXiv:2001.09128 (2020)
6. D'Haro, L.F., Banchs, R.E.: Automatic correction of ASR outputs by using machine translation. In: INTERSPEECH (2016)
7. Higuchi, Y., Karube, K., Ogawa, T., Kobayashi, T.: Hierarchical conditional end-to-end ASR with CTC and multi-granular subword units. In: ICASSP (2022)
8. Javanmardi, F., Tirronen, S., Kodali, M., Kadiri, S.R., Alku, P.: Wav2Vec-based detection and severity level classification of dysarthria from speech. In: ICASSP (2023)
9. Korkut, C., Haznedaroglu, A., Arslan, L.: Comparison of deep learning methods for spoken language identification. In: SPECOM (2020)
10. Kreyssig, F.L., Shi, Y., Guo, J., Sari, L., Mohamed, A., Woodland, P.C.: Biased self-supervised learning for ASR. CoRR (2022)
11. Liu, M., Ke, Y., Zhang, Y., Shao, W., Song, L.: Speech emotion recognition based on deep learning. In: TENCON (2022)
12. Long, Y., Li, Y., Wei, S., Zhang, Q., Yang, C.: Large-scale semi-supervised training in deep learning acoustic model for ASR. IEEE Access (2019)
13. Nguyen, Q., Valizadegan, H., Hauskrecht, M.: Learning classification models with soft-label information. J. Am. Med. Inf. Assoc. (2014)
14. Nguyen, T.N., Pham, N.-Q., Waibel, A.: Accent conversion using pre-trained model and synthesized data from voice conversion. In: Interspeech (2022)
15. Novotney, S., Schwartz, R.M., Ma, J.Z.: Unsupervised acoustic and language model training with small amounts of labelled data. In: ICASSP (2009)
16. Daniel, S., et al.: Improved noisy student training for automatic speech recognition. In: INTERSPEECH (2020)
17. Radford, A., Kim, J.W., Xu, T., Brockman, G., McLeavey, C., Sutskever, I.: Robust speech recognition via large-scale weak supervision. In: ICML (2023)
18. Rouhe, A., Virkkunen, A., Leinonen, J., Kurimo, M., et al.: Low resource comparison of attention-based and hybrid ASR exploiting Wav2Vec 2.0. In: Interspeech (2022)
19. Schneider, S., Baevski, A., Collobert, R., Auli, M.: Wav2Vec: unsupervised pre-training for speech recognition. In: INTERSPEECH (2019)

20. Shan, C., Zhang, J., Wang, Y., Xie, L.: Attention-based end-to-end speech recognition on voice search. In: ICASSP (2018)
21. Vyas, A., Madikeri, S.R., Bourlard, H.: Comparing CTC and LFMMI for out-of-domain adaptation of Wav2Vec 2.0 acoustic model. In: Interspeech (2021)
22. Weninger, F., Mana, F., Gemello, R., Andrés-Ferrer, J., Zhan, P.: Semi-supervised learning with data augmentation for end-to-end ASR. In: INTERSPEECH (2020)
23. Xu, Q., Likhomanenko, T., Kahn, J., Hannun, A.Y., Synnaeve, G., Collobert, R.: Iterative pseudo-labeling for speech recognition. In: INTERSPEECH (2020)
24. Zhao, X., et al.: Disentangling content and fine-grained prosody information via hybrid ASR bottleneck features for voice conversion. In: ICASSP (2022)

MM-PhyQA: Multimodal Physics Question-Answering with Multi-image CoT Prompting

Avinash Anand(✉), Janak Kapuriya, Apoorv Singh, Jay Saraf, Naman Lal, Astha Verma, Rushali Gupta, and Rajiv Shah

Indraprastha Institute of Information Technology, Delhi, India
{avinasha,kapuriya22032,apoorv17027,jay20438,asthav, rajivratn}@iiitd.ac.in

Abstract. While Large Language Models (LLMs) can achieve human-level performance in various tasks, they continue to face challenges when it comes to effectively tackling multi-step physics reasoning tasks. To identify the shortcomings of existing models and facilitate further research in this area, we curated a novel dataset, **MM-PhyQA**, which comprises well-constructed, high school-level multimodal physics problems. By evaluating the performance of contemporary LLMs that are publicly available, both with and without the incorporation of multimodal elements in these problems, we aim to shed light on their capabilities. For generating answers for questions consisting of multimodal input (in this case, images and text) we employed Zero-shot prediction using GPT-4 and utilized LLaVA (LLaVA and LLaVA-1.5), the latter of which were fine-tuned on our dataset. For evaluating the performance of LLMs consisting solely of textual input, we tested the performance of the base and fine-tuned versions of the Mistral-7B and LLaMA2-7b models. We also showcased the performance of the novel **Multi-Image Chain-of-Thought (MI-CoT)** Prompting technique, which when used to train **LLaVA-1.5 13b** yielded the best results when tested on our dataset, with superior scores in most metrics and the highest accuracy of 71.65% on the test set.

Keywords: Large Language Models · Large Multimodal Models · Prompt Engineering · Chain-of-Thought

1 Introduction

Recent advances in Large Multimodal Models (LMMs) show impressive capabilities in handling multiple modalities, excelling in tasks like zero-shot generalization, visual reasoning, and instruction-following. Models like LLaMA-2 [1] and Mistral-7b [2] have displayed decent performance on famous textual mainstream question-answering benchmarks. SciPhyRAG [3] used retrieval augmentation to solve physics questions. However, the challenge of effectively handling queries combining textual and visual components persists, especially in subjects like Math and Physics, a problem that is exemplified by state-of-the-art models like GPT-4 [4] being proprietary. Fine-tuning

D.-N. Yang et al. (Eds.): PAKDD 2024, LNAI 14649, pp. 53–64, 2024.
https://doi.org/10.1007/978-981-97-2262-4_5

general-purpose LLMs to perform well at a singular task has been effective in a variety of complex scenarios [5,6]. Hence, developing open-source domain-specific chatbots with multimodal capabilities is promising. These chatbots can empower students with interactive question sessions, providing instant clarifications and guidance, and revolutionizing exam preparation.

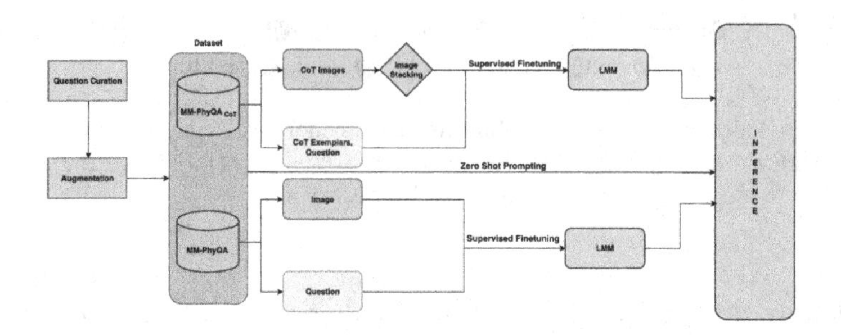

Fig. 1. Schematic Pipeline of Multimodal Question Answering

To evaluate the capabilities of Large Multimodal Models (LMMs) for question-answering we have created a novel multimodal multiple-choice high school physics question-answering dataset. Physics questions require a good understanding of the underlying concepts and construction of steps with reasoning to reach the correct solution, hence not solvable by simply memorizing certain facts. High-school physics numerical questions are often accompanied by diagrams, which adds additional complexity that models should be able to interpret and understand for effective problem-solving, therefore acting as a valuable benchmark for evaluating the performance of LMMs. Given the dearth of multimodal physics datasets containing complex, high-quality questions, our dataset facilitates the study performance of LMMs and LLMs in a challenging setting.

Introduction of techniques like Chain-of-Thought (CoT) Prompting [7] has further enhanced the performance of LLMs, and subsequent experiments using the technique in a multimodal context [8,9] have been fruitful. CoT-Prompting involves providing the necessary prompts to a model to steer it toward the correct solution. It is analogous to how humans go about solving a problem, wherein we try to think of the intermediate steps that build logically toward the final answer. However, the prospect of incorporating images and figures with the prompt exemplars is yet to be explored by contemporary literature.

In this paper, we do a quantitative analysis regarding the effect of utilizing a modality other than text and the difference in the performance of LLMs and LMMs between using them out of the box (Zero Shot Prompting) and fine-tuning them for a specific purpose. We also examine the effects of using Chain-of-Thought Prompting in a multimodal setting, for which we came up with a novel method to incorporate multiple images during the CoT prompting process.

Hence, the contributions of this paper are threefold. Firstly we introduce a novel multimodal dataset, MM-PhyQA, containing challenging physics questions. We also generate its CoT-Prompting variant, providing exemplar questions during the training process. Secondly, we analyze the effects of using an additional modality other than text, the effects of utilizing techniques like CoT Prompting, and the performance gain witnessed by fine-tuning LLMs and LMMs for a specific purpose, particularly for a task like answering physics questions. Finally, we introduce an approach, **Multi-Image Chain-Of-Thought (MI-CoT)** for employing multiple images during CoT-Prompting that is novel, to the best of our knowledge.

2 Related Works

2.1 Available Datasets

Numerous educational datasets are available for math and science. GSM8k [10] offers 8500 grade school math problems, while JEEBench [11] provides 450 questions from JEE advanced exams. SciQ [12] contains 13,697 science questions, and SciBench [13] offers college-level scientific problems. MMLU [14] is a multitask test dataset with 15908 samples, and C-Eval [15] includes multiple-choice questions in Chinese across 52 disciplines.

In the realm of multimodal datasets, GeoQA [16] offers middle school geometric questions with images and text, while TQA [17] provides middle school science questions in a similar format. ChartQA [18] is a chart-based reasoning dataset, and MMQA [19] consists of questions with images, text, and tables. ScienceQA [8] is a diverse multimodal dataset with 21208 science questions spanning various topics but lacks challenging high school-level questions.

2.2 Large Multimodal Models and Chain-of-Thought

Large language models' extension into multi-modal versions has led to significant attention and successful applications. GPT4-V [20] and PaLM-E [21] are state-of-the-art multimodal models, with PaLM-E directly incorporating visual features for enhanced performance. LLaVA [22,23] is recognized for its versatility in handling various multimodal tasks, utilizing a CLIP [24] encoder with Vicuna for vision-language understanding. Shikra [25] excels in Visual Question Answering (VQA) and image-captioning tasks, particularly in multimodal conversation scenarios. Kosmos-2 [26] demonstrates strong performance across diverse multimodal tasks, including grounding, referring, learning within context, and generation.

The Chain-of-Thought paradigm has transformed how large language models process reasoning, significantly improving NLP tasks. It has evolved from vanilla CoT to more complex structures like Tree-of-Thoughts [27] and Graph-of-Thoughts [28]. Despite these advancements, the shift towards multimodal reasoning led by multimodal CoT [9], has limitations due to reliance on multiple question-answer chains from a single image during training. To overcome this, we propose the Multi-Image Chain-Of-Thought (MI-CoT) technique, ensuring each question-answer pair used in training is associated with a distinct image, enhancing diversity and robustness.

3 Novel Dataset

There is a lack of multimodal datasets that comprise physics questions and are catered to high school students. While there are a few datasets available that consist of questions at a high school level, the quality of the questions does not belong to the highest of standards. We curated a novel MM-PhyQA Dataset from publicly available resources. The resources are geared toward individuals who prepare for competitive exams throughout India, ensuring a higher difficulty level than that of an average high school physics question.

Question: A ball of mass (m) 0.5 kg is attached to the end of a string having length (L) 0.5m. The ball is rotated on a horizontal circular path about vertical axis. The maximum tension that string can bear is 324 N.The maximum possible value of angular velocity of ball (in rad/s) is:

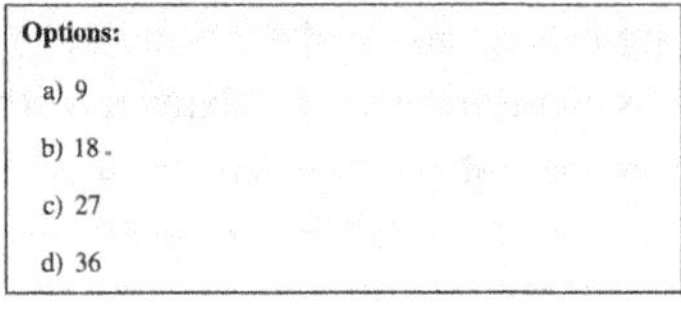

Options:

a) 9

b) 18

c) 27

d) 36

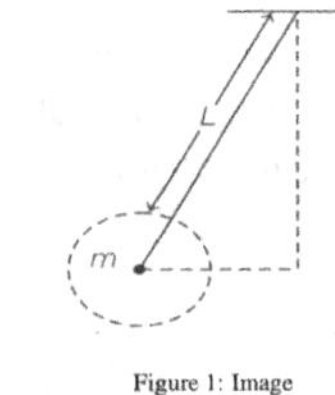

Figure 1: Image

Answer: (d)
Explanation: From the figure, $T \sin\theta = mL \sin\theta\omega^2$.

$$324 = 0.5 \times 0.5 \times \omega^2$$

$$\omega^2 = \frac{324}{0.5 \times 0.5}$$

$$\omega = \sqrt{\frac{324}{0.5 \times 0.5}} = \frac{18}{0.5} = 36\, rad/s$$

(a) Sample question of MMPhy-QA dataset

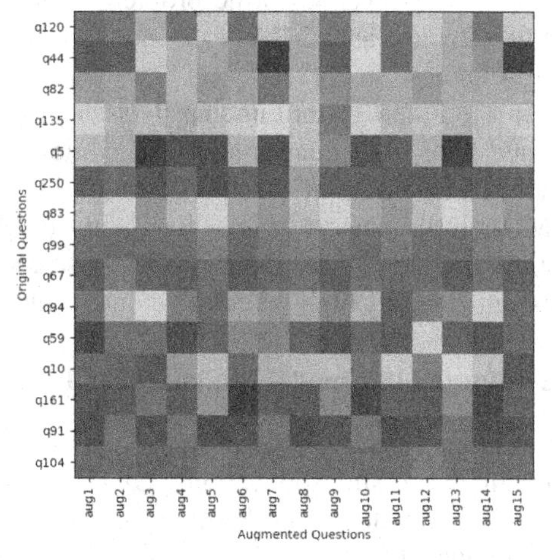

(b) Heatmap of text similarity between 15 randomly sampled original and augmented questions

Fig. 2. MMPhy-QA Dataset questions

3.1 Original Dataset Creation

Around 300 questions were manually created. As shown in Fig. 2a, each question consists of a question, four options, the correct answer to the question, and an explanation that shows the reasoning by giving steps to approach the correct answer to select the correct answer.

3.2 Data Augmentation Procedure

For augmenting the data ChatGPT [29] was given a prompt to create other variations of the text while ensuring that the meaning remained the same, bringing the total count of the questions in the dataset to 4500. Figure 2b shows the heatmap of the cosine similarity scores of the augmented questions w.r.t the original one for some of the questions. The questions were altered in two ways:

- **Numerical Value Variation:** During augmentation, numerical values in the original questions are adjusted to diversify the solutions, ensuring the model's impartiality. Python functions were developed for each question to get the correct answers after changing the values.
- **Structural Variation:** To avoid pattern memorization, the questions' structure was intentionally altered by rephrasing with ChatGPT and sometimes manual adjustments. Options were kept the same but randomly rearranged.

Initially, attempts to rephrase the entire query sometimes failed to properly shuffle the questions. Manual adjustments were made to correct these errors. While including the entire query didn't consistently result in a rephrased version, prompting ChatGPT to generate separate variations for the question and explanation improved results. However, some questions still required manual rephrasing, involving adjustments to the question, explanation, options, and correct answer.

3.3 Chain of Thought Variant

To facilitate the model to generate better reasoning, two questions were added corresponding to each question. These questions were based on the same topic and care was taken that similar concepts were utilized as seen in Fig. 2a. All three questions consist of figures.

3.4 MM-PhyQA Dataset Topics

The dataset consists of topics that are present in high school physics curricula throughout India. The topics and the corresponding subtopics are listed in Table 1.

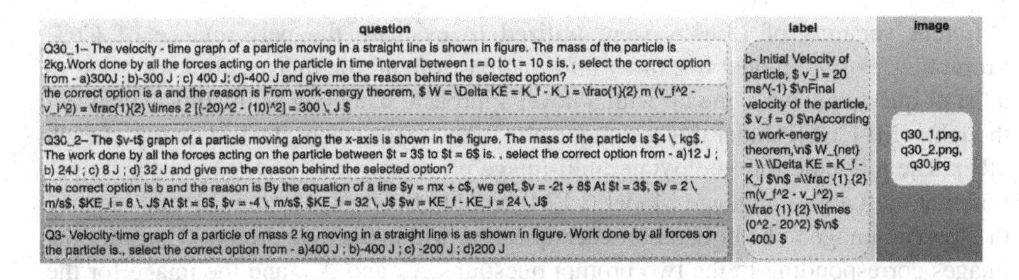

Fig. 3. Multi-Image Chain of thought (MI-CoT) Prompted text provided as input to LMMs during training. The main question to be answered is preceded by two exemplars, with the three questions separated by a delimiter. The image is a sequence of three comma-separated file names and the label is the ground truth

Table 1. Topics and subtopics in the MM-PhyQA dataset

Topic	Subtopics
Kinematics	Velocity-Time, Acceleration, Rotational Motion, Gravitation, Motion in a Straight Line, Motion in a Plane, Periodic Motion, Wave Motion
Mechanics	Law of Motion, Work, Power, Force, Law of Motion
Electrostatics and Current Electricity	Current, Voltage, Resistance, Electric Field, Ohm's Law, Kirchhoff's Laws, and Their Applications, Series and Parallel Combinations of Resistors
Thermodynamics	Laws of Thermodynamics, Thermal Equilibrium, Heat Transfer, Temperature, Reversible and Irreversible Processes, Kinetic Theory of Gases
Optics	Reflection, Mirrors, Lenses, Wave Optics, Magnification
Magnetism	Magnetic Field, Hysteresis, Permeability, Electromagnets
Electronic Devices	Semiconductors, Logic Gates, Diode
Atoms	Nuclei, Isotopes

4 Methodology

Figure 1 shows the pipeline that was utilized for data processing, input processing, and output generation. Each element in the dataset consists of the question ID, the question, the label consisting of the corresponding answer and the reasoning, and the image filename. A function was used to convert each element to a prompt which can be fed to the model for generating the answer. For the Chain of Thought variant of the dataset, the structure was modified. As shown in Fig. 3, the question was preceded by two similar questions with their correct answers and reasoning. All the three questions were separated by a delimiter consisting of hyphens. The filenames of the three images were stored in a comma-separated fashion.

4.1 Multi-image Chain-of-Thought (MI-CoT)

Different versions of LLaVA were utilized to evaluate the performance of CoT-Prompting. For the model to extract information from all the images corresponding to a list of questions, we came up with a novel approach, namely a Multi-Image chain of thoughts (MI-CoT). Under this technique, the three images were stacked on top of each other. The rationale for employing multi-image prompting was driven by the anticipation that the Large Language Model (LLM) would effectively distinguish and identify the specific image to be utilized for each question within a single prompt. Consider the images corresponding to the two prompt questions X_p and X_q, and the image for the main question X_r. LLaVA utilizes the CLIP visual encoder to get the visual feature Z_v:

$$Z_v = g(X_v) \tag{1}$$

where

$$X_v = X_p \cdot X_q \cdot X_r \tag{2}$$

The filenames were passed as a list in the same order in which they were stacked. To make sure that the dimensions were correct for feeding the resultant concatenated image X_v into the CLIP encoder, the size of the images was reduced along one dimension using an autoencoder after basic pre-processing (normalization and padding) of the images. A basic neural-network-based autoencoder was employed and was trained on the train split for this purpose.

5 Experiments

For evaluating the performance of the models, an 85/15 train-test split was used. We made sure that the percentage share of questions with options a, b, c, and d was roughly the same in both the training and testing datasets. This was especially important in the case of the training dataset to ensure no bias imposed by any option during the training process. We used accuracy as the primary metric for judging the performance of the models and rouge scores for evaluating the correctness of the reasoning.

5.1 Models

We conducted a variety of experiments with both text and multimodal LLMs to gauge the difference in performance that comes about due to the change in the modality. LLaMA2-7b and Mistral-7b are the current state-of-the-art open-source LLMs for textual input. These models were tested with text-only inputs. We use these LLMs to highlight the difference in the level of performance between fine-tuned models versus using them straight out of the box, aka through zero-shot prompting. For the ablation study, we also experimented with GPT-4, which is the current state-of-the-art model for multimodal question-answering.

LLaVA and LLaVA-1.5 being multimodal were provided with the figures along with the textual input. All the models were trained on A100 GPU and were fine-tuned for 5 epochs with a batch size of 8. Weighted Adam optimizer was utilised and the learning rate was set to 2e−4.

We also experimented with different LoRA values in the case of the LLaVA-1.5 model. LoRA or Low-Rank Adaptation [30], is a method to represent the weight changes during the training process in lower-ranked matrices. This is especially useful while fine-tuning general-purpose LLMs, as it speeds up the training process. A lower LoRA rank means fewer parameters are learned during the adaptation process, however, it results in a faster training process as well. We tested the 7b (7 billion) and 13b (13 billion) variants of LLaVA which correspond to the number of learning parameters. The different LLaVA configurations also formed the basis of our comparison of the performance of (MI-CoT) Prompting. For fine-tuning, open-source base model checkpoints from huggingface were utilized.

6 Results and Discussion

Table 2. Performance of text-only and multimodal (MM) models. Model training specifications such as LoRA Rank and whether MI-CoT Prompting was used have been mentioned. All models were fine-tuned except for GPT-4, for which the answers were extracted using zero-shot prompting

Model	MI-CoT	Modality	Accuracy	Rouge1	Rouge2	RougeL	LoRA Rank
LLaMA2-7b	×	Text Only	0.25	0.380	0.187	0.315	8
Mistral-7b	×	Text Only	0.428	0.460	0.256	0.391	8
GPT-4	×	MM	0.331	–	–	–	–
LLaVA-13b	×	MM	0.293	0.551	0.383	0.501	64
LLaVA-1.5 7b	×	MM	0.533	**0.712**	0.579	**0.676**	64
LLaVA-1.5 13b	×	MM	0.527	0.672	0.532	0.634	64
LLaVA-1.5 13b	×	MM	0.531	0.621	0.490	0.586	128
LLaVA-13b	✓	MM	0.291	0.383	0.184	0.306	64
LLaVA-1.5 7b	✓	MM	0.354	0.496	0.343	0.444	64
LLaVA-1.5 13b	✓	MM	0.653	0.686	**0.585**	0.656	64
LLaVA-1.5 13b	✓	MM	**0.716**	0.677	0.582	0.650	128

6.1 Model Performance

The results of the experiments with their accuracy scores on the test dataset are listed in Table 2. Mistral-7b and LLaMA2-7b being text-only models only take into account the textual data which means that they are bound to miss critical information in some questions. We observed an accuracy score of 25.95% and 42.83% for LLaMA2-7b and Mistral-7b, respectively. Thus, we conclude that text-only LLMs are not capable of providing the right answers for a large number of multimodal questions which require multiple steps with complex reasoning to reach the final answer.

LLaVA is a model that can potentially answer complex questions due to its ability to process images. While the older LLaVA version with 13 billion parameters exhibited a lower accuracy than Mistral-7b, LLaVA-1.5 was able to perform significantly better than Mistral-7b. The best performance was seen when LLaVA-1.5, trained with 13 billion parameters, was fine-tuned with a LoRA rank of 128 and employed Chain of Thought Prompting with an accuracy score of 71.65%. A higher LoRA rank means that the model can learn more parameters during fine-tuning which makes it ideal for task-specific situations, such as answering complex physics questions. LLaVA-1.5 13b performs better than the 7b variant with an equal LoRA rank of 64 when multi-image prompting was utilized. This is because the larger number of trainable parameters allowed the model to learn and generalize better.

Table 3. Performance of text-only LLMs using zero-shot prompting and fine-tuning

Model	Task	Modality	Accuracy (in %)	Rouge 1	Rouge 2	Rouge L
LLaMA2-7b	Zero Shot Prompting	Text Only	14.22	0.301	0.096	0.201
	Supervised Fine-tuning	Text Only	25.95	0.380	0.187	0.315
Mistral-7b	Zero Shot Prompting	Text Only	23.32	0.259	0.083	0.180
	Supervised Fine-tuning	Text Only	**42.83**	**0.460**	**0.256**	**0.391**

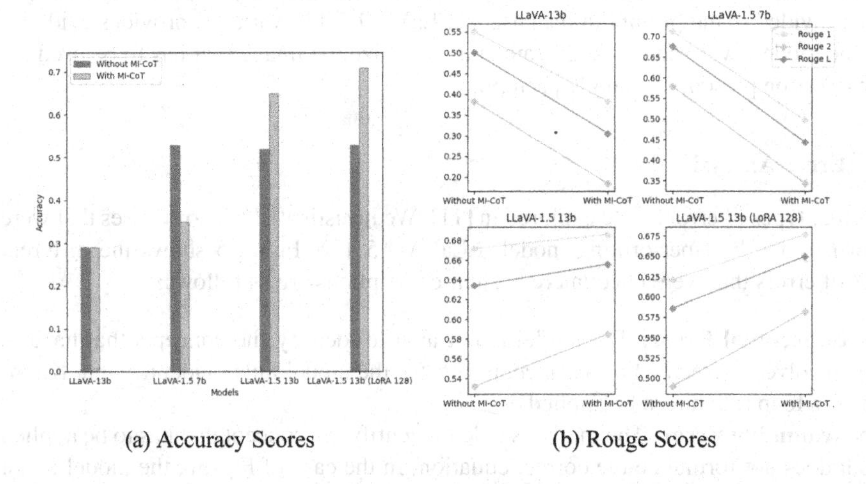

(a) Accuracy Scores (b) Rouge Scores

Fig. 4. Comparison of the accuracy and rouge scores of different LLaVA variants when trained using (MI-CoT) Prompting vs their non-CoT prompted supervised fine-tuned (SFT) counterparts

6.2 Zero Shot Prompting Vs Supervised Fine-Tuning

Table 3 shows the performance of LLaMA2-7b and Mistral-7b with zero-shot prompting and supervised fine-tuning. There is a marked improvement in the accuracy, Rouge1, Rouge2, and RougeL scores for both the models when fine-tuned on the dataset. This proves the assertion that current LLM models in their out-of-the-box configurations are not able to answer physics questions satisfactorily, and there is a need to fine-tune the models on domain-specific datasets to get better performance.

Zero-shot inferencing was done using the GPT-4 model. In most instances, GPT-4 failed to give correct answers and was not able to extract the entire information from the image. In some failure cases, GPT-4 needed more context than questions to make progress toward the solution.

6.3 Effect of Chain of Thought Prompting

For all variants of LLaVA-1.5 that were tested, there was an increase in the accuracy score when MI-CoT Prompting was employed as seen in Fig. 4a except in the case of LLaVA-1.5 7b model. A smaller number of trainable parameters meant that the model was not able to process the more complex multi-image input, leading to a sharp dip in the performance. The difference was the most significant in the case of LLaVA-1.5 13b trained with LoRA as 128, which also gave the best performance out of all the models tested when trained using MI-CoT Prompting. The MI-CoT Prompting trained version also exhibited high rouge scores as seen in Table 2. It can be observed from Fig. 4b that the rouge scores were higher in the LLaVA-1.5 13b CoT variants, showcasing the fact that models that were able to leverage the MI-CoT prompt also showed a bump in the reasoning capabilities. A marked improvement in all metrics, when multiple images were provided in the prompt in the case of LLaVA-1.5 13b variants, provides evidence that the models were able to segregate and recognize the image that has to be used for each question present in a single prompt.

6.4 Error Analysis

Different types of errors were explored in [11]. We investigated the error cases that were thrown by the best-performing model, LLaVA-1.5 13b. Figure 5 shows the different types of errors that were encountered. Their descriptions are as follows:

◇ **Conceptual Error:** The model is not able to identify the concepts that have to be involved correctly. For instance, in Fig. 5a, the model fails to identify that Kirchhoff's loop rule has to be applied.
◇ **Grounding Error:** The model is able to identify the concept that has to be applied but does not formulate the correct equation. In the case of Fig. 5b, the model is not able to apply the correct equation to get the centripetal acceleration of the cyclist.
◇ **Computational Error:** The model makes an algebraic mistake. In Fig. 5c, the concept and the equations are correct, but the computation of the final answer is incorrect.

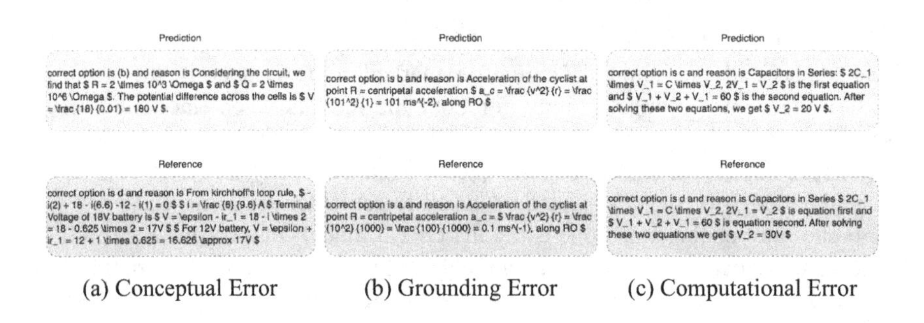

(a) Conceptual Error (b) Grounding Error (c) Computational Error

Fig. 5. Types of errors encountered by LLaVA-1.5 13b

7 Conclusion

This paper introduces the MM-PhyQA dataset, comprising high-quality problems solved by tested LLMs, serving as a benchmark for LLM performance in education. From our experiments, we concluded that the base configurations of Mistral-7b, LLaMA2, LLaVA-1.5, and GPT-4 struggled with complex reasoning tasks, but fine-tuning, particularly with MI-CoT prompting, showed promise, notably with the LLaVA-1.5 13b model. LLaVA's image extraction abilities yielded high metric scores, and leveraging multimodality and MI-CoT Prompting, improved performance significantly. Future work may explore incorporating Reinforcement Learning from Human Feedback (RLHF) for model alignment and extending MI-CoT Prompting to other multimodal tasks.

Acknowledgements. Rajiv Ratn Shah is partly supported by the Infosys Center for AI, the Center of Design and New Media, and the Center of Excellence in Healthcare at IIIT Delhi.

Disclosure of Interests. The authors have no competing interests to declare that are relevant to the content of this article.

References

1. Touvron, H., et al.: Llama 2: open foundation and fine-tuned chat models. arXiv preprint arXiv:2307.09288 (2023)
2. Jiang, A.Q., et al.: Mistral 7B. arXiv preprint arXiv:2310.06825 (2023)
3. Anand, A., et al.: SciPhyRAG-retrieval augmentation to improve LLMs on physics Q & A. In: Goyal, V., Kumar, N., Bhowmick, S.S., Goyal, P., Goyal, N., Kumar, D. (eds.) BDA 2023. LNCS, vol. 14418, pp. 50–63. Springer, Cham (2023). https://doi.org/10.1007/978-3-031-49601-1_4
4. OpenAI: GPT-4 technical report. ArXiv abs/2303.08774 (2023). n. pag
5. Anand, A., et al.: Context-enhanced language models for generating multi-paper citations. In: Goyal, V., Kumar, N., Bhowmick, S.S., Goyal, P., Goyal, N., Kumar, D. (eds.) BDA 2023. LNCS, vol. 14418, pp. 80–94. Springer, Cham (2023). https://doi.org/10.1007/978-3-031-49601-1_6
6. Anand, A., et al.: KG-CTG: citation generation through knowledge graph-guided large language models. In: Goyal, V., Kumar, N., Bhowmick, S.S., Goyal, P., Goyal, N., Kumar, D. (eds.) BDA 2023. LNCS, vol. 14418, pp. 37–49. Springer, Cham (2023). https://doi.org/10.1007/978-3-031-49601-1_3
7. Wei, J., et al.: Chain-of-thought prompting elicits reasoning in large language models. Adv. Neural. Inf. Process. Syst. **35**, 24824–24837 (2022)
8. Lu, P., et al.: Learn to explain: multimodal reasoning via thought chains for science question answering. Adv. Neural. Inf. Process. Syst. **35**, 2507–2521 (2022)
9. Zhang, Z., Zhang, A., Li, M., Zhao, H., Karypis, G., Smola, A.: Multimodal chain-of-thought reasoning in language models. arXiv preprint arXiv:2302.00923 (2023)
10. Cobbe, K., et al.: Training verifiers to solve math word problems. arXiv preprint arXiv:2110.14168 (2021)
11. Arora, D., Singh, H.G.: Have LLMs advanced enough? A challenging problem solving benchmark for large language models. arXiv preprint arXiv:2305.15074 (2023). pag

12. Welbl, J., Liu, N.F., Gardner, M.: Crowdsourcing multiple choice science questions. arXiv preprint arXiv:1707.06209 (2017)
13. Wang, X., et al.: SciBench: evaluating college-level scientific problem-solving abilities of large language models. ArXiv abs/2307.10635 (2023). n. pag
14. Hendrycks, D., et al.: Measuring massive multitask language understanding. arXiv preprint arXiv:2009.03300 (2020)
15. Huang, Y., et al.: C-eval: a multi-level multi-discipline Chinese evaluation suite for foundation models. arXiv preprint arXiv:2305.08322 (2023)
16. Chen, J., et al.: GeoQA: a geometric question answering benchmark towards multimodal numerical reasoning. arXiv preprint arXiv:2105.14517 (2021)
17. Jin, N., Siebert, J., Li, D., Chen, Q.: A survey on table question answering: recent advances. In: Sun, M., et al. (eds.) CCKS 2022. CCIS, vol. 1669, pp. 174–186. Springer, Singapore (2022). https://doi.org/10.1007/978-981-19-7596-7_14
18. Masry, A., Long, D.X., Tan, J.Q., Joty, S., Hoque, E.: ChartQA: a benchmark for question answering about charts with visual and logical reasoning. arXiv preprint arXiv:2203.10244 (2022)
19. Deepak, G., Kumari, S., Ekbal, A., Bhattacharyya, P.: MMQA: a multi-domain multi-lingual question-answering framework for English and Hindi. In: Proceedings of the Eleventh International Conference on Language Resources and Evaluation (LREC 2018) (2018)
20. https://openai.com/research/gpt-4v-system-card
21. Driess, D., et al.: PaLM-E: an embodied multimodal language model. arXiv preprint arXiv:2303.03378 (2023)
22. Liu, H., Li, C., Wu, Q., Lee, Y.J.: Visual instruction tuning. arXiv preprint arXiv:2304.08485 (2023)
23. Liu, H., Li, C., Li, Y., Lee, Y.J.: Improved baselines with visual instruction tuning. arXiv preprint arXiv:2310.03744 (2023)
24. Radford, A., et al.: Learning transferable visual models from natural language supervision. In: International Conference on Machine Learning, pp. 8748–8763. PMLR (2021)
25. Chen, K., Zhang, Z., Zeng, W., Zhang, R., Zhu, F., Zhao, R.: Shikra: unleashing multimodal LLM's referential dialogue magic. arXiv preprint arXiv:2306.15195 (2023)
26. Peng, Z., et al.: Kosmos-2: grounding multimodal large language models to the world. arXiv preprint arXiv:2306.14824 (2023)
27. Yao, S., et al.: Tree of thoughts: deliberate problem solving with large language models. arXiv preprint arXiv:2305.10601 (2023)
28. Besta, M., et al.: Graph of thoughts: solving elaborate problems with large language models. arXiv preprint arXiv:2308.09687 (2023)
29. https://chat.openai.com/
30. Hu, E.J., et al.: LoRA: low-rank adaptation of large language models. CoRR abs/2106.09685 (2021)

Adversarial Text Purification: A Large Language Model Approach for Defense

Raha Moraffah(✉), Shubh Khandelwal, Amrita Bhattacharjee, and Huan Liu

Arizona State University, Tempe, AZ, USA
{rmoraffa,skhand15,abhatt43,huanliu}@asu.edu

Abstract. Adversarial purification is a defense mechanism for safeguarding classifiers against adversarial attacks without knowing the type of attacks or training of the classifier. These techniques characterize and eliminate adversarial perturbations from the attacked inputs, aiming to restore purified samples that retain similarity to the initially attacked ones and are correctly classified by the classifier. Due to the inherent challenges associated with characterizing noise perturbations for discrete inputs, adversarial text purification has been relatively unexplored. In this paper, we investigate the effectiveness of adversarial purification methods in defending text classifiers. We propose a novel adversarial text purification that harnesses the generative capabilities of Large Language Models (LLMs) to purify adversarial text without the need to explicitly characterize the discrete noise perturbations. We utilize prompt engineering to exploit LLMs for recovering the purified samples for given adversarial examples such that they are semantically similar and correctly classified. Our proposed method demonstrates remarkable performance over various classifiers, improving their accuracy under the attack by over 65% on average.

Keywords: Textual Adversarial Defenses · Adversarial Purification · Textual Adversarial Defenses · Large Language Model

1 Introduction

Despite the tremendous success of text classification models [9,20], studies have exposed their susceptibility to adversarial examples, i.e., carefully crafted sentences with human-unrecognizable changes to the inputs that are misclassified by the classifiers [13]. The dependability and integrity of NLP applications are seriously threatened by the vulnerability of text classification models to these attacks. Thus, developing stronger defenses against adversarial attacks is crucial in improving the classification model's robustness.

Adversarial purification is a type of defense mechanism against adversarial attacks. It characterizes and removes the adversarial perturbations from the attacked inputs to generate purified samples that are similar to the attacked

D.-N. Yang et al. (Eds.): PAKDD 2024, LNAI 14649, pp. 65–77, 2024.
https://doi.org/10.1007/978-981-97-2262-4_6

ones and are classified correctly by the classifier [25,30,31,36]. These methods have demonstrated efficacy in the field of image classification without making assumptions on the form of an attack and a classification model, thus being able to defend pre-existing classifiers against unseen threats. The potential of adversarial purification, however, has not been explored for text classification, due to the challenges of characterizing the adversarial perturbations for discrete data. In particular, contrary to images, where perturbations can be generated based on continuous gradients, for text data, adversarial perturbations are generated by manipulating combinations of words in the input text [13]. Therefore, identifying these perturbations is also a combinatorial problem.

An ideal solution to adversarial purification for text is to generate the purified example without explicitly characterizing the noise perturbations. In an attempt to achieve this, Li et al. [18] propose a greedy approach that randomly masks the adversarial examples and uses their reconstructed versions by the Masked Language Models (e.g., BERT [9]) as benign purified examples. However, due to its greedy nature, this defense can be ineffective for defending text classifiers.

The exponential growth of the sheer size of LLMs has expedited their generative applications in various fields [27]. To study the effectiveness of adversarial purification for texts, we investigate if LLMs can be exploited to directly generate the purified examples from their adversarial counterparts, eliminating the need for the characterization of adversarial perturbation. To this end, we utilize the generative power of instruction-based LLMs, particularly GPT-3.5, and design a prompt to exploit the contextual understanding and capacity of LLMs to recover purified samples.

Compared to the greedy approach of selecting random combinations of tokens iteratively to remove adversarial perturbations, our proposed method exploits the comprehension and contextual understanding of LLMs to effectively reverse the adversarial perturbations, while utilizing their extensive generation power and capacity to produce cohesive, fluent texts. Our method demonstrates the effective use of adversarial purification methods for text classification, improving the performance of the classifier under attack by over 65%, and improving the performance of the existing text purification defense by over 25% in most cases. Our results open a new avenue for future research in textual adversarial defense based on purification. Our contributions are summarized as follows:

- We study if it is possible to effectively implement the adversarial text purification defense for text.
- We are the first to utilize the contextual understanding and capacity of LLMs for effective text-based adversarial purification defense.
- We conduct extensive experiments on two state-of-the-art transformer-based text classifiers and demonstrate the effectiveness of our proposed adversarial purification method in defending the pre-trained classifiers against strong attacks without any knowledge of the attack.

2 Related Work

Adversarial Attacks on Text Classifiers: Over the years there have been various types of adversarial attacks for text, with varying degrees of success on different types of model architectures. Adversarial attacks, broadly categorized into black box and white box [32], manipulate textual data through insertion, deletion, or swapping of characters and words. The substitution-based strategies to craft adversarial examples employ techniques like genetic algorithms, greedy-search, or gradient-based methods for word replacement [2,13,29]. Recent works involving word-level perturbations include TextFooler [13], BERT-Attack [16], TextHoaxer [35]. Alongside the vast body of work on word-level attacks, there is also significant amount of works in character-level and sentence-level attacks [32].

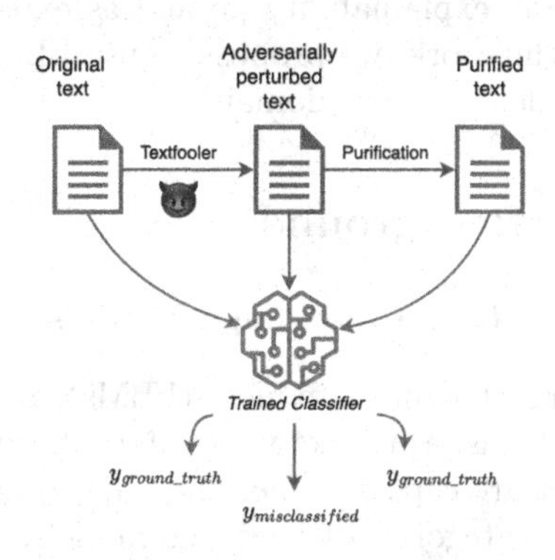

Fig. 1. Our Proposed LLM-guided Adversarial Text Purification Framework.

Adversarial Purification and Other Defenses: Influenced by the rapid development of various adversarial attacks in text, there has also been an increasing number of defense mechanisms to ensure robustness of models against different types of attacks. Some of these defense methods introduce certified robust models to create a defensive range within which substitutions cannot perturb the model [12]. Gradient-based adversarial training strategies have shown effectiveness in defending attacks with no prior knowledge and improving defense [8,10,17,22,23,38]. Adversarial purification is a particularly desirable type of defense since it does not require prior knowledge of the type of attack. Prior work in adversarial purification has traditionally focused on continuous inputs [18] such as images, exploring generative models such as GANs [30], EBMs [15], and diffusion models [25,33]. However, the field of creating better adversarial defenses and improving robustness in NLP has experienced considerable interest in recent years. Adversarial purification has been explored, however, it is comparably uncommon in NLP. [18] aims to utilize the contextual and masking capabilities of pre-trained masked language models (such as BERT [9]) in order to create a defense against adversarial attacks. However, in this work, we aim to use the power of generative AI, in particular, recent state-of-the-art Large Language Models (LLMs) to perform adversarial purification in the context of capabilities to explore the possibility of improving the robustness of the models.

LLMs as Pseudo-oracles: Alongside the impressive performance of LLMs on a variety of natural language tasks [7], LLMs are also being increasingly used as pseudo-oracles, such as in data annotation [1,11,14], as detectors [4], for model explainability [5] and as experts in general [34]. Inspired by such works, in this work, we propose to use LLMs to perform adversarial purification in the challenging text domain.

3 Background

3.1 Large Language Models

Large language models (LLMs) are essentially deep networks that are based on transformer networks. Transformer-based LLMs are highly effective models that are capable of learning and generating natural language. Broadly there are two categories of language models: (i) Autoregressive language models and (ii) Masked language models. Autoregressive language models are simply trained to predict the next token in a sentence, thereby learning how to generate fluent text when pre-trained on a large corpora of data. Such models include GPT-2 [28], GPT-3 [6], etc. Masked language models (MLMs) are bi-directional models that learn by first masking some fraction of tokens in the sentence and then predicting appropriate tokens to fill the masked slots. Examples of such models include BERT [9], RoBERTa [20], etc. The bidirectional nature of MLMs help the models to have higher language understanding capabilities, and thereby better performance on NLU tasks. More recently, autoregressive models such as the GPT-3 family of models are also being further trained via instruction-tuning [26] with (instruction, response) text pairs, whereby the model learns to generate text to follow user-specified instructions and perform tasks. Some of these instruction-tuned models undergo further training steps (e.g., via RLHF [3]) to align their responses with human preferences. State-of-the-art LLMs such as GPT-3.5 and GPT-4 from OpenAI demonstrate impressive performance when it comes to understanding long and complex human-written instructions in the prompts, as well as editing and generating text. Therefore, we use one of these models in an off-the-shelf manner for our framework.

3.2 Adversarial Text Purification

Adversarial purification is an adversarial defense mechanism that is relatively newer in the natural language domain. As we elaborated in the previous section, this method has been well explored in the domain of computer vision, whereby generative models are used to perform the purification. In the image domain, the standard method is to inject random noise into a perturbed input image, and then use a generative model i.e., the purification algorithm to reconstruct the original *clean* image from the noisy image over multiple rounds. The generated image would now be free of the adversarial perturbations. However, in the domain of text, the discrete nature of the input makes it infeasible to apply the standard computer vision methods directly. One recent attempt at adversarial text purification [18] uses masked language models to randomly mask multiple copies of the perturbed text, and then recovering the text by filling in the mask using the masked language model. This method essentially is somewhat similar to the standard process of injecting noise and iteratively reconstructing the input, as followed in the image domain. However, there is no other method for performing adversarial text purification. To fill this gap, we propose to directly leverage the instruction understanding and text generation capabilities of recent state-of-the-art LLMs and use these LLMs to perform the text purification.

4 LLM-Guided Adversarial Text Purification

In this section, we describe our purification framework and explain the necessary design choices.

We show our overall framework in Fig. 1. As mentioned previously, in this work we focus on the task of text classification and we use fine-tuned pre-trained language models (such as BERT [9]), denoted by $f(\cdot)$ as the classifier. During inference, we evaluate such a classifier on the test set of our task dataset (X_{test}, Y_{test}) where X_{test} and Y_{test} are the sequence of input texts and associated ground truth labels respectively. For an input text $x_i \in X_{test}$, say the classifier correctly predicted $f(x_i) = y_i$, or y_{ground_truth} for ease of reference. Now, say this text is perturbed by an adversarial attack method such that the perturbed text x'_i now gets misclassified to a different label, say $y_{misclassified}$. While many defense mechanisms train the model i.e., the classifier to be adversarially robust to some specific categories of perturbations, purification methods enable simply editing the text, ideally removing the adversarial perturbation from the text and thereby enabling the model to correctly classify the text. Following this, we collect this set of adversarially perturbed input texts X'_{test} and attempt to purify them by using off-the-shelf large language models. In order to do this, we carefully design prompts, as elaborated in the following paragraph. After the purification step, we obtain $\tilde{X}_{test}$ which then is correctly classified by the classifier in majority of the cases.

We use an instruction-tuned LLM which is capable of following human-written instructions in the prompt, in order to generate the purified samples. To enable this, we carefully design the following prompt:

'Human: You are a teacher tasked with grading a quiz. The quiz consists of a sentence (the question) and a classification label (the student's answer). Unfortunately, the sentence has been manipulated by an adversarial attack, leading to a misclassification.
Given the altered sentence and its incorrect label, your job is to generate a new sentence that is semantically similar to the altered one but will be classified correctly according to the correct label.
The categories for classification are: [list of classification categories]
ALTERED SENTENCE (QUESTION): [altered sentence]
MISCLASSIFIED LABEL (STUDENT ANSWER): [misclassified label]
CORRECT LABEL (TRUE ANSWER): [correct label]
Please create a new sentence that conveys the same meaning as the altered sentence but will be classified under the CORRECT LABEL when graded.
Even if there is not a misclassification, provide/construct the sentence to the best of your capability. The output format must be json:
"Original Sentence": "[New sentence here]" Begin!'

In the prompt above, [altered sentence] refers to the adversarially perturbed input text x_i', [misclassified label] refers to $y_{misclassified}$, [correct label] refers to y_{ground_truth} and [list of classification categories] refer to the list of possible labels for the particular classification task. As evident in the prompt, we 'prime' the LLM to enable it to act like a knowledgeable teacher, thereby guiding the editing process. This is the prompt we use for eliciting the purified version of the text from the LLM, and we denote this prompt as P0.

To investigate the efficacy of this carefully designed prompt, we further design and test out two variants of this prompt: P1: which removes the instruction regarding generating text that would correct the misclassified label, and P2: which essentially prompts the LLM to generate a paraphrased version of the input text. The prompt P1 is created by simply removing the text highlighted in pink from P0. Finally, the prompt P2 is:

'Human: Please generate a paraphrased sentence version of the following sentence.
SENTENCE: [altered sentence]
The output format must be json:
"Original Sentence": "[Paraphrased sentence here]" Begin!'

5 Experiments

We conduct comprehensive experiments to evaluate the effectiveness of our proposed LLM-guided adversarial purification method. Our experiments are designed to examine the three main aspects of our method: **(i)** Effectiveness of the proposed method; **(ii)** Ablation study of the components of the designed prompt; and, **(iii)** case study of the purified examples. In the following, we first explain our experimental setting and then discuss our experimental results.

5.1 Experimental Setting

In this section, we describe the datasets, the adversarial attack and the LLM we used in our experiments. We also describe the relevant defense baselines we compare our method to and provide information on our experimental setup to ensure reproducibility. Note that our experimental settings closely follow the ones in the state-of-the-art methods [18].

Datasets. We conduct experiments on two commonly-used benchmark NLP datasets: (1) **IMDb** [21]: for sentiment classification of movie reviews where each review is labeled with a *positive* or *negative* label, and (2) **AG News** [37]: news topic classification where each article is labeled with one of the four categories of $\{$*science, business, world, sports*$\}$.

Adversarial Attack and Defense Baselines. For all our experiments we use the one of the strongest textual attacks named TextFooler [13]. Similar to our baselines, we use the open-source implementation of TextAttack library [24]. The TextFooler attack is selected due to its efficient generation of strong and highly successful adversarial examples, making it an ideal attack to assess the effectiveness of the defense mechanisms. Following previous work, for the size of candidate list we choose $K = \{12, 50\}$ in our experiments.

Following the previous work on adversarial purification [18], we compare the performance of our method with two types of adversarial defense, namely (1) *Textual adversarial training methods:* these methods are based on adversarial training of the classifiers using the adversarial examples generated based on the gradients of the latent space. We use **Adv-HotFlip** [10] and **FreeLB** [17], two state-of-the-arts in this category that do not require For the choice of baseline defenses, as well as the **FreeLB++** [19], which requires the candidate list; and (2) *Textual adversarial purification methods:* methods based on purifying the adversarial examples to generate correctly-classified benign examples. To the best of our knowledge, only one text adversarial purification method exists as in [18]. We include this method as our baseline.

Classifier. Following work in [18] we use classifiers based on two pre-trained masked language models: BERT [9] and RoBERTa [20]. For each dataset, we use BERT and RoBERTa models from Huggingface Transformers (bert-base-uncased[1] and roberta-base[2]), fine-tuned on that specific dataset. Note that our proposed method does not require any further fine-tuning or adversarial training of the model and we can simple query the fine-tuned BERT and RoBERTa

[1] https://huggingface.co/bert-base-uncased.
[2] https://huggingface.co/roberta-base.

Table 1. Comparison of our LLM-guided purification methods with baselines as described in Sect. 5.1. Post-attack accuracy numbers are as reported in [18]. **Bold** denotes the best performance in terms of recovered accuracy, and underline implies the second-best performance.

Defense ↓	Original Accuracy	TextFooler ($K = 12$)	TextFooler ($K = 50$)
IMDb ↓			
fine-tuned BERT	94.1	20.4	2.8
Adv-HotFlip (BERT)	95.1	36.1	8.0
FreeLB (BERT)	96.0	30.2	7.3
FreeLB++ (BERT)	93.2	–	45.3
Text purification (BERT) [18]	93.0	<u>81.5</u>	51.0
Text purification (RoBERTa) [18]	96.1	**84.2**	54.3
(*Ours*) LLM-guided purification (BERT)	94.54	79.34	<u>73.52</u>
(*Ours*) LLM-guided purification (RoBERTa)	95.06	78.9	**76.16**
AG News ↓			
fine-tuned BERT	92.0	32.8	19.4
Adv-HotFlip (BERT)	91.2	35.3	18.2
FreeLB (BERT)	90.5	40.1	20.1
Text purification (BERT) [18]	90.6	61.5	34.9
Text purification (RoBERTa) [18]	90.8	59.1	34.2
(*Ours*) LLM-guided purification (BERT)	95.12	**83.58**	<u>81.3</u>
(*Ours*) LLM-guided purification (RoBERTa)	94.76	<u>82.84</u>	**81.4**

models in an off-the-shelf manner. For evaluating our framework, we report the post-attack accuracy, with and without the purification method, along with the original classifier accuracy without any attack.

Implementation Details. We use OpenAI's GPT-3.5 (version as of November 2023) and use our carefully designed prompts to obtain purified versions of adversarially altered texts. The process involved crafting prompts that guide the model to generate semantically similar but unperturbed versions of the input texts. We chose GPT-3.5 for its advanced contextual understanding and generative capabilities as indicated in [6]. We automated this process using the OpenAI API[3] and LangChain[4]. Our experiments were implemented in Pytorch and were

[3] https://platform.openai.com/docs/api-reference.

[4] https://www.langchain.com/.

run on two systems: (1) Linux system with one A30 and (ii) Linux system with four A100s. All code and links to data will be made available.

Effectiveness of Proposed Purification Method

5.2 Results and Discussion

In this section, we aim to answer if our proposed LLM-based adversarial text purification method is able to effectively purify the adversarial examples. For the sake of comparison, we also report the accuracy under attack for vanilla fine-tuned classifiers. We apply our defense and the state-of-the-art adversarial defenses on the IMDB and AG News datasets and report the results in Table 1. Our results demonstrate that our proposed method effectively defends the state-of-the-art transform-based text classifiers, improving their accuracy under attack by more than 60% in most cases. We elaborate on our observations in the following: (1) The adversarial training-based defenses, i.e., Adv-HotFlip, FreeLB, and FreeLB++, are constantly outperformed by our method based on purification by a large margin (more than 30%). This is because these models are robustified against continuous gradient-based adversarial perturbations and not the discrete word-level perturbations used by text adversarial attacks; (2) the state-of-the-art purification-based defense, namely Text purification, has remarkably lower performance compared to our method. This is because the Text purification method is based on a greedy approach and iteratively selects and perturbs random words. Our method, on the other hand, utilizes the power of LLMs to directly generate purified examples; and (3) finally, our proposed method (LLM-guided purification) achieves the highest after attack accuracy, which is comparable to the accuracy of the model before the attack. For instance, for the BERT trained on the AG News dataset, the original accuracy before the attack is 95.06%, whereas the accuracy after the attack is 83.58%, which is more that 20% better than the accuracy under attack for the second best-performing defense (Text purification (BERT)).

Table 2. Effectiveness of our full prompt as described in Sect. 4 (denoted by P0).

Prompt Type	AG News
Original (BERT)	95.12
Full prompt P0	**81.3**
P1	78
P2	52.7

Ablation: Effectiveness of Prompt Components. In this section, we conduct experiments with two additional prompts namely P1 and P2 as explained in Sect. 4, and compare their results with the results obtained using the main prompt (P0). Specifically, P1 is designed to understand the effect of the explicit instruction to ensure the purified text is classified as the correct label. The goal of designing P2 is to assess the effectiveness of our proposed prompt to ensure the purified samples retain semantic similarity to the adversarial counterparts. To this end P2 simply asks the LLM to paraphrase the adversarial example.

Table 3. Examples from the AG News dataset with TextFooler perturbations (with both $K = 12$ and $K = 50$) along with LLM-purified versions of the perturbed input. Portions of the input text altered by the TextFooler method are shown in teal. Labels in blue are correctly classified, labels in red are misclassified. We see that our methods successfully retains the original label after attack, while maintaining semantics of the original input.

	Texts	Label
Original	E-mail scam targets police chief Wiltshire Police warns about "phishing" after its fraud squad chief was targeted.	science
Adv. Perturbed ($K = 12$)	E-mail scam targets gendarmerie chief Wiltshire Police warns about "phishing" after its deception battalion massa was targeted.	the world
LLM-purified	Wiltshire Police issues warning about phishing email scam targeting their deception battalion massa.	science (conf.: 0.994)
Adv. Perturbed ($K = 50$)	E-mail scam targets police chief Wiltshire Police warns about "phishing" after its hoax battalion leiter was targeted.	the world
LLM-purified	Wiltshire Police alerts about a scam email targeting their police chief, warning about phishing after their hoax battalion leiter was targeted.	science (conf.: 0.984)
Original	Consumer Prices Down, Industry Output Up WASHINGTON (Reuters) - U.S. consumer prices dropped in July for the first time in eight months as a sharp run up in energy costs reversed, the government said in a report that suggested a slow rate of interest rate hikes is likely.	business
Adv. Perturbed ($K = 12$)	Eaters Pricing Down, Departments Product Arriba WASHINGTON (Reuters) - U.S. consuming prices declined in July for the first time in eight months as a ferocious manage up in energy costs quashed, the government tell in a notification that recommendations a sluggish cadence of relevance pace hiking is possible.	science
LLM-purified	Consumers face lower prices as government report suggests slower pace of interest rate hikes due to decrease in energy costs.	business (conf.: 0.954)
Adv. Perturbed ($K = 50$)	User Charging Down, Industry Product Up WASHINGTON (Reuters) - U.S. clients prices dwindled in July for the first time in eight months as a sharp run up in energy costs quashed, the government tell in a report that recommendation a slow rate of interest rate hikes is likely.	science
LLM-purified	U.S. consumer prices fell in July for the first time in eight months due to a significant increase in energy costs, as reported by the government. This suggests that the pace of interest rate hikes is likely to slow down.	business (conf.: 0.999)

Our results reported in Table 2 indicate the effectiveness of our main prompt. The accuracy under attack for purification based on P1 is about 4% less than the full prompt P0. This indicates that even though the full prompt is useful to achieve higher performance, our proposed methodology can obtain similar performance, even when the original correct label of the sample is unknown. However, the performance achieved with P2 is remarkably lower compared to the main prompt, indicating that our proposed prompt is indeed necessary for a successful adversarial purification.

Case Study. We showcase some examples from the AG News dataset in Table 3. We can observe that our purified examples are semantically similar to the adversarial examples while being classified to the original correct class before the attack. This shows that our method can successfully remove the adversarial perturbation and does not change the original benign content of the example. It is important to note that our method can effectively remove adversarial perturbations of any length with only one prompt. Additionally, our generated examples are fluent and grammatically correct, due to the generative power of the LLMs.

6 Conclusion

In this paper, we propose a novel text adversarial purification method, that can effectively remove the adversarial perturbations of any lengths from the adversarial examples and generate purified examples that are semantically similar but are classified to the original correct class. Overcoming the challenges of characterizing adversarial perturbations for discrete inputs (i.e., text), our proposed method utilizes the advanced contextual understanding and generative capabilities of the LLMs to effectively purify the adversarial examples. Our method results in an average accuracy improvement of over 65% under attack.

Acknowledgements. This work is supported by Army Research Office (ARO) W911NF2110030 and Army Research Laboratory (ARL) W911NF2020124. Opinions, interpretations, conclusions, and recommendations are those of the authors' and should not be interpreted as representing the official views or policies of the Army Research Office or the Army Research Lab.

References

1. Alizadeh, M., et al.: Open-source large language models outperform crowd workers and approach ChatGPT in text-annotation tasks. arXiv preprint arXiv:2307.02179 (2023)
2. Alzantot, M., Sharma, Y., Elgohary, A., Ho, B.J., Srivastava, M., Chang, K.W.: Generating natural language adversarial examples. arXiv preprint arXiv:1804.07998 (2018)
3. Bai, Y., , et al.: Training a helpful and harmless assistant with reinforcement learning from human feedback. arXiv preprint arXiv:2204.05862 (2022)
4. Bhattacharjee, A., Liu, H.: Fighting fire with fire: can ChatGPT detect AI-generated text? arXiv preprint arXiv:2308.01284 (2023)
5. Bhattacharjee, A., Moraffah, R., Garland, J., Liu, H.: LLMS as counterfactual explanation modules: can ChatGPT explain black-box text classifiers? arXiv preprint arXiv:2309.13340 (2023)
6. Brown, T., et al.: Language models are few-shot learners. Adv. Neural. Inf. Process. Syst. **33**, 1877–1901 (2020)
7. Chang, Y., et al.: A survey on evaluation of large language models. arXiv preprint arXiv:2307.03109 (2023)
8. Cheng, Y., Jiang, L., Macherey, W.: Robust neural machine translation with doubly adversarial inputs. arXiv preprint arXiv:1906.02443 (2019)

9. Devlin, J., Chang, M.W., Lee, K., Toutanova, K.: BERT: pre-training of deep bidirectional transformers for language understanding. arXiv preprint arXiv:1810.04805 (2018)
10. Ebrahimi, J., Rao, A., Lowd, D., Dou, D.: HotFlip: white-box adversarial examples for text classification. arXiv preprint arXiv:1712.06751 (2017)
11. Flamholz, Z.N., Biller, S.J., Kelly, L.: Large language models improve annotation of viral proteins. Res. Sq. (2023)
12. Jia, R., Raghunathan, A., Göksel, K., Liang, P.: Certified robustness to adversarial word substitutions. arXiv preprint arXiv:1909.00986 (2019)
13. Jin, D., Jin, Z., Zhou, J.T., Szolovits, P.: Is BERT really robust? A strong baseline for natural language attack on text classification and entailment. In: Proceedings of the AAAI Conference on Artificial Intelligence, vol. 34, pp. 8018–8025 (2020)
14. Latif, S., Usama, M., Malik, M.I., Schuller, B.W.: Can large language models aid in annotating speech emotional data? uncovering new frontiers. arXiv preprint arXiv:2307.06090 (2023)
15. LeCun, Y., Chopra, S., Hadsell, R., Ranzato, M., Huang, F.: A tutorial on energy-based learning. Predicting Struct. Data **1**(0) (2006)
16. Li, L., Ma, R., Guo, Q., Xue, X., Qiu, X.: BERT-attack: adversarial attack against BERT using BERT. arXiv preprint arXiv:2004.09984 (2020)
17. Li, L., Qiu, X.: Token-aware virtual adversarial training in natural language understanding. In: Proceedings of the AAAI Conference on Artificial Intelligence, vol. 35, pp. 8410–8418 (2021)
18. Li, L., Song, D., Qiu, X.: Text adversarial purification as defense against adversarial attacks. arXiv preprint arXiv:2203.14207 (2022)
19. Li, Z., et al.: Searching for an effective defender: benchmarking defense against adversarial word substitution. arXiv preprint arXiv:2108.12777 (2021)
20. Liu, Y., et al.: RoBERTa: a robustly optimized BERT pretraining approach. arXiv preprint arXiv:1907.11692 (2019)
21. Maas, A., et al.: Learning word vectors for sentiment analysis. In: Proceedings of the 49th Annual Meeting of the Association for Computational Linguistics: Human Language Technologies, pp. 142–150 (2011)
22. Madry, A., Makelov, A., Schmidt, L., Tsipras, D., Vladu, A.: Towards deep learning models resistant to adversarial attacks. arXiv preprint arXiv:1706.06083 (2017)
23. Miyato, T., Dai, A.M., Goodfellow, I.: Adversarial training methods for semi-supervised text classification. arXiv preprint arXiv:1605.07725 (2016)
24. Morris, J.X., Lifland, E., Yoo, J.Y., Grigsby, J., Jin, D., Qi, Y.: TextAttack: a framework for adversarial attacks, data augmentation, and adversarial training in NLP. arXiv preprint arXiv:2005.05909 (2020)
25. Nie, W., Guo, B., Huang, Y., Xiao, C., Vahdat, A., Anandkumar, A.: Diffusion models for adversarial purification. arXiv preprint arXiv:2205.07460 (2022)
26. Ouyang, L., et al.: Training language models to follow instructions with human feedback. Adv. Neural. Inf. Process. Syst. **35**, 27730–27744 (2022)
27. Peng, C., et al.: A study of generative large language model for medical research and healthcare. arXiv preprint arXiv:2305.13523 (2023)
28. Radford, A., et al.: Language models are unsupervised multitask learners. OpenAI Blog **1**(8), 9 (2019)
29. Ren, S., Deng, Y., He, K., Che, W.: Generating natural language adversarial examples through probability weighted word saliency. In: Proceedings of the 57th Annual Meeting of the Association for Computational Linguistics, pp. 1085–1097 (2019)

30. Samangouei, P., Kabkab, M., Chellappa, R.: Defense-GAN: protecting classifiers against adversarial attacks using generative models. arXiv preprint arXiv:1805.06605 (2018)
31. Shi, C., Holtz, C., Mishne, G.: Online adversarial purification based on self-supervision. arXiv preprint arXiv:2101.09387 (2021)
32. Shreya, G., Khapra, M.M.: A survey in adversarial defences and robustness in NLP. arXiv preprint arXiv:2203.06414 (2022)
33. Song, Y., Sohl-Dickstein, J., Kingma, D.P., Kumar, A., Ermon, S., Poole, B.: Score-based generative modeling through stochastic differential equations. arXiv preprint arXiv:2011.13456 (2020)
34. Xu, B., et al.: Expertprompting: instructing large language models to be distinguished experts. arXiv preprint arXiv:2305.14688 (2023)
35. Ye, M., Miao, C., Wang, T., Ma, F.: TextHoaxer: budgeted hard-label adversarial attacks on text. In: Proceedings of the AAAI Conference on Artificial Intelligence, vol. 36, pp. 3877–3884 (2022)
36. Yoon, J., Hwang, S.J., Lee, J.: Adversarial purification with score-based generative models. In: International Conference on Machine Learning. pp. 12062–12072. PMLR (2021)
37. Zeng, J., Xu, J., Zheng, X., Huang, X.: Certified robustness to text adversarial attacks by randomized [mask]. Comput. Linguist. **49**(2), 395–427 (2023)
38. Zhu, C., Cheng, Y., Gan, Z., Sun, S., Goldstein, T., Liu, J.: Freelb: Enhanced adversarial training for natural language understanding. arXiv preprint arXiv:1909.11764 (2019)

lil'HDoC: An Algorithm for Good Arm Identification Under Small Threshold Gap

Tzu-Hsien Tsai, Yun-Da Tsai(✉), and Shou-De Lin

National Taiwan University, Taipei, Taiwan
{f08946007,sdlin}@csie.ntu.edu.w

Abstract. Good arm identification (GAI) is a pure-exploration bandit problem in which a single learner outputs an arm as soon as it is identified as a good arm. A good arm is defined as an arm with an expected reward greater than or equal to a given threshold. This paper focuses on the GAI problem under a small threshold gap, which refers to the distance between the expected rewards of arms and the given threshold. We propose a new algorithm called lil'HDoC to significantly improve the total sample complexity of the HDoC algorithm. We demonstrate that the sample complexity of the first λ output arm in lil'HDoC is bounded by the original HDoC algorithm, except for one negligible term, when the distance between the expected reward and threshold is small. Extensive experiments confirm that our algorithm outperforms the state-of-the-art algorithms in both synthetic and real-world datasets.

1 Introduction

The stochastic multi-armed bandit problem (MAB) is a well-known task with various applications. In this problem, there are K arms in the environment, and in each round, denoted by t, the learner selects an action $a_t \in [K]$ based on a policy and then pulls the corresponding arm. The selected arm generates an independent and identically distributed (i.i.d) reward $X_{a_t}(t)$ from an unknown distribution v_i with an unknown expectation u_i. Subsequently, the learner observes the exact reward and updates its policy to achieve a specific objective. The classical objective of MAB is to minimize cumulative regret, where the learner aims to maximize the cumulative reward over a fixed number of trials [2]. In this setting, the learner faces the exploration-exploitation dilemma, where exploration involves pulling seemingly sub-optimal arms to discover the arm with the highest expected reward, while exploitation involves selecting the arm with the highest empirical reward to increase the cumulative reward.

One well-studied variant of MAB is best arm identification (BAI), a pure exploration problem where the learner aims to find the arm with the highest expected reward with as few samples as possible [10,12]. In 2016, Locatelli et al. [13] proposed the threshold bandit problem (TBP), a specific instance of the pure

Supplementary Information The online version contains supplementary material available at https://doi.org/10.1007/978-981-97-2262-4_7.

D.-N. Yang et al. (Eds.): PAKDD 2024, LNAI 14649, pp. 78–89, 2024.
https://doi.org/10.1007/978-981-97-2262-4_7

exploration bandit framework [3]. TBP divides all arms into two groups based on whether their expected reward is above a given threshold or not, using minimal samples. However, in some real-world applications, neither identifying the best arm (BAI) nor correctly partitioning all the arms (TBP) is necessary. Instead, it is more desirable to quickly identify a set of reasonably good arms. For example, in recommendation tasks, the goal is not necessarily to identify the most popular item or all popular items, but rather to quickly identify a set of items that are popular enough to recommend. To address this, Kano et al. [11] proposed the good arm identification (GAI) framework. GAI has the same goal as TBP, but it aims to minimize the sample complexity not only of identifying all good arms but also of identifying the first λ good arms, where $\lambda \in \mathbb{N}$. GAI faces a new type of exploration-exploitation dilemma, the exploration-exploitation dilemma of confidence, where exploration means the learner samples suboptimal arms to identify a good arm, and exploitation means the learner continues to sample the currently best arm to increase the confidence in its goodness.

HDoC [11] has been proposed as the SOTA method to solve the GAI task. However, the sample complexity shown in Table 1 suggests that HDoC can be quite expensive to use when the threshold gap is small. This is because it requires a large number of samples on a single arm in order to determine whether it is a good or bad arm. For example, consider a recommendation system that uses bandit to determine which items to recommend, with rewards associated with click-through-ratio (CTR). In such systems, many items may have very small CTR (e.g. close to zero), while the good arms (e.g. CTR much larger than zero) are sparse. In this scenario, Δ is likely to be small, which can significantly hurt the GAI performance. Therefore, we propose to decrease the confidence bound in the identification method of HDoC.

In this paper, we consider the GAI framework with i.i.d Bernoulli reward when the threshold gap is small (<0.01). Inspired by the challenging situation, we propose a new algorithm called "lil'HDoC" where "lil" stands for the Law of Iterated Logarithm. In the bandit problem, the operation of sampling (or pulling the arm) is considered as acquiring information about the sampled arm. HDoC suggests that at the beginning of the algorithm, each arm should be sampled once. However, if the threshold gap is small, sampling each arm only once might not be sufficient to make the right decision in the subsequent sampling algorithm because we have less confidence in the goodness/badness of each arm. One change we propose in our algorithm is to sample each arm more than once in the beginning, denoted as $T > 1$. We will show that the value of T can be determined based on the acceptance error rate and the number of arms. Sampling each arm for T times in the beginning can harm the sample complexity when it is easy to identify the goodness/badness of arms. However, in a challenging situation where a large number of samples is needed to identify one arm, the effect of sampling each arm T times is negligible. By sampling each arm more than once in the beginning of the algorithm, we have more confidence in the goodness/badness of arms, and therefore can obtain a tighter confidence bound than HDoC. However, T cannot be too large, otherwise, it can hurt the overall performance. Thus, determining a suitable value for T while still maintaining the theoretical performance guarantee is the main challenge we address. We need to adjust the confidence bound for identification to reach the theoretical guarantee.

Our contribution is as follow:

- Applying the law of iterative logarithm to design a new algorithm, named lil'HDoC that improves the total sample complexity of the HDoC algorithm in the context of Good Arm Identification (GAI).
- Exhibiting a PAC bounded sample complexity, particularly when the distance between the expected reward and the threshold is small.
- Providing various experiments to show that the lil'HDoC algorithm surpasses state-of-the-art algorithms in both synthetic and real-world datasets.

This paper is organized as follow. Section 2 reviews the related works on GAI and threshold bandit problem. Section 3 and 4 provide the basics about GAI. Our algorithm and the proof of the theoretical guarantee is provided in Sect. 5. Section 6 describes the experiment results before the conclusion in Sect. 7.

2 Background

2.1 Good Arm Identification

Kano [11] proposed a formulation of the good arm identification problem, derived from the threshold bandit problem. This formulation addresses a new type of exploration-exploitation dilemma, called the dilemma of confidence. In this dilemma, exploration refers to the agent pulling other arms, rather than the currently best one, to increase confidence in identifying whether it is good or bad. Exploitation involves pulling the currently best arm to increase confidence in its goodness. The algorithm is decomposed into two parts: the sampling method and the identifying method. The former is responsible for selecting the arm to sample in each round, and the latter decides whether an arm is good or bad. For the sampling method, Kano et al. proposed a Hybrid algorithm for the Dilemma of Confidence (HDoC) and two baseline algorithms: the Lower and Upper Confidence Bounds algorithm for GAI (LUCB-G), based on the LUCB algorithm for best arm identification [10], and the Anytime Parameter-free Thresholding algorithm for GAI (APT-G), based on the APT algorithm for the thresholding bandit problem [13]. The lower bound on the sample complexity for GAI is $\Omega(\lambda log\frac{1}{\delta})$, and HDoC can find λ good arms within $O(\frac{\lambda \log \frac{1}{\delta} + (K-\lambda) \log\log \frac{1}{\delta} + K \log \frac{K}{\Delta}}{\Delta^2})$ samples.

3 Problem Setting

Let K denote the number of arms, ξ the threshold, and δ the acceptable error rate. The reward of each arm $i \in [1, \ldots, K]$ follows a Bernoulli distribution with mean μ_i, which is unknown to the learner. We define "good" arms as those whose means are larger than the threshold ξ. Without loss of generality, we can assume that the means of the arms are ordered such that:

Table 1. Sample Complexity of HDoC

	HDoC
First λ arm	$O(\frac{\lambda \log \frac{1}{\delta} + (K-\lambda) \log \log \frac{1}{\delta} + K \log \frac{K}{\Delta}}{\Delta^2})$
Total	$O(\frac{K \log \frac{1}{\delta} + K \log K + K \log \frac{1}{\Delta}}{\Delta^2})$

Table 2. Notation list

K	Number of arms
A	Set of arms, where $\|A\| = K$
ξ	Threshold determining whether an arm is good or not
δ	Acceptance error rate. $\delta \leq 1/e$
μ_i	The true mean of ith arm
$\hat{\mu}_{i,t}$	The empirical mean of ith arm at time t
τ_λ	Round that learner identifies λ good arms
τ_{stop}	Round that learner identifies every arms
$N_i(t)$	The number of samples of arm i by the end of round t
$T_i(t)$	The number of times of arm i be pulled at t
$\overline{\mu}_{i,t}$	Upper confidence bound of arm i at time t
$\underline{\mu}_{i,t}$	Lower confidence bound of arm i at time t
Δ_i	$\|\mu_i - \xi\|$
$\Delta_{i,j}$	$\mu_i - \mu_j$
Δ	$\min(\min_{i \in K} \Delta_i, \min_{j \in [K-1]} \frac{\Delta_{j,j+1}}{2})$

$$\mu_1 \geq \mu_2 \geq \cdots \geq \mu_m \geq \xi \geq \mu_{m+1} \geq \cdots \geq \mu_K \tag{1}$$

Note that the learner is unaware of the number of "good" arms and their indexing.

In each round t, the agent selects an arm $a(t)$ to pull and receives a reward that is i.i.d. generated from distribution $v_{a(t)}$. The agent identifies one arm as the good arm and the rest as bad arms based on the rewards received from them in previous rounds. The agent stops at time τ_{stop} when all the good arms are identified. The objective is to minimize the upper bound of τ_λ (i.e., the number of sample times to identify λ good arms) and τ_{stop} (i.e., the number of samples required to identify all good arms) with an acceptance error rate of δ.

4 Preliminary

The notation is listed in Table 2. Two important lemmas are stated below. The first is the finite form of the Law of Iterated Logarithm [8] which is the kernel of our new confidence bound. The second is the inequality involving the operation of iterated logarithm.

Table 3. Sample Complexity of lilHDoC

	lil'HDoC
First λ arm	$O\big(K\log(K+1)\log[\max(\frac{1}{\delta},e)] + \frac{\lambda\log\frac{1}{\delta}+(K-\lambda)\log\log\frac{1}{\delta}+K\log\frac{K}{\Delta}}{\Delta^2}$
Total	$O(\frac{K\log\frac{1}{\delta}+K\log K+K\log\log\frac{1}{\Delta}}{\Delta^2}) + O\big(K\log(K+1)\log(\max(\frac{1}{\delta},e))\big)$

Lemma 1 (Finite form of LIL). *Let $X_1, X_2, \ldots X_n$ be i.i.d. σ–sub–gaussian random variables. Then for algorithm parameters $\epsilon \in (0,1)$ and $\rho \in (0, \frac{\log(1+\epsilon)}{e})$, with probability at least $1 - c_\epsilon \rho^{1+\epsilon}$,*

$$\frac{1}{t}\sum_{s=1}^{t} X_s \leq U(t,\rho) \tag{2}$$

for all $t > 0$. Here, U is the upper confidence bound

$$U(t,\omega) = (1+\sqrt{\epsilon})\sqrt{\frac{2\sigma^2(1+\epsilon)}{t}\log[\frac{\log((1+\epsilon)t)}{\omega}]} \tag{3}$$

$$c_\epsilon = \frac{2+\epsilon}{\epsilon}\left[\frac{1}{\log(1+\epsilon)}\right]^{1+\epsilon} \tag{4}$$

Proof. See Section 4, Lemma 1 in [8].

We will use $U(t,\omega)$ and c_ϵ throughout the paper for brevity. It is worth noting that the Bernoulli distribution is $\frac{1}{2}$-sub-gaussian.

Lemma 2. *[9] For $t \geq 1$, $\epsilon \in (0,1)$, $c > 0$, and $0 < \omega < 1$,*

$$\frac{1}{t}\log\left[\frac{\log[(1+\epsilon)t]}{\omega}\right] \geq c \Rightarrow t \leq \frac{1}{c}\log\left[\frac{2\log[\frac{(1+\epsilon)}{c\,\omega}]}{\omega}\right] \tag{5}$$

Proof. Direct calculation. For details, see [9].

5 Algorithm

We propose a new algorithm, lil'HDoC. This algorithm improve the complexity of HDoC by sampling every arms more than one time in the beginning and redesigning the confidence bound in the identifying method. The algorithm is provided in Algorithm 1, and its sample complexity is listed in Table 3. In lil'HDoC, the confidence bound in the identification method has a faster convergence rate than HDoC algorithm, thanks to the term related to $N_i(t)$ being in the form of

Algorithm 1. lil'HDoC

Input: K, δ ($\leq 1/e$)
1: $T_i(K) = 1, \forall i \in [K]$
2: $A = [K]$
3: $B = K + 1$
4: $C = \max(\frac{1}{\delta}, e)$
5: $r = (1+\sqrt{\epsilon})^2(1+\epsilon)$
6: $\epsilon = \text{argmax}_{\epsilon\in[0,1]}\left(r - 1 \leq \min\left(\frac{\log\log B}{\log B}, \frac{\log\log C}{\log C}\right)\right)$
7: $T = \text{argmin}_{t>0}\left(\frac{t^2}{[\log[(1+\epsilon)t]]^r} \geq \frac{1}{4}K^{r-1}(\frac{1}{\delta})^{r-1}(c_\epsilon)^r\right)$
8: **for** $s \in A$ **do**
9: Pull arm s T times
10: $T_i(s) = T$
11: **end for**
12: **while** $A \neq \phi$ **do**
13: **for** $s \in A$ **do**
14: $u_{i,T_i(t)} \leftarrow \sqrt{\frac{\log t}{2N_i(t)}}$
15: **end for**
16: $h_t = \text{argmax}_{i\in[K]}\left(\hat{\mu}_{i,T_i(t)} + u_{i,T_i(t)}\right)$
17: Pull h_t
18: **if** $\hat{\mu}_{i,T_i(t)} - U(N_{h_t}(t), \frac{\delta}{c_\epsilon K}) \geq \xi$ **then**
19: Output h_t as a good arm
20: Delete h_t from A
21: **else if** $\hat{\mu}_{i,T_i(t)} + U(N_{h_t}(t), \frac{\delta}{c_\epsilon K}) \leq \xi$ **then**
22: Delete h_t from A
23: **end if**
24: **end while**

$\sqrt{\frac{\log\log N_i(t)}{N_i(t)}}$, instead of $\sqrt{\frac{\log N_i(t)}{N_i(t)}}$. Consequently, the required number of samples decreases. It can be observed that for any $c_1, : c_2 \in \mathbb{R}^+$, there exists a value T such that for all $t > T$,

$$c_1\sqrt{\frac{\log t}{t}} \geq c_2\sqrt{\frac{\log\log t}{t}} \tag{6}$$

This implies that after sampling arm i T times, lil'HDoC can identify it as well as HDoC.

Moreover, the explicit form of T is easy to handle, and we apply binary search to determine its value in lil'HDoC. We can consider sampling each arm for T times as a way to gain sufficient confidence in the goodness or badness of the arms, which enables us to develop a more precise confidence bound for identifying the arms in lil'HDoC. In the complexity analysis of the first λ arm in Table 3, the first term corresponds to T, while the second term corresponds to the sample complexity of HDoC. In challenging situations, the terms divided by Δ dominate the sample complexity, making the $K\log(K+1)\log[\max(\frac{1}{\delta}, e)]$ negligible. Thus, the sample complexity of the first λ arms is the same under big-O notation in challenging situations. Additionally, when an arm is sampled more than T times, lil'HDoC requires fewer samples to identify that arm. While the total complexity in Table 3 could potentially be $O(K\log(K+1)\log(\max(\frac{1}{\delta},e)))$, we will demonstrate in Sect. 5.3 that this scenario can be easily avoided. We will now establish the theoretical guarantee for the lil'HDoC algorithm by proving the correctness of the algorithm, providing the sample complexity for identifying the first λ good arms and obtaining the explicit form of T, and lastly proving the sample complexity for identifying all arms.

5.1 Correctness of lil'HDoC

Theorem 1. *With probability at least* $1-\delta$, *lil'HDoC correctly identifies every arms.*

Proof (Proof of Theorem 1*).* Let D as an event that $\forall\, i \in [K]$ and $t \geq 1$,

$$|\hat{\mu}_{i,t} - \mu_i| \leq U(t, \frac{\delta}{c_\epsilon K}) \tag{7}$$

By Lemma 1 and the union bound, we know that event D happen with probability at least

$$1 - Kc_\epsilon(\frac{\delta}{c_\epsilon K})^{1+\epsilon} \geq 1 - Kc_\epsilon(\frac{\delta}{c_\epsilon K}) \geq 1-\delta \tag{8}$$

Inequality 8 is because $\epsilon \in [0,1]$ and $c_\epsilon \geq 1$. So event D happens with probability at least $1-\delta$. Therefore, if

$$\hat{\mu}_{i,T_i(t)} - U(T_i(t), \frac{\delta}{c_\epsilon K}) \geq \xi \tag{9}$$

holds under event D,

$$\hat{\mu}_{i,T_i(t)} - U(T_i(t), \frac{\delta}{c_\epsilon K}) \geq \xi \tag{10}$$

$$\Rightarrow \hat{\mu}_{i,T_i(t)} - |\hat{\mu}_{i,T_i(t)} - \mu_i| \geq \xi \tag{11}$$

$$\Rightarrow \mu_i \geq \xi \tag{12}$$

and if

$$\hat{\mu}_{i,T_i(t)} + U(T_i(t), \frac{\delta}{c_\epsilon K}) < \xi \tag{13}$$

holds under event D,

$$\hat{\mu}_{i,T_i(t)} + U(T_i(t), \frac{\delta}{c_\epsilon K}) < \xi \tag{14}$$

$$\Rightarrow \hat{\mu}_{i,T_i(t)} + |\hat{\mu}_{i,T_i(t)} - \mu_i| < \xi \tag{15}$$

$$\Rightarrow \mu_i < \xi \tag{16}$$

So our algorithm output the correct answer when event D holds, so the error rate of our algorithm is at most δ.

5.2 First λ Arms Sampling Complexity

Theorem 2 (First λ Arms Sample Complexity). *After conducting at most $O\big(\log(K+1)\log(\max(\frac{1}{\delta}, e))\big)$ samples on each arm, our confidence bound will be less than that of HDoC.*

Proof. See Appendix 2.

5.3 Total Sample Complexity

Theorem 3 (Sample Complexity). *Let $T = 1$, then with probability at least $1-\delta$, lil'HDoC identifies arm i with at most*

$$\frac{2(1+\epsilon)(1+\sqrt{\epsilon})^2}{\Delta_i^2}\log\left[\frac{2c_\epsilon K\log\left[\frac{2c_\epsilon K(1+\sqrt{\epsilon})^2(1+\epsilon)^2}{\delta\Delta_i^2}\right]}{\delta}\right]$$

times of sampling.

Proof. See Appendix 3.

6 Experiment

The goal of our experiments is to learn (1) lil'HDoC improves the sample complexity of identifying first λ good arms in practice, (2) lil'HDoC improves the sample complexity of identifying all arms in practice. In the following experiment, we will focus on the datasets given the challenging situation.

6.1 Dataset

Syntactic Dataset. One syntactic dataset is provided. It has six arms with expected reward 0.007, 0.006, 0.005, 0.003, 0.002, 0.001 respectively, and the threshold is 0.004. We conduct the experiment over 10 independent runs.

Real World Dataset. We generate the experiment data from three real-world datasets: Covertype [1], Jester [6], and MovieLens [7]. We conduct the experiment over 25 independent runs. The Covertype dataset classifies the cover type of northern Colorado forest areas in 7 classes. For Covertype, we use the method similar to [4,5,14] to transform multi–class dataset to bandit dataset, and we divide the mean by 10 to make the dataset more challenging. The threshold is set as the arithmetic mean of the reward of the 3rd best arm and the 4th best arm. We conduct the experiment over 25 independent runs. Jester dataset provides continuous ratings in [−10, 10] for 100 jokes from 73421 users. For Jester, we create a recommendation system bandit problem as follows. We first count the average rating of 100 jokes and divide then by 10 in order to increase difficulty, and scale the rating from $[-10, 10]$ to $[0, 1]$. The threshold is set as the arithmetic mean of the reward of the 25-th best arm and the 26-th best arm. We conduct the experiment over 25 independent runs. MovieLens dataset provides 100,000 ratings ranging from [0, 5], from 1000 users on 1682 movies. For MovieLens, we first average the rating of each movie, and divide the average ratings by 100. The threshold is set as the arithmetic mean of the reward of the 168-th best arm and the 169-th best arm. Here we provide the sample complexity of identifying the first 25 good movies. We conduct the experiment over 20 independent runs.

6.2 Baseline

We choose HDoC and LUCB–G as the baseline since they are the top-2 models compared here [11]. All algorithms including ours consist of two stages, sampling and identifying. The former decides which arm to pull in the next round, and the latter decides whether the pulled arm can be identified as a good or bad arm.

Sampling Method

1. HDoC : Pull arm $\hat{a}^* = \text{argmax}_{i \in A} \overline{\mu}_{i,t}$, where $\overline{\mu}_{i,t} = \hat{\mu}_{i,t} + \sqrt{\frac{\log(t)}{2N_i(t)}}$
2. LUCB-G : Pull arm $\hat{a}^* = \text{argmax}_{i \in A} \overline{\mu}_{i,t}$, where $\overline{\mu}_{i,t} = \hat{\mu}_{i,t} + \sqrt{\frac{\log 4KN_i^2(t)/\delta}{2N_i(t)}}$
3. lil'HDoC : Pull arm $\hat{a}^* = \text{argmax}_{i \in A} \overline{\mu}_{i,t}$, where $\overline{\mu}_{i,t} = \hat{\mu}_{i,t} + \sqrt{\frac{\log t}{2N_i(t)}}$

Identifying Method. Here we show the confidence bound of algorithms

1. HDoC : $\sqrt{\frac{\log 4KN_i^2(t)/\delta}{2N_i(t)}}$
2. LUCB-G : $\sqrt{\frac{\log 4KN_i^2(t)/\delta}{2N_i(t)}}$
3. lil'HDoC : $(1 + \sqrt{\epsilon})\sqrt{\frac{(1+\epsilon)}{2N_i(t)} \log \frac{c_\epsilon K \log((1+\epsilon)N_i(t))}{\delta}}$

6.3 Results

The experimental results are listed from Fig. 1a to Fig. 1d. In these figures, the x axis indicates the number of identified good arms, and the y axis indicate the number of sample times. The detailed information of Fig. 1a to Fig. 1d is in Table 4a to Table 4d. In addition, we only list the τ_{stop} in the table since it is often very large with respect to τ_λ.

In every experiments, lil'HDoC outperforms HDoC, especially when an arm required more sample times to identify it. We can see this from Table 4a, Table 4b, Table 4c, and Table 4d. From the last row of the three table, we can also see that the sampling times of identifying all arms in lil'HDoC outperforms the sampling times of identifying all arms in HDoC and LUCB–G. The effect of lil'HDoC is not obvious when the required sample times is small. It is because that although the rate of convergence is faster in the confidence bound of lil'HDoC, the constant in the confidence bound of lil'HDoC is larger than that of HDoC, so when the sample times is not large, the effect will be diluted. The

Table 4. Experiment results of 4 dataset. Best results are in bold fonts.

	HDoC	LUCB–G	lil'HDoC
τ_1	3.92 ± 0.24	**1.98 ± 0.06**	2.67 ± 0.18
τ_2	9.30 ± 0.21	22.91 ± 0.48	**6.00 ± 0.24**
τ_3	30.10 ± 0.78	55.56 ± 1.01	**17.60 ± 0.51**
τ_{stop}	551.13 ± 10.89	555.58 ± 10.12	**315.85 ± 6.49**

(a) Average and standard deviation of sampling times over synthetic dataset. All number is divided by 100000.

	HDoC	LUCB–G	lil'HDoC
τ_1	**0.14 ± 0.02**	**0.13 ± 0.02**	**0.15 ± 0.01**
τ_2	**0.29 ± 0.03**	104.64 ± 3.50	**0.25 ± 0.02**
τ_3	137.55 ± 3.24	317.71 ± 0.71	**83.26 ± 2.13**
τ_{stop}	3225.9 ± 53.1	3177.1 ± 71.4	**1860.3 ± 38.2**

(b) Average and standard deviation of sample times over Covertype dataset. All numbers in this table is divided by 10000.

	HDoC	LUCB–G	lil'HDoC
τ_3	**1.61 ± 0.23**	**5.92 ± 64.6**	**1.42 ± 0.24**
τ_6	**2.69 ± 0.30**	95.79 ± 72.5	**2.35 ± 0.25**
τ_9	**4.18 ± 0.33**	217.78 ± 93.3	**3.57 ± 0.30**
τ_{12}	6.26 ± 0.42	275.79 ± 55.1	**4.85 ± 0.49**
τ_{15}	8.31 ± 0.50	301.77 ± 32.4	**6.55 ± 0.50**
τ_{18}	18.30 ± 1.97	320.13 ± 55.9	**13.94 ± 1.18**
τ_{21}	51.87 ± 4.47	478.79 ± 79.2	**36.18 ± 2.92**
τ_{24}	95.10 ± 4.75	551.42 ± 36.2	**63.47 ± 4.24**
τ_{stop}	563.63 ± 21.1	560.40 ± 31.6	**324.70 ± 19.5**

(c) Average and standard deviation of sampling times over Jester dataset. All number is divided by 100000.

	HDoC	LUCB–G	lil'HDoC
τ_3	27.02 ± 0.83	NA	**24.39 ± 0.70**
τ_6	29.16 ± 0.93	NA	**26.07 ± 0.90**
τ_9	31.38 ± 1.11	NA	**28.47 ± 1.01**
τ_{12}	85.36 ± 4.17	NA	**76.49 ± 4.56**
τ_{15}	98.47 ± 3.19	NA	**87.14 ± 3.85**
τ_{18}	107.14 ± 4.31	NA	**94.71 ± 3.09**
τ_{21}	123.67 ± 7.59	NA	**106.60 ± 3.61**
τ_{24}	**162.94 ± 11.58**	NA	**146.25 ± 5.88**

(d) Average and standard deviation of sampling times over MovieLens dataset. All number is divided by 1000000. NA in the entries means that the sample times is more than 1.5e+8

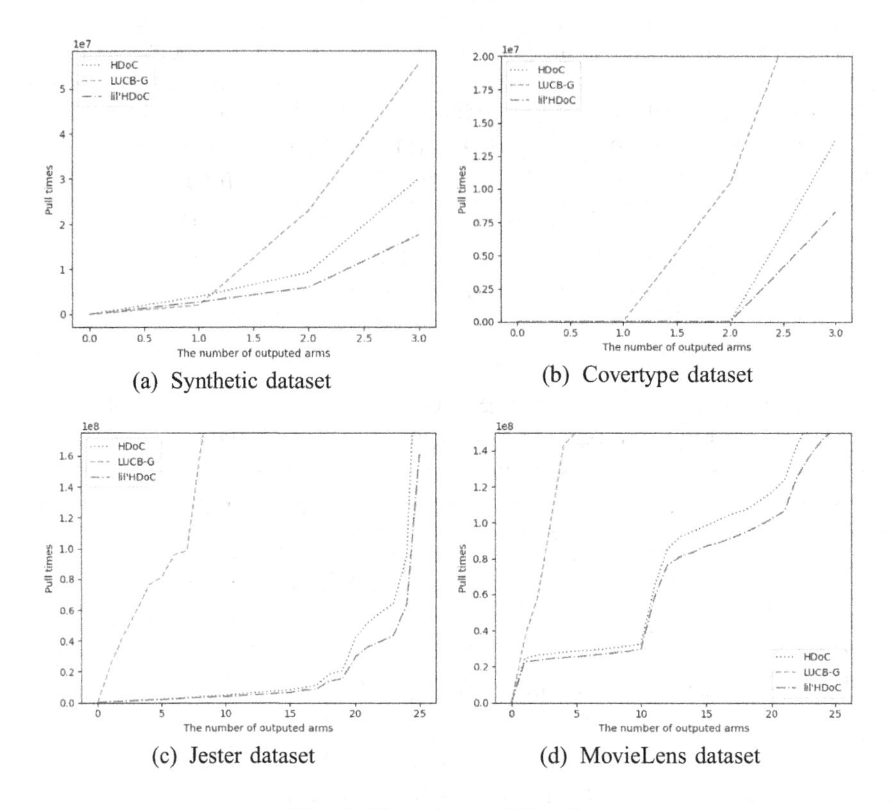

(a) Synthetic dataset

(b) Covertype dataset

(c) Jester dataset

(d) MovieLens dataset

Fig. 1. Experimental Results

other issue is that in Jester and MovieLens, LUCB–G may identify many bad arms before all good arms is identified. This is because the sampling method of LUCB–G doesn't put the total sampling times t into consideration, so its exploration part would be less than HDoC and lil'HDoC. Thus, LUCB–G will exploit the suboptimal arms.

7 Conclusion

In this paper, we propose a new algorithm lil'HDoC, based on the HDoC algorithm in GAI [11]. Intuitively, we leverage the fact that under challenging situation (when the threshold gap is small < 0.01) every arms require a huge number of sample times. Thus, we can sample each arm for a rather small number of times in the beginning to obtain more confidence on the goodness/badness of arms, which can lead to a tighter confidence bound in the identifying method. From the theoretical perspective, the first λ good arms of lil'HDoC is bounded by the original HDoC algorithm except for one negligible term under challenging situation, and the total sample complexity of lil'HDoC is less than HDoC by decrease the $\frac{1}{\Delta}\log\frac{1}{\Delta}$ term to $\frac{1}{\Delta}\log\log\frac{1}{\Delta}$, which makes a conspicuous improvement when Δ is small. From the practical perspective, on both synthetic and read world dataset, lil'HDoC outperforms HDoC under the same acceptance rate. Therefore, we conclude that lil'HDoC outperform HDoC and LUCB–G from theoretical and empirical performance under challenging situation.

References

1. Asuncion, A., Newman, D.: UCI machine learning repository (2007)
2. Auer, P., Cesa-Bianchi, N., Fischer, P.: Finite-time analysis of the multiarmed bandit problem. Mach. Learn. **47**(2), 235–256 (2002)
3. Chen, S., Lin, T., King, I., Lyu, M.R., Chen, W.: Combinatorial pure exploration of multi-armed bandits. Adv. Neural. Inf. Process. Syst. **27**, 379–387 (2014)
4. Da Tsai, Y., De Lin, S.: Fast online inference for nonlinear contextual bandit based on generative adversarial network. arXiv preprint arXiv:2202.08867 (2022)
5. Dudík, M., Langford, J., Li, L.: Doubly robust policy evaluation and learning. arXiv preprint arXiv:1103.4601 (2011)
6. Goldberg, K., Roeder, T., Gupta, D., Perkins, C.: Eigentaste: a constant time collaborative filtering algorithm. Inf. Retrieval **4**(2), 133–151 (2001)
7. Harper, F.M., Konstan, J.A.: The movielens datasets: history and context. ACM Trans. Interact. Intell. Syst. (TIIS) **5**(4), 1–19 (2015)
8. Jamieson, K., Malloy, M., Nowak, R., Bubeck, S.: lil'ucb: an optimal exploration algorithm for multi-armed bandits. In: Conference on Learning Theory, pp. 423–439. PMLR (2014)
9. Jiang, H., Li, J., Qiao, M.: Practical algorithms for best-k identification in multi-armed bandits (2017)
10. Kalyanakrishnan, S., Tewari, A., Auer, P., Stone, P.: PAC subset selection in stochastic multi-armed bandits. In: ICML, vol. 12, pp. 655–662 (2012)
11. Kano, H., Honda, J., Sakamaki, K., Matsuura, K., Nakamura, A., Sugiyama, M.: Good arm identification via bandit feedback. Mach. Learn. **108**(5), 721–745 (2019)
12. Kaufmann, E., Cappé, O., Garivier, A.: On the complexity of best-arm identification in multi-armed bandit models. J. Mach. Learn. Res. **17**(1), 1–42 (2016)
13. Locatelli, A., Gutzeit, M., Carpentier, A.: An optimal algorithm for the thresholding bandit problem. In: International Conference on Machine Learning, pp. 1690–1698. PMLR (2016)
14. Tsai, Y.D., Tsai, T.H., Lin, S.D.: Differential good arm identification. arXiv preprint arXiv:2303.07154 (2023)

Recommender Systems

ScaleViz: Scaling Visualization Recommendation Models on Large Data

Ghazi Shazan Ahmad[1], Shubham Agarwal[2], Subrata Mitra[2(✉)], Ryan Rossi[3], Manav Doshi[4], Vibhor Porwal[2], and Syam Manoj Kumar Paila[5]

[1] IIIT B, Bengaluru, India
[2] Adobe Research, Bangalore, India
subrata.mitra@adobe.com
[3] Adobe Research, San Jose, USA
[4] IIT B, Mumbai, India
[5] IIT KGP, Kharagpur, India

Abstract. Automated visualization recommendation (Vis-Rec) models help users to derive crucial insights from new datasets. Typically, such automated Vis-Rec models first calculate a large number of statistics from the datasets and then use machine-learning models to score or classify multiple visualizations choices to recommend the most effective ones, as per the statistics. However, state-of-the-art models rely on a very large number of expensive statistics and therefore using such models on large datasets becomes infeasible due to prohibitively large computational time, limiting the effectiveness of such techniques to most large real-world datasets. In this paper, we propose a novel reinforcement-learning (RL) based framework that takes a given Vis-Rec model and a time budget from the user and identifies the best set of input statistics, specifically for a target dataset, that would be most effective while generating accurate enough visual insights. We show the effectiveness of our technique as it enables two state of the art Vis-Rec models to achieve up to 10X speedup in *time-to-visualize* on four large real-world datasets.

1 Introduction

As more and more data is being collected from various sources, users often encounter data that they are not familiar with. A dataset can contain numerous columns, both numerical and categorical, with multiple categories. It is a daunting task to even decide how to dissect and plot such data to reveal any *interesting* insights. Visualization recommendation (Vis-Rec) techniques [6,13] help to automatically generate, score, and recommend the most relevant visualizations for a dataset and can improve productivity by reducing the time required by analysts to first find interesting insights and then visualize them.

Automated Vis-Rec techniques typically work through the following steps. First, they calculate various statistics, as much as up to 1006 number of statistics as used by Qian et al. in [13] per column from the data to capture overall

G. S. Ahmad, M. Doshi and S. M. K. Paila—Work done during internship at Adobe Research.

D.-N. Yang et al. (Eds.): PAKDD 2024, LNAI 14649, pp. 93–104, 2024.
https://doi.org/10.1007/978-981-97-2262-4_8

statistical landscape of the data. Second, these statistics are used as features to score prospective visualization configurations (i.e. combination of columns, aggregates, plot types etc.) in a supervised learning setup. Finally, queries are issued against the actual data to populate the top recommended visualization charts. A significant number of prior works [4–7, 15] focused on perfecting the visualization recommendation technique, which evolved from initial algorithmic approaches to most recent deep-learning based approaches [6, 11, 13]. Further, Qian et al. [13] extended these techniques to address the problem of how to generalize these models on unseen datasets having completely different schema structure and data distributions. The way such generalization work is that the neural network learns the importance of different visualizations at much abstract level by extracting a large number of higher order statistical features extracted from the data. However, prior works did not address another very important problem, which is the scalability of these algorithms on datasets with large number of columns and/or rows. Real world datasets can be huge, having several hundreds of millions or even several hundreds of billions of rows. Calculating large number of statistical features on such large datasets is intractable by the state-of-the-art (SOTA) visualization recommendation algorithms. For example, in Qian et al. [13] collects 1006 various higher-order statistics *per column* of the dataset, which itself can have a large number of columns. On top of that, they calculate multi-column statistics to capture dependency, correlation and other such properties. In Fig. 1 we show the CDF of computation time for different statistical features that are needed by SOTA `Vis-Rec` models MLVR [13] and VizML [6] for 4 datasets of different sizes. Table 1 lists the number of rows and columns for each dataset and total number of statistical features that needs to be computed for MLVR and VizML for each dataset. Calculating so many statistics on large datasets makes the very first step of the typical visualization recommendation pipeline infeasible and unscalable.

Now, there can be two ways to overcome this problem:

- First option could be to drop certain statistics to reduce the computation. *But which ones?* These statistics are basically the features to the core visualization recommendation model. Which statistics are important and carry important signals that would make a particular combination of columns and visualization style interesting - is very dataset dependent. A statistics that is very computationally intensive might carry significant importance for one dataset and might not be relevant for another dataset. Indiscriminate dropping of certain statistics or identifying the important statistics based on few datasets and extending that decision to other datasets, can lead to poor quality output.
- Second option could be to take a small sample of the data, on which calculation of large number statistics is tractable, and then generate the visualization recommendations based on that sample. However, for massive amounts of data, such sample has to be a tiny fraction for the existing `Vis-Rec` pipeline to work and such a tiny sample may not be representative of the complete data. Therefore, the visualization recommendations generated on the sample can be completely misleading or inaccurate.

To overcome these drawbacks of naive ways to speed-up visualization recommendation generation, in this paper we present a framework, called ScaleViz,

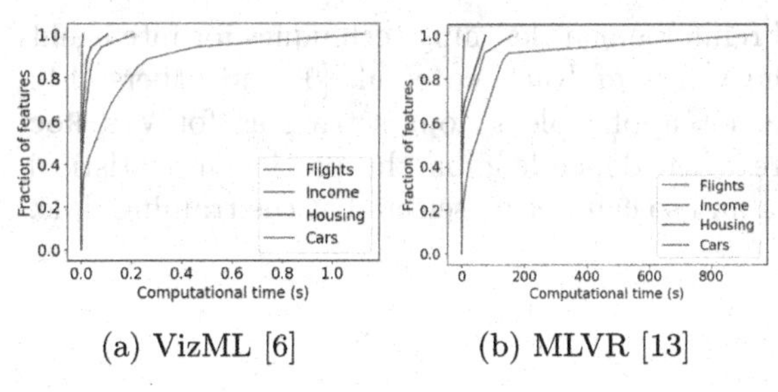

(a) VizML [6] (b) MLVR [13]

Fig. 1. CDF of computation time (in seconds) of various features on datasets of different sizes. With increase in dataset size, the computation time of complex features drastically increases.

Table 1. Number of statistical features that are needed by MLVR and VizML for 4 different datasets with (# rows, # columns).

Datasets	VizML	MLVR
Flights ($1M$, 12)	972	12072
Income ($200k$, 41)	3321	41246
Housing ($20k$, 10)	810	10060
Cars ($10k$, 9)	729	9054

that takes such a generalized `Vis-Rec` model and *customizes* it for a given dataset so that we can produce visualization recommendations at large scale, for that dataset. ScaleViz does this by through the following steps: (1) It profiles the computational cost of calculating statistics for each statistics that are needed by the generalized model on a few samples of data of different sizes. (2) It uses regression models to extrapolate that cost to the size of the full dataset. (3) It uses a *budget-aware* Reinforcement-Learning (RL) based technique to identify the most crucial features from the original `Vis-Rec` model — using multiple samples containing a very small fraction from the original dataset. (4) Finally, ScaleViz only calculates these selected statistical features from the full dataset and produces the visualization recommendations using the given model. In summary, we make the following contributions:

1. We propose a framework that enables a `Vis-Rec` model to generate accurate enough insights for a target large-scale dataset within a chosen time budget.
2. We formulate the problem as a budget-aware Reinforcement Learning problem that incrementally learns the most useful statistical features from a large-scale dataset for the target model.
3. Our evaluations with 2 recent ML-based `Vis-Rec` models [6,13] and with 4 large public datasets show that ScaleViz can provide upto 10X speedup in producing accurate enough visualization recommendations.

2 Related Works

Several prior works [2,4–8,13,15] targeted visualization recommendations to help insight discovery. But scalability of such technique when handling large datasets were not addressed and is the focus of this paper. As recent `Vis-Rec` models use large number of statistical features from the data, feature selection literature is also related to our work. Prior works by Li et al. [10], Deng et al. [1], and Farahat et al. [3], have employed decision trees or greedy-based approaches.

Some researchers have explored reinforcement learning techniques for intelligent feature selection, as seen in the works of Kachuee et al. [9] and others [14]. However, these approaches are not applicable to our setting as for `Vis-Rec` models, the crucial features are often dependent on the particular statistical characteristics of the target data and so can not be selected at the training time.

3 Problem Formulation

In this section, we first formally define the problem of budget-aware visualization recommendation generation.

Let $\mathcal{P}$ be a target `Vis-Rec` model that a user wants to apply on a large tabular dataset $\mathcal{D}$. Let $\mathcal{D}$ consists m columns and r rows. Let $\mathcal{F}$ be the feature space for dataset $\mathcal{D}$ based on statistical features used in the model $\mathcal{P}$. As `Vis-Rec` models calculate a large number of different statistics from each column, let us denote the number of statistical features computed from each column be n. Let f_{ij} denote the j-th feature for i-th column, where $i \in \{1, \ldots, m\}$ and $j \in \{1, \ldots, n\}$.

We introduce the cost function $c^k : \mathcal{F} \rightarrow \mathbb{R}^{m \times n}$, quantifying the *computational time* required to calculate each of the features based on a k fraction from $\mathcal{D}$ (i.e. such fraction will consist of $1/k$ rows of $\mathcal{D}$). Notably, c^1 serves as the cost function for the entire dataset $\mathcal{D}$, and for brevity, we use c^1 denoted as c throughout the paper.

To formalize the problem, we frame the statistical feature selection as an optimization task. Let $\theta : \mathcal{F} \rightarrow \{0, 1\}^{m \times n}$ be a function mapping features to binary acquisition decisions. $\theta(f) \odot f$ gives a subset of features by ignoring the masked features, where $f \in \mathcal{F}$ and $\odot$ is the Hadamard operator which calculates the product of two matrices of the same dimension. $\mathcal{L}$ is a loss function which compares the output of the model on two different input feature set. The objective is to find the feature mask minimizing the error in the model's prediction while ensuring the total cost of selected features adheres to the budget $\mathcal{B}$:

$$\min_{\theta} \mathcal{L}[\mathcal{P}(\theta(f) \odot f) - \mathcal{P}(f)], \text{ subject to: } \sum_{i,j} \theta(f) \odot c(f) \leq \mathcal{B} \tag{1}$$

Here, the budget $\mathcal{B}$ is constrained by the total computational cost of features calculated on the complete dataset:

$$\mathcal{B} \leq \sum_{i,j} c(f) \tag{2}$$

Note, in this formulation, we use $\mathcal{B}$, that is *time-to-compute* visualization recommendations as the constraint, because it is intuitive for users to specify a time-budget. Alternatively, we could also make this constraint relative to the size of $\mathcal{D}$. In that case, $\mathcal{B} \leq r \times \sum_{i,j} c(f)$ where r is a particular user-specified fraction of the statistical feature computation time for the base `Vis-Rec` model.

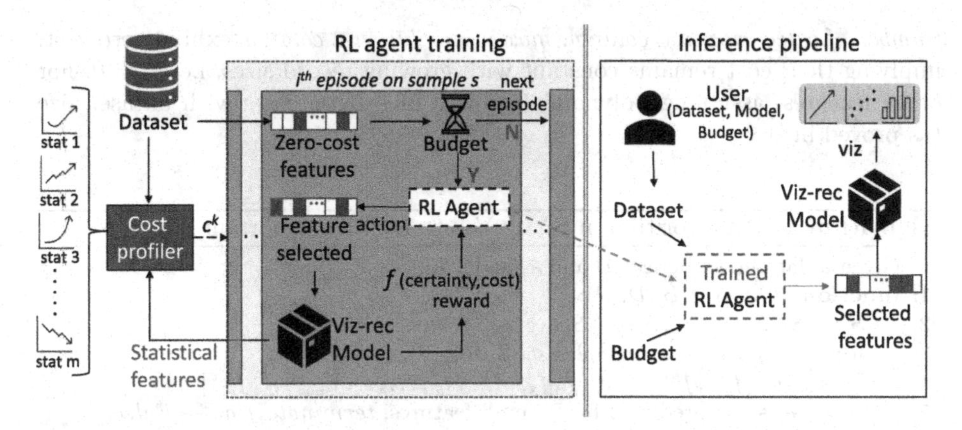

Fig. 2. Pipeline overview: The cost profiler estimates the computational cost for computing statistics. RL agent training begins with a set of zero-cost features. Within each episode, the agent dynamically acquires features until the budget is exhausted. The Q-value is estimated using rewards which is based on increased certainty in recommendations with newly acquired features, considering acquisition costs. This iterative process continues until error converges below a certain error value. Once trained, in the inference pipeline, the RL agent now selects features for the specified budget, tailored to the dataset and model.

4 Proposed Solution

We approach the above problem as a scenario where decisions are made sequentially over time and model the problem as a reinforcement learning problem. The overall pipeline, as shown in Fig. 2, consists of a **Cost profiler**, which employs polynomial regression to estimate the computational cost of computing statistics across varying dataset sizes. This estimation is crucial for predicting costs without actually computing them. Subsequently, the **RL agent training** module teaches the agent to acquire features under budget constraints across increasing data samples. Once trained, the **Inference pipeline** utilizes the RL agent to select features for the given budget, computing only the learned subset of features on the entire dataset to obtain model predictions. We provide a detailed description of the two main components and also describe the RL agent training algorithm.

4.1 Cost Profiling

The Cost Profiler module profiles the computation time (cost) of each statistical feature across varying dataset sizes. It collects data points to estimate the computation cost for each feature on larger datasets without actual computation.

Given the dataset $\mathcal{D}$, the cost function c^k is obtained for $|k|$ fractions of the dataset, denoted as $\{c^{k_1}, c^{k_2}, \ldots, c^{|k|}\}$. For each feature f_{ij}, the goal is to predict its cost c_{ij} on the full dataset. Some features, such as *column types, number of categories in a column, max-min value in a column*, exhibit zero-cost, implying their cost remains constant with growing record sizes, i.e $c_{ij} = 0$. For other features, assuming polynomial growth of feature costs with dataset size (as proved in [16]).

Algorithm 1. RL algorithm in ScaleViz to identify important features

Given a dataset $\mathcal{D}$, budget $\mathcal{B}$ and a model $\mathcal{P}$

1: **function** ScaleViz($\mathcal{B}$, $\mathcal{D}$, $\mathcal{P}$):
2: $S \leftarrow [S_1, S_2, S_3, ..., S_{|S|}]$ $(|S_{k+1}| > |S_k| \; \forall k \in [1, |S| - 1])$
3: **for** sample S_k *in the samples set* S **do**
4: $x^{k,t} \leftarrow [f_{11}^{k,t}, f_{12}^{k,t}, \ldots, f_{1n}^{k,t}, f_{21}^{k,t}, \ldots f_{2n}^{k,t}, \ldots, \ldots f_{m1}^{k,t}, \ldots, f_{mn}^{k,t}]$
5: $\tilde{y_k} \leftarrow$ score predicted by $\mathcal{P}$ on all features, $terminate\_flag \leftarrow False$
6: **while** not $terminate\_flag$ **do**
7: **if** $random$ in $[0, 1) \leq Pr_{\text{rand}}$ **then**
8: $ij \leftarrow$ index of a randomly selected unknown feature
9: **else**
10: $ij \leftarrow Q(x^{k,t})$ (index of the feature with the maximum Q value)
11: **end if**
12: $x^{k,t+1} \leftarrow$ acquire f_{ij} and unmask it
13: $\mathcal{P}(x^{k,t}) \leftarrow$ score predicted using the feature set $x^{k,t}$
14: $total\_cost \leftarrow total\_cost + \boldsymbol{c}_{ij}$
15: $r_{ij}^{k,t} \leftarrow \frac{\|\mathcal{P}(x^{k,t}) - \mathcal{P}(x^{k,t+1})\|}{c_{ij}}$
16: push $\left(\boldsymbol{x}^{k,t}, ij, \boldsymbol{x}^{k,t+1}, r_{ij}^{k,t}\right)$ into the replay memory
17: $t \leftarrow t + 1$
18: **if** $total\_cost \geq B$ **then**
19: $terminate\_flag \leftarrow True$
20: $\hat{y_k} \leftarrow \mathcal{P}(x^{k,t})$predicted score on subset of features $x^{k,t}$
21: **end if**
22: loss $\leftarrow \mathcal{L}(\tilde{y_k}, \hat{y_k})$
23: **if** $update\_condition$ **then**
24: train _batch $\leftarrow$ random mini-batch from the replay memory
25: update (Q, target Q) networks using train batch
26: **end if**
27: **end while**
28: **if** $\epsilon > loss$ **then** terminate loop
29: **end if**
30: **end for**
31: **end function**

4.2 RL Agent

We use an RL agent based framework to learn feature acquisition under budget constraints. Each episode consists of the agent choosing the important subset of features for a sample S_k. We define the state of the agent for an episode k as the feature set acquired by it so far in an episode (i.e. $x^{k,t} = \theta^{k,t}(f) \odot f$), where $\theta^{k,t}$ is the mask of the features at time t. The action $a_{k,t}$ of the agent is to select a feature which has not been masked in the feature set (i.e. $x^{k,t}$).

At every step t, the agent selects a feature ij and masks that feature as selected. The agent moves to the next state, which is $x^{k,t+1} = \theta^{k,t+1}(f) \odot f$). A cost of c_{ij} is deducted from the remaining budget for choosing the feature. The reward for an action $a_{k,t}$ is calculated as the absolute change in the score before and after acquiring the feature, ij with a penalty of c.

$$r_t = \frac{\left\| \mathcal{P}\left(\boldsymbol{x}^{k,t}\right) - \mathcal{P}\left(\boldsymbol{x}^{k,t+1}\right) \right\|}{c_{ij}} \tag{3}$$

We use the technique of double deep Q-learning with experience replay buffer [12] to train the RL agent. The agent explores the feature space with a ϵ-greedy approach, with the probability of exploration decaying exponentially. The architecture of the Q-networks is a feed-forward neural network with three layers of sizes $[512, 128, 64]$.

Algorithm 1 describes the training procedure for the RL agent, designed for cost-effective feature acquisition. The process initiates with the agent receiving a dataset $\mathcal{D}$, a pre-defined budget $\mathcal{B}$ and a `Vis-Rec` model $\mathcal{P}$. The dataset is sequentially explored through a series of samples. The algorithm initializes by setting an initial exploration probability Pr_{rand} and a termination threshold ϵ. In each episode, the agent learns the important subset of features for a particular sample S_k. Every episode starts with the same budget, $\mathcal{B}$ and the size of the samples keeps increasing with the number of episodes. The RL agent starts with a *zero-cost* feature set and keeps acquiring features till it runs out of budget. At every step of an episode, the agent chooses to explore randomly or exploit the current knowledge by selecting the feature with the maximum Q-value. The tuple (state, action, next state, reward) is pushed into the experience replay buffer. The Q and the target-Q networks are periodically updated using the tuples from the buffer. The process is terminated when the loss for an episode falls below the threshold ϵ. The increasing size of the samples across episodes helps the agent to exploit the learned behavior of the model on a larger sample. This is particularly important because, we ultimately want the agent to predict the important features on the full dataset which it has not been trained on.

The RL agent ultimately selects the important and highly sensitive statistical features for the target base `Vis-Rec` model $\mathcal{P}$ from a given dataset $\mathcal{D}$.

5 Evaluations

5.1 Experimental Setup

We use an NVIDIA GeForce RTX 3090 GPU and a 32-core processor for all experiments. PyTorch, scikit-learn, and pandas were used for both training of the agent and running `Vis-Rec` models. For `Vis-Rec` input features, non-selected features were imputed based on a smaller 0.01% sample. We use an exponential decay exploration probability which starts with probability of 1.0 and eventually reaches 0.1. A batch size of 128 is used to randomly sample experiences from the replay buffer. The agent's action space was normalized to facilitate efficient Q-network training. The training of both the Q and target Q networks employed the Adam optimization algorithm and Mean Squared Error (MSE) loss for effective convergence.

`Vis-Rec` Models

1. **VizML** [6]**:** VizML provides visualization-level and encoding-level prediction tasks using 81 column-level features, which are aggregated using 16 functions for predicting visualizations. In our experiments, the RL agent selects column-level features during training, which are then aggregated and fed into the VizML model. We use cross-entropy loss to calculate the error introduced due to selection of subset of features.
2. **MLVR** [13]**:** MLVR recommends the top-k visualizations for a given dataset and set of visualizations. It predicts visualization probabilities by leveraging 1006 column-level features. This approach becomes computationally challenging with a high number of columns. In our experiments, we use the mean-squared loss of the prediction scores on the top-k visualization configurations given by the model using the full feature set to calculate the error.

Datasets

1. **Flights:**[1] On-time performance of domestic flights operated by large air carriers in the USA. It comprises approximately 1 million rows and 12 columns.
2. **Income:**[2] USA Census Income data, with 200k rows and 41 columns.
3. **Cars:**[3] Features of vehicles, including mileage, transmission, price, etc. The dataset consists of 10k rows and 9 columns.
4. **Housing:**[4] Home prices based on factors like area, bedrooms, furnishings, proximity to the main road, etc. It contains around 20k rows and 10 columns.

Baselines. We compare ScaleViz with the following baselines.

1. `Random`**:** Features are randomly selected through uniform sampling until the budget is exhausted, forming the set of features for the `Vis-Rec` model.
2. `Greedy`**:** Features are chosen using a greedy technique inspired by [17] until the budget is exhausted, and then passed to the `Vis-Rec` model.
3. `Sample`**:** Features are computed on 1%, 2%, 3%, and 5% uniform samples of the dataset. This baseline approach allows calculation of all the statistics using a small sample, which can be passed to the `Vis-Rec` model.

---

[1] https://www.kaggle.com/datasets/mexwell/carrier-dataset.
[2] https://www.kaggle.com/datasets/manishkc06/usa-census-income-data.
[3] https://www.kaggle.com/datasets/mrdheer/cars-dataset.
[4] https://www.kaggle.com/datasets/ashydv/housing-dataset.

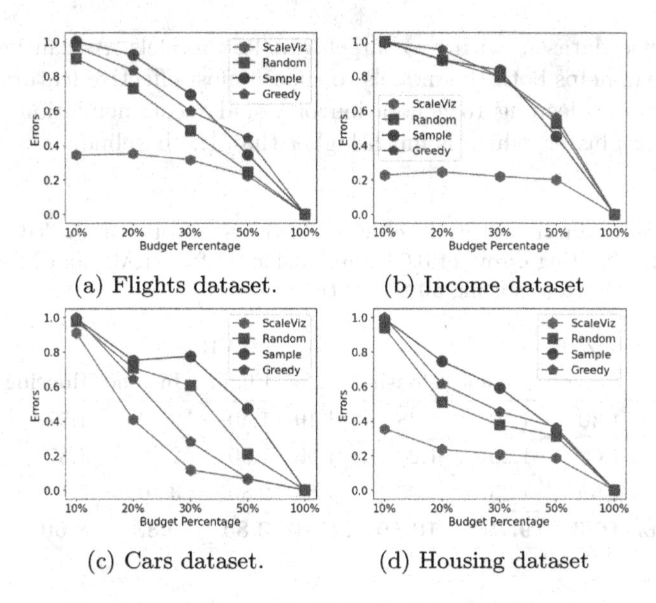

(a) Flights dataset. (b) Income dataset

(c) Cars dataset. (d) Housing dataset

Fig. 3. Evaluation of VizML on different datasets

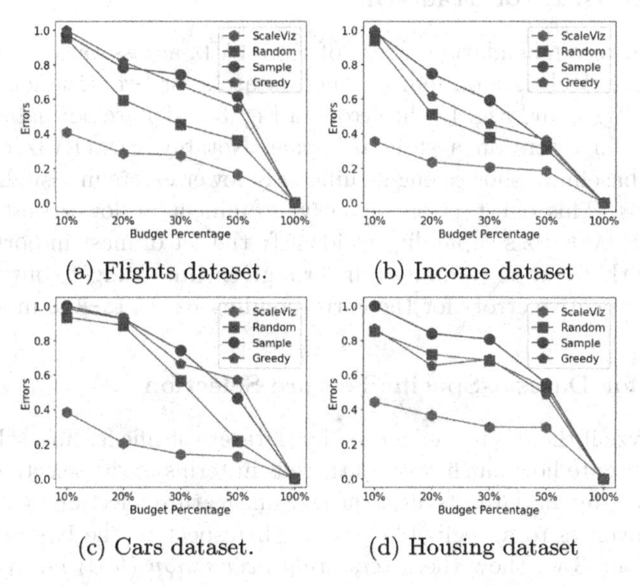

(a) Flights dataset. (b) Income dataset

(c) Cars dataset. (d) Housing dataset

Fig. 4. Evaluation of MLVR on different datasets

5.2 Speed-Up in Visualization Generation

We first evaluate the speed-up achieved by ScaleViz compared to the baselines approaches when we ensure that the resulting error due to use of less features (for ScaleViz, `Random`, and `Greedy`) or less data (for `Sample`) is less than $0.0002, 3.43e^{-05}$ for VizML and MLVR respectively. Table 2 presents the speedup for four diverse datasets with two target `Vis-Rec` models. As can be observed that ScaleViz helps both the models to choose most effective features, tailored for each datasets, leading to generation of visual recommendation generation upto 10.3 times faster, which is much higher than the baseline models.

Table 2. Speedup in visualization recommendation generation provided by different techniques with limiting errors of 0.0002 and $3.43e^{-05}$ for VizML and MLVR respectively, compared to results using all the features.

Method	ViZML				MLVR			
	Flights	Income	Housing	Cars	Flights	Income	Housing	Cars
`Sample`	1.30	1.50	1.38	1.10	1.40	2.80	1.52	2.20
`Greedy`	1.60	1.20	1.20	1.90	1.90	2.76	1.37	1.40
`Random`	2.50	1.25	1.96	1.63	2.80	3.10	2.10	2.80
ScaleViz	**10.3**	**9.70**	**10.10**	**8.10**	**9.84**	**9.88**	**8.60**	**9.94**

5.3 Budget vs. Error Trade-Off

We assess the recommendation errors of ScaleViz across various budget percentages of the total time on four distinct datasets for two `Vis-Rec` models, as illustrated in Fig. 3 and Fig. 4. The errors in Fig. 3 and 4 are normalized to show the difference in errors on a standard scale. Notably, ScaleViz consistently outperforms baselines, showcasing significantly lower errors in visualization recommendations. This effect is particularly prominent at lower budget ranges, highlighting ScaleViz's capability to identify the set of most important statistical features that can be computed under a given time-budget constraint while minimizing respective errors for the corresponding base `Vis-Rec` models.

5.4 Need for Dataset-Specific Feature Selection

We now analyze if there is indeed a need for dataset specific feature-selection. For this, we investigate how much overlap there is in terms of the selected statistical features from different datasets after the runtime feature selection by ScaleViz's RL-agent converges to a negligible error with respect to the baseline `Vis-Rec` models. In Table 3 we show the *intersection over union* (IoU) between the sets of features important features selected by ScaleViz for all pairs from the 4 real world datasets. It can be observed that IoU values ranges from 10% to a maximum of 22% for VizML. Similarly, for MLVR, the overlap varies from 3% to a maximum of 14%. This emphasizes the design choice of ScaleViz highlighting the fact that feature selection is highly dependent on both the choice of `Vis-Rec` model ($\mathcal{P}$) and the target dataset ($\mathcal{D}$) and a dataset agnostic pruning of features (even when done in a computation cost-aware manner) would remain suboptimal.

Table 3. Intersection over Union (IoU) of features selected by ScaleViz for different datasets. It can be observed that the important statistical features identified for each dataset has very low overlap with other datasets, highlighting the importance of runtime and data-specific feature selection by ScaleViz and the fact that a generic feature selection technique would be sub-optimal.

	Flights	Income	Cars	Housing
Flights	1.00	0.22	0.10	0.12
Income	0.22	1.00	0.12	0.12
Cars	0.10	0.12	1.00	0.19
Housing	0.12	0.12	0.19	1.00

(a) VizML model

	Flights	Income	Cars	Housing
Flights	1.00	0.03	0.14	0.11
Income	0.03	1.00	0.03	0.05
Cars	0.14	0.03	1.00	0.12
Housing	0.11	0.05	0.12	1.00

(b) MLVR model

5.5 Scalability with Increasing Data Size

We now show how ScaleViz's benefit keeps on increasing as the size of the dataset (in terms of number of rows) increases. We define a saturation budget $\mathcal{B}'$ as the computation time taken by the selected features by ScaleViz where the resulting visualization recommendations has insignificant error ($\leq \epsilon$) compared to the base `Vis-Rec` model. For VizML $\epsilon = 0.0002$ and for MLVR $\epsilon = 3.43e^{-05}$. We us $\mathcal{B}_{MAX}$ to denote the time taken by the base `Vis-Rec` model to produce the visualizations. Table 4 shows the values of $\mathcal{B}_{MAX}$ and $\mathcal{B}'$ for both VizML and MLVR models for increasing sizes of a dataset (Flights). As can be observed ScaleViz saturated at around **half** the budget for a 1k dataset, saturated at around **one-fifth** of the budget for 100k, and its efficiency scales even more impressively with larger datasets, reaching about **one-tenth** of the budget for a dataset size of 1M. This scalability advantage positions ScaleViz as an efficient and cost-effective solution to boost `Vis-Rec` models for large datasets.

Table 4. Analysis of the minimum budget $\mathcal{B}'$ and $\mathcal{B}_{MAX}$ (*milliseconds* for VizML, *seconds* for MLVR) required to achieve specified errors on the flights dataset. VizML to achieve an error $\epsilon = 0.0002$ MLVR to achieve an error $\epsilon = 3.43e^{-05}$

`Vis-Rec`	VizML				MLVR			
Records	1k	10k	100k	1M	1k	10k	100k	1M
$\mathcal{B}_{MAX}$	160	233	1311	13259	592	928	1950	11161
$\mathcal{B}'$	79	112	262	1285	458	583	771	1134

6 Conclusion

In this paper, we identify an important drawback of the state-of-the-art visualization recommendation (`Vis-Rec`) models that these models sacrificed the scalability in order to make them generalize over unknown datasets. Such models compute a very large number of statistics from the target dataset, which becomes infeasible at larger dataset sizes. In this paper, we propose ScaleViz- a scalable and time-budget-aware framework for visualization recommendations on large datasets. Our approach can be used with existing `Vis-Rec` models to tailor them for a target dataset, such that visual insights can be generated in a timely manner with insignificant error compared to alternate baseline approaches.

References

1. Deng, H., Runger, G.: Feature selection via regularized trees. In: The 2012 International Joint Conference on Neural Networks (IJCNN), pp. 1–8. IEEE (2012)
2. Ding, R., Han, S., Xu, Y., Zhang, H., Zhang, D.: QuickInsights: quick and automatic discovery of insights from multi-dimensional data. In: ICMD (2019)
3. Farahat, A.K., Ghodsi, A., Kamel, M.S.: An efficient greedy method for unsupervised feature selection. In: ICDM, pp. 161–170. IEEE (2011)
4. Godfrey, P., Gryz, J., Lasek, P.: Interactive visualization of large data sets. IEEE TKDE (2016). https://doi.org/10.1109/TKDE.2016.2557324
5. Harris, C., et al.: Insight-centric visualization recommendation. arXiv:2103.11297 (2021)
6. Hu, K., Bakker, M.A., Li, S., Kraska, T., Hidalgo, C.: VizML: a machine learning approach to visualization recommendation. In: CHI, pp. 1–12 (2019)
7. Hulsebos, M., Demiralp, C., Groth, P.: Gittables: a large-scale corpus of relational tables. Proc. ACM Manag. Data **1**, 1–17 (2023)
8. Idreos, S., Papaemmanouil, O., Chaudhuri, S.: Overview of data exploration techniques. In: SIGMOD (2015)
9. Kachuee, M., et al.: Opportunistic learning: budgeted cost-sensitive learning from data streams. arXiv preprint arXiv:1901.00243 (2019)
10. Li, J., et al.: Feature selection: a data perspective. ACM Comput. Surv. (CSUR) **50**, 1–45 (2017)
11. Luo, Y., Qin, X., Tang, N., Li, G.: DeepEye: towards automatic data visualization. In: ICDE, pp. 101–112. IEEE (2018)
12. Mnih, V., Kavukcuoglu, K., Silver, D., Rusu, E.A.: Human-level control through deep reinforcement learning. Nature **518**, 529–533 (2015)
13. Qian, X., et al.: Learning to recommend visualizations from data. In: KDD 2021. ACM (2021)
14. Sali, R., Adewole, S., Akakpo, A.: Feature selection using reinforcement learning. CoRR **abs/2101.09460** (2021). https://arxiv.org/abs/2101.09460
15. Vartak, M., Huang, S., Siddiqui, T., Madden, S., Parameswaran, A.: Towards visualization recommendation systems. ACM SIGMOD Rec. **45**, 34–39 (2017)
16. Wang, C., Chen, M.H., Schifano, E., Wu, J., Yan, J.: Statistical methods and computing for big data. Stat. Interf. **9**(4), 399 (2016)
17. Xu, Z., Weinberger, K., Chapelle, O.: The greedy miser: learning under test-time budgets. arXiv preprint arXiv:1206.6451 (2012)

Collaborative Filtering in Latent Space: A Bayesian Approach for Cold-Start Music Recommendation

Menglin Kong[1], Li Fan[2], Shengze Xu[3], Xingquan Li[4], Muzhou Hou[1], and Cong Cao[1(✉)]

[1] School of Mathematics and Statistics, Central South University, Changsha, China
{212112025,congcao}@csu.edu.cn
[2] School of Advanced Technology, Xi'an Jiaotong-Liverpool University, Suzhou, China
hmzw@csu.edu.cn
[3] Department of Mathematics, The Chinese University of Hong Kong, Hong Kong, China
[4] Peng Cheng Laboratory, Shenzhen, China

Abstract. Personalized music recommendation technology is effective in helping users discover desired songs. However, accurate recommendations become challenging in cold-start scenarios with newly registered or limited data users. To address the accuracy, diversity, and interpretability challenges in cold-start music recommendation, we propose CFLS, a novel approach that conducts collaborative filtering in the space of latent variables based on the Variational Auto-Encoder (VAE) framework. CFLS replaces the standard normal distribution prior in VAE with a Gaussian process (GP) prior based on user profile information, enabling consideration of user correlations in the latent space. Experimental results on real-world datasets demonstrate the effectiveness and superiority of our proposed method. Visualization techniques are employed to showcase the diversity, interpretability, and user-controllability of the recommendation results achieved by CFLS.

Keywords: Music Recommendation · Bayesian Inference · Variational Auto-Encoder · Gaussian Process

1 Introduction

Music is a popular leisure and entertainment activity in people's daily lives. However, with the increasing problem of information overload, it has become challenging for users to efficiently discover songs of their interest from a vast music library [18]. To address this issue, personalized music recommendation technology has emerged as the most effective method to help users quickly find their desired songs [15]. Personalized music recommendation is a service that

Menglin Kong and Li Fan contributed equally to this work.

D.-N. Yang et al. (Eds.): PAKDD 2024, LNAI 14649, pp. 105–117, 2024.
https://doi.org/10.1007/978-981-97-2262-4_9

utilizes deep learning (DL) and machine learning (ML) technology to offer music suggestions based on users' music preferences, interests, and behavioral data [20]. Its primary objective is to enhance users' music experience, increase user engagement on music platforms, and drive the growth of the music industry [3].

As a typical representative of the more generalized recommender system task of Sequential Recommendation (SR), personalized music recommendation systems encounter similar challenges as SR in practice, including (i) **data sparsity and cold starts** [22,25]. Recommender systems face the cold-start problem when dealing with newly registered users or users with limited historical data. Making accurate personalized recommendations without sufficient information about their preferences becomes difficult. (ii) **balancing diversity and personalization** [7]. Personalized music recommendation systems aim to provide users with recommendations that align with their preferences. While recommendations should match users' interests, it is also important to introduce new music that may differ from their previous preferences, allowing them to explore and discover fresh content. (iii) **interpretability and user controllability** [24]. Complex ML/DL algorithms used for recommendations often hinder users' understanding of why certain recommendations are given. Simultaneously, users desire control over the recommendation process, including the ability to customize or adjust the results. Users and music platforms are most concerned about the cold start problem because it is directly related to the user experience and platform revenue [8].

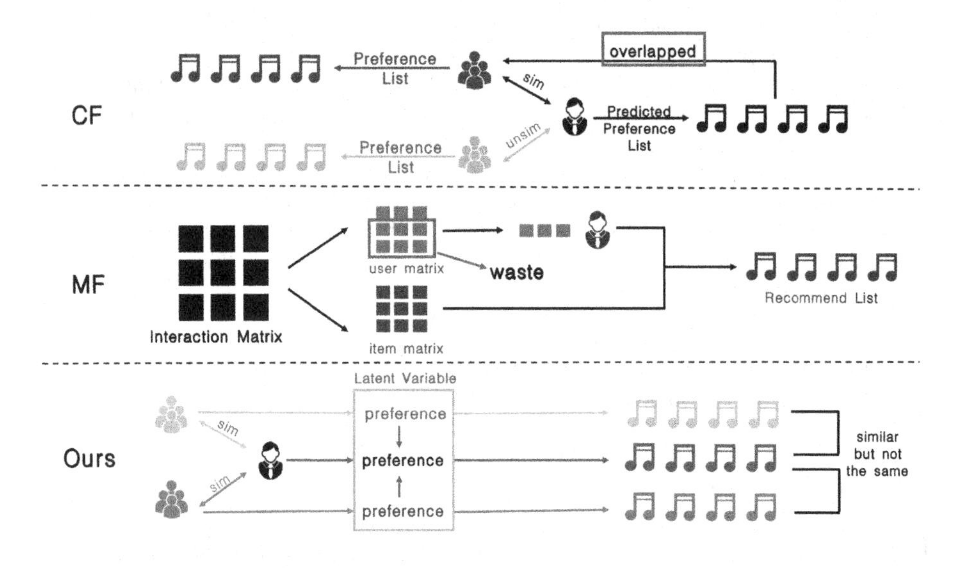

Fig. 1. The basic idea of different cold-start music recommendation methods.

As shown in Fig. 1, existing personalized music recommendation systems can be classified into two categories in solving the cold-start problem: **collaborative filtering (CF)**-based approach [15,20] and **matrix factorization (MF)**-based approach [3,18]. The former utilizes the similarity between users to recommend songs from neighboring users' song lists to cold-start users. However, it suffers from poor accuracy in scenarios where data is sparse and users have limited interaction records. Furthermore, it lacks the capability to provide diverse recommendation results. On the other hand, the latter predicts the probability of user-item interactions by employing latent vectors derived from matrix decomposition for users and items. It then selects the item with the highest score as the recommendation. However, this approach fails to fully exploit the rich user behavior information available in the dataset to enhance the recommendation process. Moreover, the latent vectors used for prediction lack explicit semantic information and are not interpretable.

In this paper, we propose a new approach based on Bayesian inference to solve the cold-start problem in music recommendation, taking into account the diversity and interpretability of the recommendation results. The approach involves introducing a latent variable of user preferences, which is determined by the user's profile information that is available to all users in the dataset. Collaborative filtering in the latent space (CFLS) is then performed to infer the preferences of cold-start users. This inference process samples the conditional distributions of the user's behaviors to obtain a candidate set of recommendation results with diversity. Specifically, CFLS employs a neural network (NN) architecture similar to a variational autoencoder (VAE). The NN serves as an encoder, mapping the user's behavioral sequences into the hidden space representing user preference. The decoder NN reconstructs the sampled user preference back into user behavioral sequences. A key difference is that CFLS incorporates a Gaussian Process (GP)-based latent variable prior to account for the similarity among users in the preference hidden space. The covariance function of the GP prior is calculated on the user's profile information. The parameters of the overall model can be optimized using gradient-based methods, enabling accurate and simple out-of-sample predictions. The contributions of this paper are as follows:

- We introduce a new perspective for solving the cold-start problem in music recommendation with additional benefits: Bayesian inference. Sample-based prediction can provide more diversity in recommendation results.
- We specify the prior distribution of user preferences as a GP based on user profile information, which enables us to consider user similarity in the preference latent space and enhances the interpretability of its results.
- By changing the form of the kernel function in the GP prior, different populations can introduce their a priori knowledge about user preferences into the model, leading to user-controllable recommender systems.

2 Related Work and Problem Formulation

Sequential Recommendation (SR): The significance of sequential behaviors in reflecting user preferences has been underscored by studies such as [6,25], leading to a surge in interest from both academia and industry in the field of SR. With the emergence of DL, Hidasi et al. [6] pioneered the use of Gated Recurrent Units (GRU) to model sequential behaviors in recommendation systems. Subsequently, researchers have explored a diverse range of deep learning techniques, including Transformers [8], and Large Language Models (LLM) [7,24], to encode interaction sequences. Nevertheless, despite these advancements, it is worth noting that the majority of research in SR has primarily focused on improving recommendation performance, often overlooking the challenging cold-start problems that are inherent in SR [22,25], that is, when the available user-item interactions are very limited, the performance of the SR model decreases dramatically. In this paper, for the cold-start challenge in SR (specifically, music recommendation), we propose a novel method based on Bayesian inference to generate precise recommendation lists for cold-start users by collaborative filtering in the latent space based on the user's profile information.

Cold-Start Recommendation: To address the cold-start problems in recommendations, several primary techniques are commonly employed. Content-based methods aim to incorporate features extracted from side information into CF-based frameworks [21]. Transfer learning methods leverage shared features learned from a source domain to enhance recommendation quality in a target domain [7]. Meta-learning approaches involve a learning-to-learn process across multiple training tasks, optimizing global knowledge to enable rapid adaptation to new recommendation tasks [13]. In this paper, we primarily focus on addressing the cold-start problem in music recommendation, a specific scenario of SR that has been relatively underexplored due to its inherent complexity. While content methods have been proven effective in a large number of recommendation tasks, they struggle to smoothly integrate historical interactions with user profile information in the cold-start scenario [21]. Consequently, researchers have predominantly turned to meta-learning-based methods [25] and LLM-based methods [22] to tackle cold-start challenges in SR, relying on multi-source available user data. However, meta-learning-based methods have high requirements on data volume and data diversity and are more sensitive to the choice of hyperparameters, which increases implementation difficulties [4]; LLM-based methods, on the other hand, lack interpretability of recommendation results [7]. To this end, from the perspective of Bayesian statistics, we propose an improved VAE-based method for generating recommendation lists for cold-start in music recommendation, which effectively fills the gap in existing research by eliminating the need for a large amount of data and possessing good interpretability.

2.1 Problem Formulation

Assume there are $|\mathcal{U}|$ users with historical interaction and profile information (listening habits, length of use, etc.), $|\mathcal{I}|$ songs with item attributes in the dataset, where $\mathcal{U}$ and $\mathcal{I}$ denote the user and item set respectively. For a user $u \in \mathcal{U}$, her profile information is $\mathbf{x}_u \in \mathbb{R}^L$ and interacted sequence is $\mathbf{s}_u = (s_1, s_2, \cdots, s_T)$, where L is the number of available user-related features and T is the length of the sequences. Given a cold-start user u^* with $\mathbf{x}_{u^*}$ and $\mathbf{s}_{u^*}$ (specifically, a full cold-start user when $\mathbf{s}_{u^*} = \emptyset$), we would like to obtain the set of candidate songs $\hat{\mathbf{s}}_{u^*}$ that accurately matches the user's u^* musical preferences, based on $\mathbf{X}$ and $\mathbf{S}$, and $\mathbf{s}_{u^*}, \mathbf{x}_{u^*}$. Where $\mathbf{X} \in \mathbb{R}^{|\mathcal{U}| \times L}$ and $\mathbf{S} = \bigcup_{u \in \mathcal{U}} \mathbf{s}_u$ are the collective information in the dataset.

3 Methodology

3.1 Overview

Our proposed CFLS is similar to the VAE in terms of model structure. Due to the presence of aligned multi-view data (user-item interactions, user profile information, song attributes), in order to fully utilize the information in the dataset to perform cold-start recommendation, we first train a matrix factorization (MF) [11] model on the original dataset and obtain the pre-trained latent vectors of songs $\mathbf{v}_i \in \mathbb{R}^M$ as the representation of song i. Thus, the interaction sequence of user u can be represented as $\mathbf{s}_u = \frac{1}{|\mathcal{I}_u|} \sum_{j=1}^{|\mathcal{I}_u|} \mathbf{v}_j \in \mathbb{R}^M$, where $\mathcal{I}_u$ denotes the song sets the user u once interacted with.

As can be seen from Fig. 2(c), CFLS consists of two parts: first, an MLP parameterized by ψ (i.e., the encoder $E_\psi(\cdot)$) is used to map user u's sequence of interactions $\mathbf{s}_u$ into the latent space where the user's preference latent variable $\mathbf{z}_u \in \mathbb{R}^K$ are distributed, where $K \ll M$. This process is equivalent to applying a variational posterior $q_\psi(\mathbf{z}_u)$ to approximate the intractable true posterior $p(\mathbf{z}_u|\mathbf{s}_u)$. Subsequently, based on the $\mathbf{z}_u$ sampled from the variational posterior $q_\psi(\mathbf{z}_u)$, a decoder $D_\phi(\cdot)$ parameterized by ϕ is used to map $\mathbf{z}_u$ back to the predicted sequence $\hat{\mathbf{s}}_u$ of user u. Unlike the original VAE, which specifies the prior distribution of the latent variable $\mathbf{z}_u$ as a standard Gaussian distribution, i.e., the samples are assumed to be independent of each other in the latent space, we

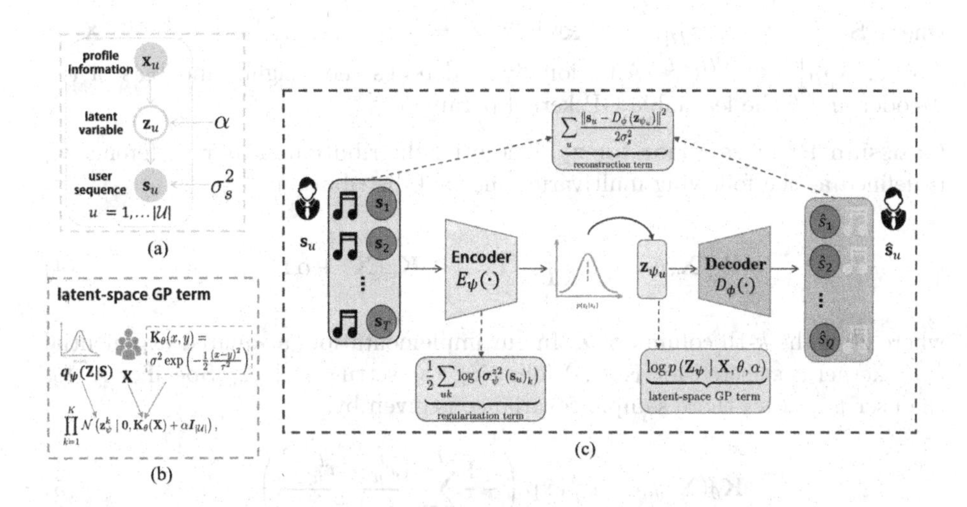

Fig. 2. The framework of CFLS. (a) The generative model underlying the proposed CFLS. (b) The formula of latent-space GP term with diagrams. (c) Forward process based on VAE model and loss function in CFLS.

specify the prior distribution of $\mathbf{z}_u$ as a GP whose covariance function $\mathbf{K}_\theta(\cdot,\cdot)$'s independent variables are the user's profile information, thus introducing similarity among users into the latent space. Finally, the overall model parameters $\Theta = \{\psi, \phi, \theta\}$ can be optimized by the Stochastic Gradient Variational Bayesian (SGVB) method [23] in an end-to-end manner.

3.2 Statistical Model in CFLS

In this section, we describe the data generation process in CFLS in detail from the perspective of a statistical model, including the latent variable of user preferences $\mathbf{z}_u$ based on profile information $\mathbf{x}_u$ and the generation of user behaviors $\mathbf{s}_u$ based on user preferences (Fig. 2 (a)).

For a user u, her preference $\mathbf{z}_u$ is generated from profile information $\mathbf{x}_u$:

$$\mathbf{z}_u = f(\mathbf{x}_u) + \boldsymbol{\eta}_u, \text{ where } \boldsymbol{\eta}_u \sim \mathcal{N}(\mathbf{0}, \alpha \boldsymbol{I}_K), \tag{1}$$

and sequence $\mathbf{s}_u$ is generated from its preference $\mathbf{z}_u$ as:

$$\mathbf{s}_u = g(\mathbf{z}_u) + \boldsymbol{\epsilon}_u, \text{ where } \boldsymbol{\epsilon}_u \sim \mathcal{N}\left(\mathbf{0}, \sigma_s^2 \boldsymbol{I}_M\right). \tag{2}$$

CFLS uses a GP prior on f, which enables it to model sample covariances in the latent space as a function of profile information. In this paper, we use an MLP (i.e., decoder $D_\phi(\cdot)$) for g to output the distribution hyperparameters for the user sequence $\mathbf{s}_u$ (i.e., $p_\phi(\mathbf{s}_u|\mathbf{z}_u)$). The formulation is as follows:

$$p\left(\mathbf{S} \mid \mathbf{X}, \phi, \sigma_s^2, \theta, \alpha\right) = \int p\left(\mathbf{S} \mid \mathbf{Z}, \phi, \sigma_s^2\right) p(\mathbf{Z} \mid \mathbf{X}, \theta, \alpha) d\mathbf{Z}, \tag{3}$$

where $\mathbf{S} = \left[\mathbf{s}_1, \ldots, \mathbf{s}_{|\mathcal{U}|}\right]^T \in \mathbb{R}^{|\mathcal{U}| \times M}, \mathbf{Z} = \left[\mathbf{z}_1, \ldots, \mathbf{z}_{|\mathcal{U}|}\right]^T \in \mathbb{R}^{|\mathcal{U}| \times K}, \mathbf{X} = \left[\mathbf{x}_1, \ldots, \mathbf{x}_{|\mathcal{U}|}\right]^T \in \mathbb{R}^{|\mathcal{U}| \times L}$. Additionally, ϕ denotes the weights and bias of the decoder and θ the learnable GP kernel parameters.

Gaussian Process Prior for $\mathbf{z}_u$. The prior distribution of user preference $\mathbf{z}_u$ is defined as the following multivariate normal distribution:

$$p(\mathbf{Z} \mid \mathbf{X}, \theta, \alpha) = \prod_{k=1}^{K} \mathcal{N}\left(\mathbf{z}^k \mid \mathbf{0}, \mathbf{K}_\theta(\mathbf{X}) + \alpha \boldsymbol{I}_{|\mathcal{U}|}\right), \tag{4}$$

where $\mathbf{z}^k$ is the k-th column of $\mathbf{Z}$. In our implementation, a squared exponential (SE) kernel is selected for $\mathbf{K}_\theta(\cdot,\cdot)$ following the setting in [1,2]. Specifically, for the user u and u', there sample covariance is given by:

$$\mathbf{K}_\theta(\mathbf{X})_{uu'} = \sigma_f^2 \exp\left(-\frac{1}{2}\sum_{l=1}^{L} \frac{\left(x_u^l - x_{u'}^l\right)^2}{\tau_l^2}\right), \tag{5}$$

where $\theta = [\sigma_f, \tau_1, \cdots, \tau_L]$, σ_f is the standard hyperparameter and τ_l is the lengthscale along each individual input direction. When x^l is a sparse feature, we first train a factorization machine (FM) model [17] to obtain the dense low-dimensional hidden vector $\tilde{\mathbf{x}}^l$ of feature l, and then compute the sample covariance based on $\tilde{\mathbf{x}}_u^l$ and $\tilde{\mathbf{x}}_{u'}^l$.

Variational Posterior for $\mathbf{z}_u$. As with a standard VAE, we employ an Encoder $E_\psi(\cdot)$ to output the hyperparameters of the variational posterior $q_\psi(\mathbf{z}_u|\mathbf{s}_u)$, which is optimized to approximate the intractable true posterior $p(\mathbf{s}_u|\mathbf{z}_u)$:

$$q_\psi(\mathbf{Z} \mid \mathbf{S}) = \prod_u \mathcal{N}\left(\mathbf{z}_u \mid \boldsymbol{\mu}_\psi^z\left(\mathbf{s}_u\right), \operatorname{diag}\left(\boldsymbol{\sigma}_\psi^{z2}\left(\mathbf{s}_u\right)\right)\right), \tag{6}$$

where ψ denotes the weights and bias of the MLP for encoder $E_\psi(\cdot)$, $\boldsymbol{\mu}_\psi^z\left(\mathbf{s}_u\right)$ and $\operatorname{diag}\left(\boldsymbol{\sigma}_\psi^z\left(\mathbf{s}_u\right)\right)$ are the hyperparameters of the variational distribution and output by $E_\psi(\mathbf{s}_u)$. Specifically, latent user preference $\mathbf{z}_{\psi_u}$ are sampled using the re-parameterization trick in [10], that is:

$$\mathbf{z}_{\psi_u} = \mu_\psi^z\left(\mathbf{s}_u\right) + \boldsymbol{\delta}_u \odot \sigma_\psi^z\left(\mathbf{s}_u\right), \boldsymbol{\delta}_u \sim \mathcal{N}\left(\mathbf{0}, \boldsymbol{I}_{K\times K}\right), \tag{7}$$

where $\odot$ denotes the Hadamard product.

3.3 Optimization

Based on the model and notations defined above, we have the following evidence lower bound (ELBO) for the likelihood in Eq. (3):

$$\begin{aligned}\log p\left(\mathbf{S} \mid \mathbf{X}, \phi, \sigma_s^2, \theta\right) \geq & \mathbb{E}_{Z\sim q_\psi}\left[\sum_u \log \mathcal{N}\left(\mathbf{s}_u \mid D_\phi\left(\mathbf{z}_u\right), \sigma_s^2 I_M\right) + \log p(\mathbf{Z} \mid \mathbf{X}, \theta, \alpha)\right] \\ & + \frac{1}{2}\sum_{uk} \log\left(\sigma_\psi^{z2}\left(\mathbf{s}_u\right)_k\right) + \text{const.}\end{aligned} \tag{8}$$

Inspired by [23], we employ stochastic backpropagation to maximize the above ELBO. By sampling latent variable $\mathbf{z}_{\psi_u}$ from a reparameterized variational posterior $q_\psi(\mathbf{z})$, we approximate the expectation in (8) and obtain a loss function as follows:

$$\begin{aligned}&\mathcal{L}\left(\phi, \psi, \theta, \alpha, \sigma_s^2\right) \\ &= |\mathcal{U}| M \log \sigma_s^2 + \underbrace{\sum_u \frac{\left\|\mathbf{s}_u - D_\phi\left(\mathbf{z}_{\psi_u}\right)\right\|^2}{2\sigma_s^2}}_{\text{reconstruction term}} - \underbrace{\log p\left(\mathbf{Z}_\psi \mid \mathbf{X}, \theta, \alpha\right)}_{\text{latent-space GP term}} + \underbrace{\frac{1}{2}\sum_{uk} \log\left(\sigma_\psi^{z2}\left(\mathbf{s}_u\right)_k\right)}_{\text{regularization term}}\end{aligned} \tag{9}$$

By minimizing (9) via mini-batch stochastic gradient descent (SGD), the optimal parameters $\Theta^* = \left\{\phi^*, \psi^*, \theta^*, \alpha^*, \sigma_s^{2*}\right\}$ can be obtained by end-to-end.

3.4 Prediction

Given a cold-start user u^* with $\mathbf{x}_{u^*}$, we would like to obtain the set of candidate songs $\hat{\mathbf{s}}_{u^*}$ that accurately matches the user's u^* musical preferences, based on $\mathbf{X}$ and $\mathbf{S}$, and Θ^*. The predictive posterior for $\hat{\mathbf{s}}_{u^*}$ is given by:

$$p\left(\hat{\mathbf{s}}_{u^*} \mid \mathbf{x}_{u^*}, \mathbf{S}, \mathbf{X}\right) \approx \int p_{\phi^*}\left(\hat{\mathbf{s}}_{u^*} || \mathbf{z}_{u^*}\right) p_{\Theta^*}\left(\mathbf{z}_{u^*} \mid \mathbf{x}_{u^*}, \mathbf{Z}_{\psi^*}, \mathbf{X}\right) q_{\psi^*}(\mathbf{Z} \mid \mathbf{S}) d\mathbf{z}_{u^*} d\mathbf{Z}. \tag{10}$$

The candidate songs $\hat{\mathbf{s}}_{u^*}$ can be obtained in the prediction stage by the following procedure: (i) encode training data to get $\mathbf{Z}_{\psi^*}$ via encoder E_{ψ^*}, (ii) sample $\mathbf{z}_{u^*}$ from the GP predictive posterior $p_{\theta^*}(\mathbf{z}_{u^*} \mid \mathbf{x}_{u^*}\mathbf{Z}_{\psi^*}, \mathbf{X})$, (iii) decode $\mathbf{z}_{u^*}$ to get the item embedding of $\hat{\mathbf{s}}_{u^*}$, (iv) search for top Q nearest neighbors of $\hat{\mathbf{s}}_{u^*}$ in the item set $\mathcal{I}$ based on their location in the embedding space, return the final recommendation list for u^*.

4 Experiments

In this section, we present the industrial dataset used for the experiments, the settings for the experiments, including the evaluation protocols and baseline models, and finally, the results of the experiments and related analysis. Moreover, we demonstrate the diversity, interpretability, and user controllability of CFLS's recommendation results through data visualization.

4.1 Dataset

In this study, we validate the effectiveness of CFLS based on the **#nowplaying-RS** dataset [16]. This dataset comprises listening data from Twitter collected in 2014, including 138,150 users, 346,646 songs, and 11,606,689 listening records. Each dataset's listening record contains rich song features and user profile information. For a more comprehensive understanding of the dataset and its features and detailed statistical information, please refer to [16].

4.2 Experimental Settings

Evaluation Protocol. Following previous cold-start recommendation and SR works [25], we utilize the leave-one-out method to evaluate the recommendation performance, and set Precision@K, Recall @K, MAP@K and NDCG@K as the evaluation metrics. In order to verify the superiority and robustness of CFLS in cold-start recommendation scenarios, we introduce the hyperparameter r to denote the proportion of completely cold-start users in the test set to evaluate the performance variation of CFLS and baseline models when r takes different values. We repeated each set of experiments 10 times to ensure the stability of the results and reported their mean values as the final results.

Baseline Models. We selected a bunch of recent DL-based recommendation models as baseline models, including DSSM [9], YoutubeDNN [5], MIND [12], SDM [14] and Bert4Rec [19]. The hyperparameters of all baseline models are consistent with those reported in their paper. In addition, in order to verify the validity of the proposed GP prior, we introduce the CFLS w/o GP prior, a variant that uses the standard normal distribution as the prior distribution of the latent variables of user preferences.

Implementation and Hyperparameters. For CFLS, the hyperparameters are as follows: the training epoch is fixed as 10, the early stopping patience is fixed as 3, the item embedding dimension M is fixed as 128, the latent variable dimension K is fixed as 64, the mini-batch size is 64, the Adam [23] is used for optimization and the learning rate is 0.001. Both encoder and decoder are two-layer MLP with ReLU as an activation function.

Table 1. Performance comparisons of different methods

Model	Precision@K				Recall@K				MAP@K				NDCG@K			
	K = 20	K = 40	K = 60	K = 80	K = 20	K = 40	K = 60	K = 80	K = 20	K = 40	K = 60	K = 80	K = 20	K = 40	K = 60	K = 80
DSSM [9]	8.65%	7.75%	7.42%	6.93%	17.30%	31.01%	44.52%	55.44%	9.88%	9.08%	8.47%	8.18%	10.42%	9.32%	8.82%	8.35%
YoutubeDNN [5]	8.83%	7.96%	7.60%	7.05%	17.66%	31.85%	45.61%	56.40%	10.14%	9.42%	8.72%	8.41%	10.60%	9.58%	8.90%	8.54%
MIND [12]	9.11%	8.29%	7.97%	7.23%	18.22%	33.16%	47.82%	57.84%	10.35%	9.63%	8.95%	8.64%	10.82%	9.79%	9.15%	8.75%
SDM [14]	9.23%	8.45%	8.13%	7.37%	18.46%	33.82%	48.78%	58.94%	10.48%	9.77%	9.16%	8.76%	11.08%	10.02%	9.39%	8.93%
Bert4Rec [19]	9.17%	8.35%	8.03%	7.34%	18.35%	33.38%	48.17%	58.70%	10.38%	9.60%	9.00%	8.55%	10.80%	9.85%	9.32%	8.82%
CFLS w/o GP prior	9.05%	8.32%	7.99%	7.33%	18.11%	33.26%	47.92%	58.61%	10.36%	9.56%	8.96%	8.55%	10.85%	9.83%	9.27%	8.74%
CFLS	**9.46%**	**8.62%**	**8.28%**	**7.50%**	**18.92%**	**34.48%**	**49.68%**	**59.99%**	**10.72%**	**10.01%**	**9.28%**	**8.91%**	**11.21%**	**10.12%**	**9.63%**	**9.09%**

4.3 Performance Comparisons

In this section, we specify the percentage of completely cold-start users in the test set $r = 0.7$ and give the evaluation results for the four evaluation metrics for $K = 20, 40, 60, 80$. The specific values are shown in Table 1. Bolded numbers are optimal representations, and underlined numbers are sub-optimal. We use the DSSM as a baseline for calculating relative lift to facilitate a comparison of model performance. The following observations can be obtained from Table 1.

The DSSM model gives the worst performance in all the comparison experiments. This is due to the fact that DSSM only considers user profile information while modeling user preferences and does not utilize user behavioral information, resulting in the inability to achieve personalized music recommendations. YoutubeDNN introduces a heterogeneous subnetwork to model user behavioral information, thereby obtaining more expressive representations of user interests, which improves the NDCG@80 metric by 2.3%. MIND introduces a dynamic routing mechanism to model the user's interest evolution, while SDM uses a session-based approach to separately model the user's interest into long and short periods. Both achieve finer-grained user interest modeling by introducing a new induction bias, resulting in relative performance improvements of 4.79% and 6.95% in NDCG@80, respectively. Among them, SDM achieves the sub-optimal performance in all comparison experiments Bert4Rec utilizes pre-trained Bert to encode information about users and items and introduce knowledge from other domains to represent user interests better. However, when the pre-training is based on data unrelated to the target task, it may lead to performance degradation. This is demonstrated by the increase of 5.63% of NDCG@80, lower than SDM and CFLS.

Comparing the results of CFLS w/o GP prior and CFLS can verify the validity of our proposed GP prior: replacing the standard normal prior for the latent variable of user preferences in the naive VAE with the profile information-based GP prior, our method achieves a 4% improvement on NDCG@80. CFLS relative to DSSM improves by 8.86% and achieves optimal performance among all compared models, demonstrating the superiority of modeling user similarity and performing collaborative filtering in the latent space.

4.4 Influence of Different Cold-Start Levels

In this section, in order to verify the robustness of the performance of CFLS to the percentage r of fully cold-started users in the test set, we set up four sets of experiments by setting the values of the hyperparameter r to 0.7, 0.8, 0.9, and 1.0, i.e., gradually increasing the percentage of fully cold-started users in the test set.

As can be seen in Fig. 3(a), with the gradual improvement of r, the performance of YoutubeDNN, Bert4rec, and CFLS on all 4 metrics decreases to different degrees. However, compared with the other two methods, CFLS has a smaller decrease, and the overall change trend is relatively stable and always maintains the optimal performance. This verifies the robustness of our proposed CFLS method, which is less affected by the percentage of completely cold-start users r in the test set.

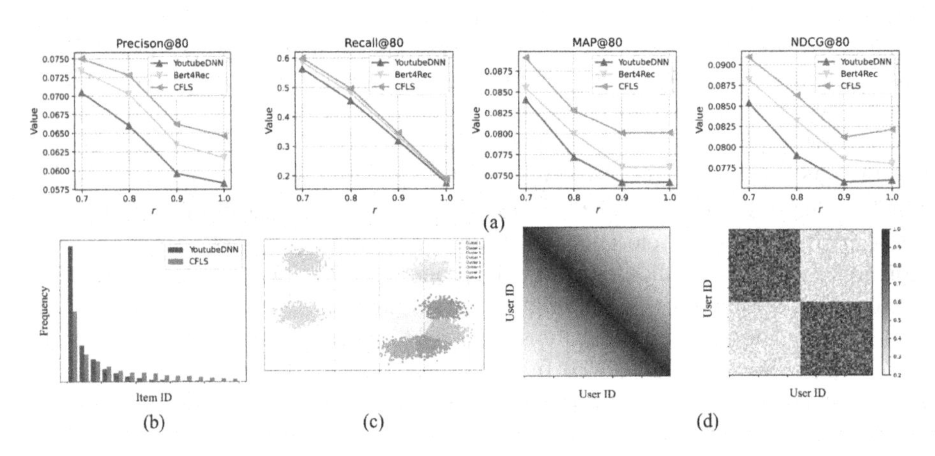

Fig. 3. (a) Robustness of different methods to variations in of r, the percentage of fully cold-started users in the test set. (b) Occurrence frequency of items with different popularity in YoutubeDNN and CFLS recommendation results. (c) Distribution of user preference latent variables encoded by CFLS in 2D space. (d) The covariance matrix of user preference latent variables was obtained by using different kernel functions in the GP prior.

4.5 Diversity, Interpretability and User Controllability

In this section, we first evaluate the diversity of CFLS's recommendation results. Specifically, we select a part of popular items and a part of cold items, arrange them in descending order of their frequency of appearance in the test set, and then count the frequency of their appearances in the final recommendation results of YoutubeDNN and CFLS, respectively, and the results are shown in Fig. 3(b). It can be seen that compared with YoutubeDNN, CFLS, which obtains recommendation results based on sampling, has more significant diversity, which is specifically reflected in the fact that cold items can also get some exposure opportunities, and the exposure opportunities of popular items are suppressed, which reduces the popularity bias to a certain extent.

To illustrate the good interpretability of user preference latent variables encoded in CFLS, we first cluster users based on their profile information. The user preference latent variables $\mathbf{z}_\psi$ obtained from the encoder are then subjected to PCA dimensional reduction to visualize their distribution in 2D space. It can be seen from Fig. 3(c) that $\mathbf{z}_\psi$ reflects the similarity between users very well, which justifies our collaborative filtering in the latent space.

Finally, we compare the user covariance matrices obtained using the two kernel functions mentioned in the literature [1] (SE kernel and automatic relevance determination (ARD) kernel) as covariance functions for the GP prior. We choose the SE kernel in our implementation because we believe that the preferences of neighboring users in the feature space should have greater correlation in the latent space and that the change in this correlation is smooth (as shown on the left in Fig. 3(d)); however, if others believe that there are two groups of preference in the users that are more different and have greater intra-group correlation and less inter-group correlation, they can choose the ARD kernel (as in the right of Fig. 3(d)). This validates the user-controllability of CFLS, i.e., users can flexibly adapt the model in both training and prediction stage based on their prior knowledge.

5 Conclusions

In this paper we present a novel Bayesian inference-based approach to address the cold-start problem in music recommendation. Our method incorporates diversity and interpretability considerations, utilizing the VAE framework. We introduce a Gaussian Process (GP) prior for user preferences latent variable, leveraging user profile information to account for user similarity in the preference latent space. Experimental results on real-world datasets demonstrate the efficacy and superiority of our proposed method. Additionally, we employ visualization techniques to showcase the diversity, interpretability, and user-controllability of the recommendation results achieved by our approach.

Acknowledgements. This study was supported by Natural Science Foundation of Hunan Province (grant number 2022JJ30673) and by the Graduate Innovation Project of Central South University (2023XQLH032, 2023ZZTS0304).

References

1. Botteghi, N., Guo, M., Brune, C.: Deep kernel learning of dynamical models from high-dimensional noisy data. Sci. Rep. **12**(1), 21530 (2022)
2. Casale, F.P., Dalca, A., Saglietti, L., Listgarten, J., Fusi, N.: Gaussian process prior variational autoencoders. In: Advances in Neural Information Processing Systems, vol. 31 (2018)
3. Chen, K., Liang, B., Ma, X., Gu, M.: Learning audio embeddings with user listening data for content-based music recommendation. In: ICASSP 2021–2021 IEEE International Conference on Acoustics, Speech and Signal Processing (ICASSP), pp. 3015–3019. IEEE (2021)
4. Chu, Z., Wang, H., Xiao, Y., Long, B., Wu, L.: Meta policy learning for cold-start conversational recommendation. In: Proceedings of the Sixteenth ACM International Conference on Web Search and Data Mining, pp. 222–230 (2023)
5. Covington, P., Adams, J., Sargin, E.: Deep neural networks for youtube recommendations. In: Proceedings of the 10th ACM Conference on Recommender Systems, pp. 191–198 (2016)
6. Hidasi, B., Karatzoglou, A., Baltrunas, L., Tikk, D.: Session-based recommendations with recurrent neural networks. arXiv preprint arXiv:1511.06939 (2015)
7. Hou, Y., He, Z., McAuley, J., Zhao, W.X.: Learning vector-quantized item representation for transferable sequential recommenders. In: Proceedings of the ACM Web Conference 2023, pp. 1162–1171 (2023)
8. Hou, Y., Mu, S., Zhao, W.X., Li, Y., Ding, B., Wen, J.-R.: Towards universal sequence representation learning for recommender systems. In: Proceedings of the 28th ACM SIGKDD Conference on Knowledge Discovery and Data Mining, pp. 585–593 (2022)
9. Huang, P.-S., He, X., Gao, J., Deng, L., Acero, A., Heck, L.: Learning deep structured semantic models for web search using clickthrough data. In: Proceedings of the 22nd ACM International Conference on Information & Knowledge Management, pp. 2333–2338 (2013)
10. Kingma, D.P., Welling, M.: Auto-encoding variational bayes. arXiv preprint arXiv:1312.6114 (2013)
11. Koren, Y., Bell, R., Volinsky, C.: Matrix factorization techniques for recommender systems. Computer **42**(8), 30–37 (2009)
12. Li, C., et al.: Multi-interest network with dynamic routing for recommendation at Tmall. In: Proceedings of the 28th ACM International Conference on Information and Knowledge Management, pp. 2615–2623 (2019)
13. Lu, Y., Fang, Y., Shi, C.: Meta-learning on heterogeneous information networks for cold-start recommendation. In: Proceedings of the 26th ACM SIGKDD International Conference on Knowledge Discovery & Data Mining, pp. 1563–1573 (2020)
14. Lv, F., et al.: SDM: sequential deep matching model for online large-scale recommender system. In: Proceedings of the 28th ACM International Conference on Information and Knowledge Management, pp. 2635–2643 (2019)
15. Mao, Y., Zhong, G., Wang, H., Huang, K.: Music-CRN: an efficient content-based music classification and recommendation network. Cogn. Comput. **14**(6), 2306–2316 (2022)
16. Poddar, A., Zangerle, E., Yang, Y.-H.: nowplaying-RS: a new benchmark dataset for building context-aware music recommender systems. In: Proceedings of the 15th Sound & Music Computing Conference, pp. 21–26 (2018)

17. Rendle, S., Freudenthaler, C., Schmidt-Thieme, L.: Factorizing personalized Markov chains for next-basket recommendation. In: Proceedings of the 19th International Conference on World Wide Web, pp. 811–820 (2010)
18. Shen, J., Tao, M., Qu, Q., Tao, D., Rui, Y.: Toward efficient indexing structure for scalable content-based music retrieval. Multimedia Syst. **25**, 639–653 (2019)
19. Sun, F., et al.: BERT4Rec: sequential recommendation with bidirectional encoder representations from transformer. In: Proceedings of the 28th ACM International Conference on Information and Knowledge Management, pp. 1441–1450 (2019)
20. Van den Oord, A., Dieleman, S., Schrauwen, B.: Deep content-based music recommendation. In: Advances in Neural Information Processing Systems, vol. 26 (2013)
21. Wu, L., Quan, C., Li, C., Wang, Q., Zheng, B., Luo, X.: A context-aware user-item representation learning for item recommendation. ACM Trans. Inf. Syst. (TOIS) **37**(2), 1–29 (2019)
22. Wu, Y., et al.: Personalized prompts for sequential recommendation. arXiv preprint arXiv:2205.09666 (2022)
23. Yu, H., Nghia, T., Low, B.K.H., Jaillet, P.: Stochastic variational inference for Bayesian sparse Gaussian process regression. In: 2019 International Joint Conference on Neural Networks (IJCNN), pp. 1–8. IEEE (2019)
24. Yuan, Z., et al.: Where to go next for recommender systems? ID-vs. modality-based recommender models revisited. arXiv preprint arXiv:2303.13835 (2023)
25. Zheng, Y., Liu, S., Li, Z., Wu, S.: Cold-start sequential recommendation via meta learner. In: Proceedings of the AAAI Conference on Artificial Intelligence, vol. 35, pp. 4706–4713 (2021)

On Diverse and Precise Recommendations for Small and Medium-Sized Enterprises

Ludwig Zellner[1(✉)], Simon Rauch[1,2], Janina Sontheim[1], and Thomas Seidl[1,2,3]

[1] Database Systems and Data Mining, LMU Munich, Munich, Germany
{zellner,rauch,sontheim,seidl}@dbs.ifi.lmu.de

[2] Fraunhofer IIS, Fraunhofer Institute for Integrated Circuits IIS, Division Supply Chain Services, Nuremberg, Germany

[3] Munich Center for Machine Learning, Munich, Germany

Abstract. Recommender Systems are a popular and common means to extract relevant information for users. Small and medium-sized enterprises make up a large share of the overall amount of business but need to be more frequently considered regarding the demand for recommender systems. Different conditions, such as the small amount of data, lower computational capabilities, and users frequently not possessing an account, require a different and potentially a more small-scale recommender system. The requirements regarding quality are similar: High accuracy and high diversity are certainly an advantage. We provide multiple solutions with different variants solely based on information contained in event-based sequences and temporal information. Our code is available at GitHub (https://github.com/lmu-dbs/DP-Recs). We conduct experiments on four different datasets with an increasing set of items to show a possible range for scalability. The promising results show the applicability of these grammar-based recommender system variants and leave the final decision on which recommender to choose to the user and its ultimate goals.

Keywords: Recommender System · Rule Mining · Temporal Diversity

1 Introduction

The mass of collected data is rapidly increasing, fueled by digitization. Especially in current years, contingent on the pandemic, many companies read the signs of the times and digitally mapped their processes. Roles such as salesman or consultant, which humans formerly took, are now integrated into an automated system. Recommender Systems provide means to successfully filter the overwhelming mass of data and present an excerpt considered a good fit for the user. These systems are popular and common among social media platforms, search engines, entertainment systems, and many more [1]. Research for platforms providing big data for recommender systems to train on has been the focus for a long time [14]. Small and medium-sized enterprises (SMEs), however, are often overlooked. Nevertheless, they benefit as well but pose other challenges,

D.-N. Yang et al. (Eds.): PAKDD 2024, LNAI 14649, pp. 118–130, 2024.
https://doi.org/10.1007/978-981-97-2262-4_10

such as a small amount of provided data, sparsity of data, lower computational capabilities, and lack of additional data since users in most cases do not own accounts. Additionally, with a smaller product portfolio the amount of one-time users is higher. Beyond that, there is a limited size of shopping history records because of that very reason. Therefore, recommendation systems like user-based or item-based collaborative filtering, which are considered traditional rating-based personalization techniques, are not applicable [9,10]. In general, methods such as matrix factorization that excel with expanded user feedback have limited applicability in this context. With plain item sequences there is no such feedback except a binary indicator. Additionally, matrix factorization techniques do not inherently capture the sequential nature of item sequences. Thus, sequence-based recommendation algorithms are more appropriate in this context.

Popular grammar-based approaches to receiving recommendation candidates comprise mining sequential rules from a sequence database. Among these approaches, totally-ordered sequential rule mining [19] and partially-ordered sequential rule (POSR) mining [6] using frequency and confidence thresholds have been extensively analyzed. In particular, POSR mining significantly impacts the number of rules generated. It combines many totally-ordered sequential rules by joining items in the antecedent and items in the consequent of many rules.

The requirements of potential recommender systems for SMEs are similar, e.g., the demand for accuracy and diversity in the recommendation process. In this work, we want to provide multiple POSR-based solutions. With SMEs' restrictions in mind, we solely build on information contained in sequences of items [5]. One of our variants extends this scope by using temporal information.

Our approaches can be used on sparse datasets from different environments. Here, we test our solutions on four different datasets, where three stem from clickstream data of news websites and one from an online retailer. These datasets have an increasing set of unique items to show a possible range for scalability.

The contributions of our paper are summarized as follows: We develop six variants of recommender systems based on POSR (1), designed to satisfy both accuracy and diversity measures (2). The benchmarking process reveals advantages and disadvantages among these variants compared to a naïve baseline (3).

The remainder of this work is organized in the following manner. In Sect. 2, we provide the context of our approach and describe ideas of related areas. Section 3 contains all definitions required to understand the methods of our approaches. Section 4 explicitly outlines our approach, part of our experiments, and evaluation of the results in Sect. 5. Finally, we summarize our work and describe the prospects in Sect. 6.

2 Related Work

Below, we review existing research in this area for recommender systems.

Wang et al. [18] emphasize the high relevancy of sequential recommender systems. They define several future directions with open issues, such as the inclusion of user preferences or more contextual factors. Among them, *Interactive Session-Based Recommendations* are mentioned, where our paper is settled. Kim et al. [12] propose the *RF-Miner* algorithm to extract rules that prioritize events close to a rule's consequent. They aim to find more diverse and relevant patterns by accounting for gap information. In contrast, we base our approaches on integrating further information like the size of gaps between antecedent and consequent. Another approach, by Fahed et al. [4], introduces two algorithms *D-SR-post-Mining* and *D-SR-in-Mining* that integrate a minimal gap constraint into sequential rule mining. By prioritizing distant events and omitting recent ones, the authors ensure an early prediction of events. In our work, we combine distant, recent, and other events that frequently occur with a specific gap that circumvents the exclusion of certain relevant patterns if they meet the given condition. Lathia et al. [13] are concerned with the area of fairness. Especially, they developed the concept of temporal diversity, which is concerned with preferring recommendations of not yet selected items.

Gharahighehi et al. [7] examine diversity-aware session-based recommender systems based on the session-based k-nearest-neighbour method. However, their approach does not optimally grasp diversity as it inherently prioritizes similar query results. We adapt the recommended items based on the gap between a previous action and the frequency of item combinations, leading to more diverse recommendations. Nikookar et al. [15] minimize *intra-diversity* and maximize *inter-diversity* as the diversity within and between consecutive sessions, respectively. In contrast, we target the diversity of a single recommendation list as *inter-list-diversity* and the relation between each recommendation list regarding successive recommendation steps as the *inter-list-diversity*. Also more recent approaches [3,11] make use of rule mining algorithms as a base for recommendations. However they are not discussing the potential of their approach with respect to diversity metrics. While there are existing approaches relying on the availability of various and plenty of user-item interaction data, the need for algorithms that produce reliable recommendation results with only a limited amount of that data, as is often the case in SMEs, is also addressed by existing contributions in the field of recommender systems [9].

Kaminskas et al. [9] formulate an approach based on association rules and text mining to perform *item-centric* recommendations in a small-scale retailer setting. However, they do not account for a sequential order in their identified rules and, most importantly, do not emphasize diversity. With our approach, we leverage ideas of the papers as mentioned earlier - particularly those that are rule-based and based on recency - but, additionally, increase diversity and relevancy in a successive-item-recommendation task. Thus, we extend partially-ordered rule-mining algorithms into a fully-fledged recommender system.

3 Definitions and Problem Statement

Definition 1 (Sequence/Sequence Database). *A sequence* $s = \langle e_1, \ldots, e_n \rangle$ *contains items* $e_i, 1 \leq i \leq n$, $n \in \mathbb{N}$, *whereby the order of items is essential. The*

items may have optional further attributes. We only consider items with a timestamp: $t(s) = \langle t(e_1), t(e_2), \ldots, t(e_n)\rangle$ *where* $t(e_i) < t(e_{i+1})$ *for* $i \in 1, \ldots, n-1$.
A sequence database $\mathcal{DB}$ *is a set of non-empty, finite sequences* s_p *with* $1 \leq p \leq |\mathcal{DB}|$.

Definition 2 (Partially-Ordered Sequential Rule (POSR)). *A partially-ordered sequential rule* $r : X \rightarrow Y$ *is a totally-ordered relation between two partially-ordered sets of items* $X = \{e_1^X, \ldots, e_m^X\}$ *and* $Y = \{e_1^Y, \ldots, e_n^Y\}$*, with* $e^X \prec e^Y \ \forall e^X \in X, \forall e^Y \in Y$ *where the relation* $e_k \prec e_l$ *for two items* e_k*,* e_l *in a given sequence* s *implies* $t(e_k) < t(e_l)$.
In a POSR $r : X \rightarrow Y$*,* X *is called the antecedent, and* Y *is the consequent.*

Definition 3 (Rule Occurrence). *A rule* $r : X \rightarrow Y$ *occurs in a sequence* $s = \langle e_1, \ldots, e_n\rangle$*, iff an integer* k *exists, with* $1 \leq k < n$*,* $X \subseteq \bigcup_{i=1}^{k}\{e_i\}$ *and* $Y \subseteq \bigcup_{i=k+1}^{n}\{e_i\}$ *[20]. We refer to such rules as occurring or matching rules.*

Definition 4 (Gaps and Discrete/Continuous Gap Sizes). *The gap* $g_r(s)$ *of a sequence* $s = \langle e_1, \ldots, e_n\rangle$ *regarding a POSR* $r : X \rightarrow Y$ *is a list of items of* s *occurring between the latest occurring item of the antecedent* X*,* e_i*, and the first occurring item of the consequent* Y*,* e_j*, of* r *on* s*, with* $i < j, 1 \leq i, j \leq n$*:* $g_r(s) = [e_{i+1}, \ldots, e_{j-1}]$.
The discrete gap size $|g_r^d(s)|$ *of gap* $g_r(s)$ *is the number of events held in* $g_r(s)$.
The continuous gap size $|g_r^c(s)|$ *takes the time perspective into account and is the time difference between the first occurring item of the consequent,* e_j*, and the last occurring item of the antecedent,* e_i*:* $|g_r^c(s)| = t(e_j) - t(e_i)$.

Definition 5 (Rule Ranking Methods). *Let* $\mathcal{R}^s = \{r_1, \ldots, r_n\} \subseteq \mathcal{R}^{\mathcal{DB}}$ *be the resulting set of rules generated by a partially-ordered sequential rule mining algorithm that occur in* $\mathcal{DB}$ *with a sufficient frequency. A rule ranking method is a function with total order* $\pi : \mathcal{R} \rightarrow \mathbb{N}$ *that creates a permutation of* $\mathcal{R}^s$ *representing the actual ranking of the rule set* $\mathcal{R}^s$ *regarding sequence* s*. For rule ranking, further information and criteria regarding the respective rules and/or the historical sequence information can be considered.*

Further information for ranking rules we propose are gap sizes – either in terms of item indices or their exact temporal information (see Definition 4) – and individual item occurrences in the sequence up to the time of recommendation.

In the context of recommending users meaningful items, e.g., news articles or retail goods, we use the past information about the users' behaviour. For each user we focus on individually recorded sessions of all users with the respective set of possible items to choose from for the item recommendation task. The formal problem we tackle with our approach is described as follows:

Suppose we have a database of user sessions $\mathcal{S} = \{s_1, \ldots, s_n\}$ and a set of items $\mathcal{E} = e_1, \ldots, e_m$ where each user session s_i contains information on past item interactions $I_i = \langle e_1^i, \ldots, e_k^i\rangle$. A session-based sequential recommendation model generates a ranked list of recommended items $l_i = \{\hat{e}_1, \ldots, \hat{e}_{|l_i|}\}$ as the next item interaction e_{k+1}^i for each ongoing user session $s_i \in \mathcal{S}$ based on I_i. For single item recommendation the highest-ranked item of l_i is chosen as the most relevant item for session s_i.

4 Variants of a Session-Based Recommender System

In this section, we introduce six variants of a Session-Based Recommender System, which we compare against a naïve baseline approach and each other. Each variant is created to serve the notion of diversity, especially temporal diversity (cf. Lathia et al. [13]). The rule mining algorithm detects all suitable rules on the training set of our sequence database. In this paper, we execute it with the application of the POSR mining algorithm ER-Miner [6]. This algorithm yields POSRs which are then used for the successive recommendation. We decide to focus on rules with consequents of the cardinality of one. Thus, each recommendation contains exactly one item. In the following, we describe the various variants, namely *DGap, DGap-Acc, CGap, CGap-Acc, UniqueC, UniqueC-TW.*

The baseline methods, called *Naïve-AR* and *Naïve-MR*, operate as follows: *Naïve-AR* selects randomly from all available rules, including those with antecedents not in the sequence prefix, ensuring strong diversity. *Naïve-MR* works similar but excludes rules without antecedents matching items in the sequence prefix. Both approaches can be implemented by randomly selecting an element from a list, resulting in a constant computational complexity of $\mathcal{O}(1)$.

For the DGap approach, Definition 4 yields, a discrete gap $g_r(s)$ are items that occur between the last item of the antecedent of a rule r and the first item of a consequent of r given r occurs in a sequence s (see Definition 3). The gap-size $|g_r(s)|$ is the number of items contained by $|g_r(s)|$. Hence, for each occurrence, o, a specific number $g_r(s)$ describes the gap size between the antecedent and consequent of r. Thereby, we can iterate over our training set and create a histogram as a *proximity signature* for each rule in the rule set. With each rule having its signature, we can rank those rules based on the current recommendation step.

We introduce another alternative of this approach, called *DGap-Acc.* Here, we do not look up the position in the proximity signature regarding the specific gap size but accumulate every occurrence frequency in the proximity signature up to that gap size. This variant is reasonable as the unfulfilled needs of a user could intensify up to the point when they are satisfied.

The continuous definition of a gap is related to the aforementioned discrete gap. Instead of using the number of elements between the antecedent and consequent of a rule, we rely on the timestamp $t(e_i)$ of the last item of the occurring antecedent and the timestamp $t(e_j)$ of the first item of the occurring consequent in a sequence (see Definition 4). As a result, we can calculate the duration $|g_r^c(s)| = t(e_j) - t(e_i)$ between the occurring antecedent and consequent. Equal to calculating the histogram for *DGap*, we iterate over the training set and save every duration of occurring rules. We can not assume that the consequent occurrence follows a statistical parametric model. Therefore, we have to utilize a non-parametric approach. Hence, the timestamp data is the input for a kernel density estimator (KDE). Per rule r, we are interested in the occurrence probability of its consequent after a specific timespan, given that the antecedent occurred in the sequence. We calculate the ranking value of a rule by using the distance d_{t_1,t_2} where t_1 denotes the timestamp of the last item of the antecedent and t_2 the timestamp of the item position which has to be recommended. As an analogy, this could correspond to a recommendation button clicked by the user. We look up d_{t_1,t_2} in the KDE, which yields the occurrence probability for ranking.

Additionally, we also provide an alternative approach called *CGap-Acc*, which works analogously to *DGap-Acc* on timestamps. Here, we calculate the integral of the KDE up to d_{t_1,t_2} and use it as the ranking value. Due to the similarity of the four variants regarding the ranking procedure, we jointly illustrate it in Fig. 1. The computational complexity remains consistent across these variations. Traversing the training set once to generate the proximity signature results in a time complexity of $\mathcal{O}(n)$, where n denotes the size of the training set. Subsequently, sorting the rule set during each recommendation step incurs a computational complexity of $\mathcal{O}(m \log m)$, where m represents the number of rules. When diversity in sequential recommendation is a priority, it stands to reason that recommended items should not be repeated. Repetition would contribute to the dissatisfaction of users. For example, in online retailers or news platforms, the recommendation of an item a user already possesses, i.e., a user has already read, does not create added value. A fundamental approach to accomplishing this goal is to exclude any rule whose consequent already exists in the sequence. Hence, no item can exist twice in the recommended suffix, and diversity within the sequence can be maximized. We implement this approach as *UniqueC*. This

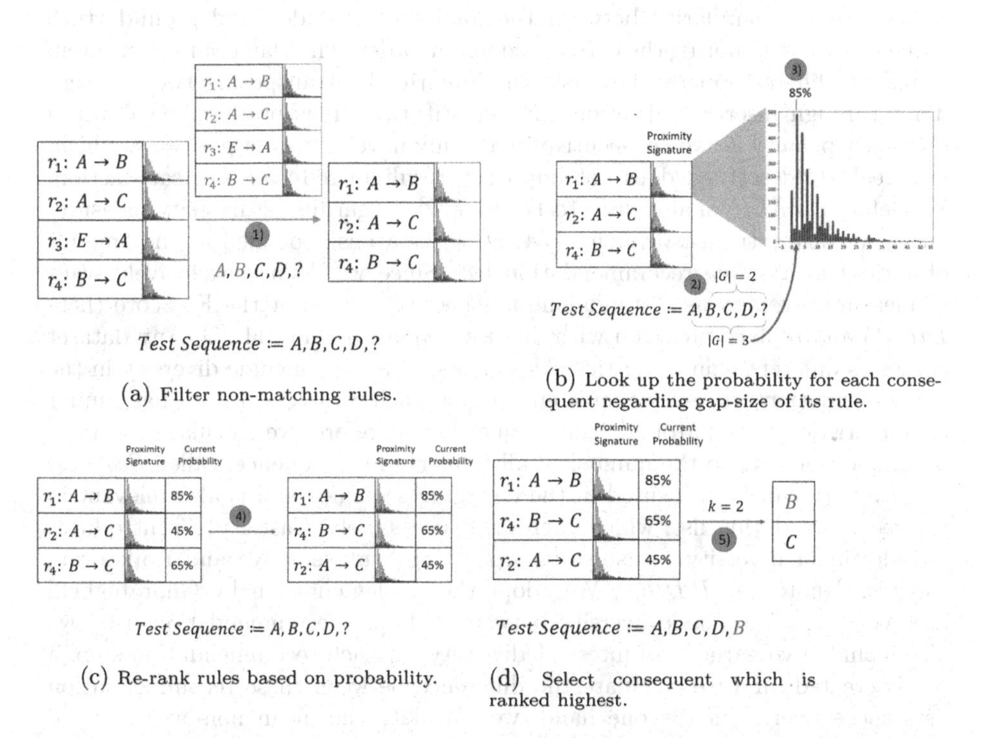

(a) Filter non-matching rules.

(b) Look up the probability for each consequent regarding gap-size of its rule.

(c) Re-rank rules based on probability.

(d) Select consequent which is ranked highest.

Fig. 1. Stepwise procedure for our recommendation system (DGap/CGap).

approach has a computational complexity of $\mathcal{O}(m)$. A variant of this method is called *UniqueC-TW* where *TW* stands for *Time Window*. Here, we do not adduce the complete sequence for rule deletion but implement a sliding window of size Δt. It is used to delete rules which hold consequents that already exist in this time window. Thus, items can be recommended more than once, but a certain delay is satisfied. This approach is helpful for subjects where items are consumed repeatedly, such as food products. Even when we need to handle the addition and removal of items entering and leaving the time window, this variant can be executed with a computational complexity of $\mathcal{O}(m)$. It is worth noting that the computational complexity of the rule mining phase varies depending on the specific rule mining algorithm utilized.

4.1 Quality Metrics

In comparing the different approaches with each other and a naïve baseline, we focus on both precision and diversity metrics and adduce established quality metrics in this regard. We use three metrics considering precision and four metrics concerning diversity. One of the former is Recall@k. We implement it as the mean number of hits over all recommendation lists. The second metric is Accuracy, which denotes the ratio of all correct recommendations.

We additionally include the Damerau-Levenshtein (*DamLev*) similarity to measure the similarity between the final recommended and ground truth sequences while honoring the correct sequential order. The Damerau-Levenshtein similarity differs from the Levenshtein similarity by transposing two adjacent items through insertion, deletion, and substitution. This property results in a less strict penalty if two consecutive items only have to switch positions, which, in a real-world setting, does not implicitly result in reduced user satisfaction. We define *DamLev* analogously to Boytsov [2]. Regarding a diversity measurement, we take intra-list diversity (*IALD@k*) as a basis, defined as the amount of variety in a single recommendation list. Since we do not exclusively want to measure diversity as a standalone perspective, we adapt the F1-Score (here *DivF*1) and replace precision with diversity equal to Hu et al. [8]. Any dataset possesses different characteristics. As soon as we aim to include diversity in the recommendation process, we must ensure that each created sequence has similar characteristics to the ground truth sequence. Therefore, we calculate the ratio of unique elements to the number of all elements in a sequence, called *SeqDiv*.

Diversity can be measured in the emerging sequence and is also relevant in the recommendation list, which is created successively. Castells [17] introduces an additional diversity measure, incorporating rank-sensitive and rank-aware aspects, denoted as *ILD-RR*. We adopt the implementation by Gharahighehi and Vens [7][1]. To assess diversity between different recommendation lists, we also include two variants of inter-list diversity. At each recommendation step, a list is created, and we calculate the difference between these recommendation lists successively. On the one hand, we calculate the mean non-overlap ratio

[1] https://github.com/alirezagharahi/d_SBRS/blob/main/performance_measures.py

between two recommendation lists R_1' and R_2' of successive recommendation steps ($IELD_p@k$). We compare recommendation lists pairwise, matching items from the first list at position i against those from the second list at the corresponding position i. On the other hand, we adopt the variant from Lathia et al. [13]. We treat each successive recommendation list as a set of items R_1'' and R_2''. With these sets, we calculate the set-theoretic difference and define the inter-list diversity based on the set-theoretic difference ($IELD_s@k$). Using partially-ordered sequential rules, we can not ensure that we have a matching rule for each case. Therefore, we abort the recommendation procedure of a sequence once we have no matching rule. The complement number of successfully executed sequence recommendations is tracked, and we include the value as the proportion of all successful recommendations concerning all recommendations, namely *SR-Ratio*.

5 Experiments and Evaluation

5.1 Selection of Real-World Datasets

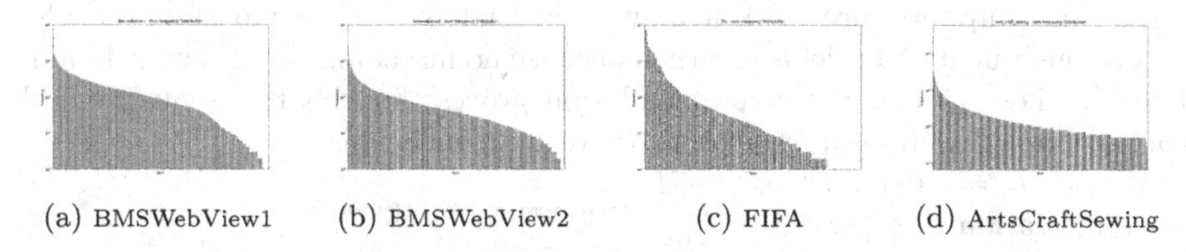

(a) BMSWebView1 (b) BMSWebView2 (c) FIFA (d) ArtsCraftSewing

Fig. 2. Item-Frequency-Distribution of the datasets revealing potential long-tails. The x-axis denotes each item and the y-axis represents frequency using logarithmic scale. The y-axis is limited to 10^4.

We conduct experiments on four real-world datasets, namely *BMSWebView1* and *BMSWebView2*, a small and a comparatively large version of the same dataset from the KDD Cup 2000. It represents a clickstream of an e-commerce platform. Additionally, we include a clickstream dataset from the website of the FIFA World Cup in 1998. All three datasets can be found at SPMF Website by Fournier-Viger[2]. A fourth dataset is *Amazon Arts, Crafts and Sewing*[3].

Figure 2 provides insights regarding the frequency distribution of the datasets. A common characteristic is a long-tail plot, which is the case for both BMSWebView1 and BMSWebView2. However, FIFA has a much more distinctive long-tail, which shows that few articles have been read very often, whereas many have been clicked only a few times. ArtsCraftSewing contrasts with this pattern, as many items within it have been purchased at similar frequencies.

[2] https://www.philippe-fournier-viger.com/spmf/index.php?link=datasets.php.
[3] https://nijianmo.github.io/amazon/index.html.

5.2 Task Definition and Parameter Configuration

We conduct experiments using an 80/20 random training-test-split based on the number of sequences in the database. Each test sequence is split into a prefix and a remainder for experimentation across all four datasets.

The prefix serves as a knowledge basis, which is used to filter the first recommendation list and to not run into a cold start problem. Several design decisions for the recommendation process have been based on practical experience and observation. We set the cut-off at e_3, meaning we start with the first three items as a-priori knowledge. Furthermore, we assume the sequence length to be known beforehand. This conforms with a real-world assumption where a recommender system is not used once the desired activity terminates, e.g., reading news articles or buying items on a retail platform. In this experimental setting, we recommend items until the sequence length is reached. Since our approaches do not have access to the whole item space, due to the rule's support threshold, it may occur that in a particular step, no rule, i.e., no consequent of a rule, is returned because no matching antecedent is found. Here, we chose to halt the recommendation process for this test sequence and monitor both the aborted and successful recommendations.

Table 1. Parameter values of minimum support and confidence thresholds and their respective number of rules.

Dataset	Min. Support	Min. Confidence	#Rules
BMSWebView1	31	0.5	8 784
FIFA	2 782	0.6	387
BMSWebView2	31	0.6	2 663
ArtsCraftSewing	3	0.5	2 425

We set $k = 10$ for each recommendation list, ensuring that the approaches utilize their full potential without exceeding the limits of rule inclusion. For UniqueC-TW, the time window size is an additional parameter where items occurring within a sliding window after the latest Δt items are excluded from recommendations.

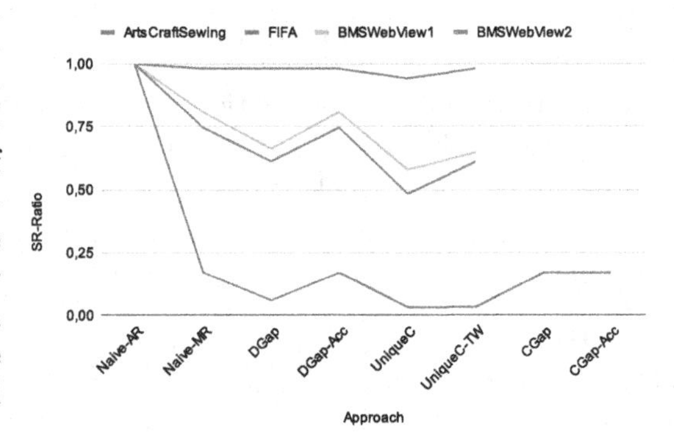

Fig. 3. Comparison of *SR-Ratio* on all four datasets. CGap and CGap-Acc are only available on datasets with timestamps.

Our experiments show that a $\Delta t = 3$ accomplishes an appropriate trade-off between precision and diversity. For setting the minimum support and confidence thresholds, experiments show that the values in Table 1 result in a suitable trade-off. Regarding CGap and CGap-Acc, we use a Gaussian kernel for KDE and a bandwidth calculated via Silverman's rule of thumb [16].

5.3 Evaluation of Experimental Results

The following section contains a comprehensive analysis of the results (cf. Table 2). These results are an expressive excerpt of conducted experiments that reflect the tendency of each approach. We start the evaluation by confirming evident attributes and edge cases. Remarkably, DGap achieves a relatively high $SeqDiv_Y$ outperforming Naïve-AR at BMSWebView2. Here, $DivF1$ validates this result as many different items in the recommendation list seemingly contribute to a high IALD@k. Thus, the chance of recommending various items at each step is high. Additionally, $IELD_s$ for DGap shows that successive recommendation lists differ significantly, amplifying this effect. Except for ArtsCraftSewing DGap shows a significantly higher $IELD_p$ value than the other approaches. This is useful as it suggests to the user that the recommendation system directly depends on the input given. Hence, it is evident that the ranking order is successfully influenced.

Figure 3 shows that the approaches perform rather well on FIFA. A reason for that is the median sequence length, which sets itself apart from other datasets. Thus, the rule mining algorithm can learn relations between items extensively. The low number of rules does not decrease this effect and shows that a model relying on rules can also be utilized. ArtsCraftSewing shows the opposite effect. This can be traced back to the fact that the dataset consists of many items with a similar frequency (observable in Fig. 2d). Hence, the minimum support threshold is difficult to set, yielding a very high number of rules on the one hand or hardly any rules on the other.

The approaches show similar behaviour on BMSWebView1 and BMSWebView2. The frequency distributions (cf. Fig. 2a and Fig. 2b) can be adduced as an explanation, as well. A moderate long-tail distribution leads to a set of items not represented in the rule mining model, causing sequence recommendations to be aborted. In the latter three datasets, there is a slight increase in *SR-Ratio* regarding DGap-Acc. DGap-Acc not only focuses on rules in which a gap occurs at least once in the training set but also allows an accumulation of buckets for every gap up to the specific size. Hence, it filters fewer rules than DGap in its unmodified version. This, again, leads to a higher number of potential rules to be selected in a recommendation step.

Proceeding to the Damerau-Levenshtein similarity, UniqueC leads in almost every dataset. This is interesting as UniqueC is an approach designed to increase diversity and does not aim for more precision than the other approaches. Regarding $IELD_s$, UniqueC leads at every dataset except FIFA. Here, DGap outperforms the other methods by far, as explained before. A reason for the low value concerning UniqueC is that it, indeed, filters already occurring items, but the rest of the recommendation list stays the same. Since FIFA has a low number of rules available, this has a substantial impact on that metric. In general, DGap and UniqueC, i.e., UniqueC-TW, outperform other approaches regarding $IELD_p$.

Table 2. Overall comparison of results (BMSWebView1, BMSWebView2, FIFA and ArtsCraftSewing top-down in this order). ArtsCraftSewing dataset was added because of the availability of timestamps which are necessary for the approaches CGap and CGap-Acc. Highest values are underlined. In case of highest value being 1.0 the second highest values are highlighted in bold.

Approach	Recall	Accuracy	DamLev	$SeqDiv_Y$	DivF1	ILD-RR	$IELD_p$	$IELD_s$	SR-RATIO
Naïve-AR	0,09539	0,01223	0,01634	0,83246	0,17309	0,00936	0,00000	0,00000	1,00000
Naïve-MR	0,23398	0,07692	0,09518	0,67159	0,36493	0,00744	0,21805	0,11185	**0,80783**
DGap	0,41348	0,13430	0,09931	0,81738	0,54068	0,00763	0,76730	0,24678	0,66201
DGap-Acc	0,11690	0,07082	0,08593	0,55865	0,20477	0,00816	0,23607	0,06355	**0,80783**
UniqueC	0,39580	0,16419	0,10722	1,00000	0,54139	0,00775	0,49767	0,28855	0,57920
UniqueC-TW	0,39181	0,12745	0,11053	**0,97019**	0,53863	0,00765	0,35002	0,19871	0,64718
Naïve-AR	0,01971	0,00692	0,00727	0,57360	0,03817	0,00615	0,00000	0,00000	1,00000
Naïve-MR	0,13415	0,05702	0,03396	0,55122	0,22859	0,00674	0,17246	0,08748	**0,74622**
DGap	0,23995	0,10851	0,03799	0,65188	0,34780	0,00620	0,55955	0,11903	0,61219
DGap-Acc	0,11937	0,05820	0,02963	0,39092	0,20770	0,00798	0,23829	0,04948	**0,74622**
UniqueC	0,29799	0,12913	0,03936	1,00000	0,43659	0,00712	0,43431	0,31738	0,48360
UniqueC-TW	0,19679	0,09086	0,03982	**0,88879**	0,31428	0,00660	0,27154	0,16864	0,61030
Naïve-AR	0,20442	0,01071	0,01244	0,34847	0,33679	0,00944	0,00000	0,00000	1,00000
Naïve-MR	0,31030	0,01870	0,01980	0,56219	0,46909	0,00865	0,19163	0,04606	**0,98141**
DGap	0,33438	0,02708	0,02903	0,38813	0,49259	0,00933	0,78072	0,17146	0,98136
DGap-Acc	0,18124	0,01922	0,02161	0,11073	0,30472	0,00958	0,26651	0,03727	**0,98141**
UniqueC	0,48722	0,02414	0,03796	1,00000	0,64767	0,00873	0,41538	0,07831	0,94256
UniqueC-TW	0,31402	0,01846	0,02160	**0,61674**	0,47333	0,00863	0,20083	0,04646	0,98081
Naïve-AR	0,01976	0,00493	0,00607	0,56724	0,03834	0,00605	0,00000	0,00000	1,00000
Naïve-MR	0,26198	0,11331	0,00644	0,35526	0,39580	0,00773	0,20280	0,02749	**0,17084**
DGap	0,55221	0,24860	0,00584	0,66178	0,59964	0,00644	0,58867	0,06476	0,06047
DGap-Acc	0,26225	0,11331	0,00612	0,32282	0,39576	0,00802	0,21561	0,02731	**0,17084**
UniqueC	0,63987	0,23200	0,00592	1,00000	0,73142	0,00825	0,83346	0,12246	0,03231
UniqueC-TW	0,63197	0,21374	0,00588	**0,99593**	0,72411	0,00812	0,80089	0,11658	0,03381
CGap	0,26352	0,11646	0,00642	0,33518	0,39499	0,00784	0,23477	0,02713	**0,17084**
CGap-Acc	0,26298	0,11511	0,00641	0,33231	0,39658	0,00802	0,21925	0,02708	**0,17084**

6 Conclusion and Future Work

We compare all proposed approaches regarding four diversity and three precision metrics to analyze different aspects of these two areas. Surprisingly, UniqueC is a strong competitor as it outperforms the other methods, not only in diversity, which is the primary goal of its design. It also proves to be beneficial regarding precision when recommending items.

In future investigations, we aim for an alternative regarding the preprocessing step. Zellner et al. [20] published a rule mining algorithm that already uses the gap information when analyzing patterns for rules in a dataset. Unfortunately, it runs slow on large datasets because the a-priori principle can not be used on that approach. Once this performance issue is improved, it supports methods like DGap or CGap by already aiming for that goal in a preprocessing step. This work explores diverse recommendation approaches and variants, each offering a complete recommendation system requiring minimal input from SMEs, thus adding value at low costs. Grammar-based recommendation models emerge as valuable alternatives for small- and medium-sized companies, presenting viable options in settings with limited information.

References

1. Bennett, J., Lanning, S., et al.: The netflix prize. In: Proceedings of KDD Cup and Workshop, New York, vol. 2007 (2007)
2. Boytsov, L.: Indexing methods for approximate dictionary searching: Comparative analysis. J. Exper. Algorithmics (JEA) **16** (2011)
3. Dogan, O.: A recommendation system in e-commerce with profit-support fuzzy association rule mining (p-farm). J. Theor. Appl. Electron. Commer. Res. **18**(2), 831–847 (2023)
4. Fahed, L., Lenca, P., Haralambous, Y., Lefort, R.: Distant event prediction based on sequential rules. Data Sci. Pattern Recogn. **4**(1) (2020)
5. Felfernig, A., et al.: Persuasive recommendation: serial position effects in knowledge-based recommender systems. In: de Kort, Y., IJsselsteijn, W., Midden, C., Eggen, B., Fogg, B.J. (eds.) PERSUASIVE 2007. LNCS, vol. 4744, pp. 283–294. Springer, Heidelberg (2007). https://doi.org/10.1007/978-3-540-77006-0_34
6. Fournier-Viger, P., Gueniche, T., Zida, S., Tseng, V.S.: ERMiner: sequential rule mining using equivalence classes. In: Blockeel, H., van Leeuwen, M., Vinciotti, V. (eds.) IDA 2014. LNCS, vol. 8819, pp. 108–119. Springer, Cham (2014). https://doi.org/10.1007/978-3-319-12571-8_10
7. Gharahighehi, A., Vens, C.: Diversification in session-based news recommender systems. Pers. Ubiq. Comput. **27**, 5–15 (2021)
8. Hu, L., Cao, L., Wang, S., Xu, G., Cao, J., Gu, Z.: Diversifying personalized recommendation with user-session context. In: IJCAI (2017)
9. Kaminskas, M., Bridge, D., Foping, F., Roche, D.: Product recommendation for small-scale retailers. In: Stuckenschmidt, H., Jannach, D. (eds.) EC-Web 2015. LNBIP, vol. 239, pp. 17–29. Springer, Cham (2015). https://doi.org/10.1007/978-3-319-27729-5_2
10. Kaminskas, M., Bridge, D., Foping, F., Roche, D.: Product-seeded and basket-seeded recommendations for small-scale retailers. J. Data Semant. **6**, 3–14 (2017)
11. Karimi, M., Cule, B., Goethals, B.: Leveraging sequential episode mining for session-based news recommendation. In: Zhang, F., Wang, H., Barhamgi, M., Chen, L., Zhou, R. (eds.) WISE 2023. LNCS, pp. 594–608. Springer, Singapore (2023). https://doi.org/10.1007/978-981-99-7254-8_46
12. Kim, H., Choi, D.W.: Recency-based sequential pattern mining in multiple event sequences. Data Min. Knowl. Disc. **35**(1), 127–157 (2021)
13. Lathia, N., Hailes, S., Capra, L., Amatriain, X.: Temporal diversity in recommender systems. In: Proceedings of the 33rd International ACM SIGIR Conference on Research and Development in Information Retrieval (2010)
14. Nasir, M., Ezeife, C.: A survey and taxonomy of sequential recommender systems for e-commerce product recommendation. SN Comput. Sci. **4**(6), 708 (2023)
15. Nikookar, S., Esfandiari, M., Borromeo, R.M., Sakharkar, P., Amer-Yahia, S., Basu Roy, S.: Diversifying recommendations on sequences of sets. VLDB J. **32**, 283–304 (2022)

16. Silverman, B.W.: Density Estimation for Statistics and Data Analysis, vol. 26. CRC Press, Boca Raton (1986)
17. Vargas, S., Castells, P.: Rank and relevance in novelty and diversity metrics for recommender systems. In: Proceedings of the Fifth ACM Conference on Recommender Systems (2011)
18. Wang, S., Cao, L., Wang, Y., Sheng, Q.Z., Orgun, M.A., Lian, D.: A survey on session-based recommender systems. ACM Comput. Surv. (CSUR) **54**(7), 1–38 (2021)
19. Zaki, M.J.: Spade: an efficient algorithm for mining frequent sequences. Mach. Learn. **42**(1), 31–60 (2001)
20. Zellner, L., Sontheim, J., Richter, F., Lindner, G., Seidl, T.: Scorer-gap: sequentially correlated rules for event recommendation considering gap size. In: 2021 International Conference on Data Mining Workshops (ICDMW). IEEE (2021)

HMAR: Hierarchical Masked Attention for Multi-behaviour Recommendation

Shereen Elsayed[1(✉)], Ahmed Rashed[2], and Lars Schmidt-Thieme[1]

[1] Information Systems and Machine Learning Lab (ISMLL) and VWFS Data Analytics Research Center (VWFS DARC), University of Hildesheim, Hildesheim, Germany
{elsayed,schmidt-thieme}@ismll.uni-hildesheim.de

[2] Data Analytics and AI Engineering Volkswagen Financial Services AG, Braunschweig, Germany
ahmed.rashed@vwfs.io

Abstract. In the context of recommendation systems, addressing multi-behavioral user interactions has become vital for understanding the evolving user behavior. Recent models utilize techniques like graph neural networks and attention mechanisms for modeling diverse behaviors, but capturing sequential patterns in historical interactions remains challenging. To tackle this, we introduce Hierarchical Masked Attention for multi-behavior recommendation (HMAR). Specifically, our approach applies masked self-attention to items of the same behavior, followed by self-attention across all behaviors. Additionally, we propose historical behavior indicators to encode the historical frequency of each item's behavior in the input sequence. Furthermore, the HMAR model operates in a multi-task setting, allowing it to learn item behaviors and their associated ranking scores concurrently. Extensive experimental results on four real-world datasets demonstrate that our proposed model outperforms state-of-the-art methods. Our code and datasets are available here (https://github.com/Shereen-Elsayed/HMAR).

Keywords: Sequential Recommendation · Multi-behavior Recommendation · Multi-task Learning

1 Introduction

Multi-behavioural recommendation aims to integrate all user behaviors into the next-item recommendation task. Such kind of approaches leverage all user behaviors, yielding more insightful recommendations. For example, in retail platforms, clicking on an item and adding it to favorites can indicate an intent to purchase.

Recommender systems approaches have evolved over time, from traditional matrix factorization to modern neural network-based methods like Neural Collaborative Filtering (NCF) [4], SASRec [7] and BERT4Rec [11] for sequential recommendations, CARCA [10], SSE-PT [20] and TiSASRec [9] for context and attribute-aware recommendations. However, these techniques primarily rely only

D.-N. Yang et al. (Eds.): PAKDD 2024, LNAI 14649, pp. 131–143, 2024.
https://doi.org/10.1007/978-981-97-2262-4_11

on the user's purchase interactions within retail platforms. Recently, newly developed models aim to harness additional user-item interaction data, particularly in multi-behavior scenarios. Graph-based approaches like MB-GCN [6] and MB-GMN [16] construct user-item and item-item graphs to model diverse behaviors. More recent methods, like MB-STR [19] and MBHT [18], employ attention mechanisms. Yet, they often struggle to capture comprehensive behavioral representations and capture complex sequential patterns in the user history.

To bridge this gap, we introduce HMAR, a hierarchical masked attention model that captures latent behavioral representations while preserving temporal sequences in the data. Our contributions can be summarized as follows:

- We present the HMAR model for the multi-behavior recommendation, incorporating hierarchical masked attention to learn comprehensive behavioral representations. Furthermore, we employ multi-task learning to understand both user intentions and interaction types.
- Within the HMAR model, we introduce hierarchical masked attention that models deep behavioral relationships within and across behaviors while preserving input sequence dynamics. We also include historical behavior indicators (HBI) for tracking item behavior frequency.
- Extensive experiments on four real-world datasets show that our proposed model HMAR outperforms previous state-of-the-art multi-behavioral recommendation models.

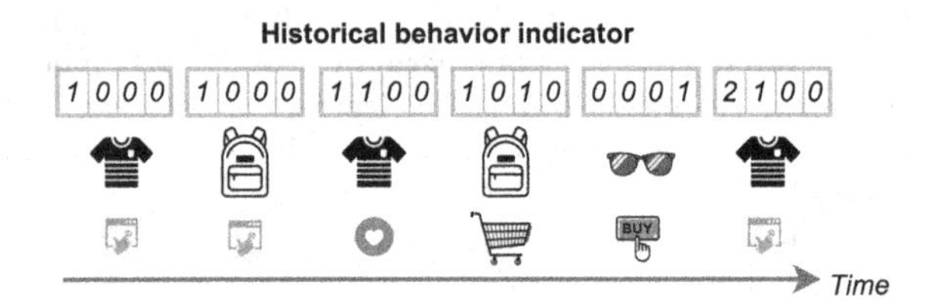

Fig. 1. Historical behavior indicator example

2 Methodology

In this section, we present the HMAR model, employing masked attention historical behavior indicator features to capture complex multi-behavioral sequential representations. Additionally, we present the multi-task learning aspect, where HMAR learns both item scores and behaviors through an auxiliary classification task.

2.1 Problem Formulation

In multi-behavior settings, given a set of items $\mathcal{I} := \{1, \ldots, I\}$ and a set of users $\mathcal{U} := \{1, \ldots, U\}$, each user u has a historical ordered sequence S of item interactions $S^u := \{v_1^u, \ldots, v_{|S_t^u|}^u\}$. Given K behaviors, each item in the sequence has a corresponding behavior type $b \in \mathcal{B} := \{1, \ldots, K\}$, such as buy and add to favorite. Our goal is to predict the next item to be interacted with based on the primary behavior, while other behaviors serve as auxiliary information to better model the user behavior over time.

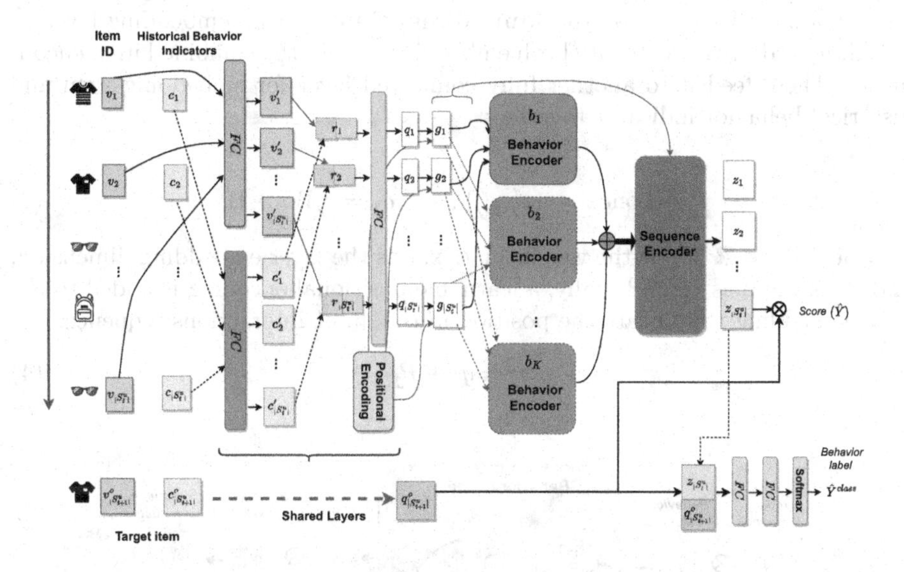

Fig. 2. Hierarchical Masked Attention Model Architecture

2.2 HMAR

Figure 2 illustrates the model architecture. It consists of Items and Behaviors Encoding, Hierarchical masked attention, and Multi-task learning.

Items and Behaviors Encoding. As discussed earlier, each user has an input sequence of historical item interactions S^u and a corresponding behaviors sequence $B^u = [b_1, b_2, ..., b_{|S^u|}]$. To construct the latent item representation, We first embed each item in the input sequence using a fully connected layer to obtain the initial embeddings as:

$$v'_j = v_j W_v + \mathrm{B}_v \tag{1}$$

where $W_v \in \mathbb{R}^{I \times d}$ is the weight matrix, I is the number of items, and d is the items embedding dimension and B_v is the bias term. To encode item behaviors, we use historical behavior indicators c_j that track the frequency of each behavior type on an item throughout the sequence. Figure 1 shows an example of the historical behavior indicators in a sequence. We also embed these indicators using a fully connected layer. This allows the model to learn the order and the frequency of user behaviors on items.

$$c_j' = c_j W_c + \mathrm{B}_c \tag{2}$$

where $W_c \in \mathbb{R}^{N \times d}$ is the weight matrix, N is the number of different historical behavior indicators, d is the context embedding dimension and B_c is the bias term. For simplicity, we use the **same dimension** d for all embedding layers.

Afterward, we concatenate both embeddings to get the combined item encoding r_j. Then, feed it to another fully connected layer for embedding item and historical behavior indicator together:

$$r_j = \text{concat}_{col}\left(v_j', c_j'\right), \qquad q_j = r_j W_r + \mathrm{B}_r \tag{3}$$

where $W_r \in \mathbb{R}^{2d \times d}$ is the weight matrix, d is the layer embedding dimension, and B_r is the bias term. Finally, a learnable positional encoding is added to the final embedding to indicate the position in the input interactions sequence:

$$g_j = q_j + P_j \tag{4}$$

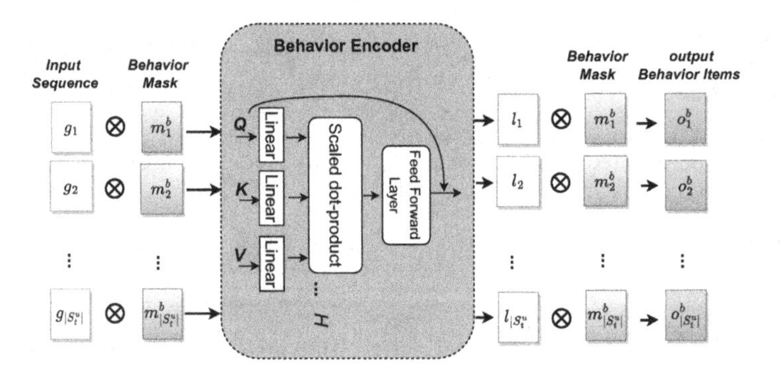

Fig. 3. Behavior Encoder Architecture

Hierarchical Masked Attention. In sequential multi-behavior recommendation, sequence order is crucial for an accurate next-item recommendation. In our work, we model behavioral interactions using two-stage self-attention to preserve input sequence order. The first stage is the **Behavior Encoder**, which encodes the items with the same behavior in the sequence, followed by a **Sequence Encoder**, which encodes the whole sequence and allows the modeling of different behaviors in the input sequence.

Behavior Encoder. Recent approaches use graphs to gather multi-behavior information, potentially sacrificing the input sequence's sequential pattern. To maintain the items' order while obtaining better representations, we employ a behavior mask M_b^u, created from the sequence's items, to filter out items not associated with a specific behavior b. Accordingly, we multiply the item sequence embedding $G^u := [g_1, g_2, ..., g_{|S_t^u|}]$ and with the corresponding mask of the specified behavior $M_b^u := [m_1^b, m_2^b, ..., m_{|S_t^u|}^b]$ as shown in Fig. 3.

This yields a number of behavior encoder blocks equivalent to the number of behaviors K. The masked latent embeddings are then fed into a fully connected layer to get $E_b^u := \{e_1^b, \ldots, e_{|S_t^u|}^b\}$ followed by a multi-head self-attention component to sequentially encode items with the same behavior as follows:

$$e_j^b = (g_j * m_j^b) W_e + \mathrm{Be}, \quad W_e \in \mathbb{R}^{d \times d}, \quad \mathrm{B}_e \in \mathbb{R}^d \tag{5}$$

$$X_b^u = \mathrm{SA}(E_b^u) = \mathrm{concat}_{col}\left(\mathrm{Att}(E_b^u \mathbf{W}_h^{Q^b}, E_b^u \mathbf{W}_h^{K^b}, E_b^u \mathbf{W}_h^{V^b})\right)_{h=1:H} \tag{6}$$

where $\mathbf{W}_h^{Q^b}$, $\mathbf{W}_h^{K^b}$, $\mathbf{W}_h^{V^b} \in \mathbb{R}^{d \times \frac{d}{H}}$ represent the linear projection matrices of the head at index h, and H is the number of heads. X_b^u represents the column-wise concatenation of the attention heads. Finally, we have the point-wise feed-forward layers to obtain the component's final output representations $F_b^u \in \mathbb{R}^{|S_t^u| \times d}$, $F_b^u = \mathrm{FFN}(X_b^u)$ as follows:

$$\mathrm{FFN}(X_b^u) = \mathrm{concat}_{row}\left(ReLU(X_{b,j}^u \mathbf{W}^{(1)^b} + \mathrm{B}^{(1)^b})\mathbf{W}^{(2)^b} + \mathrm{B}^{(2)^b}\right)_{j=1:|S_t^u|} \tag{7}$$

where $\mathbf{W}^{(1)^b}$, $\mathbf{W}^{(2)^b} \in \mathbb{R}^{d \times d}$ are the weight matrices of the two feed-forward layers, and $\mathrm{B}^{(1)^b}$, $\mathrm{B}^{(2)^b} \in \mathbb{R}^d$ are their bias vectors. The initial embedding sequence is added via a ReZero [1] residual connection to the behavior encoder output to obtain as:

$$L_b^u = F_b^u + \gamma (G^u * M_b^u) \tag{8}$$

where γ is a learnable weight initialized with zero to adjust the contribution of $(G^u * M_b^u)$. The output is multiplied by the behavior mask again to mask the positions of the other behaviors as follows:

$$O_b^u = L_b^u * M_b^u \tag{9}$$

Finally, we combine the output sequences of each behavior $O_b^u := [o_1^b, o_2^b, ..., o_{|S_t^u|}^b]$ by summing them element-wise to generate the first stage multi-behavioral latent sequence embeddings $\mathcal{O}^u$ as follows:

$$\mathcal{O}^u = \sum_{b=0}^{K} O_b^u \tag{10}$$

Sequence Encoder. In the second phase of hierarchical attention, we utilize a multi-head self-attention component across all behaviors while maintaining the original temporal order of the sequence. Specifically, once we've constructed the sequence comprising all behaviors, denoted as $\mathcal{O}^u$, the sequence encoder employs attention mechanisms over all items within the sequence. This enables the model to capture dependencies among various behaviors within the sequence, yielding an intra and inter-multi-behavioral latent representation for each item:

$$A^u = \text{SA}(\mathcal{O}^u) = \text{concat}_{col}\left(\text{Att}(\mathcal{O}^u\mathbf{W}_h^Q, \mathcal{O}^u\mathbf{W}_h^K, \mathcal{O}^u\mathbf{W}_h^V)\right)_{h=1:H} \tag{11}$$

where $\mathbf{W}_h^Q$, $\mathbf{W}_h^K$, $\mathbf{W}_h^V \in \mathbb{R}^{d\times\frac{d}{H}}$ represent the linear projection matrices of the head at index h, and H is the number of heads. A^u represents the column-wise concatenation of the attention heads. Additionally,for the model stability, we add a residual connection between sequence attention output A^u and sequence items G^u:

$$A^{u\prime} = A^u + G^u \tag{12}$$

Finally, we have the point-wise feed-forward layers to obtain the component's output representations $Z^u \in \mathbb{R}^{|S_t^u|\times d}$ as follows:

$$Z^u = \text{FFN}(A^{u\prime}) = \text{concat}_{row}\left(ReLU(A^{u\prime}\mathbf{W}^{(1)^Z} + \text{B}^{(1)^Z})\mathbf{W}^{(2)^Z} + \text{B}^{(2)^Z}\right)_{j=1:|S_t^u|} \tag{13}$$

where $\mathbf{W}^{(1)^Z}$, $\mathbf{W}^{(2)^Z} \in \mathbb{R}^{d\times d}$ are the weight matrices of the two feed-forward layers, and $\text{B}^{(1)^Z}$, $\text{B}^{(2)^Z} \in \mathbb{R}^d$ are their bias vectors.

2.3 Multi-task Learning

In order to train the model, we utilize a multi-task learning approach to learn item ranking and behavior type simultaneously. For efficient training, we adopt the SASRec [7] process, which uses autoregressive training, considering only historical items for each target item recommendation. The input sequence, S^u, has a fixed length, achieved by padding or truncating. The target sequence is constructed by right-shifting the input sequence and including the last item interaction. On the other hand, negative items are sampled using random items not in the user's interaction history $v \notin S^u$.

For item ranking, we calculate the final score by taking the dot product of the last item embedding $z_{|S_t^u|}$ from the sequence encoder and the target item embedding $q^o_{|S_{t+1}^u|}$ as follows:

$$\hat{Y}_{t+1} = \sigma(z_{|S_t^u|} \cdot q^o_{|S_{t+1}^u|}) \tag{14}$$

where σ is a sigmoid function. In multi-behavior datasets, interactions have varying importance for next-item recommendation. To handle this, we introduce weighting factors α_b in the loss function, one for each behavior. The number and values of α_b may differ based on the dataset's behavior count and their importance. Thus, our weighted binary cross-entropy objective for multi-behavior recommendation is as follows:

$$\mathcal{L}_{rank} = - \sum_{S^u \in \mathcal{S}} \sum_{t=0}^{|S^u|} [\alpha_b log(\hat{Y}_t^{O^{(+)}}) + \beta(log(1 - \hat{Y}_t^{O^{(-)}}))] \tag{15}$$

where $\hat{Y}_t^{O^{(+)}}$ are the output scores for the positive samples and $\hat{Y}_t^{O^{(-)}}$ are the output scores for the negative samples, $\mathcal{S}$ is the set of all sequences, α_b is the behaviors weights, and lastly, β is the sampling weight.

In this work, we also propose learning target interaction behavior type as an auxiliary task enhancing the model's ability to understand user behavior patterns and anticipate the behavior type of the next interaction. The behavior-type task involves a two-layer neural network that takes concatenated item embeddings $z_{|S_t^u|}$ and $q^o_{|S^u_{t+1}|}$ as input, as follows:

$$\hat{Y}^{class} = \text{Softmax}((\text{concat}_{col}(z_{|S_t^u|}, q^o_{|S^u_{t+1}|})\mathbf{W}^{(1)} + \text{B}^{(1)})\mathbf{W}^{(2)} + \text{B}^{(2)}) \tag{16}$$

where $\mathbf{W}^{(1)} \in \mathbb{R}^{2d \times d}$ and $\mathbf{W}^{(2)} \in \mathbb{R}^{d \times d}$ are the weight matrices of the two fully connected layers, and $\text{B}^{(1)}, \text{B}^{(2)} \in \mathbb{R}^d$ are their bias vectors. Then, we apply softmax function on the output to get the labels scores of each behavior and utilize the cross entropy classification loss as follows:

$$\mathcal{L}_{class} = - \sum_{S^u \in \mathcal{S}} \sum_{t=0}^{|S^u|} \sum_{j=0}^{K} [Y_{t_j}^{class} log(\hat{Y}_{t_j}^{class})] \tag{17}$$

where Y_{t_j} is the true behavior labels and $\hat{Y}_{t_j}^{class}$ is the predicted classification scores. Finally, the model is optimized using ADAM optimizer [8]. The final model loss can be defined as a weighted sum of the two losses:

$$\mathcal{L}_{model} = \mathcal{L}_{rank} + \theta \mathcal{L}_{class} \tag{18}$$

where θ is a hyperparameter controlling the contribution of the classification task to the final loss.

Table 1. Datasets Statistics

Dataset	Interactions	User#	Item#	Behaviors
Taobao	7,658,926	147,894	99,037	Page View, Fav., Cart, Buy
Yelp	1,400,000	19,800	22,734	Tip, Dislike, Neutral, Like
MovieLens	9,922,036	677,88	8704	Dislike, Neutral, Like
Tianchi	4,619,389	25,000	500,900	Page View, Fav., Cart, Buy

Table 2. Model performance and comparison against baselines on Taobao and Yelp datasets. Published results are indicated by the † symbol.

Method	Taobao		Yelp	
	HR@10	NDCG@10	HR@10	NDCG@10
Sequential and Context-Aware Recommendation Methods				
SASRec [7]	$0.390 \pm 4.2E{-3}$	$0.249 \pm 8E{-4}$	$0.853 \pm 1.7E{-3}$	$0.5601 \pm 5.1E{-3}$
SSE-PT [13]	$0.393 \pm 2.7E{-3}$	$0.232 \pm 3E{-4}$	$0.857 \pm 1.5E{-2}$	$0.572 \pm 1.8E{-2}$
CARCA [10]	$0.769 \pm 4E{-3}$	$0.662 \pm 5E{-3}$	$0.854 \pm 2E{-3}$	$0.589 \pm 4E{-3}$
Multi-Behavior Recommendation Methods				
MATN [14]	0.354†	0.209†	0.826†	0.530†
MB-GCN [6]	0.369†	0.222†	0.796†	0.502†
MB-GMN [16]	0.491†	0.300†	$0.861 \pm 8E{-3}$	$0.570 \pm 1.2E{-2}$
KHGT [15]	0.464†	0.278†	0.880†	0.603†
KMCLR [17]	0.4557 †	0.2735 †	0.8897 †	0.6038 †
MB-STR [19]	0.768†	0.608†	0.882†	0.624†
MBHT [18]	$0.745 \pm 6.1E{-3}$	$0.559 \pm 8.4E{-3}$	$0.885 \pm 5E{-4}$	$0.618 \pm 1.4E{-3}$
HMAR (ours)	$\mathbf{0.8515} \pm 1.8E{-3}$	$\mathbf{0.7294} \pm 1.2E{-3}$	$\mathbf{0.9015} \pm 5.7E{-5}$	$\mathbf{0.6374} \pm 8.08E{-4}$
Improv.(%)	10.663%	10.120%	1.349%	2.083%

3 Experiments

In this section, we aim to address the following research questions:

- **RQ1:** How does the HMAR model's performance compare to that of state-of-the-art multi-behavior recommendation models?
- **RQ2:** What impact do auxiliary behaviors have on the model's performance?
- **RQ3:** What influence does each individual model component have on the overall model performance?

3.1 Experimental Settings

Datasets. We assess our model on four multi-behavioral datasets:
Taobao[1] [15,19]**:** Data from the popular Chinese online shopping platform Taobao with behaviors like buy, add-to-cart, add-to-favorite, and pageview. We consider buy behavior as the target behavior. **Tianchi**[2]**:** Collected from Tmall.com, similar to Taobao, it includes buy, add-to-cart, add-to-favorite, and pageview behaviors. **Yelp**[3] [15,19]**:** Data gathered from the Yelp challenge, categorizing behaviors based on user ratings. Behaviors are split into dislike (rating ≤ 2), natural ($2 < \text{rating} < 4$), and like (rating ≥ 4). Users can also write

[1] https://github.com/akaxlh/MB-GMN/tree/main/Datasets/Tmall.
[2] https://tianchi.aliyun.com/competition/entrance/231576/information.
[3] https://github.com/akaxlh/KHGT/tree/master/Datasets/Yelp.

venue tips. **MovieLens**[4] [15]: Similar to Yelp, but with movie ratings, categorizing behaviors as dislike, neutral, and like. The main behavior considered for both MovieLens and Yelp datasets is the "like" behavior. Datasets statistics are summarized in Tables 1.

Table 3. Model performance and comparison against baselines on MovieLens and Tianchi datasets. Published results are indicated by the † symbol.

Method	MovieLens		Tianchi	
	HR@10	NDCG@10	HR@10	NDCG@10
Sequential and Context-Aware Recommendation Methods				
SASRec [7]	0.911 ± 1E−3	0.668 ± 5.1E−3	0.659 ± 3E−3	0.495 ± 2E−3
SSE-PT [13]	0.911 ± 7.1E−3	0.657 ± 4.5E−3	0.663 ± 1.2E−2	0.468 ± 1.3E−2
CARCA [10]	0.906 ± 2E−3	0.665 ± 1E−3	0.713 ± 4E−4	0.500 ± 1E−3
Multi-Behavior Recommendation Methods				
MATN [14]	0.847†	0.569†	0.714 ± 7E−4	0.485 ± 2E−3
MB-GCN [6]	0.826†	0.553†	-	-
MB-GMN [16]	0.820 ± 1.1E−3	0.530 ± 9E−4	<u>0.737 ± 4.3E−3</u>	0.502 ± 1.9E−3
KHGT [15]	0.861†	0.597†	0.652 ± 1E−4	0.443 ± 1E−4
MBHT [18]	<u>0.913 ± 5.9E−3</u>	<u>0.695 ± 7E−3</u>	0.725 ± 6.3E−3	<u>0.554 ± 4.8E−3</u>
HMAR (ours)	**0.9412** ± 5.5E−4	**0.7370** ± 1.7E−3	**0.7842** ± 1.0E−3	**0.5974** ± 1.7E−3
Improv.(%)	3.066%	6.043%	6.377%	7.761%

3.2 Evaluation Protocol

For a fair comparison, we follow the same evaluation process as recent state-of-the-art methods [15,19]. We use a leave-one-out mechanism, training and validating the model with the entire sequence except the last interaction, which is used for testing. We generate 99 negative items for each positive item and assess model performance with Hit Ratio (HR@N) and Normalized Discounted Cumulative Gain (NDCG@N). Higher HR and NDCG values indicate better performance. We report the mean and standard deviation of results from three separate runs to ensure statistical robustness.

Baselines. We compare our proposed method against various sequential and multi-behavioral recommendation methods.

Sequential and Context-Aware Recommendation Methods

- **SASRec** [7]: A model that utilizes multi-head self-attention to capture the sequential pattern in the users' history, then applies dot product for calculating the items scores.

[4] https://github.com/akaxlh/KHGT/tree/master/Datasets/MultiInt-ML10M.

- **SSE-PT** [13]: A state-of-the-art model which incorporates the user embedding into a personalized transformer with stochastic shared embedding regularization for handling extremely long sequences.
- **CARCA** [10]: A context-aware sequential recommendation approach employing cross-attention between user profiles and items for score prediction.

Multi-behavior Recommendation Models

- **MATN** [14]: A memory-augmented transformer network that utilizes a transformer-based encoder that jointly models behavioral dependencies.
- **MB-GCN** [6]: A model that represents the data as a unified graph, then employs a graph-convolutional network to learn the node's representation.
- **MB-GMN** [16]: Learns the multi-behavior dependencies through graph meta-network.
- **KHGT** [15]: This model employs a knowledge graph hierarchical transformer to capture the behavior dependencies in the recommender model.
- **MB-STR** [19]: Employs multi-behavior sequential transformers to learn sequential patterns across various behaviors.
- **MBHT** [18]: Utilizes a hypergraph-enhanced transformer model with low-rank self-attention to model short and long behavioral dependencies.
- **KMCLR** [17]: A knowledge graph-based approach that utilizes contrastive learning to capture the commonalities between different behaviors.

3.3 Model Performance (RQ1)

We conducted experiments comparing HMAR to various state-of-the-art models. As shown in Tables 2 and 3 SASRec and SSE-PT performed well, even against multi-behavior models, due to their ability to capture sequential patterns. In contrast, MATN, MB-GCN, MB-GMN, and KHGT had weaker results, especially on the Taobao dataset. HMAR outperformed all these baselines, showing significant HR improvements on Taobao and Tianchi datasets. In comparison to the context-aware method CARCA, HMAR achieved better results with a reasonable margin on all datasets. CARCA showed competitive performance due to its use of sequential and contextual information.

3.4 Effect of Auxiliary Behaviors and Individual Model Components (RQ2 & RQ3)

As shown in Table 4, excluding auxiliary behaviors and relying solely on target behavior interactions has a notable impact on the Tianchi, Taobao, and Yelp datasets. This shows that auxiliary behaviors enhance the model's understanding of user behavior. Particularly, the Historical Behavior Indicator (HBI) significantly benefits the Tianchi dataset, given the influence of behaviors like page views and favorites on performance. However, there is no observable effect on the MovieLens and Yelp datasets. Multitask learning demonstrates its significance in datasets where auxiliary behaviors play a more substantial role but has a less pronounced impact on others, such as the MovieLens dataset. Additionally, eliminating the behavior encoder component, responsible for capturing deep behavioral dependencies, and relying solely on the sequence encoder, has an impact on all datasets, highlighting the advantages of the two-stage attention procedure for modeling multi-behavior sequences.

Table 4. Ablation study on model components (HR@10)

Model	Dataset			
	Taobao	Yelp	Tianchi	MovieLens
w/o Aux. behaviors	0.665	0.878	0.666	0.939
w/o Multitask	0.846	0.897	0.781	**0.943**
w/o HBI	0.851	0.901	0.780	0.942
w/o Behavior Encoder	0.840	0.893	0.779	0.936
HMAR	**0.851**	**0.901**	**0.784**	0.941

4 Related Work

Sequential Recommendation aims to predict the next user interaction by leveraging users' historical interactions. Early methods used Convolutional Neural Networks (CNNs) [12] and Gated Recurrent Units (GRUs) [5] for encoding interactions. SASRec [7] was among the pioneers introducing the transformer architecture to sequential recommendation. Later models like BERT4Rec [11] improved this approach with Bidirectional Self-Attention. Notable innovations include SSE-PT [20] and TiSASRec [9]. Contemporary models like CARCA [10] incorporate contextual information and item attributes, utilizing a cross-attention mechanism for scoring. Recent approaches, like ICLRec [2], employ contrastive techniques to boost model performance.

Multi-behavior Recommendation has gained traction recently due to the increased availability of diverse data sources in online applications, leading to improved recommendation accuracy. Models like MB-GCN [6] use graph convolutional networks to model multi-behavior relationships, while MB-GMN [16] employs meta-learning to address behavior heterogeneity. Recent advancements like KHGT [15] utilize graph attention layers for modeling relationships. Temporal aspects are considered with models like MBHT [18] and MB-STR [19], using attention mechanisms to incorporate sequential patterns. Cascaded graph convolutional networks [3] further enhance understanding of behavior dependencies, and KMCLR [17] introduces contrastive learning to tackle data sparsity.

In contrast to the previously mentioned models, our model combines the benefits of the two families of models, achieving superior recommendation performance.

5 Conclusion

In this paper, we introduced HMAR, a model designed to capture deep behavioral dependencies and sequential patterns in multi-behavioral data. It encodes items of the same behavior with a behavior encoding component, followed by a sequence encoder to capture cross-behavior dependencies. We also applied multi-task learning for behavior types. Our experiments on four multi-behavioral datasets demonstrate that HMAR outperforms state-of-the-art methods.

References

1. Bachlechner, T., Majumder, B.P., Mao, H., Cottrell, G., McAuley, J.: Rezero is all you need: fast convergence at large depth. In: Uncertainty in Artificial Intelligence, pp. 1352–1361. PMLR (2021)
2. Chen, Y., Liu, Z., Li, J., McAuley, J., Xiong, C.: Intent contrastive learning for sequential recommendation. In: Proceedings of the ACM Web Conference 2022, pp. 2172–2182 (2022)
3. Cheng, Z., Han, S., Liu, F., Zhu, L., Gao, Z., Peng, Y.: Multi-behavior recommendation with cascading graph convolution networks. In: Proceedings of the ACM Web Conference 2023, pp. 1181–1189 (2023)
4. He, X., Liao, L., Zhang, H., Nie, L., Hu, X., Chua, T.S.: Neural collaborative filtering. In: Proceedings of the 26th International Conference on World Wide Web, pp. 173–182 (2017)
5. Jannach, D., Ludewig, M.: When recurrent neural networks meet the neighborhood for session-based recommendation. In: Proceedings of the Eleventh ACM Conference on Recommender Systems, pp. 306–310 (2017)
6. Jin, B., Gao, C., He, X., Jin, D., Li, Y.: Multi-behavior recommendation with graph convolutional networks. In: Proceedings of the 43rd International ACM SIGIR Conference on Research and Development in Information Retrieval, pp. 659–668 (2020)
7. Kang, W.C., McAuley, J.: Self-attentive sequential recommendation. In: 2018 IEEE International Conference on Data Mining (ICDM), pp. 197–206. IEEE (2018)
8. Kingma, D.P., Ba, J.: Adam: a method for stochastic optimization. arXiv preprint arXiv:1412.6980 (2014)
9. Li, J., Wang, Y., McAuley, J.: Time interval aware self-attention for sequential recommendation. In: Proceedings of the 13th International Conference on Web Search and Data Mining, pp. 322–330 (2020)
10. Rashed, A., Elsayed, S., Schmidt-Thieme, L.: Context and attribute-aware sequential recommendation via cross-attention. In: Proceedings of the 16th ACM Conference on Recommender Systems, pp. 71–80 (2022)
11. Sun, F., et .: Bert4rec: Sequential recommendation with bidirectional encoder representations from transformer. In: Proceedings of the 28th ACM International Conference on Information and Knowledge Management, pp. 1441–1450 (2019)
12. Tang, J., Wang, K.: Personalized top-n sequential recommendation via convolutional sequence embedding. In: Proceedings of the Eleventh ACM International Conference on Web Search and Data Mining, pp. 565–573 (2018)
13. Wu, L., Li, S., Hsieh, C.J., Sharpnack, J.: SSE-PT: sequential recommendation via personalized transformer. In: Fourteenth ACM Conference on Recommender Systems, pp. 328–337 (2020)

14. Xia, L., Huang, C., Xu, Y., Dai, P., Zhang, B., Bo, L.: Multiplex behavioral relation learning for recommendation via memory augmented transformer network. In: Proceedings of the 43rd International ACM SIGIR Conference on Research and Development in Information Retrieval, pp. 2397–2406 (2020)
15. Xia, L., et al.: Knowledge-enhanced hierarchical graph transformer network for multi-behavior recommendation. In: Proceedings of the AAAI Conference on Artificial Intelligence, vol. 35, pp. 4486–4493 (2021)
16. Xia, L., Xu, Y., Huang, C., Dai, P., Bo, L.: Graph meta network for multi-behavior recommendation. In: Proceedings of the 44th International ACM SIGIR Conference on Research and Development in Information Retrieval, pp. 757–766 (2021)
17. Xuan, H., Liu, Y., Li, B., Yin, H.: Knowledge enhancement for contrastive multi-behavior recommendation. In: Proceedings of the Sixteenth ACM International Conference on Web Search and Data Mining, pp. 195–203 (2023)
18. Yang, Y., Huang, C., Xia, L., Liang, Y., Yu, Y., Li, C.: Multi-behavior hypergraph-enhanced transformer for sequential recommendation. In: Proceedings of the 28th ACM SIGKDD Conference on Knowledge Discovery and Data Mining, pp. 2263–2274 (2022)
19. Yuan, E., Guo, W., He, Z., Guo, H., Liu, C., Tang, R.: Multi-behavior sequential transformer recommender. In: Proceedings of the 45th International ACM SIGIR Conference on Research and Development in Information Retrieval, pp. 1642–1652 (2022)
20. Zhou, K., et al.: S3-rec: self-supervised learning for sequential recommendation with mutual information maximization. In: Proceedings of the 29th ACM International Conference on Information & Knowledge Management, pp. 1893–1902 (2020)

Residual Spatio-Temporal Collaborative Networks for Next POI Recommendation

Yonghao Huang(✉), Pengxiang Lan, Xiaokang Li, Yihao Zhang, and Kaibei Li

School of Artificial Intelligence, Chongqing University of Technology, Chongqing, China
{huangyonghao,lixiaokang,likaibei}@stu.cqut.edu.cn, pengxianglan@2020.cqut.edu.cn, yhzhang@cqut.edu.cn

Abstract. As location-based services become increasingly integrated into users' lives, the next point-of-interest (POI) recommendation has become a prominent area of research. Currently, many studies are based on Recurrent Neural Networks (RNNs) to model user behavioral dependencies, thereby capturing user interests in POIs. However, these methods lack consideration of discrete check-in information, failing to comprehend the complex motivations behind user behavior. Moreover, the information collaboration efficiency of existing methods is relatively low, making it challenging to effectively incorporate the numerous collaborative signals within the historical trajectory sequences, thus limiting improvements in recommendation performance. To address the issues mentioned above, we propose a novel Residual Spatio-Temporal Collaborative Network (RSTCN) for improved next POI recommendation. Specifically, we design an encoder-decoder architecture based on residual linear layers to better integrate spatio-temporal collaborative signals by feature projection at each time step, thus improving the capture of users' long-term dependencies. Furthermore, we have devised a skip-learning algorithm to construct discrete data in a skipping manner, aiming to consider potential relationships between discrete check-ins and thus enhance the modeling capacity of short-term user dependencies. Extensive experiments on two real-world datasets demonstrate that our model significantly outperforms state-of-the-art methods.

Keywords: Next Point-of-interest · User dependency · Spatio-Temporal · Recommendation

1 Introduction

In recent years, due to the thriving development of location-based social services such as Foursquare and Uber, the field of the next POI recommendation has made unprecedented progress. The substantial check-in behaviors of users have provided us with abundant data to recommend next POI for users. The next POI recommendation, as an extension of traditional POI recommendation, aims to provide users with more personalized content recommendations. Consequently, it has garnered significant attention from researchers [15,17,18].

D.-N. Yang et al. (Eds.): PAKDD 2024, LNAI 14649, pp. 144–155, 2024.
https://doi.org/10.1007/978-981-97-2262-4_12

Previous research has covered various aspects and proposed numerous personalized next POI recommendation models. Early studies primarily focused on short-term user dependencies, such as using Markov chains [2]. Subsequently, methods based on RNN and its variants were introduced to better capture short-term trajectory features, such as ST-RNN [7] and CARA [10]. Meanwhile, in addition to the short-term dynamic dependencies exhibited by recent user behavior, people's actions also display long-term periodicity. For instance, a user might visit a supermarket every weekend, which demonstrates long-term cyclic behavior. Nevertheless, in the task of next POI recommendation, fewer studies have simultaneously considered both long and short-term user behaviors. Notable examples of such studies are STGN [17] and LSTPM [11]. STGN employs time gates and distance gates to account for long and short-term interests, while LSTPM utilizes nonlocal operations and geo-dilated Long Short-Term Memory (LSTM) to model users' long and short-term preferences. Recent research, ST-PEGD, employs a gated-deep network and a position-extended algorithm to respectively capture users' long- and short-term dependencies. Besides, Some studies also incorporate extensive collaborative signals and employ enhanced Transformer-based models to accurately recommend the next POI [13,14].

Despite the continuous emergence of new models, there are still several key issues yet to be addressed. Firstly, existing models fail to effectively leverage spatio-temporal collaborative signals. Specifically, long-term trajectories typically contain rich spatio-temporal factors, but models based on RNNs struggle to integrate these signals effectively due to their sequential data processing nature [11]. Some attention-based methods also lack consideration of collaborative information [8]. Furthermore, some Transformer-based methods do not account for users' long-term dependencies and solely focus on short-term trajectory modeling, thereby failing to utilize valuable spatio-temporal collaborative signals present in long-term trajectory sequences [14]. Secondly, in daily life, users' activities may exhibit periodic patterns. This implies that even if certain check-ins are temporally discontinuous, there may still exist some form of relationship between them. RNN-based methods primarily emphasize sequential patterns, overlooking correlations between discrete check-ins [9,17].

To address these issues, we propose the Residual Spatio-Temporal Collaborative Network (RSTCN) for next POI recommendation. RSTCN enhances the model's ability to capture long-term user dependencies by incorporating a novel residual structure that strengthens the combination of spatio-temporal collaborative signals at each time step. Furthermore, RSTCN employs a skip learning algorithm to consider the correlation between discrete check-ins, overcoming the sequential regularity constraints found in RNN. The primary contributions of this paper are outlined as follows:

- We employ an encoding-decoding structure based on residual linear layers and a global residual connection to better harness spatio-temporal collaborative signals, thus enhancing long-term dependency modeling.
- We have designed a skip learning algorithm to learn relationships between discrete check-ins and model users' short-term dynamic dependencies.

– We introduce the RSTCN model and validate its effectiveness through extensive experiments on two datasets. The experimental results demonstrate that our model outperforms the existing state-of-the-art models.

2 Related Works

Different from traditional POI recommendation, next POI recommendation [1] focuses on predicting the user's next visited location within a trajectory, rather than offering general location recommendations. Early research in this field incorporated methods from other sequential recommendation studies, such as Markov chains [1,2,16]. For instance, Cheng et al. [1] utilized matrix factorization and the embedding of personalized Markov chains to recommend the next POI. However, due to their limitation in modeling only static user dependencies, most of these early methods lacked the modeling capacity that was found in deep learning approaches.

Currently, most researchers have turned to exploring deep learning methods. Many methods based on RNN and its variants have been proposed for capturing dynamic user dependencies [4,6,7,11,17]. For instance, Liu et al. [7] utilizes distance- and time-specific transition matrices to model spatio-temporal context information. Zhao et al. [17] introduced the Spatio-Temporal Gated Network (STGN), which consists of two temporal gates and two spatial gates to learn user's long and short-term dependencies. LSTPM [11] designed a nonlocal network and a geo-dilated RNN to model the long- and short-term preference. Some studies also employ attention mechanisms for next POI recommendation, such as STAN [8] and ST-PEGD [6]. ST-PEGD generates auxiliary binary gates and uses the position-extended algorithm to capture users' behavior dependency. Furthermore, some methods utilize Transformers [12] for next POI recommendations [13,14]. For example, STTF-Recommender [13] utilizes the Transformer [12] aggregation layer to capture spatio-temporal relationships between any two locations. However, in the aforementioned methods, most of them struggle to effectively harness the spatio-temporal collaborative information within historical trajectory sequences, resulting in suboptimal modeling of long-term dependencies. Additionally, there is a lack of attention to the issue of discrete check-ins. As a result, we constructed RSTCN to address these issues.

3 Method

In this section, we provide a detailed description of our proposed RSTCN model. Figure 1 illustrates the overview of our architecture. Our model comprises three key parts, namely, short-term preference part, long-term preference part and the sample balancer. We first generate multi-feature embeddings to represent various information. We then define two modules to achieve the recent user dependency and historical user dependency respectively. Lastly, the sample balancer weighs the output of the two modules mentioned above and generates a POI prediction.

3.1 Problem Formulation

Let $P = \{p_1, p_2, ..., p_{|M|}\}$ and $U = \{u_1, u_2, ..., u_{|U|}\}$. Each POI $p \in P$ is geocoded by a (Lon, Lat) tuple, i.e., (Lon_p, Lat_p). Each user $u \in U$ has an available trajectory sequence, which can be represented by $T = \{T_1, T_2, ..., T_{|N|}\}$, where N denotes the index of the current sequence. Each trajectory $T_m \in T$ contains a chronological check-in sequence of user $u \in U$, i.e., $T_m = \{p_1, p_2, ..., p_{|T_m|}\}$ and $p \in P$. Each user's historical trajectory sequence can be represented by $T_h = \{T_1, T_2, ..., T_{|N-1|}\}$. The current chronological trajectory sequence is defined as $T_{|N|} = \{p_1, p_2, ..., p_{|P_{t-1}|}\}$, where P_{t-1} denotes the most recent POI checked in by user u at time $t-1$.

3.2 Long-Term Dependence Module

Long-term preference provides information about how users evolve and adapt over time, revealing a persistent pattern of behavior. Inspired by, LSTPM [11] and TiDE [3], we designed an encoder-decoder structure to capture long-term preferences of users. The encoder maps $e_t^{(T_{i-1})}$ at each time step to a low-dimensional projection of size r, significantly smaller than temporal width R. We can describe the process as,

$$\overline{e}_t^{(T_{i-1})} = Encoder(e_t^{(T_{i-1})}) \tag{1}$$

where $e_t^{(T_{i-1})} \in \mathbb{R}^r$ denotes r-dimensional embedding vector for the historical trajectory sequence $T_{i-1} \in T_h$ at time t and $i \in [2, N]$.

Residual Linear Layer. We use residual linear layers (RLLs) as the underlying processing layer of the architecture, which means most data will flow through this layer. As shown in Fig. 1, in one RLL, we employ two multilayer perceptrons (MLP) and one is for shortcut operation. The hidden layer of the other ReLU-activated MLP is mapped to the output by dropout operation and we use layer norm before output. We use one RLL both as an encoder and decoder. We can describe the residual linear layer as follows,

$$\begin{aligned} dc_t^{(T_{i-1})} &= MLP(\overline{e}_t^{(T_{i-1})}), \\ \widetilde{e}_t^{(T_{i-1})} &= MLP(ReLU(\overline{e}_t^{(T_{i-1})})), \\ I_t^{(T_{i-1})} &= LayerNorm(dc_t^{(T_{i-1})} \oplus Dropout(\widetilde{e}_t^{(T_{i-1})})) \end{aligned} \tag{2}$$

where $dc_t^{(T_{i-1})}$ denote the r-dimensional dynamic covariates of trajectory sequence T_{i-1} at time t, $\widetilde{e}_t^{(T_{i-1})}$ denote the dense expression of historical trajectory, $\oplus$ denotes the shortcut connection and $I_t^{(T_{i-1})}$ is the intermediate expression of the historical trajectories before we obtain the output of RLLs.

Overall, we stack and flatten all past and future dynamic covariates and map them into embeddings using multiple stacked RLLs, which means the RLLs project the encoding $\overline{e}_t^{(T_{i-1})}$ to a vector for each time step. This operation can be described as,

$$E_t^{(T_{i-1})} = n \times RLL(\overline{e}_t^{(T_{i-1})}) \tag{3}$$

where n denotes the number of residual linear layers.

The long-term interests of users often exhibit a certain degree of periodicity, therefore, it is necessary to compute the saliency of different POIs. Consequently, we introduce a multi-head self-attention mechanism. It can be described as,

$$OA_t^{\gamma} = MultiheadSA(e_t^{\gamma}) \tag{4}$$

where e_t^{γ} denotes the target historical or current trajectory embedding encoded by RLLs, which means γ could be h or c, respectively denoting the output of RLLs in short- and long-term module. We then let OA_t^h and OA_t^c denote the output of multi-head self-attention for e_t^h and e_t^c. After the attention mechanism, we designed a decoder to project OA_t^h to OD_t^h, restoring its dimension to the same as $e_t^{(T_{i-1})}$:

$$OD_t^h = Decoder(OA_t^h) \tag{5}$$

We then introduce a global residual link to combine the look-back $e_t^{(T_{i-1})}$ with the decoded historical trajectory OL_t^h :

$$OL_t^h = LayerNorm(OD_t^h \oplus e_t^{(T_{i-1})}) \tag{6}$$

where $\oplus$ denotes the shortcut connection.

The above algorithm is particularly useful in cases where certain POIs have a strong direct impact on the actual values at specific time steps. This implies that lost collaborative signals can be reutilized. In the absence of such residual connections, the model may lose these spatio-temporal collaborative signals or require a longer time to learn them.

The temporal factor largely determines the popularity of a POI at the current time point. Accordingly, we construct a temporal collaborative matrix. Specifically, we divide each week into 48 time steps proportionally, with 24 time steps for weekdays and 24 time steps for weekends. We consider a POI set $D_i = \{p_1, p_2, ..., p_{D_i}\}$, where there are no POIs that users have not visited. We then achieve the similarity $\alpha_{i,j}$ between the i-th and j-th time steps by:

$$\alpha_{i,j} = \frac{|D_i \bigcap D_j|}{\sqrt{|D_i| * |D_j|}} \tag{7}$$

The more identical POIs between two time steps, the higher their similarity. When providing recommendations to users, we place greater emphasis on the POIs visited by users who are more similar to the users within the same time step. This approach involves calculating temporal collaborative weights, represented as W_t, and applying them to target historical trajectories:

$$t_h = \sum_{t=1}^{|T_h|} W_t OL_t^h, W_t = \frac{\exp(\alpha_{c,D_t})}{\sum_{i=1}^{|T_h|} \exp(\alpha_{c,D_i})} \tag{8}$$

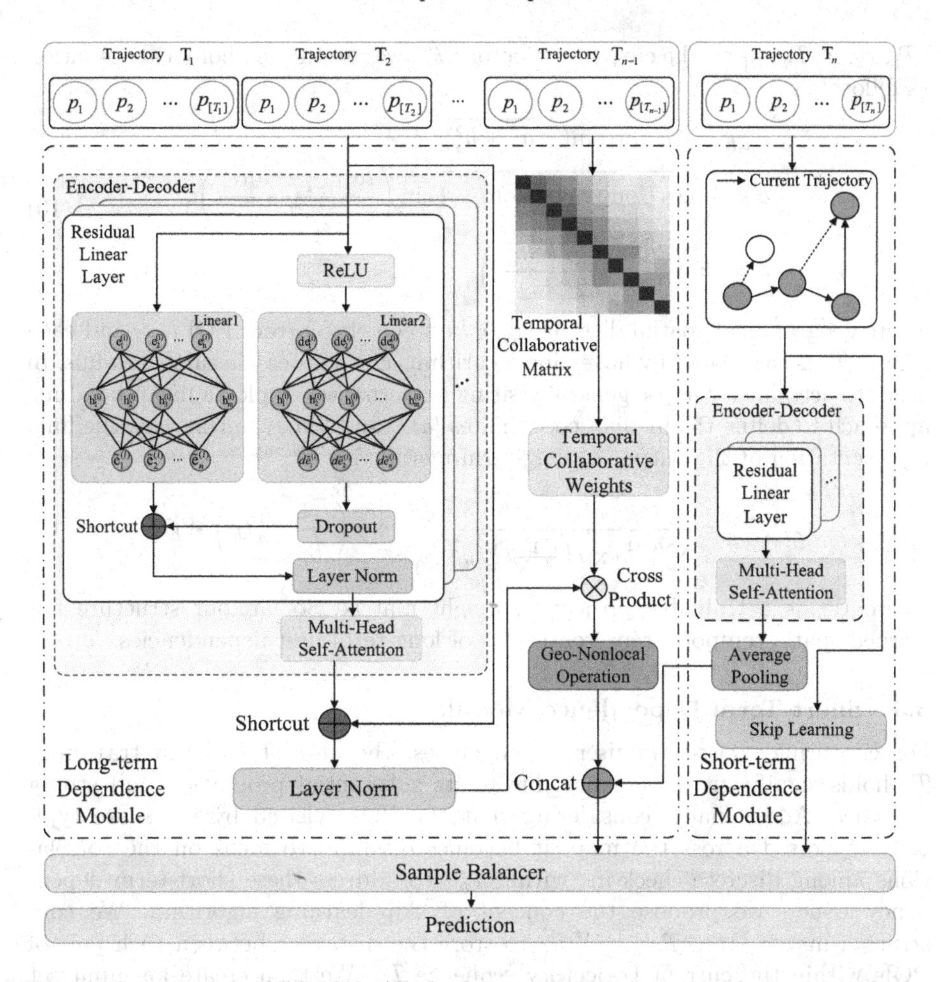

Fig. 1. The overview of RSTCN. All check-in trajectories were used to capture long-term user dependencies. Capturing short-term user dependencies only used the current check-in trajectory. The Sample Balancer balances the outputs of the long- and short-term dependency modules and derives the prediction results.

after considering temporal collaborative information, historical check-in trajectory can now be represented by $t_h \in \{t_1, t_2, ..., t_{n-1}\}$.

We then preserve all information with average pooling to better reflect the real-time behavior dependencies of users:

$$t_n = \frac{1}{|T_n|} \sum_{t=1}^{|T_n|} OL_t^c \tag{9}$$

where the target current check-in sequence is represented as t_n. Clearly, the real-time location information of the target user has a certain impact on its spatial dimension. Accordingly, to consider the impact of each historical trajectory $T_h \in$

$\{T_1, T_2, ..., T_{n-1}\}$ on the current trajectory T_n, we use the geo-nonlocal operation as follows,

$$\begin{aligned} d_{n,h} &= 2Er \arcsin(\sqrt{a^2+b^2}), \\ a &= \sqrt{\cos(\text{lon}_{t_h})\cos(\text{lon}_{l_{t-1}})}\sin\left(\frac{lat_{l_{t-1}} - lat_{t_h}}{2}\right), \\ b &= \sin\left(\frac{lon_{l_{t-1}} - lon_{t_h}}{2}\right) \end{aligned} \tag{10}$$

where geographical spatial distance $d_{n,h}$ between the current POI l_{t-1} and each POI in T_h is calculated by haversine algorithm. Er denotes the earth's radius. In order to account for users' general visiting patterns, we employ a mean-handling approach to define the median coordinates lat_{t_h} and lon_{t_h}. In this way, the final representation of the output of long-term module is OP_n^{long}:

$$OP_n^{long} = \frac{1}{\sum_h^{n-1}\exp\left(\frac{1}{d_{n,h}}\widetilde{t_n^\top} t_h\right)}\sum_h^{n-1}\exp\left(\frac{1}{d_{n,h}}\widetilde{t_n^\top} t_h\right)W_h t_h \tag{11}$$

where W_h is a trainable projection weight matrix. So far, our structure has learned spatio-temporal representations of long-term user dependencies.

3.3 Short-Term Dependence Module

For capturing short-term user dependencies, the current check-in trajectory T_n holds significant importance due to its substantial proportion and strong influence. Additionally, considering that the POIs visited by users are typically dispersed across the map, it becomes essential to focus on the correlations among discrete check-ins within T_n. To address these short-term dependency issues, we propose the concept of skip learning algorithm. We construct a matrix $M \in \mathbb{R}^{(t-1)\times(t-1)}$ to store the distances between each pair of POIs within the current trajectory sequence T_n. We then create an input set $T_n^{Skip} = \{\{s_1, s_\sigma\}, ..., \{s_\theta, s_{t-1}\}\}$ (where $1 < \sigma < ... < \theta < t-1$) and the computation of skip learning contains the iteration of a two-step process,

$$h_{t-1} = n \times RLL(h_\theta) \tag{12}$$

where n denotes number of RLLs and h_{t-1} is calculated from the last sequence $\{s_\theta, s_{t-1}\} \in T_n^{Skip}$. Finally, we use an attention mechanism in the skip learning algorithm,

$$\begin{aligned} N_{t-1} &= \tanh(W_q h_{t-1}) + b_q \\ A_{t-1} &= \sum_t \frac{\exp(N_t^\top u_q)}{\sum_t \exp(N_t^\top u_q)} h_{t-1} \end{aligned} \tag{13}$$

with

$$OP_n^{short} = ReLU(A_{t-1}W' + b')W'' + b'' \tag{14}$$

where W_q and u_q are randomly initialized matrices, A_{t-1} is the weighted vectors and W', W'', b', b'' are learnable parameters. OP_n^{short} denotes the output of the short-term dependence module.

3.4 Sample Balancer

Up to this point, we have captured both the long-term and short-term dependencies of users. We design a sample balancer to balance and integrate the information obtained so far:

$$\mathbf{p} = \text{softmax}\left(W_s\left(\epsilon OP_n^{long} \oplus (1-\epsilon)OP_n^{short}\right)\right) \tag{15}$$

where $\oplus$ denotes concatenation operation, $\mathbf{p}$ is the probability set of each POI and trainable matrix $W_s \in \mathbb{R}^{|L|\times 2d}$ can project each POI. Therefore, the probability of the target user visiting a POI at the next time step t is maximized. We then denote the probability of true POI as $p \in \mathbf{p}$ and describe the objective function as follows,

$$\mathcal{L} = -\sum_{j=1}^{K} \log(p_j) \tag{16}$$

where p_j is the probability that the model correctly identifies the true POI according to the j-th training sample. K denotes the total training samples number.

4 Experiments

In this section, we evaluate the performance of RSTCN by comparing it with state-of-the-art approaches using two real-world datasets.

4.1 Experimental Settings

Datasets and Evaluation Metrics. To evaluate our model correctly, we conducted rigorous experiments on two standard datasets, including Gowalla (collected worldwide from February 2009 to October 2010) [4] and Foursquare (collected in New York from February 2010 to January 2011) [11]. We utilize the initial 80% of each user's trajectory for training, while the remaining 20% is allocated for testing, according to [5]. We also filtered out POIs with fewer than 10 visits and users with fewer than 5 trajectories to enhance the data density. We compute the *Recall@K* and Normalized Discounted Cumulative Gain ($NDCG@K$), which are common metrics in POI recommendation and $K = \{5, 10\}$. *Recall@K* indicates whether the true POI is in the top-k recommended POIs and $NDCG@K$ evaluates the quality of the top-k POIs list.

Baselines and Settings. We compare the following methods with our RSTCN.

- **DRCF** [9] captures both users' dynamic and static dependencies with a pairwise ranking function.
- **DeepMove** [4] separates user trajectory sequences into historical and current trajectory sequences, and utilizes attention mechanisms and RNN to capture users' long-term and short-term dependencies separately.

- **STGN** [17] incorporates time gates and space gates into LSTM to capture user preferences in both temporal and spatial dimensions.
- **LSTPM** [11] employs geo-dilated LSTM and LSTM with nonlocal operation to model long-and short-term user preferences.
- **STAN** [8] captures the interactions between non-adjacent check-ins with spatio-temporal information on trajectories and a self-attention layer.
- **ST-PEGD** [6] captures the long-term dependencies with position-extended algorithm and short-term dependencies with a gated-deep network.
- **STTF-Recommender** [13] employs the Transformer aggregation layer to explore spatio-temporal relationships between two visited locations and an attention matcher to match the most likely candidates.

According to DeepMove [4], we set the learning rate to 0.00005 and use Adam to optimize our parameters in our model with the batch size of 32. The number of layers for the residual linear layer is set to 8 in the long- and short-term dependence module and 2 in the skip learning algorithm. The head number of the multi-head self-attention is set to 4.

4.2 Recommendation Performance

Table 1 presents the comparative performance of our RSTCN model against other state-of-the-art models. We compared the results of these methods in terms of $Recall@K$ and $NDCG@K$ ($K = 5, 10$).

Firstly, from these results, it can be observed that the performance of RSTCN surpasses all the baselines, indicating that our model indeed enhances recommendation performance. RSTCN exhibits an average improvement of 15.45% on Foursquare and 41.27% on Gowalla. Particularly noteworthy is the 49.34% improvement over the second best result in $NDCG@5$ on Gowalla. Therefore, these results demonstrate the effectiveness of our RSTCN.

Secondly, ST-PEGD and STTF-Recommender employ distinct methods for modeling long-term user dependencies and outperform all baselines. This underscores the importance of modeling long-term user dependencies in the task of next POI recommendation. Furthermore, several methods (such as LSTPM, ST-PEGD, and DeepMove) focus on modeling short-term dependencies to capture dynamic user preferences, affirming the indispensability of short-term modeling. In contrast, DRCF and DeepMove, which do not incorporate spatio-temporal collaborative signals, exhibit suboptimal performance. Evidently, owing to the intricate yet inefficient long-term modeling techniques used by ST-PEGD and STTF-Recommender, RSTCN significantly outperforms them.

Finally, RNN-based models like LSTPM and STGN did not address the issue of discrete check-ins within sequences and only focused on periodicity between sequences, which is a key reason for their poor performance. Therefore, RSTCN designed skip learning algorithm to model discrete check-ins, which is one of the significant factors contributing to the superior performance of RSTCN over these models.

Table 1. Recommendation performance of our model compared to baselines on Gowalla and Foursquare datasets. The best results are highlighted in bold. The underlines indicate the second-best performance. STTF-Recommender is abbreviated as STTF in the table.

Model	Gowalla				Foursquare			
	Rec@5	Rec@10	NDCG@5	NDCG@10	Rec@5	Rec@10	NDCG@5	NDCG@10
DRCF	0.1709	0.2192	0.1249	0.1406	0.2743	0.3466	0.1971	0.2206
DeepMove	0.1330	0.1656	0.1023	0.1128	0.3110	0.3831	0.2323	0.2556
STGN	0.1600	0.2041	0.1191	0.1333	0.2730	0.3547	0.1951	0.2217
LSTPM	0.1922	0.2442	0.1430	0.1597	0.3294	0.4065	0.2441	0.2688
STAN	0.3016	0.3998	0.2473	0.2621	0.4318	0.5186	0.3450	0.3671
ST-PEGD	0.3715	0.4191	0.3012	0.3166	0.4072	0.4856	0.3146	0.3387
STTF	0.3444	0.4294	0.2890	0.3358	0.5097	0.5934	0.4105	0.4365
RSTCN	**0.5385**	**0.5732**	**0.4498**	**0.4611**	**0.5969**	**0.6551**	**0.4864**	**0.5055**
%Improv.	44.95%	33.49%	49.34%	37.31%	17.11%	10.40%	18.49%	15.81%

Table 2. Analysis on Key Components in RTSCN.

Model	Gowalla				Foursquare			
	Rec@5	Rec@10	NDCG@5	NDCG@10	Rec@5	Rec@10	NDCG@5	NDCG@10
noRLLs	0.3851	0.4283	0.3211	0.3351	0.4733	0.5436	0.3758	0.3987
STCN	0.5260	0.55923	0.4401	0.4509	0.5454	0.6160	0.4404	0.4634
noSL	0.5140	0.5504	0.4314	0.4432	0.5162	0.5833	0.4170	0.4388
RSTCN-L	0.4875	0.5118	0.4145	0.4224	0.5540	0.6066	0.4526	0.4697
RSTCN-S	0.2411	0.2820	0.2031	0.2168	0.2757	0.3565	0.1916	0.2177
RSTCN	**0.5385**	**0.5732**	**0.4498**	**0.4611**	**0.5969**	**0.6551**	**0.4864**	**0.5055**

4.3 Ablation Study

As shown in Table 2, comparative experiments were conducted to evaluate the performance of RSTCN and its various variants. These results reflect the contributions of the different key components of RSTCN to its overall performance.

noRLLs is a variant that removes RLLs. **STCN** is a variant that removes the global residual connection. **noSL** is a variant that removes skip learning algorithm. **RSTCN-L** is a variant that only considers long-term user dependency. **RSTCN-S** is a variant that only considers short-term user dependency.

The results indicate that our RSTCN outperforms any of its variants, thus demonstrating the contributions of these key components to the overall performance. It is evident that the removal of RLLs results in poorer performance compared to other variants, emphasizing the crucial role of the RLL module. This is because RLLs can flatten the dynamical covariates, providing more detailed modeling of each time step, enhancing the temporal modeling capacity, and better utilizing the spatio-temporal collaborative signals.

Furthermore, we illustrate the importance of simultaneously considering long-term and short-term user dependencies by setting up variant models RSTCN-S

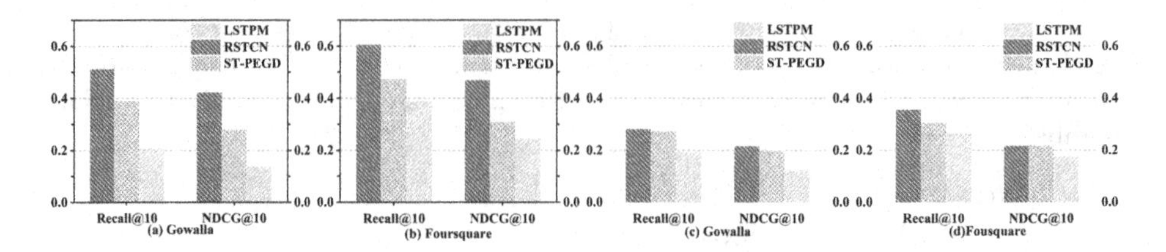

Fig. 2. Experimental results on two standard datasets. (a) and (b) is a comparison of LSTPM, ST-PEGD, and RSTCN for modeling long-term user dependency only. (c) and (d) is a comparison of LSTPM, ST-PEGD, and RSTCN for modeling short-term user dependency only.

and RSTCN-L. As shown in Fig. 2, we can observe that, concerning the modeling of long-term user dependencies, the improvements far surpass those related to short-term dependencies. To be more specific, in comparison to LSTPM [11], our long-term dependency module exhibits a significant enhancement in *Recall*@10. This increase reaches 143.71% on the Gowalla dataset and 55.06% on the Foursquare dataset. However, the *Recall*@10 improvement for the short-term dependency module is more modest, with a rate of 45.29% on Gowalla and 34.94% on Foursquare. These findings unequivocally establish the outstanding capability of RSTCN in capturing long-term user dependencies. The efficacy of RSTCN in modeling long-term dependencies is attributed to the encoding-decoding structure based on RLLs that we have designed. Conversely, the less effective performance of RSTCN-S can be ascribed to the substantially reduced spatio-temporal collaborative signals in the current trajectory compared to historical trajectories. In summary, RSTCN, a fusion of RSTCN-L and RSTCN-S, underscores the utility of considering both long- and short-term user dependencies for next POI recommendation.

RSTCN also takes into account geographical factors to model discrete check-ins, and the decrease in the effectiveness of the NoSL variant compared to RSTCN underscores the efficacy of the skip learning algorithm. This also demonstrates the necessity of considering discrete check-ins. Furthermore, the absence of a global residual connection in STCN may lead to the potential loss of signals, resulting in a decrease in recommendation performance.

5 Conclusions

In this paper, we introduce a novel RSTCN model. Specifically, we employ an attention network based on residual linear layers to model long-term user dependencies. Furthermore, we propose a skip learning algorithm that combines spatio-temporal collaborative signals to focus on discrete check-ins and capture short-term user dependencies. Through extensive experimental validation, our RSTCN model demonstrates outstanding recommendation performance, which is validated on two standard datasets. In the future, we will further investigate how to achieve superior results with simpler model designs to alleviate the computational burden on researchers.

References

1. Cheng, C., et al.: Where you like to go next: successive point-of-interest recommendation. In: IJCAI (2013)
2. Cheng, H., et al.: What's your next move: User activity prediction in location-based social networks. SIAM (2013)
3. Das, A., et al.: Long-term forecasting with tide: time-series dense encoder. arXiv preprint arXiv:2304.08424 (2023)
4. Feng, J., et al.: Deepmove: predicting human mobility with attentional recurrent networks. In: WWW, pp. 1459–1468 (2018)
5. Feng, S., et al.: Personalized ranking metric embedding for next new poi recommendation. In: IJCAI (2015)
6. Lan, P., et al.: Spatio-temporal position-extended and gated-deep network for next poi recommendation. In: Wang, X., et al. (eds.) DASFAA, pp. 505–520. Springer, Heidelberg (2023). https://doi.org/10.1007/978-3-031-30672-3_34
7. Liu, Q., et al.: Predicting the next location: A recurrent model with spatial and temporal contexts. In: AAAI, vol. 30 (2016)
8. Luo, Y., et al.: Stan: spatio-temporal attention network for next location recommendation. In: WWW, pp. 2177–2185 (2021)
9. Manotumruksa, J., et al.: A deep recurrent collaborative filtering framework for venue recommendation. In: CIKM, pp. 1429–1438 (2017)
10. Manotumruksa, J., et al.: A contextual attention recurrent architecture for context-aware venue recommendation. In: SIGIR, pp. 555–564 (2018)
11. Sun, K., et al.: Where to go next: modeling long-and short-term user preferences for point-of-interest recommendation. In: AAAI, vol. 34, pp. 214–221 (2020)
12. Vaswani, A., et al.: Attention is all you need. In: NIPS, vol. 30 (2017)
13. Xu, S., et al.: Spatio-temporal transformer recommender: next location recommendation with attention mechanism by mining the spatio-temporal relationship between visited locations. ISPRS Int. J. Geo-Inf. **12**(2), 79 (2023)
14. Yang, S., et al.: Getnext: trajectory flow map enhanced transformer for next poi recommendation. In: SIGIR, pp. 1144–1153 (2022)
15. Yin, H., et al.: Joint modeling of user check-in behaviors for real-time point-of-interest recommendation. TOIS **35**(2), 1–44 (2016)
16. Zhang, J.D., Chow, C.Y., Li, Y.: Lore: Exploiting sequential influence for location recommendations. In: Proceedings of the 22nd ACM SIGSPATIAL International Conference on Advances in Geographic Information Systems, pp. 103–112 (2014)
17. Zhao, P., et al.: Where to go next: a spatio-temporal gated network for next poi recommendation. IEEE Trans. Knowl. Data Eng. **34**(5), 2512–2524 (2020)
18. Zhou, X., et al.: Topic-enhanced memory networks for personalised point-of-interest recommendation. In: SIGKDD International Conference on Knowledge Discovery & Data Mining, pp. 3018–3028 (2019)

Conditional Denoising Diffusion for Sequential Recommendation

Yu Wang[1(✉)], Zhiwei Liu[2], Liangwei Yang[1], and Philip S. Yu[1]

[1] University of Illinois Chicago, Chicago, USA
{ywang617,lyang84,psyu}@uic.edu
[2] Salesforce AI Research, Palo Alto, USA
zhiweiliu@salesforce.com

Abstract. Contemporary attention-based sequential recommendations often encounter the oversmoothing problem, which generates indistinguishable representations. Although contrastive learning addresses this problem to a degree by actively pushing items apart, we still identify a new *ranking plateau* issue. This issue manifests as the ranking scores of top retrieved items being too similar, making it challenging for the model to distinguish the most preferred items from such candidates. This leads to a decline in performance, particularly in top-1 metrics. In response to these issues, we present a conditional denoising diffusion model that includes a stepwise diffuser, a sequence encoder, and a cross-attentive conditional denoising decoder. This approach streamlines the optimization and generation process by dividing it into simpler, more tractable sub-steps in a conditional autoregressive manner. Furthermore, we introduce a novel optimization scheme that incorporates both cross-divergence loss and contrastive loss. This new training scheme enables the model to generate high-quality sequence/item representations while preventing representation collapse. We conduct comprehensive experiments on four benchmark datasets, and the superior performance achieved by our model attests to its efficacy. We open-source our code at https://github.com/YuWang-1024/CDDRec.

Keywords: Sequential Recommendation · Diffusion Models · Generative Models

1 Introduction

Sequential Recommendation (SR) [10,13,24–26,28] has been intensively investigated because of its scalability and efficacy in capturing user temporal trends from histories. Recent research in SR focuses on attention-based methods for their promising results. Early attempts e.g., SASRec [10] and Bert4Rec [22] utilize the attention-based transformer structure. However, the attention mechanism tends to lead to a condition known as oversmoothing [5,6], which results in generating indistinguishable representations. Current methods predominantly

D.-N. Yang et al. (Eds.): PAKDD 2024, LNAI 14649, pp. 156–169, 2024.
https://doi.org/10.1007/978-981-97-2262-4_13

address this complication from the item representation perspective, utilizing contrastive learning. These methods effectively counter the collapse of item representation learning [17,26,27,29]. The DuoRec model [17] proposes to use sequences with the same predicted item as augmented views and implements the noise contrastive estimation objective for regularization. ContrastVAE [26] incorporates variational augmentation and the contrastELBO objective into the attention-based variational autoencoder for SR.

Despite the success of the above methods, we still observe a phenomenon we term the *ranking plateau*, characterized by indistinguishable ranking scores, even when the quality of item representation is commendable. Specifically, as shown in Sect. 4.1, the ranking scores of the top-40 retrieved items are too similar to allow the recommender to differentiate the best candidates among them. These oversmoothed ranking scores result in performance degradation, especially in top-1 metrics. For instance, the score assigned by DuoRec to the second-best item is only 1% lower than that of the top item. (We will discuss such phenomenon in detail in Sect. 4.1.) This suggests that the under-performance might not be solely due to item representation degeneration, but also to the complicated reasons beyond the user-item engagement. During our experiments, we observe significant shifts in some user intents from their historical records. For example, if a user regularly purchases *collection kits* but suddenly transitions to buying *printing-related* items, the representation of such sequences can be easily skewed by the user's past behaviors. This results in continuously recommending *collection-related* items, as the sequence representations are essentially a weighted sum of past item representations, regardless of the quality of candidate item representations.

Intuitively, if such dynamic intent transitions cannot be captured during the one-step dot-product, one might question whether it would be feasible to divide the transition process into easier and more tractable multi-steps such that the model could potentially correlate the intermediate transition steps and produce high-fidelity results progressively. For these desiderata, we turn to diffusion models [3,8,18] for solutions, as they break the higher-order complicated transitions into feasible sub-steps by removing certain noise stepwisely. Generally, the diffuser of diffusion models gradually adds a certain scale of Gaussian Noise to the data in the forward diffusion process, and the denoiser reconstructs such intermediate states by learning to remove the added noise in the reverse denoising process. In this way, the denoiser is able to learn fine-grained intermediate transitions from these multi-step generations.

However, it is rather challenging to incorporate such a learning paradigm into SR. One primary reason is that traditional diffusion models are designed for continuous spaces like image generation, where input features are fixed and contain substantial information. They are optimized by reconstructing **original** images. In contrast, SR involves item input information that is randomly initialized based on item IDs and dynamically optimized. Original reconstruction objective in discrete spaces could be adversely affected by representation collapse that all embeddings collapse to a trivial solution [3,4,11]. Furthermore, the SR is a retrieval task, which aims to generate user preferences reflecting the **next** item engagement. Merely reconstructing original item representations within a sequence (Gaussian Noise vector from the beginning) does not contribute to ranking performance but exacerbates the collapse issues of diffusion models.

To address these challenges, we propose **C**onditional **D**enoising **D**iffusion Models for Sequential **Rec**ommendation (CDDRec), including a stepwise diffuser, sequence encoder, cross-attentive conditional denoising decoder, and cross-divergence objective. The stepwise diffuser introduces noise into target item representations to construct corrupted targets, simulating the small stepwise noise in the sequences. The sequence encoder learns sequence representations from historical interactions, used as the conditioned information for a stepwise generation of next-engagement preference. The conditional denoising decoder aims to generate high-quality next-engagement representation by removing the noise of historical sequence representations step-by-step. To enhance the denoising decoder's awareness of each denoising step, we adopt the cross-attention mechanism with the denoising step as input. Additionally, We introduce a cross-divergence loss, enabling the model to construct high-fidelity sequence/item representations while being attuned to next-engagement preferences and preventing learning collapse. Furthermore, we leverage the In-view and Cross-view contrastive optimization to prevent item representation degeneration. Our contribution can be summarized as follows:

- To the best of our knowledge, we are the first to propose the novel conditional denoising diffusion models for sequential recommendation CDDRec in the conditional autoregressive generation paradigm.
- We first observe the *ranking plateau* issue and propose the multi-step next-engagement generation to address this issue.
- We introduce cross-divergence to equip the CDDRec with ranking capability.
- We conduct comprehensive experiments on the SR dataset, the substantial improvement across all metrics in four datasets indicates the effectiveness of CDDRec. We also conduct ablation studies to examine each key design's effectiveness further.

2 Related Work

Denoising Diffusion Probabilistic Models (DDPMs) have shown great success in continuous spaces, such as image generation [8,9,15,18]. Recently, several attempts have been made to apply DDPMs to discrete tasks, such as text generation. SUNDAE [21] is one of pioneers that use DDPMs for text generation. They introduce a step-unrolled denoising autoencoder that reconstructs corrupted sequences in a non-autoregressive manner. Diffusion-LM [11] gradually reconstructs word vectors from Gaussian noise guided by attribute classifiers and introduces a rounding process that maps continuous word embeddings to discrete words. DiffSeq [4] introduces a forward process with partial noise that uses the question of a dialog as the uncorrupted part and the answer of the dialog as the corrupted part and adds partial noise to the answer part during the forward pass. The backward pass reconstructs the answer in a non-autoregressive way.

There are also several concurrent attempts to introduce DDPMs for recommender systems. [23] introduces noise and reconstructs information for user interactions, and introduces L-DiffRec resembling latent diffusion, and T-DiffRec to encode temporal information to reweight user interactions respectively. [12] adds Gaussian noise on target items and reconstructs them through an approximator, which inputs the corrupted target item representations and historical interactions. [2] introduces the partial noise only on the target items that resemble DiffSeq and reconstructs them in a non-autoregressive way. Unlike the above methods, we introduce conditional generation in an autoregressive manner, which equips the model with the ability to generate high-fidelity sequence/item representations without too many generation steps.

3 Methodology

In this section, we present the methodology of CDDRec and illustrate it in Fig. 1. CDDRec consists of 1) stepwise diffuser that gradually corrupts target item embeddings via adding Gaussian noise; 2) sequence encoder that learns historical sequence representation, serving as conditional information for stepwise preference generation; 3) cross-attentive conditional denoising decoder that learns the stepwise user preference transition from conditional historical sequence representation to next target preference via stepwise removing noise from conditional sequence embeddings; 4) cross-divergence objective that enables the model with ranking capability while preventing model from collapsing.

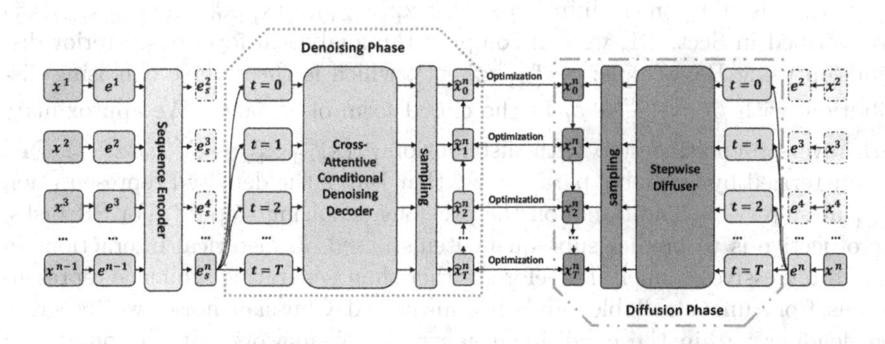

Fig. 1. Framework of CDDRec. Orange dots from top to bottom indicate the diffusion phase that gradually adds Gaussian noise to target item embeddings $\mathbf{x}_t^n$, while blue dots from the bottom up illustrate the reverse denoising phase that stepwisely removes noise from estimated user preference $\hat{\mathbf{x}}_t^n$ at step t. (Color figure online)

3.1 Stepwise Diffuser

As shown in Fig. 1 with the orange dot box from top to bottom, the stepwise diffuser is designed to incrementally introduce Gaussian noise to target item embeddings. This process creates the corrupted target for each step, thereby facilitating the denoise learning of the denoiser. Given the predefined noise scale added at diffusion step t: β_t and the corresponding diffusion transition distribution $q(\mathbf{x_t^n}|\mathbf{x_{t-1}^n}) \sim \mathcal{N}(\mathbf{x_t^n}; \sqrt{1-\beta_t}\mathbf{x_{t-1}^n}, \beta_t\mathbf{I})$, the distribution of the current diffusion step has the analytical form conditional on the first diffusion step:

$$q(\mathbf{x_t^n}|\mathbf{x_0^n}) = \mathcal{N}(\mathbf{x_t^n}; \sqrt{\bar{\alpha}_t}\mathbf{x_0^n}, (1-\bar{\alpha}_t)\mathbf{I}), \quad \alpha_t = 1-\beta_t, \quad \bar{\alpha}_t = \prod_{i=1}^{t}\alpha_i, \tag{1}$$

where $\mathbf{x_0^n} = \mathbf{e^n}$ is target item representation as the initialization of the diffusion phase. Thus, we can sample the corrupted target $\mathbf{x_t^n}$ at any step t using $\mathbf{x_0^n}$:

$$\mathbf{x_t^n} = \sqrt{\bar{\alpha}_t}\mathbf{x_0^n} + \sqrt{1-\bar{\alpha}_t}\epsilon, \quad \epsilon \sim \mathcal{N}(0, \mathbf{I}). \tag{2}$$

3.2 Sequence Encoder

Previous methods predominantly concentrate on denoising from a randomly initialized Gaussian noise and generating sentences non-autoregressively [3,4,11]. However, predicting the next item based on historical interaction records necessitates a conditional autoregressive generation in the SR. Consequently, in this paper, we utilize SASRec as our sequence encoder to learn hidden representations of historical interactions $\mathbf{e_s}$. These are used as the condition of the subsequent conditional denoising decoder for the multi-step preference generation.

3.3 Cross-Attentive Conditional Denoising Decoder

Given the distribution of diffusion step $q(\mathbf{x_t^n}|\mathbf{x_{t-1}^n})$, $q(\mathbf{x_t^n}|\mathbf{x_0^n})$, and $q(\mathbf{x_{t-1}^n}|\mathbf{x_0^n})$ as described in Sect. 3.1, we can compute the analytical form of posterior distribution using Bayes' rule: $q(\mathbf{x_{t-1}^n}|\mathbf{x_t^n}, \mathbf{x_0^n})$, which is the reverse denoising distribution, with $\hat{\beta}_t = \frac{1-\bar{\alpha}_{t-1}}{1-\bar{\alpha}_t}\beta_t$ as the closed form of variance. We approximate such reverse denoising step with distribution $p_\theta(\hat{\mathbf{x}}_{\mathbf{t-1}}^{\mathbf{n}}|\hat{\mathbf{x}}_{\mathbf{t}}^{\mathbf{n}}) \sim \mathcal{N}(\mu_\theta(\hat{\mathbf{x}}_{\mathbf{t}}^{\mathbf{n}}, t), \hat{\beta}_t\mathbf{I})$, parameterized by learnable parameter θ that learns the denoised representation $\hat{\mathbf{x}}_{\mathbf{t-1}}^{\mathbf{n}}$ at step $t-1$ conditional on the previous denoising step $\hat{\mathbf{x}}_{\mathbf{t}}^{\mathbf{n}}$. For SR tasks, the objective is to predict subsequent items based on historical interactions in an autoregressive manner. Therefore, rather than generating sequence representations from uncontrollable randomly initialized Gaussian noise, we integrate the denoiser within the conditional generation framework with a conditional denoising decoder. Consequently, as shown in Fig. 1 blue dot box from bottom up, we condition the reverse denoising phase on the preceding sequence representations, expressed as $p_\theta(\hat{\mathbf{x}}_{\mathbf{t}}|\mathbf{e_s}, t) \sim \mathcal{N}(\mu_\theta(\mathbf{e_s}, t), \hat{\beta}_{t+1}\mathbf{I})$, where $\mathbf{e_s}$ denotes the encoded historical interactions using the sequence encoder. Given such probabilistic modeling, we will discuss the corresponding model design to learn the denoised mean $\mu_\theta(\mathbf{e_s}, t)$.

In contrast to earlier methods [19] that only maintain the final position's representation as the sequence representation, we strive to preserve as much information as possible due to the sparse nature of SR. Hence, we select a cross-attention architecture as the denoising decoder instantiation, which is capable of taking the entire sequence representation and corresponding step indicator as input. Formally, given a sequence embedding $\mathbf{e_s^{1:n}}$ and the corresponding denoising step t, the conditional denoising decoder is designed to predict the denoised mean of corrupted target item embedding at the corresponding diffusion step. Initially, we acquire a learnable embedding $\mathbf{e_t}$ for the indicator t from a step lookup embedding table and expand it to the dimension of $\mathcal{R}^{(n-1)\times d}$, ensuring that every previously hidden embedding is conscious of the same denoising step. We define the cross-attention (CA) as follows:

$$\mu_\theta^{\mathbf{1:n}}(\mathbf{e_s^{1:n}}, t) = CA(\mathbf{e_s^{1:n}}, \mathbf{e_t}) = \text{Softmax}\left(\frac{(\mathbf{e_t}\mathbf{W}^Q)(\mathbf{e_s^{1:n}}\mathbf{W}^K)^\top}{\sqrt{d}}\right)(\mathbf{e_s^{1:n}}\mathbf{W}^V). \tag{3}$$

Given the predicted denoised mean and the precomputed posterior variance, we can sample the generated user preference at step t:

$$\hat{\mathbf{x}}_\mathbf{t}^\mathbf{n} = \mu_\theta^\mathbf{n} + \hat{\beta}_{t+1}\epsilon, \quad \epsilon \sim \mathcal{N}(0, \mathbf{I}). \tag{4}$$

3.4 Optimization

Traditional DDPM is designed to reconstruct an image by removing the Gaussian noise added to it. Consequently, the objective is to learn the denoising function $p_\theta(x_{t-1}|x_t)$ for the corresponding step, minimizing the KL divergence $D_{KL}[q(x_{t-1}|x_t, x_0)||p_\theta(x_{t-1}|x_t)]$ at each step. Such an objective maximizes the similarity between the predicted and corrupted input data. However, since all item embeddings are randomly initialized and optimized dynamically in SR, the model may learn trivial item representations, where every pair of item embeddings is highly similar, resulting in high-ranking scores for all items. Furthermore, the SR is a retrieval task requiring the model to effectively rank items, giving higher scores to target items over non-interest items. Merely reconstructing the input sequence does not contribute to the next-engagement prediction.

To circumvent these issues, we require the KL divergence between the predicted and target item embeddings to be smaller than that between the predicted and negative item embeddings. Consequently, we introduce the cross-divergence loss using KL-divergence as a dissimilarity metric at each denoising step t:

$$\begin{aligned}\mathcal{L}_{cd}^t = &\frac{1}{N}\sum_n \log(\sigma(-D_{KL}[q(\mathbf{x_t^n}|\mathbf{x_{t+1}^n}, \mathbf{x_0^n})||p_\theta(\hat{\mathbf{x}}_\mathbf{t}^\mathbf{n}|\mathbf{e_s}, t)])) \\ &+ \log(1 - \sigma(-D_{KL}[q(\mathbf{x_t'^n}|\mathbf{x_{t+1}'^n}, \mathbf{x_0'^n})||p_\theta(\hat{\mathbf{x}}_\mathbf{t}^\mathbf{n}|\mathbf{e_s}, t)]))],\end{aligned} \tag{5}$$

where $\mathbf{x_0'^n}$ is the embedding of a randomly sampled negative item that has never appeared in the user history. We sample both corrupted target item embeddings $\mathbf{x_t^n}$ and generated user preferences $\hat{\mathbf{x}}_\mathbf{t}^\mathbf{n}$ according to Eq. 2 and Eq. 4.

Contrastive Loss. To endow the model with robustness against the noisy interactions and prevent item representation from collapsing, we incorporate a simple yet effective in-view and cross-view contrastive learning using InfoNCE loss [16].

The in-view InfoNCE minimizes the distance between user preferences and target item embedding while enlarging inter-users/inter-item distance:

$$\mathcal{L}_{in}^{t} = \frac{1}{N}\sum_{i=1}^{N} \log \frac{\exp(\hat{\mathbf{x}}_t^{i\top}\mathbf{x}_t^i/\tau)}{\sum_j \exp(\hat{\mathbf{x}}_t^{i\top}\mathbf{x}_t^j/\tau) + \sum_j \mathbb{1}_{[j\neq i]} \exp(\hat{\mathbf{x}}_t^{i\top}\hat{\mathbf{x}}_t^j/\tau)}, \tag{6}$$

where $\hat{\mathbf{x}}_\mathbf{t}^\mathbf{i}$ is the output of conditional denoising decoder, $\mathbf{x}_\mathbf{t}^\mathbf{i}$ is the output of stepwise diffuser at diffusion step t of position i in the sequence.

The cross-view InfoNCE ensures the sequence encoder generates reasonable sequence representation. It achieves this by minimizing the distance between the same input sequence with a slight noise interpolation while pushing the in-batch sequence representation away from each other:

$$\mathcal{L}_{cross}^{t} = \frac{1}{N}\sum_{i=1}^{N} \log \frac{\exp(\hat{\mathbf{x}}_t^{i\top}\tilde{\mathbf{x}}_t^i/\tau)}{\sum_j \exp(\hat{\mathbf{x}}_t^{i\top}\tilde{\mathbf{x}}_t^j/\tau) + \sum_j \mathbb{1}_{[j\neq i]} \exp(\hat{\mathbf{x}}_t^{i\top}\hat{\mathbf{x}}_t^j/\tau)}, \tag{7}$$

where $\tilde{\mathbf{x}}_t^i$ is the output of conditional denoising decoder at diffusion step t, position i of augmented view.

Step-Adaptive Objective. Since the noise added to target item embeddings increases as the diffusion phase progresses, more information is lost at higher diffusion steps. Intuitively, to avoid focusing too much on reconstructing non-informative noise, we rescale the loss term of each diffusion step by dividing it by the corresponding step indicator. Furthermore, unlike previous methods that randomly sample step indicators for optimization, we explicitly calculate the loss term for every diffusion step. The final optimization objective is formalized as:

$$\mathcal{L} = \sum_{t=0}^{T} \frac{1}{t+1}(\mathcal{L}_{cd}^{t} + \lambda(\mathcal{L}_{in}^{t} + \mathcal{L}_{cross}^{t})). \tag{8}$$

4 Experiments

Dataset. In this paper, we conduct experiments on four Amazon datasets [14]: *Office*, *Beauty*, *Tools and Home*, and *Toys and Games*. In line with common practice [10,22,29], for each user, we sort the interactions chronologically. We use the penultimate, last records as validation and test datasets, while all preceding records as train datasets. We report the statistics of the datasets in Table 1.

Table 1. Statistics of datasets.

Dataset	#Users	#Items	#Interactions	#Ints/item	Avg. seq. len.
Beauty	22,363	12,101	198,502	16.40	8.3
Toys	19,412	11,924	167,597	14.06	8.6
Tools	16,638	10,217	134,476	13.16	8.1
Office	4,905	2,420	53,258	22.00	10.8

Baseline Models. We compare CDDRec with these three related types of state-of-the-art (SOTA) methods: Generative Models: **SAVE** [20], **ACVAE** [28], **ContrastVAE** [26]. SVAE first introduces VAE into the SR. ACVAE introduces the concept of adversarial variational Bayes and mutual information maximization to optimize the VAE. ContrastVAE introduces the objective named ContrastELBO to maximize the mutual information among latent variables. Contrastive Models: **CL4Rec** [27], **DuoRec** [17], **CBiT** [1]. CL4Rec introduces the data augmentation strategies: mask, shuffle, and crop, and optimizes the model via InfoNCE [16] loss. DuoRec improves the performance via semantic augmentation considering sequences with the same target items as the positive views. CBiT improves Bert4Rec by introducing the additional InfoNCE objective. Encoder Models: **GRU4Rec** [7], **SASRec** [10], **FMLP** [30]. GRU4Rec first attempts the RNN for the SR, while SASRec first employs a transformer-based encoder for SR. FMLP replaces the multi-head self-attention layer of SASRec with the denoising Fourier layer.

Metrics. To evaluate the performance of our model, we employ ranking-related evaluation metrics, including Recall@N, NDCG@N, and MRR, following common practice [10,26,27].

4.1 Plateau of Ranking Prediction

As previously mentioned, traditional generative models often encounter *ranking plateau*, where the ranking scores of top-40 candidate items are too similar. This smoothness makes it difficult for models to distinguish the best from these candidates, resulting in degraded top-1 metrics. To study such a phenomenon, we conduct experiments comparing the average absolute percentage change (Avg.Change) of the top-40 ranking scores. The metric is defined as follows:

$$\text{Avg.Change} = \sum_{i=1}^{N} \frac{1}{N-1} \frac{|rank_{i+1} - rank_i|}{rank_i} \times 100, \tag{9}$$

where *rank* is the ranking score calculated using dot-product between predicted and candidate item embeddings.

We utilize this metric to evaluate the descending speed of the ranking scores, which can reflect the smoothness of the ranking prediction. We report the results in Fig. 2. In general, baseline models tend to provide more similar ranking scores among the top-40 candidates. Interestingly, we observe a positive correlation between Recall@1 and Avg.Change when comparing Avg.Change on 'ALL' sequences. Specifically, the Recall@1 is 0.012, 0.0194, 0.0224, and 0.0271 for DuoRec, ContrastVAE, FMLP, and CDDRec, and the Avg.Change is 0.99%, 1.0928%, 1.2269%, and 5.7765% respectively. The Avg.Change of CDDRec decreases with increasing sequence length, indicating that the model is less certain for longer sequences. We also evaluate the Avg.Change w.r.t. the denoising stage (inverse process w.r.t diffusion step t, i.e., denoising stage is $T-t$). One observation is that the Avg.Change increases with the denoising stage. At the beginning of the denoising stage, the model shows uncertainty in ranking predictions, but it gradually gains clarity as the denoising phase progressively removes noise from the preference predictions. This also reveals the relationship between the noise level in the generated user preference and the ranking score smoothness. Specifically, a noisier preference correlates with a smoother ranking score, which supports our intuition of adding a denoiser after the sequence encoder.

4.2 Overall Experiments

In this paper, we conduct a comprehensive comparison between CDDRec and SOTA models, reporting the numerical results in Table 2. Our model CDDRec consistently outperforms others on these four datasets, demonstrating the effectiveness of CDDRec. Specifically, in terms of Recall@1, CDDRec shows substantial improvements with gains of 20.98%, 16.67%, 17.59%, and 18.42% compared to the second-best models on Office, Beauty, Tools, and Toys, respectively. We attribute these improvements to the high-quality next-item-engagement representations generated by CDDRec. The sequence representation generated from the multi-step denoising process can reveal the user preference from a fine-grained level, thus, being more distinguishable among top-rated candidate items. This approach avoids the *ranking plateau* phenomenon, and obtaining high performance w.r.t top-1 metrics and MRR. On the contrary, baseline methods

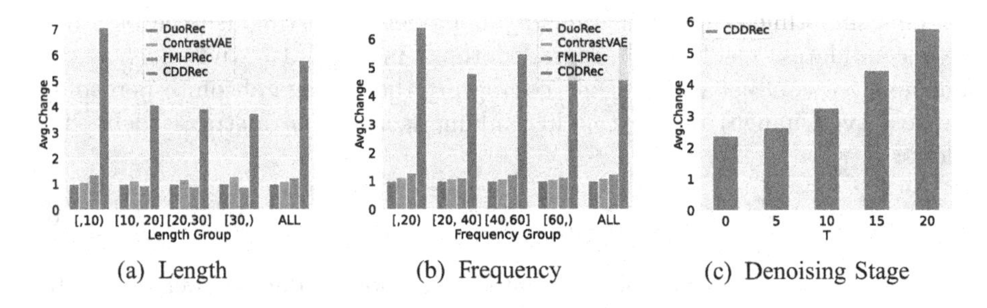

(a) Length (b) Frequency (c) Denoising Stage

Fig. 2. The Avg.Change for CDDRec and baseline methods across various subset sequences and denoising stage on Office dataset.

Table 2. Overall Comparison. The best is bolded, and the runner-up is underlined

Dataset	Metric	SVAE	ACVAE	ContrastVAE	CL4Rec	DuoRec	CBiT	GRU4Rec	SASRec	Bert4Rec	FMLP	CDDRec	Imp
Office	R@1	0.0088	0.0139	0.0194	0.0094	0.0120	0.0198	0.0051	0.0198	0.0137	0.0224	**0.0271**	20.98%
	R@5	0.0316	0.0457	0.0642	0.0294	0.0330	0.0593	0.0241	0.0656	0.0485	0.0593	**0.0765**	16.62%
	R@10	0.0597	0.0742	0.1052	0.0430	0.0559	0.0917	0.0510	0.0989	0.0848	0.0901	**0.1091**	3.71%
	N@5	0.0202	0.0300	0.0411	0.0194	0.0223	0.0396	0.0149	0.0428	0.0309	0.0414	**0.0521**	21.73%
	N@10	0.0292	0.0392	0.0544	0.0237	0.0296	0.0500	0.0234	0.0534	0.0426	0.0513	**0.0627**	15.26%
	MRR	0.0249	0.0351	0.0463	0.0207	0.0264	0.0437	0.0204	0.0457	0.0408	0.0455	**0.0548**	18.36%
Beauty	R@1	0.0014	0.0167	0.0161	0.0045	0.0107	0.0174	0.0079	0.0129	0.0119	0.0154	**0.0203**	16.67%
	R@5	0.0068	0.0428	0.0491	0.0160	0.0278	0.0512	0.0266	0.0416	0.0396	0.0433	**0.0542**	5.86%
	R@10	0.0127	0.0606	0.0741	0.0250	0.0403	0.0762	0.0421	0.0633	0.0595	0.0627	**0.0770**	1.05%
	N@5	0.0041	0.0299	0.0327	0.0103	0.0193	0.0343	0.0172	0.0274	0.0257	0.0297	**0.0376**	9.62%
	N@10	0.0060	0.0356	0.0407	0.0131	0.0233	0.0424	0.0222	0.0343	0.0321	0.0360	**0.0447**	5.42%
	MRR	0.0046	0.0310	0.0345	0.0111	0.0201	0.0359	0.0191	0.0291	0.0294	0.0305	**0.0387**	7.80%
Tools	R@1	0.0055	0.0090	0.0108	0.0060	0.0058	0.0066	0.0047	0.0103	0.0059	0.0089	**0.0127**	17.59%
	R@5	0.0118	0.0242	0.0315	0.0189	0.0182	0.0214	0.0154	0.0284	0.0189	0.0251	**0.0359**	13.97%
	R@10	0.0204	0.0364	0.0483	0.0293	0.0361	0.0347	0.0242	0.0427	0.0319	0.0359	**0.0522**	8.07%
	N@5	0.0086	0.0166	0.0212	0.0123	0.0120	0.0139	0.0102	0.0194	0.0123	0.0170	**0.0244**	15.09%
	N@10	0.0114	0.0206	0.0266	0.0156	0.0148	0.0182	0.0129	0.0240	0.0165	0.0204	**0.0297**	11.65%
	MRR	0.0098	0.0178	0.0227	0.0132	0.0128	0.0154	0.0113	0.0207	0.0160	0.0174	**0.0253**	11.45%
Toys	R@1	0.0022	0.0156	0.0228	0.0067	0.0099	0.0195	0.0066	0.0193	0.0110	0.0189	**0.0270**	18.42%
	R@5	0.0057	0.0349	0.0591	0.0180	0.0258	0.0525	0.0226	0.0551	0.0300	0.0516	**0.0665**	12.52%
	R@10	0.0098	0.0492	0.0823	0.0259	0.0360	0.0747	0.0363	0.0797	0.0466	0.0674	**0.0935**	13.61%
	N@5	0.0038	0.0255	0.0414	0.0124	0.0179	0.0364	0.0148	0.0377	0.0206	0.0357	**0.0472**	14.01%
	N@10	0.0038	0.0301	0.0489	0.0149	0.0212	0.0435	0.0192	0.0456	0.0260	0.0408	**0.0559**	14.31%
	MRR	0.0044	0.0270	0.0422	0.0132	0.0182	0.0373	0.0165	0.0385	0.0244	0.0347	**0.0479**	13.51%

encounter *ranking plateau*, where the ratings of top-rate items are indistinguishable. From another perspective, metrics like NDCG@5, NDCG@10, and MRR, which take the ranking position of target items into account, show impressive improvements. This indicates that our model CDDRec ranks target items relatively higher than other models.

4.3 Ablation Study

In this section, we conduct experiments to examine the contributions of the denoising and diffusion phases. Notably, the model is designed to predict the denoised mean of the corrupted target item embedding. By employing this sampling procedure (Eq. 4) with the predicted mean, the model is able to mimic the corrupted target items. Accurate prediction of the mean for these perturbed items, devoid of noise, confers the denoising ability upon the model. On the other hand, the diffusion step also follows a sampling process as described by Eq. 2. To scrutinize the impact of these dual processes, we substitute the two sampling steps with the predicted mean and the original item embedding, respectively, and present the experimental findings in Table 3.

Firstly, the diffusion and denoising processes generally contribute positively to the overall performance across all datasets, as evidenced by the performance decline in comparison to our model CDDRec. Moreover, the significance of diffusion and denoising varies among datasets. Specifically, the denoising process demonstrates greater importance in the Office and Toys datasets, while the diffusion phase is more crucial for the Beauty and Tools datasets, when we compare the performance drop within each row. An explanation for this variation could be attributed to the differences in sequence length across the datasets. Office and Toys datasets exhibit relatively longer sequences, which could result in a higher likelihood of noisy interactions, thereby rendering the sampling process on the target sequence less effective. Conversely, when dealing with shorter sequences, the diffusion phase that introduces noise may serve as an augmentation strategy, bolstering the model's robustness to noisy interactions.

Table 3. Ablation Study

	-Diffusion			-Denoising			CDDRec	
	R@1	MRR	avg.drop	R@1	MRR	avg.drop	R@1	MRR
Office	0.0236	0.0496	11.20%	0.0222	0.0490	14.33%	**0.0271**	**0.0548**
Beauty	0.0154	0.0342	17.88%	0.0179	0.0365	8.75%	**0.0203**	**0.0387**
Tools	0.0112	0.0226	11.24%	0.0116	0.0243	6.31%	**0.0127**	**0.0253**
Toys	0.0267	0.0464	2.12%	0.0264	0.0462	2.89%	**0.0270**	**0.0479**

Table 4. Case Study

Denoising Stage	10	15	20
Rank 1	Swivel Tower Sorter	Swivel Tower Sorter	**Wristbands**
Rank 2	Desk Tray	Paper Clip Holder	Erase Markers
Rank 3	Paper Clip Holder	Desk Sorter	Pencil Sharpener
Rank 4	Desk Sorter	Desk Tray	Graphite Pencils
Rank of target item	>40	7	1

4.4 Hyperparameter Sensitivity

In this section, we investigate the sensitivities of CDDRec's hyperparameters. Due to space constraints, we focus on reporting the experimental results for key hyperparameters, including the maximum diffusion step and maximum noise schedule. As depicted in Fig. 3, the optimal maximum diffusion steps are 10 and 30 for Office and Beauty datasets, respectively. The possible reason is that the Beauty dataset has a higher number of items, presenting a greater challenge for CDDRec to learn meaningful item embeddings. Consequently, a greater number of denoising steps is required to refine intermediate states of item representations. We observe that the optimal maximum noise levels for Office and Beauty are 0.04 and 0.1, respectively. A possible explanation is that longer sequences may inherently contain noisy interactions, thus necessitating less added noise.

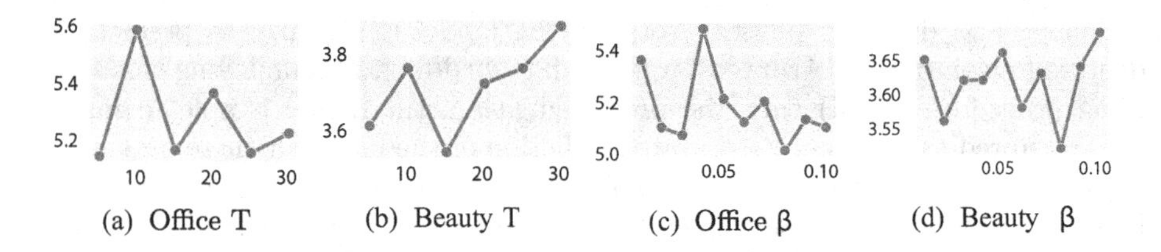

(a) Office T (b) Beauty T (c) Office β (d) Beauty β

Fig. 3. The evaluation of CDDRec on MRR through two datasets with different maximum diffusion step T and noise schedule β.

4.5 Case Study for Stepwise Generation

In the effort to demystify the intermediate generation of CDDRec, we execute item retrieval utilizing the intermediate-generated preference and item representations. The top-4 retrieved items along with the rank of the actual target item are presented in Table 4. Prior interacted items encompass *Desk Organizer, Organization Cube, Pencil Cup, Wristbands*, with the target item also being wristbands. A notable observation from history is that the sequence is dominated by collection-related kits, with a sudden shift in user engagement towards wristbands. After the initial ten denoising stages, the system continues to suggest collection kits, and the target item's rank falls outside the top 40. As the system undergoes more stepwise denoising, CDDRec begins to acknowledge the significance of the most recent purchase behavior, subsequently improving the ranking of the target items. After 20 stages of denoising, CDDRec manages to place the target item at the top position, and the recommended items display greater diversity, including items such as *pencil sharpener, graphite pencil*, etc.

5 Conclusion

In summary, we highlight the *ranking plateau* issue and underline the importance of stepwise generation as an effective solution. We introduce CDDRec, a model characterized by a cross-attentive conditional denoising decoder. This decoder makes use of the denoising step indicator and sequence encoder output as input, predicting the denoised mean at each denoising step. We also propose the cross-divergence objective with contrastive loss, tailored for sequence recommendation. These objectives guard against representation collapse while enabling the model to exhibit ranking capacity. Consequently, CDDRec can generate high-fidelity sequence/item representations and provide fine-grained ranking predictions, thus addressing *ranking plateau* issues. Thorough experimental results indicate CDDRec's superior performance, outshining contemporary SOTA methods, especially in top-1 metrics.

Acknowledgement. This work is supported in part by NSF under grant III-2106758.

References

1. Du, H., et al.: Contrastive learning with bidirectional transformers for sequential recommendation. In: CIKM (2022)
2. Du, H., Yuan, H., Huang, Z., Zhao, P., Zhou, X.: Sequential recommendation with diffusion models. arXiv preprint arXiv:2304.04541 (2023)
3. Gao, Z., et al.: Difformer: empowering diffusion model on embedding space for text generation. arXiv preprint arXiv:2212.09412 (2022)
4. Gong, S., Li, M., Feng, J., Wu, Z., Kong, L.: DiffuSeq: sequence to sequence text generation with diffusion models. arXiv preprint arXiv:2210.08933 (2022)
5. Guo, X., Wang, Y., Du, T., Wang, Y.: ContraNorm: a contrastive learning perspective on oversmoothing and beyond. arXiv preprint arXiv:2303.06562 (2023)
6. He, J., Cheng, L., Fang, C., Zhang, D., Wang, Z., Chen, W.: Mitigating undisciplined over-smoothing in transformer for weakly supervised semantic segmentation. arXiv preprint arXiv:2305.03112 (2023)

7. Hidasi, B., Karatzoglou, A.: Recurrent neural networks with top-k gains for session-based recommendations. In: Proceedings of the 27th ACM International Conference on Information and Knowledge Management, CIKM 2018, Torino, Italy, 22–26 October 2018, pp. 843–852. ACM (2018)
8. Ho, J., Jain, A., Abbeel, P.: Denoising diffusion probabilistic models. Adv. Neural. Inf. Process. Syst. **33**, 6840–6851 (2020)
9. Huang, C.W., Lim, J.H., Courville, A.C.: A variational perspective on diffusion-based generative models and score matching. Adv. Neural. Inf. Process. Syst. **34**, 22863–22876 (2021)
10. Kang, W., McAuley, J.J.: Self-attentive sequential recommendation. In: IEEE International Conference on Data Mining, ICDM 2018, Singapore, 17–20 November 2018, pp. 197–206. IEEE Computer Society (2018)
11. Li, X., Thickstun, J., Gulrajani, I., Liang, P.S., Hashimoto, T.B.: Diffusion-LM improves controllable text generation. Adv. Neural. Inf. Process. Syst. **35**, 4328–4343 (2022)
12. Li, Z., Sun, A., Li, C.: DiffuRec: a diffusion model for sequential recommendation. arXiv preprint arXiv:2304.00686 (2023)
13. Liu, Z., Fan, Z., Wang, Y., Yu, P.S.: Augmenting sequential recommendation with pseudo-prior items via reversely pre-training transformer. In: Proceedings of the 44th International ACM SIGIR Conference on Research and Development in Information Retrieval, pp. 1608–1612 (2021)
14. McAuley, J.J., Targett, C., Shi, Q., van den Hengel, A.: Image-based recommendations on styles and substitutes. In: Proceedings of the 38th International ACM SIGIR Conference on Research and Development in Information Retrieval, Santiago, Chile, 9–13 August 2015, pp. 43–52. ACM (2015)
15. Nichol, A.Q., Dhariwal, P.: Improved denoising diffusion probabilistic models. In: International Conference on Machine Learning, pp. 8162–8171. PMLR (2021)
16. Oord, A.v.d., Li, Y., Vinyals, O.: Representation learning with contrastive predictive coding. arXiv preprint arXiv:1807.03748 (2018)
17. Qiu, R., Huang, Z., Yin, H., Wang, Z.: Contrastive learning for representation degeneration problem in sequential recommendation. In: WSDM 2022: The Fifteenth ACM International Conference on Web Search and Data Mining, Virtual Event/Tempe, AZ, USA, 21–25 February 2022, pp. 813–823. ACM (2022)
18. Ramesh, A., Dhariwal, P., Nichol, A., Chu, C., Chen, M.: Hierarchical text-conditional image generation with clip latents. arXiv preprint arXiv:2204.06125 (2022)
19. Rasul, K., Seward, C., Schuster, I., Vollgraf, R.: Autoregressive denoising diffusion models for multivariate probabilistic time series forecasting. In: International Conference on Machine Learning, pp. 8857–8868. PMLR (2021)
20. Sachdeva, N., Manco, G., Ritacco, E., Pudi, V.: Sequential variational autoencoders for collaborative filtering. In: Proceedings of the Twelfth ACM International Conference on Web Search and Data Mining, WSDM 2019, Melbourne, VIC, Australia, 11–15 February 2019, pp. 600–608. ACM (2019)
21. Savinov, N., Chung, J., Binkowski, M., Elsen, E., Oord, A.v.d.: Step-unrolled denoising autoencoders for text generation. arXiv preprint arXiv:2112.06749 (2021)
22. Sun, F., et al.: BERT4Rec: sequential recommendation with bidirectional encoder representations from transformer. In: Proceedings of the 28th ACM International Conference on Information and Knowledge Management, CIKM 2019, Beijing, China, 3–7 November 2019, pp. 1441–1450. ACM (2019)
23. Wang, W., Xu, Y., Feng, F., Lin, X., He, X., Chua, T.S.: Diffusion recommender model. arXiv preprint arXiv:2304.04971 (2023)

24. Wang, Y., Liu, Z., Zhang, J., Yao, W., Heinecke, S., Yu, P.S.: DRDT: dynamic reflection with divergent thinking for LLM-based sequential recommendation. arXiv preprint arXiv:2312.11336 (2023)
25. Wang, Y., et al.: Exploiting intent evolution in e-commercial query recommendation. In: Proceedings of the 29th ACM SIGKDD Conference on Knowledge Discovery and Data Mining, pp. 5162–5173 (2023)
26. Wang, Y., Zhang, H., Liu, Z., Yang, L., Yu, P.S.: ContrastVAE: contrastive variational autoencoder for sequential recommendation. In: Proceedings of the 31st ACM International Conference on Information & Knowledge Management, pp. 2056–2066 (2022)
27. Xie, X., et al.: Contrastive learning for sequential recommendation. arXiv preprint arXiv:2010.14395 (2020)
28. Xie, Z., Liu, C., Zhang, Y., Lu, H., Wang, D., Ding, Y.: Adversarial and contrastive variational autoencoder for sequential recommendation. In: WWW 2021: The Web Conference 2021, Virtual Event/Ljubljana, Slovenia, 19–23 April 2021, pp. 449–459. ACM/ IW3C2 (2021)
29. Zhou, K., et al.: S3-Rec: self-supervised learning for sequential recommendation with mutual information maximization. In: CIKM 2020: The 29th ACM International Conference on Information and Knowledge Management, Virtual Event, Ireland, 19–23 October 2020, pp. 1893–1902. ACM (2020)
30. Zhou, K., Yu, H., Zhao, W.X., Wen, J.R.: Filter-enhanced MLP is all you need for sequential recommendation. In: Proceedings of the ACM Web Conference 2022, pp. 2388–2399 (2022)

UIPC-MF: User-Item Prototype Connection Matrix Factorization for Explainable Collaborative Filtering

Lei Pan[1(✉)] and Von-Wun Soo[1,2]

[1] National Tsing Hua University, Hsinchu, Taiwan
parrotlet@gapp.nthu.edu.tw
[2] Chang Gung University, Taoyuan City, Taiwan
soo@cgu.edu.tw

Abstract. In recent years, prototypes have gained traction as an interpretability concept in the Computer Vision Domain, and have also been explored in Recommender System algorithms. This paper introduces UIPC-MF, an innovative prototype-based matrix factorization technique aimed at offering explainable collaborative filtering recommendations. Within UIPC-MF, both users and items link with prototype sets that encapsulate general collaborative features. UIPC-MF uniquely learns connection weights, highlighting the relationship between user and item prototypes, offering a fresh method for determining the final predicted score beyond the conventional dot product. Comparative results show that UIPC-MF surpasses other prototype-based benchmarks in Hit Ratio and Normalized Discounted Cumulative Gain across three datasets, while enhancing transparency.

Keywords: collaborative filtering · prototype · explainable recommender system

1 Introduction

Recommending items to users is vital in commerce, presenting challenges in both accuracy and explainability. While many collaborative filtering methods, rooted in intricate models, deliver high performance, their sophisticated architectures often fall short in providing transparent explanations for their recommendations.

Prototypes are representative examples that involve the learning of latent representations from the data space that were widely adopted in interpretable machine learning [16,24] and had extensive applications in computer vision [5,10,18]. These approaches leverage prototypes to extract significant patterns from large images and determine similarity scores between a specific photo and learned prototypes in order to facilitate effective identification. The features of the prototypes can serve as the interpretable basis for a group of typical objects.

However, selecting or identifying the proper prototypes can be critical in both accuracy and interpretability in recommendation. We propose User-Item

D.-N. Yang et al. (Eds.): PAKDD 2024, LNAI 14649, pp. 170–181, 2024.
https://doi.org/10.1007/978-981-97-2262-4_14

Prototypes Connections Matrix Factorization (UIPC-MF), a novel prototype-based collaborative filtering algorithm. UIPC-MF learns two sets of prototypes for both users and items from large interaction data, with each prototype capturing a common attribute derived from collective wisdom. A predicted logit score is computed through a linear combination of user similarities between a user representation and each user prototype, as well as item similarities between an item representation and each item prototype. The score computation is conducted based on the connection weights among the user prototypes and item prototypes. This novel approach allows transparent understanding of how the predicted logit score is composed, offering more insights than other prototype-based methods. The transparency makes the recommender systems explainable as it enables the model designers to understand the recommender system's underlying operations and effects, and thus can provide users with rationales of a recommendation [2,6,21].

Another contribution is our incorporation of an L1 normalization term with prototypes. This combination aims to mitigate most biases and enhance accuracy.

The objectives of the paper focus on the issues around learning the models of the prototype representation connection mapping between the user space and item space in order to make accurate and transparent recommendations given user interaction history data. They can be formulated as the following three research questions:

- **RQ1.**How does the performance of our recommendation model compare with the state-of-the-art baseline model?
- **RQ2.**How can we understand the underlying inference of our model and to generate explanation rationales for the recommendations?
- **RQ3.** How can we setup the loss objective functions as well as proper regularization term in avoid of the potential learning bias of the model in the optimization of the parameters of the model against the given training data?

2 Related Work

2.1 Collaborative Filtering

The collaborative filtering based models tracking the user's history or feedback (e.g. ratings or clicks) to evaluate user's preference and relations to items.The most popular technique is the matrix factorization (MF) methods [11] which the user-item interactions construct an original user-item interaction matrix.

Recently, deep learning models have been widely used in recommendation systems due to their high performance in learning implicit representations. [8,9, 13] These systems utilize deep learning models to learn user/item representations and interaction functions to estimate the user-item scores.

However, these algorithms are hard to understand the underlying inference mechanisms for the final recommendation and thus become explanation opague. In order to make the correlation calculation more explicit and interpretable, we propose prototype-based collaborative filtering method to be discussed in the following section.

2.2 Explainable and Transparent Recommender Models

Transparency of a recommender system is important to allow users or developers to understand the rationales of underlying operations or inferences of the recommender system [2,6,21]. An explainable recommender system refers to generating explanations for the predicted outcomes of the recommender system. An explainable recommender system can provide users and developers with a reason for why a particular recommendation was made. An explainable system may not necessarily be totally transparent.

The explainable recommender models can be categorized into two types based on their levels of transparency [23]. The first is the model-intrinsic approach [16], where the recommendation mechanism is transparent by design and can be utilized to generate explanations. For example, prototype-based models such as UIPC-MF learn representative examples within the model's internal workings. The latter is the model-agnostic approach, also known as a post-hoc approach [16]. In this approach, the recommendation model is a blackbox by itself that requires to implement another explanation model to explain the recommendations made by the recommendation model [23].

2.3 The Prototype-Based Collaborative Filtering

The concept of using prototypes to learn representative vectors for users and items has recently been employed in Anchor-based Collaborative Filtering (ACF) [3] and Prototype-based Matrix Factorization (ProtoMF) [15]. ACF learns shared vectors known as anchors, and users and items are represented as convex combinations of these anchor vectors. The predicted output in ACF is obtained through the dot product between the user representation and the item representation. However, this output logit does not directly reflect a linear relationship with the anchors that can limit the interpretability. The results in [3] show that the performance of ACF outperforms LightGCN [8], which is one of the popular methods using deep learning.

In ProtoMF, prototypes are learned separately for users and items. The predicted logit in ProtoMF is computed by summing the dot products between user similarities from each user prototype and item representation with linear transformation weights, as well as the dot products between item similarities from each item prototype and user representation with linear transformation weights. Although the logit output of ProtoMF can be decomposed into a linear combination of user prototype similarities and item prototype similarities, the lack of connections between user prototypes and item prototypes makes it unable to directly associate the relationship between them as well as distinguish user's preference over each item prototype.

In summary, these methods learn representative prototype vectors from user vectors and item vectors, where the user vectors and item vectors have direct relationships with the prototype vectors. However, there is a lack of clear and direct relationships between the user prototype vectors and the item prototype vectors, which limits transparency. To address this problem, UIPC-MF introduces weights between user prototype vectors and item prototype vectors. The predicted scores are then calculated as a linear combination of user prototype similarities and item prototype similarities, resulting in improved transparency.

3 Methodology

In this section, we first provide a detailed description on on our proposed model, User-Item Prototypes Connections Matrix Factorization (UIPC-MF), and then explain the objective function. UIPC-MF is specifically designed for the top-N recommendation based on implicit feedback, such as user's listening events and click history, rather than explicit feedback like user ratings.

3.1 User-Item Prototypes Connections Matrix Factorization (UIPC-MF)

The aim of UIPC-MF is to identify collaborative patterns among users and items based on interaction data in terms of prototype vectors. For example, from the perspective of user prototypes, an individual who likes metal music may have an embedding vector that is closer to a user prototype with an attribute being a metal lover. Similarly, from the perspective of item prototypes, the embedding vector of a metal music piece will be closer to an item prototype with an attribute belonging to the metal genre.

We define $\mathcal{U} = \{u_i \in \mathbb{R}^d\}_{i=1}^N$ and $\mathcal{T} = \{t_j \in \mathbb{R}^d\}_{j=1}^M$ as the set of N users and M items, respectively. Let $\mathcal{P}^u = \{p_i^u \in \mathbb{R}^d\}_{i=1}^{L^u}$ be the set of L^u learnable user prototypes, where $L^u \ll N$, and $\mathcal{P}^t = \{p_i^t \in \mathbb{R}^d\}_{i=1}^{L^t}$ be the set of L^t learnable item prototypes, where $L^t \ll M$. L^u and L^t are hyperparameters. For the implicit interaction data, $\mathcal{I} = \{(u_i, t_j)\}$, where (u_i, t_j) represents that the user u_i has interacted with the item t_j. For simplicity, we omit the indices of users or items to represent any user or item.

To obtain the collaborative patterns of user u, we adopt the method proposed in [15], which calculates the user-prototype similarity vector u^* based on the user vector and each user prototype vector. UIPC-MF first calculates the similarity scores between a user u and each user prototype p^u, resulting in $u^* \in \mathbb{R}^{L^u}$, a user-prototype similarity vector. The similarity function used is the shifted cosine similarity, which has a range of 0 to 2. Similarly, UIPC-MF calculates $t^* \in \mathbb{R}^{L^t}$, an item-prototype similarity vector, using item t and each item prototype vector p^t in the similar manner as u^*. The index i in u^* represents the similarity score between a user embedding u and the i-th user prototype vector p_i^u, allowing u^* to capture all the scores of the user's collaborative patterns. The same concept applies to t^*. The formulas for calculating u^* and t^* are shown in Eq. 1.

$$u^* = \begin{bmatrix} sim(u, p_1^u) \\ \dots \\ sim(u, p_{L^u}^u) \end{bmatrix} \in \mathbb{R}^{L^u},\ t^* = \begin{bmatrix} sim(t, p_1^t) \\ \dots \\ sim(t, p_{L^t}^t) \end{bmatrix} \in \mathbb{R}^{L^t} \tag{1}$$

$$, where\ sim(a, b) = 1 + \frac{a^T b}{\|a\| \cdot \|b\|}$$

Finally, UIPC-MF utilizes connection weights to link u^* and t^* in order to calculate the final predicted logit for (u, t). The connection weights assign a learnable weight $w \in \mathbb{R}$ to each pair $(sim(u, p^u), sim(t, p^t))$. This implies that UIPC-MF assumes each user prototype has some degree of relation with each item prototype. For example, the user prototype representing a metal lover may exhibit a strong association with the item prototype of the metal genre and a weaker association with the classical genre. It is important to note that the connection weights are global and shared by all users and items. The overall architecture of UIPC-MF is depicted in Fig. 1. The formula for generating the predicted logit of (u, t) is shown in Eq. 2.

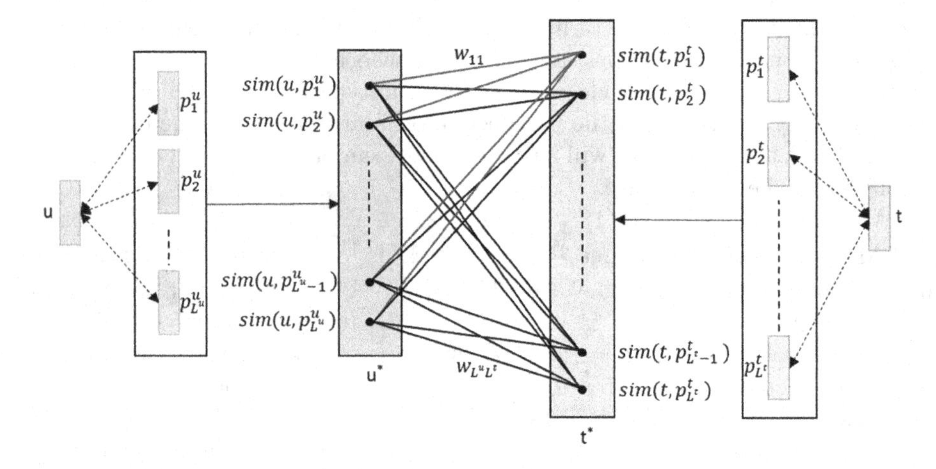

Fig. 1. UIPC-MF Model. The red lines show the used connection weights to obtain user preference value for a particular item prototype 1, r_1. (Color figure online)

$$UIPC\text{-}Score(u, t) = \sum_{i=1}^{L^u} \sum_{j=1}^{L^t} w_{ij} sim(u, p_i^u) sim(t, p_j^t) \tag{2}$$

where w_{ij} represents the connection weight between the user prototype p_i^u and item prototype p_j^t.

Equation 2 demonstrates that the predicted logit is a linear combination of $sim(u, p^u)$ and $sim(t, p^t)$. For each user u, u^* and the connection weights can be utilized to calculate the user's preference value r over an item prototype p_j^t according to Eq. 3.

$$r_j = \sum_{i=1}^{L^u} w_{ij} sim(u, p_i^u) \tag{3}$$

where, r_j represents the preference value of users toward an item prototype p_j^t. It is computed by summing the similarities between a user and all user prototypes, multiplied by the corresponding connection weights, for a specific item prototype. This calculation provides a clear insight into the user preference toward the corresponding item prototype. The red lines in Fig. 1 indicate the connection weights of users against the first item prototype as r_1. A higher absolute value of r_j indicates a stronger association between user and item prototype p_j^t. Consequently, the user preference values for all item prototypes can be represented as $\{r_1, ..., r_{L^t}\}$.

$$UIPC\text{-}Score(u,t) = \sum_{j=1}^{L^t} s_j,\ s_j = r_j \cdot sim(t, p_j^t) \tag{4}$$

In Eq. 4, s_j denotes the final score of item prototype j for the user u. The predicted logit can also be represented as the sum of all s_j. This equation provides a clear explanation of how the UIPC-Score is composed and computed as well as the roles of user preference and item prototype similarity in calculating the score.

3.2 Loss Function

To train the UIPC-MF model, we adopt the loss function $\mathcal{L}_{total}(\mathcal{I}, \Theta)$ as Eq. 5, given the data $\mathcal{I}$ with the model parameters Θ.

$$\begin{aligned}\mathcal{L}_{total}(\mathcal{I}, \Theta) = \mathcal{L}_{base}(\mathcal{I}, \Theta) + \lambda_{L2} \|\Theta\|_2 + \lambda_1 R_{\{\mathcal{P}^u \to \mathcal{U}\}} + \\ \lambda_2 R_{\{\mathcal{U} \to \mathcal{P}^u\}} + \lambda_3 R_{\{\mathcal{P}^t \to \mathcal{T}\}} + \lambda_4 R_{\{\mathcal{T} \to \mathcal{P}^t\}} + \lambda_{L1} \|r\|_1\end{aligned} \tag{5}$$

where $\mathcal{L}_{base}(\mathcal{I}, \Theta)$ can be binary cross-entropy loss (BCE Loss), Bayesian personalized ranking loss (BPR Loss), or sampled softmax loss (SSM Loss), is commonly used in implicit feedback for recommender systems [19,22]. $\|\Theta\|$ indicates the L^2-norm using the hyperparameter λ_{L2} as a regularization term and the last term is the L^1-norm for users' preference values. The terms in equations $R_{\{\mathcal{P}^u \to \mathcal{U}\}}$, $R_{\{\mathcal{U} \to \mathcal{P}^u\}}$, $R_{\{\mathcal{P}^t \to \mathcal{T}\}}$, and $R_{\{\mathcal{T} \to \mathcal{P}^t\}}$ are interpretability terms from ProtoMF [15]. These terms ensure a correlation between each user/item prototype and at least one user/item embedding, and vice versa.

The last term as shown in Eq. 6 is L^1-Norm for users' preference values that promotes the convergence of preference values r towards zero, ensuring that each preference value for an item prototype does not have a strong bias towards either a strictly positive or negative range for all users.

$$\|r\|_1 = \frac{1}{N} \sum_{i=1}^{N} \sum_{j=1}^{L^t} |r_{ij}| \tag{6}$$

Instead, we aim for each user's preference value for an item prototype to be positive or negative depending on their individual preferences. This helps prevent prototypes from becoming a bias during training.

To reduce computation time, we only calculate the users and items that appear in the sampled batch during training. Hyperparameters λ_1, λ_2, λ_3, λ_4 and λ_{L1} are used to adjust the influence of these terms.

Table 1. Statistics of the filtered datasets for collaborative filtering

	ML-1M	AMAZONVID	MO-3MON
# Users	6,034	6,950	8,385
# Items	3,125	14,494	27,696
# Interactions	574,376	132,209	921,194

4 Experiments and Discussion

In our research on collaborative filtering, we utilized three datasets: MovieLens-1M (ML-1M) [7], Amazon Video Games (AMAZONVID) [14], and Music4All-Onion-3Months (MO-3MON) [17], in accordance with the methodologies described by [15]. We considered ratings exceeding 3.5 as positive. For the ML-1M and AMAZONVID datasets, we applied a 5-core filtering approach, ensuring that both users and items appeared at least five times. The MO-3MON dataset underwent a 10-core filtering after meeting additional demographic criteria such as age and gender. The post-filtering statistics of the datasets are detailed in Table 1.

4.1 Evaluation Metrics

For evaluation, we used two commonly used accuracy metrics in recommender systems: Hit Ratio (HR) and Normalized Discounted Cumulative Gain (NDCG). The results represent the average performance across all users and were computed with a cutoff of 5 and 10.

4.2 Baseline Models

We compared the performance of three baseline methods: *Matrix Factorization (MF)* [11], a classical collaborative filtering method, and two prototype-related methods *Anchor-based Collaborative Filtering* [3] (ACF) and *Prototype-based Matrix Factorization* [15] (ProtoMF).

4.3 Training Details

We employ the leave-one-out strategy to obtain the train/validation/test data. This means we retrieve the second from last and the last interacted items sorted by timestamp as validation and test data, respectively, for each user. For evaluation, we sample 99 negative interactions for each validation and test interaction.

Hyperparameter Tuning. We followed the hyperparameter tuning method [4,12] and value range described in [15] for all models in our experiments. All hyperparameter types and ranges in UIPC-MF are the same as ProtoMF [15], except for λ_{L1} which has the same range as λ_{L2}. We repeated this entire process three times using different seeds and reported the average performance metrics on the test set.

Table 2. Recommendation evaluation results

Dataset	Metric	MF	ACF	ProtoMF	UIPC-MF	UIPC-MF-L_1
ML-1M	NDCG@5	0.230	0.275	0.308	0.325	**0.335**
	NDCG@10	0.280	0.334	0.367	0.383	**0.394**
	HR@5	0.344	0.412	0.458	0.476	**0.490**
	HR@10	0.501	0.596	0.640	0.656	**0.669**
AMAZONVID	NDCG@5	0.102	0.138	0.179	0.181	**0.187**
	NDCG@10	0.130	0.183	0.223	0.225	**0.231**
	HR@5	0.149	0.221	0.269	0.274	**0.283**
	HR@10	0.234	0.360	0.404	0.411	**0.419**
MO-3MON	NDCG@5	0.149	0.330	0.364	0.398	**0.405**
	NDCG@10	0.182	0.387	0.416	0.447	**0.455**
	HR@5	0.221	0.477	0.511	0.549	**0.558**
	HR@10	0.324	0.653	0.670	0.702	**0.710**

4.4 Evaluation Results

To empirically investigate **RQ1**, we evaluated UIPC-MF with and without the L^1-Norm. Table 2 demonstrates that UIPC-MF outperforms all baseline models on the three datasets, and UIPC-MF-L^1 further improves performance across all datasets. In-depth analysis of the results show that ACF outperforms MF on all datasets, and ProtoMF outperforms ACF. Compared with the best-performing model ProtoMF in the baselines, UIPC-MF-L^1 improves on average by 4.7% in HR@10, 6.8% in NDCG@10, 7.1% in HR@5 and 8.2% in NDCG@5 across the three datasets. UIPC-MF-L^1 significantly improves performance against all other models, especially in the cutoff 5, which means it has the ability to rank true interactions at the top in comparison to other models.

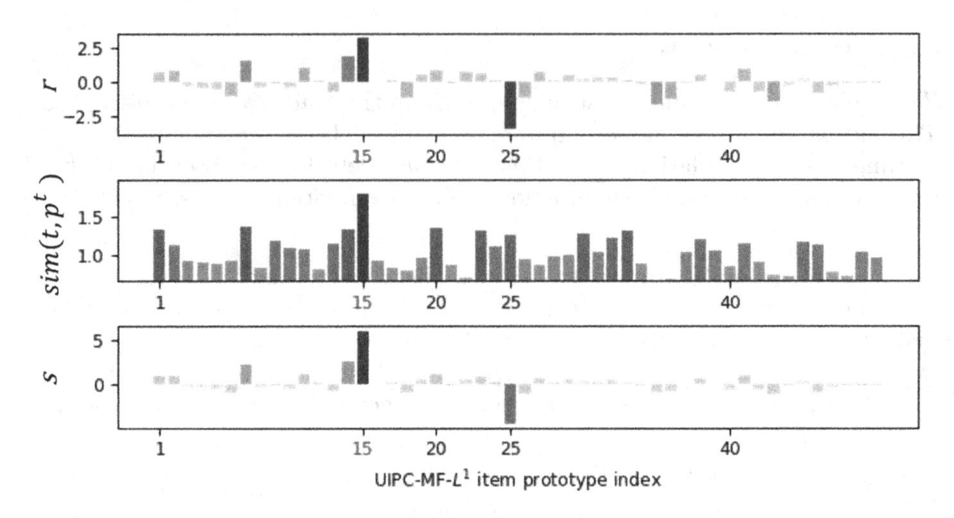

Fig. 2. The visualization of the user preference value r, $\mathrm{sim}(t, p^t)$, and the item prototype score s.

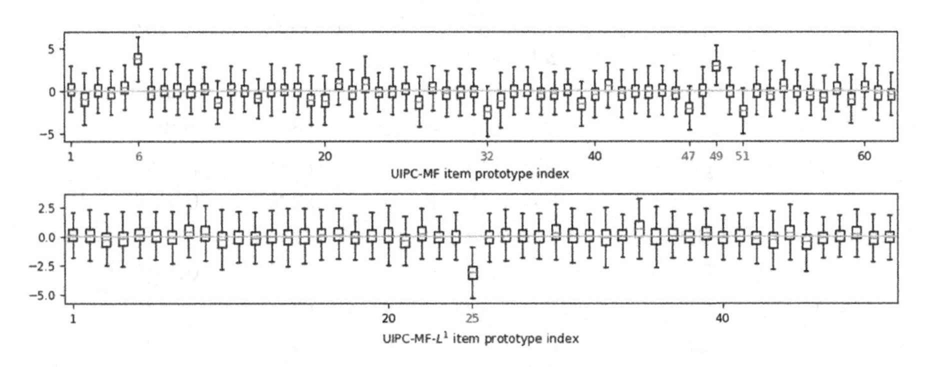

Fig. 3. The boxplots with the range of user preference values in UIPC-MF (top) and UIPC-MF-L^1 (bottom).

4.5 Explaining UIPC-MF Recommendations

To answer **RQ2**, we visualize an example of MO-3MON on Fig. 2, which corresponds to Eq. 4. In this example, the user has the highest preference for item prototype 15, and the item has the highest similarity to prototype 15. As a result, the score of item prototype 15 accounts for the majority of the weight in the predicted score.

Since the item prototypes and items are in the same space, we can calculate the top nearest items with the item prototype. The left column of Table 4 shows the top 5 nearest items to item prototype 15, including their song names, genres, and total occurrences in the dataset. All of them close to the metal genre.

In this example, where the user has the highest preference for item prototype 15, we can obtain the top songs with the highest similarity to prototype 15 that the user has listened to in the training dataset. Table 3 shows the top 5 nearest items to item prototype 15 that the user has listened to in the training dataset.

This information can be used as the explanation rationales for the recommendation. For example: *"Since you love metal music, you may also enjoy the music we are recommending."* or *"Other listeners who have listened to XXX also enjoyed the music we are recommending."* This example demonstrates the transparency of the proposed model, as it allows the recommender systems to access the underlying inference of the model in generating a recommendation item for a specific user.

Table 3. The top 5 nearest items to item prototype 15 that the user has listened to in the training dataset.

Song	Genre
Where the Slime Live	*death metal*
Grotesque Impalement	*death metal*
Severed Survival	*death metal*
Far Below	*progressive rock*
You Can't Bring Me Down	*thrash metal*

Table 4. The top 5 nearest items in item prototypes 15 and 25

Item prototype 15	Item prototype 25
Mask of the Red Death	**Low Grade Buzz**
black metal/15	*indietronica/11*
Unhallowed	**Airsick**
black metal/36	*ambient/11*
Crown	**Betrayed in the Octagon**
black metal/15	*ambient/10*
Grim and Frostbitten Kingdoms	**Hymn to Eternal Frost**
black metal/25	*black metal/9*
Release From Agony	**Blue Drive**
thrash metal/13	*ambient/12*

4.6 The Impact of L^1-Norm in Reduction of Learning Bias

To assess the impact of L^1-Norm on the model and answer **RQ3**, we present boxplots of users' preferences for item prototypes in the MO-3MON dataset.

Figure 3 shows that UIPC-MF-L^1 exhibits less bias towards extreme positive or negative values in comparison to UIPC-MF. All users in UIPC-MF exhibit strictly positive or negative preference values towards item prototypes 6, 32, 47, 49, and 51 within a specified range. However, prototype 25 in UIPC-MF-L^1 still has negative preferences for all users. We list the top 5 nearest items to item prototype 25 in the right column of Table 4. The occurrences are approximately equal to the preprocessing filtering threshold (10), and their genres lack significant consistency. This indicates a negative bias towards long tail items learned by UIPC-MF-L^1. This phenomenon aligns with common observations in collaborative filtering [1,20], and it is observed within the internal workings of UIPC-MF-L^1.

5 Conclusion

This paper presents a prototype-based collaborative filtering that offers several contributions. First, the predicted logit score is determined by a linear combination of the crossover interactions between user prototype similarities and item prototype similarities. This allows our model to reveal the preference of each user against each item prototype and the association relationships between user prototypes and item prototypes using the connection weights. Second, to enhance performance and alleviate bias, we employ the L^1-Norm on user preference to avoid bias towards positive or negative preference values for all users. Third, the transparency of the model reveals a negative bias towards long tail items for all users. Finally, through experimental results on three datasets, we demonstrate the superiority of our model in comparison to other prototype-based methods in top-N recommendation tasks, establishing it as a state-of-the-art solution.

References

1. Abdollahpouri, H., Mansoury, M., Burke, R., Mobasher, B.: The unfairness of popularity bias in recommendation. arXiv preprint arXiv:1907.13286 (2019)
2. Balog, K., Radlinski, F.: Measuring recommendation explanation quality: the conflicting goals of explanations. In: Proceedings of the 43rd International ACM SIGIR Conference on Research and Development in Information Retrieval, pp. 329–338 (2020)
3. Barkan, O., Hirsch, R., Katz, O., Caciularu, A., Koenigstein, N.: Anchor-based collaborative filtering. In: Proceedings of the 30th ACM International Conference on Information & Knowledge Management, pp. 2877–2881 (2021)
4. Bergstra, J., Bardenet, R., Bengio, Y., Kégl, B.: Algorithms for hyper-parameter optimization. In: Advances in Neural Information Processing Systems, vol. 24 (2011)
5. Chen, C., Li, O., Tao, D., Barnett, A., Rudin, C., Su, J.K.: This looks like that: deep learning for interpretable image recognition. In: Advances in Neural Information Processing Systems, vol. 32 (2019)
6. Chen, X., Zhang, Y., Wen, J.R.: Measuring "why" in recommender systems: a comprehensive survey on the evaluation of explainable recommendation. arXiv preprint arXiv:2202.06466 (2022)

7. Harper, F.M., Konstan, J.A.: The movielens datasets: history and context. ACM Trans. Interact. Intell. Syst. (TIIS) **5**(4), 1–19 (2015)
8. He, X., Deng, K., Wang, X., Li, Y., Zhang, Y., Wang, M.: LightGCN: simplifying and powering graph convolution network for recommendation. In: Proceedings of the 43rd International ACM SIGIR Conference on Research and Development in Information Retrieval, pp. 639–648 (2020)
9. He, X., Liao, L., Zhang, H., Nie, L., Hu, X., Chua, T.S.: Neural collaborative filtering. In: Proceedings of the 26th International Conference on World Wide Web, pp. 173–182 (2017)
10. Keswani, M., Ramakrishnan, S., Reddy, N., Balasubramanian, V.N.: Proto2proto: can you recognize the car, the way i do? In: Proceedings of the IEEE/CVF Conference on Computer Vision and Pattern Recognition, pp. 10233–10243 (2022)
11. Koren, Y., Bell, R., Volinsky, C.: Matrix factorization techniques for recommender systems. Computer **42**(8), 30–37 (2009)
12. Li, L., Jamieson, K., DeSalvo, G., Rostamizadeh, A., Talwalkar, A.: Hyperband: a novel bandit-based approach to hyperparameter optimization. J. Mach. Learn. Res. **18**(1), 6765–6816 (2017)
13. Liang, D., Krishnan, R.G., Hoffman, M.D., Jebara, T.: Variational autoencoders for collaborative filtering. In: Proceedings of the 2018 World Wide Web Conference, pp. 689–698 (2018)
14. McAuley, J., Targett, C., Shi, Q., Van Den Hengel, A.: Image-based recommendations on styles and substitutes. In: Proceedings of the 38th International ACM SIGIR Conference on Research and Development in Information Retrieval, pp. 43–52 (2015)
15. Melchiorre, A.B., Rekabsaz, N., Ganhör, C., Schedl, M.: ProtoMF: prototype-based matrix factorization for effective and explainable recommendations. In: Proceedings of the 16th ACM Conference on Recommender Systems, pp. 246–256 (2022)
16. Molnar, C.: Interpretable Machine Learning. 2 edn. (2022)
17. Moscati, M., Parada-Cabaleiro, E., Deldjoo, Y., Zangerle, E., Schedl, M.: Music4all-onion-a large-scale multi-faceted content-centric music recommendation dataset. In: Proceedings of the 31th ACM International Conference on Information & Knowledge Management (CIKM 2022) (2022)
18. Nauta, M., Van Bree, R., Seifert, C.: Neural prototype trees for interpretable fine-grained image recognition. In: Proceedings of the IEEE/CVF Conference on Computer Vision and Pattern Recognition, pp. 14933–14943 (2021)
19. Rendle, S.: Item recommendation from implicit feedback. In: Recommender Systems Handbook, pp. 143–171. Springer, New York (2021). https://doi.org/10.1007/978-1-0716-2197-4_4
20. Sankar, A., Wang, J., Krishnan, A., Sundaram, H.: ProtoCF: prototypical collaborative filtering for few-shot recommendation. In: Proceedings of the 15th ACM Conference on Recommender Systems, pp. 166–175 (2021)
21. Tintarev, N., Masthoff, J.: Explaining recommendations: design and evaluation. In: Ricci, F., Rokach, L., Shapira, B. (eds.) Recommender Systems Handbook, pp. 353–382. Springer, Boston, MA (2015). https://doi.org/10.1007/978-1-4899-7637-6_10
22. Wu, J., et al.: On the effectiveness of sampled softmax loss for item recommendation. arXiv preprint arXiv:2201.02327 (2022)
23. Zhang, Y., Chen, X., et al.: Explainable recommendation: a survey and new perspectives. Found. Trends® Inf. Retrieval **14**(1), 1–101 (2020)
24. Zhang, Y., Tiňo, P., Leonardis, A., Tang, K.: A survey on neural network interpretability. IEEE Trans. Emerg. Top. Comput. Intell. **5**(5), 726–742 (2021)

Towards Multi-subsession Conversational Recommendation

Yu Ji, Qi Shen, Shixuan Zhu, Hang Yu, Yiming Zhang, Chuan Cui, and Zhihua Wei(✉)

Tongji University, Shanghai, China
{2230779,1653282,2130768,2053881,2030796,cuichuan, zhihua_wei}@tongji.edu.cn

Abstract. Conversational recommendation systems (CRS) could acquire dynamic user preferences towards desired items through multi-round interactive dialogue. Previous CRS works mainly focus on the single conversation (**subsession**) that the user quits after a successful recommendation, neglecting the common scenario where the user has multiple conversations (**multi-subsession**) over a short period. Therefore, we propose a novel conversational recommendation scenario named **M**ulti-**S**ubsession **M**ulti-round **C**onversational **R**ecommendation (**MSMCR**), where the user would still resort to CRS after several subsessions and might preserve vague interests, and the system would proactively ask attributes to activate user interests in the current subsession. To fill the gap in this new CRS scenario, we devise a novel framework called **M**ulti-**S**ubsession **C**onversational Recommender with **A**ctivation **A**ttributes (**MSCAA**). Specifically, we first develop a context-aware recommendation module, comprehensively modeling user interests from historical interactions, previous subsessions, and feedback in the current subsession. Furthermore, an attribute selection policy module is proposed to learn a flexible strategy for asking appropriate attributes to elicit user interests. Finally, we design a conversation policy module to manage the above two modules to decide actions between asking and recommending. Extensive experiments on four datasets verify the effectiveness of our MSCAA framework for the proposed MSMCR setting (More details of our work are presented in https://arxiv.org/pdf/2310.13365v1.pdf).

Keywords: Conversational Recommendation · Human-in-the-Loop Learning · Recommender Systems

1 Introduction

Conversational recommendation systems (CRS) aim to elicit dynamic user preferences and make successful recommendations through a real-time multi-round conversation with the user [4]. In this work, we focus on the multi-round conversational recommendation (MCR) setting where the system takes actions based on the user's current needs in each turn, either asking questions about user preferences on attributes or recommending items, for succeeding in fewer turns.

Y. Ji and Q. Shen– Contributed equally to this research.

D.-N. Yang et al. (Eds.): PAKDD 2024, LNAI 14649, pp. 182–194, 2024.
https://doi.org/10.1007/978-981-97-2262-4_15

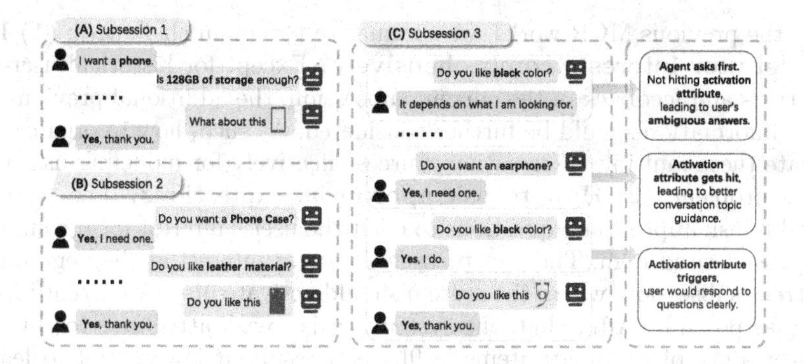

Fig. 1. A toy example of Multi-Subsession Multi-round Conversational Recommendation. The right side explains the multi-round process of the 3rd subsession in detail.

Although existing MCR studies have made significant progress, they all focus on a single conversation episode that the user quits after receiving the satisfactory item recommendation [3,8,9]. We argue that this single conversation setting overlooks the prevalence of **multiple** conversations in the real-world CRS scenario. The user would continue the dialogue with the system after a successful recommendation to browse items on other topics aimlessly or obtain additional system suggestions. Under the situation of multiple conversations, the user might also preserve possible **vague** interest rather than clearly express his/her preference, which is more realistic in the world and breaks the assumption of the previous MCR works [8,9]. For example, Fig. 1 illustrates an interaction process (**session**) that the user has three conversations (**subsessions**). In the first subsession of one session, the user successfully seeks his/her target item *Phone*. At this time, he/she might require other items related to *Phone* but be unaware of the related attributes to find them. After several system suggestions, he/she further gets satisfactory items (*Phone Case* and *Earphone*) in the subsequent subsessions. More specifically considering vague user interest shown in the 3rd subsession (C) of Fig. 1, the user would be overwhelmed by over-specific questions (e.g., *black* color) before the main attributes (e.g., *earphone*) of the target item, i.e., **activation attributes** are determined, which leads to ambiguous answers (e.g., *It depends on what I am looking for.* for attribute question *black* color). Once the activation attribute *earphone* is confirmed by the user as the topic of the current subsession, he/she would activate his/her interest and respond explicitly on system-asked questions towards the current topic, naturally including over-specific questions in the early turns.

To this end, we extend MCR [8,9,19] to a general setting named **M**ulti-**S**ubsession **M**ulti-round **C**onversational **R**ecommendation (**MSMCR**). In this scenario, the user has several subsessions with the system over a short period. And the system would proactively ask questions on attributes to elicit dynamic user interests towards desired items in the subsequent subsessions (i.e., after the first subsession of one session). Since our MSMCR scenario involves the user's previous subsessions and activation attributes of the current subsession, surpassing the previous MCR works, we summarize two main challenges: **(1) How to model user interests comprehensively?** Except for historical user-item interactions and feedback in the current subsession, the additional previous subsessions information should be further considered. As such, how to organize and aggregate these multiple information sources effectively for modeling user interests is a problem. **(2) How to ask appropriate questions?** The system is required to ask appropriate questions to elicit the user's interest for recommending his/her desired item. The user preserves unclear interest at the beginning of the current subsession, where the system should activate his/her current interest as early as possible. After that, the system needs to ask attributes to eliminate the uncertainty of candidate items [8,9]. As a result, it is essential to learn a flexible attribute selection strategy to activate user interests and then reduce the uncertainty of candidate items.

To effectively address the aforementioned challenges, we propose a novel method named **M**ulti-**S**ubsession **C**onversational Recommender with **A**ctivation **A**ttributes (**MSCAA**), which organically combines the policy modules with a recommendation module. In detail, the *Context-aware Recommendation* module models multiple information (i.e., historical user interactions, previous subsessions, and user feedback in the current subsession) to comprehensively model user interests for item and attribute prediction. The *Attribute Selection Policy* module automatically learns the strategy for asking activation attributes or others based on predicted user-like score and entropy of candidate attributes. And finally, the *Conversation Policy* module further utilizes these two modules and then decides actions between asking and recommending. We evaluate MSCAA on four adapted datasets to demonstrate the effectiveness of our method.

2 Related Works

Multi-round Conversational Recommendation (MCR) [3,8,9,17] is a common task for conversation recommendation, where the system alternates between asking attribute-based questions and recommending items several times to seek the user's desired item within fewer interaction-turns. The conversation strategy is the core design in MCR [7] for deciding whether to ask or recommend, and the strategy is typically modeled by deep reinforcement learning. For instance, CRM [14] pioneers the single-round task, and EAR [8] extends it to the MCR. SCPR [9] proposes an interactive path reasoning method to find item/attribute candidates. Based on the EAR framework, FPAN [17] utilizes user feedback to refine user embedding. UNICORN [3] contributes a dynamic weighted graph-based unified framework. Except for the above MCR task, recent studies have started to explore the new MCR scenario [2,19]. Different from these works, and [20] that utilizes the user's historical interaction sequence to improve recommendations in MCR, we focus on a **novel scenario** where the user has several dialogues with the system in short-term and might preserve unclear interests.

3 MSMCR Scenario

3.1 Definition

In this scenario, we define the set of users $\mathcal{U}$ and items $\mathcal{V}$. We collect all attributes $\mathcal{A}$ corresponding to items, and each item v is associated with a set of attributes $\mathcal{A}_v$. Unlike previous MCR works that call a conversation session, we define a conversation episode as a *subsession* s. And a *session* $S_u = [s_u^1, s_u^2, \ldots, s_u^{n-1}, s_u^n]$ consists of multiple subsessions where n is variational length for each session. Moreover, a session can be divided into two parts: the previous subsessions $P_{S_u} = [s_u^1, s_u^2, \ldots, s_u^{n-1}]$ and the current subsession s_u^n. The goal of the MSMCR task is to recommend the desired item v^n of user u in the **current** subsession s_u^n within fewer turns, based on P_{S_u} and the current subsession information.

For a user $u \in \mathcal{U}$, the workflow of MSMCR is listed as follows: **(1)** After several subsessions (i.e., P_{S_u}) successfully end, the user resorts to the subsequent conversation (i.e., s_u^n) instead of quitting the session. In this state, the user might initially have no explicit attribute query, and hence the current subsession is started from the system side. **(2)** Then, the system is free to *ask* questions about an attribute from the candidate attribute set $\mathcal{A}_{cand}$ (e.g., $\mathcal{A}_{cand} = \mathcal{A}$ at first) or to *recommend* a certain number of items (e.g., top-K) from the candidate item set $\mathcal{V}_{cand}$. **(3)** Next, user u provides feedback, i.e., accept, unknown, or reject for the asked attribute, and accept or reject for the recommended items. **(4)** After that, the system updates the sets of candidate attributes and items based on user feedback. **(5)** Within multiple iterations of step **(2)**-**(4)**, the system elicits clearer user interests and provides more accurate recommendations. The current subsession will terminate when the system recommends successfully, or the interaction turn reaches the maximum T.

Different from the user setting that has clear attribute preferences in previous MCR works [8,9], the user's preference might be vague and should be proactively guided [21] by the system in the MSMCR scenario. That is, in step (3), the user will respond "**unknown**" for the other attributes when the **activation attributes** $\mathcal{A}_{v^n}^*$ (i.e., several main attributes that enable the user to trigger current demands) have not been clarified. And after one of the activation attributes is hit by the system, the user would generate explicit feedback (accept or reject) for all attributes.

3.2 General Framework

Following existing MCR works [3,9], we formulate our MSMCR task as a Markov Decision Process (MDP) of interaction between the user and the recommendation system agent. The goal of our framework is to learn the conversation policy network π_c (cf., Sect. 4.2) to maximize the expected cumulative rewards for the overall conversations [17]. We decompose one conversation turn into four steps: **state, action, transition,** and **reward** under our framework.

State. The system maintains the conversation state s_t at each turn t in the subsession, which encodes the user feedback (e.g., accepted attributes $\mathcal{A}_{acc}^t$, rejected attributes $\mathcal{A}_{rej}^t$, and rejected items $\mathcal{V}_{rej}^t$) and the system candidate information.

Action. The agent takes actions according to the conversation state, where the action can be asking an attribute or recommending items. If the system decides to ask, it will select an attribute from the candidate attribute set $\mathcal{A}_{cand}$ via the decision of an attribute agent using policy π_a (cf., Sect. 4.2). While for recommendation action, the system will select top-K items from the candidate item set $\mathcal{V}_{cand}$ based on the recommendation score (cf., Sect. 4.1).

Transition. When the agent takes actions, and the user provides corresponding feedback to the attribute or items, the current state will transition to the next s_{t+1}. In this step, we update the sets $\mathcal{A}_{acc}$, $\mathcal{A}_{rej}$, $\mathcal{V}_{rej}$, $\mathcal{V}_{cand}$ and $\mathcal{A}_{cand}$ based on user feedback following the transition operation of [3,9,19]. Note that in our MSMCR setting, user feedback for the asked attribute will be "unknown" until an activation attribute is hit. In this case, we do not update them.

Reward. We design six kinds of reward functions in our framework: (1) r_{rec_acc} and (2) r_{rec_rej} denote positive and negative rewards when the user accepts or rejects the recommended item, respectively; (3) r_{ask_acc}, (4) r_{ask_rej} and (5) r_{ask_unk} denote positive, negative and negative rewards when the user accepts, rejects and responds "unknown" to the asked attribute, respectively; (6) r_{quit} is a strongly negative reward if the maximum subsession turn T is reached.

4 Methodology

4.1 Context-Aware Recommendation

A well-learned recommendation module will improve the performance of the system agent [7]. To this end, we carefully design the following three parts in the context-aware recommendation: *unified representation learning* to extract long-term user interest, *user interest modeling* to learn short-term interest from subsessions (i.e., context) and then integrate various interests, and finally, *prediction* to estimate scores on how the user likes items and attributes.

Unified Representation Learning. We first construct a unified heterogeneous graph [11,19] based on historical user-item interactions and static attribute-item relations, including three kinds of nodes: users, items, and attributes. This graph can be denoted as $G = \{N = \{\mathcal{U}, \mathcal{V}, \mathcal{A}\}, E = \{\mathcal{E}_{uv}, \mathcal{E}_{av}\}\}$, where the user-item edge $(u, v) \in \mathcal{E}_{uv}$ denotes that user u has interacted with item v and the attribute-item edge $(a, v) \in \mathcal{E}_{av}$ means that attribute a belongs to item v.

We introduce a L_g-layer relational graph convolutional network (RGCN) [13] to learn the node representations in the graph G. At first, each user, item, and attribute node is assigned with a unique node index and then initialized with the node embedding matrix $\boldsymbol{E}^0 \in \mathbb{R}^{|N| \times D}$, as the input embeddings of the first layer of RGCN. And for each layer until L_g, RGCN will first model different edge types based on node neighbors, and then the aggregated edge features will

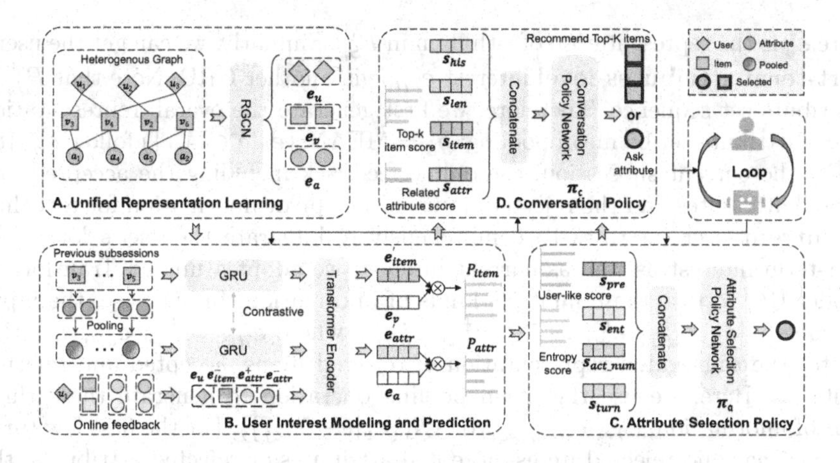

Fig. 2. The overview of MSCAA.

be integrated into node features. We take item node v which involves two kinds of edges $\mathcal{R}_v = \{r_{uv}, r_{av}\}$ as an example as follows:

$$e_v^{l+1} = \text{ReLU}(\sum\nolimits_{r\in\mathcal{R}_v}\sum\nolimits_{i\in\mathcal{N}_v^r}\frac{1}{\sqrt{|\mathcal{N}_v^r||\mathcal{N}_i^r|}}\boldsymbol{W}_r^l e_i^l + \boldsymbol{W}_0^l e_v^l), \tag{1}$$

where $\mathcal{N}_v^r$ denotes the neighbor nodes of node v under relation r. $\boldsymbol{W}_0^l$, $\boldsymbol{W}_{r_{uv}}^l$, and $\boldsymbol{W}_{r_{av}}^l \in \mathbb{R}^{D\times D}$ are trainable parameters. We can obtain the refined user and attribute embeddings $\boldsymbol{e}_u^{l+1}$, $\boldsymbol{e}_a^{l+1}$ in the same way. After L_g layers information propagation, we combine node embeddings of each layer to capture different semantics [5]: $\boldsymbol{e}_n = \frac{1}{L_g+1}\sum_{j=0}^{L_g} \boldsymbol{e}_n^j, n \in N = \{\mathcal{U}, \mathcal{V}, \mathcal{A}\}$. Finally, the contextualized node representations of users, items and attributes denote as $\boldsymbol{e}_u$, $\boldsymbol{e}_v$ and $\boldsymbol{e}_a$, where $\boldsymbol{e}_u$ can be regarded as **long-term** user interest [17].

User Interest Modeling. In our scenario, the user also has previous subsessions that occur recently with the system online, which are probably relevant, and hence crucial for user short-term interest modeling. For each subsession s_u^i in P_{S_u}, much abundant information should be considered, including the accepted item v^i and its associated attributes $\mathcal{A}_{v^i}$, which reflects short-term user preference on item and attributes. Therefore, we have two sequences from P_{S_u}: the previous desired item sequence $P_{S_u}^{\mathcal{V}} = [v^1, v^2, \ldots, v^{n-1}]$ and a sequence of the previous desired attribute sets $P_{S_u}^{\mathcal{A}} = [\mathcal{A}_{v^1}, \mathcal{A}_{v^2}, \ldots, \mathcal{A}_{v^{n-1}}]$. Following the classic sequential recommendation method [6,18], we separately model the above two sequences via gated recurrent units (GRU) [6] to capture the temporal interactions between each subsession. Specifically, for item sequence $P_{S_u}^{\mathcal{V}}$, we use the final hidden state of the GRU as the representation of the user's **short-term item-level** interest:

$$\boldsymbol{e}'_{item} = \text{GRU}(\{\boldsymbol{e}_v^1, \boldsymbol{e}_v^2, \ldots, \boldsymbol{e}_v^{n-1}\}), \tag{2}$$

where e_v^i is the representation of i-th item in $P_{S_u}^{\mathcal{V}}$. Similarly, we can get the user's **short-term attributes-level** interest e'_{attr} via another GRU. Note that $P_{S_u}^{\mathcal{A}}$ is a attribute set sequence. Therefore, we first generate the overall representation of each attribute set by mean pooling: $g_a^i = \text{MEAN}(\{e_a | a \in \mathcal{A}_{v^i}\})$ following [18].

For the current subsession, the online feedback including the accepted and rejected attributes and the rejected items, is also pivot information for revealing the **current** user interest. To comprehensively integrate the user's long- and short-term interest as well as current interest, we adopt a unified Transformer encoder [16] to fuse these multiple kinds of information. In our case, the input sequence is $[e_u, e'_{item}, e'_{attr}, e^-_{item}, e^+_{attr}, e^-_{attr}]$, where $e^-_{item}, e^+_{attr}, e^-_{attr} \in \mathbb{R}^{1\times D}$ denotes the aggregated representation of rejected items, accepted and rejected attributes. Here, we use the mean pooling operation to generate these three kinds of information, e.g., $e^-_{item} = \text{MEAN}(\{e_v | v \in \mathcal{V}_{rej}\})$. For the first few turns that may have no rejected items, accepted attributes or rejected attributes, the corresponding $e^-_{item}, e^+_{attr}, e^-_{attr}$ is masked for Transformer input. Finally, we can obtain the user's **final item-level and attribute-level interest** from different kinds of information, i.e., e'_{item} and e'_{attr} from the Transformer encoder.

Prediction. We estimate the scores that the user likes items and attributes:

$$P_{item}(u, v) = \tanh({e'_{item}}^{\top} e_v) \ ; \ P_{attr}(u, a) = \tanh({e'_{attr}}^{\top} e_a), \tag{3}$$

where v and a are candidate item in $\mathcal{V}_{cand}$ and attribute in $\mathcal{A}_{cand}$, respectively.

4.2 Policy Learning

Our framework MSCAA considers two different decision policies for attribute selection and conversation management.

Attribute Selection Policy. In our MSMCR task, due to the existence of special activation attributes, we expect these activation attributes that the user might well like should be hit as soon as possible. After that, other target attributes asked can better eliminate the uncertainty of candidate items. However, the previous attribute selection rules, like *Max Entropy* [7,9] and *Max User-like* [19], only consider one aspect of the above issues, e.g., *Max User-like* is mainly suitable for asking attributes that match user interests. Hence we employ an attribute selection policy to adaptively select the appropriate asked attribute.

Specifically, the policy network is implemented by a two-layer MLP, which can output the asked attribute based on the user's conversation state $s_{\mathfrak{a}}$. The state vector $s_{\mathfrak{a}}$ involves 4 parts of information:

$$s_{\mathfrak{a}} = s_{pre} \oplus s_{ent} \oplus s_{act\_num} \oplus s_{turn}, \tag{4}$$

where $\oplus$ is the concatenation operation, s_{pre} encodes user interests on all attributes, in which each dimension is the estimated preference score P_{attr} calculated by Equation (3). s_{ent} is the entropy of each attribute among candidate items $\mathcal{V}_{cand}$, where each dimension is the entropy w_a of attribute a calculated by $w_a = -\mathrm{p}(a)\log \mathrm{p}(a)$, $\mathrm{p}(a) = |\mathcal{V}_{cand} \cap \mathcal{V}_a| / |\mathcal{V}_{cand}|$, following [7,8]. Note that only the attributes in $\mathcal{A}_{cand}$ are preserved for the attribute selection, and all other attributes are masked in $\boldsymbol{s}_{pre}$ and $\boldsymbol{s}_{ent}$. These two vectors provide useful user-side and item-side information for attribute selection [8]. $\boldsymbol{s}_{act\_num}$ and $\boldsymbol{s}_{turn}$ denote the accepted attributes number and the current turn number, to perceive the changes during the conversation. This agent receives the ask-related rewards (r_{ask_acc}, r_{ask_rej} and r_{ask_unk}) and user-quit reward (r_{quit}) only, and updates the state with corresponding state transitions.

Conversation Policy. This policy is responsible for deciding whether to ask an attribute or recommend items with the purpose of hitting the user's desired item of the current subsession in fewer turns. We use a simple two-layer MLP to implement the interaction policy network, which maps the state vector $\boldsymbol{s}_{\mathfrak{c}}$ to two actions, i.e., ask or recommend. The state vector $\boldsymbol{s}_{\mathfrak{c}}$ is represented as:

$$\boldsymbol{s}_{\mathfrak{c}} = \boldsymbol{s}_{his} \oplus \boldsymbol{s}_{len} \oplus \boldsymbol{s}_{item} \oplus \boldsymbol{s}_{attr}, \tag{5}$$

where $\boldsymbol{s}_{his}$ encodes the conversation history in the subsession and $\boldsymbol{s}_{len}$ encodes the size of the candidate item set following previous CRS works [8,9]. $\boldsymbol{s}_{item}$ is a K-dimension vector which records the top-K candidate item score from the context-aware recommendation module. $\boldsymbol{s}_{attr}$ represents the preference and entropy score of the attribute by the attribute selection policy. Intuitively, the greater value is in vector $\boldsymbol{s}_{item}$, the more probably the agent selects the "recommend" action. The same is for $\boldsymbol{s}_{attr}$. By comparing the last two vectors, the agent can make full use of detailed information to decide the appropriate action.

4.3 Model Training

Since the performance of the online agent largely depends on the context-aware recommendation module, we first pre-train this module based on training interaction data for item and attribute prediction. The training objective is a multi-task pairwise loss (e.g., BPR [12]) to rank positive items/attributes higher than others. Moreover, we further introduce a contrastive loss (e.g., InfoNCE [10]) to constrain the consistency of the final user item-level and attribute-level interest to complement each other. Afterwards, we train two policies $\pi_{\mathfrak{a}}$ and $\pi_{\mathfrak{c}}$ meanwhile the recommendation module is fixed. Concretely, we utilize each session collected online to optimize them by Policy Gradient [15] following [17].

5 Experiments

5.1 Experimental Setup

Dataset. To evaluate the proposed method, we adopt four existing MCR benchmark datasets (**LastFM**, **LastFM***, **Yelp**, and **Yelp***) [9]. The statistics of these datasets are given in Table 1.

Table 1. Statistics of datasets used in experiments, where n is the length of a session.

Statistic	#Users	#Items	#Interactions	#Attributes	Avg. n
LastFM/LastFM*	1,801	7,432	76,693	33/8438	2.82/2.82
Yelp/Yelp*	27,675	70,311	1,368,606	29/590	2.85/2.85

User Simulator. Following previous works [3,8,9], the classical user simulator is adapted to our MSMCR scenario from the construction of a session and the generation of activation attributes.

- *The Construction of a Session.* We simulate a session composed of ordered subsessions from a chronological user-item interaction sequence. i-th ($i < n$) subsession s_u^i is constructed by i-th interacted item v^i of user u correspondingly. For each subsession s_u^i in session S_u, we regard the item v^i as the ground-truth target item, and its attribute set $\mathcal{A}_{v^i}$ as the oracle attributes preferred by the user. The session length n is defined as $\min(\text{Random}(N_{min}, N_{max}), M_u)$, where M_u is the interaction length of user u, N_{min} and N_{max} are the threshold values of session length n to align with realistic conversation session size. We report the mean value of n resulted from our simulation in Table 1.
- *The Generation of Activation Attributes.* The activation attributes $\mathcal{A}_{v^n}^*$ of the current subsession within a session are the subset of oracle attributes $\mathcal{A}_{v^n}$. Specifically, it is defined as the top-two-ranked attributes based on user-attribute affinity score: $\mathbf{U}_u^\top \mathbf{A}_a + (\sum_{j=1}^{n-1} \mathbf{V}_{v^j}^\top \mathbf{A}_a)/(n-1), \forall a \in \mathcal{A}_{v^n}$ for each session, following the design of [2]. In this formula, v^j represents the target item of the j-th subsession in session S_u, and $\mathbf{U}$,$\mathbf{V}$,$\mathbf{A}$ are pre-trained embeddings of users, items and attributes, respectively, which are obtained via TransE [1] in [3].

Baseline Models. To verify model performance, our proposed model is compared with various representative multi-round CRS methods including **Max Entropy** (**MaxE**) [8], **EAR** [8], **SCPR** [9], **FPAN** [17], and **UNICORN** (**UNI**) [3]. For a fair comparison, we adapt all the above baselines (named superscript †): (1) We utilize previous subsessions to complement the user presentation for the item and attribute scoring. Specifically, we first apply GRU to obtain the representation of user item-level interest from previous subsessions, aligned with our method (cf., Sect. 4.1). Then, the original user representation and item-level interest presentation are further fused as the updated user representation via the gating mechanism [11]. (2) We employ our user simulator.

Evaluation Metrics. Following previous studies on MCR [8,9], the cumulative ratio of successful conversational recommendation assessed by SR@T (*success rate at turn* T) and the average number of turns for all subsessions assessed by *average turn* (AT) are adopted. Similar to SR@T, we introduce AR@T (*activation rate at turn* T) to measure the cumulative ratio of hitting activation attributes in the session. Moreover, we also adopt hN@(T,K) to measure the recommendation accuracy [3], considering the rank performance in both list- and turn-level. We omit @T and K terms in the following experiments.

Implementation Details. We implement the proposed method based on PyTorch. We randomly divide interactions across each dataset into training, validation, and test parts with the ratio of 7 : 1.5 : 1.5 and then separately generate the session samples based on by user simulator described in Sect. 5.1. The maximum/minimum session length N_{max}, N_{min} is set to 4, 2 respectively. We set the maximum turn T and the size of the recommendation list K as 10. The embedding dimension D is set as 64. We employ the Adam/SGD optimizer to train the recommendation/policy module with the learning rate $5e^{-4}$ and $1e^{-3}$ separately. The heterogeneous graph G is constructed by the training dataset. The number of RGCN layers L_g and Transformer layers L_t are all set to be 2. Discount factor γ is set to be 0.7. The weight ω of InfoNCE constraint is set to be 0.01. The temperature parameter τ in InfoNCE is set with 0.5. We set six rewards $[r_{rec\_acc}, r_{rec\_rej}, r_{ask\_acc}, r_{ask\_rej}, r_{ask\_unk}, r_{quit}] = [1, -0.1, 0.01, -0.1, -0.1, -0.3]$ following reward settings in [9].

Table 2. Performance comparison. The **bolded** number indicates our improvements over the runner-up (underlined) are statistically significant ($p < 0.01$) with paired t-test. SR, hN and AR stand for SR@10, hN@(10, 10) and AR@10, respectively.

Models	LastFM				LastFM*				Yelp				Yelp*			
	SR↑	AT↓	hN↑	AR↑	SR↑	AT↓	hN↑	AR↑	SR↑	AT↓	hN↑	AR↑	SR↑	AT↓	hN↑	AR↑
MaxE†	.061	9.94	.015	.451	.067	9.91	.017	.199	.846	6.52	.278	.953	.015	9.98	.004	.324
EAR†	.117	9.78	.037	.254	.059	9.94	.015	.197	.804	6.91	.273	.956	.054	9.95	.009	.353
SCPR†	.128	9.69	.042	.273	.075	9.82	.027	.248	.826	6.85	.301	.985	.091	9.84	.023	.371
FPAN†	.264	9.54	.052	.313	.069	9.89	.028	.365	.895	5.87	.304	.972	.082	9.88	.022	.361
UNI†	.286	9.48	.068	.475	.196	9.45	.062	.178	.830	6.42	.312	.954	.162	9.76	.052	.382
Ours	**.547**	**8.88**	**.170**	**.855**	**.291**	9.47	**.089**	**.439**	**.924**	**5.81**	**.349**	.968	**.277**	**9.75**	**.081**	**.778**

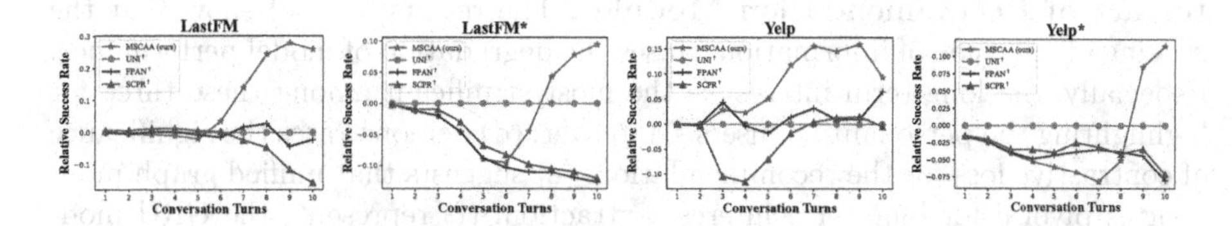

Fig. 3. Comparisons of relative success rate at different conversation turns.

5.2 Overall Performance

The comparison of overall experimental results of our MSCAA and the adapted methods are presented in Table 2. Besides the overall performance (e.g., SR), we also present a fine-grained performance comparison of success rate at each turn in Fig. 3, where the blue line of UNI† is set to $y = 0$. From the table and figures, we summarize the following observations:

(1) In most cases, our proposed method exhibits significant advantages over all the baselines across four datasets, due to the combination of three dedicated modules for activating user interests and then hitting the desired item of user.
(2) Our method's improvement is relatively limited on the Yelp dataset among all four datasets. In addition, the overall performance of both LastFM* and Yelp* is comparatively inferior to the other two datasets. These could be explained that the setting of the Yelp dataset is to ask enumerated questions [8], which makes it easier to hit the activation attributes and then sharply diminish the candidate item space for all methods. However, for the LastFM* and Yelp* datasets, asking attributes to activate user interests is harder due to the numerous candidates.
(3) UNI† may outperform MSCAA in the first few turns, while our MSCAA achieves a pretty performance at the end stage of the conversation. This phenomenon could be explained that MSCAA is more inclined to ask appropriate attributes to elicit user interest at first, and after that, can more confidently recommend items to satisfy the user's current demand. Conversely, the other baselines may recommend items successfully instead of asking more attributes in the earlier turns since they do not consider the activation of user interest. Due to lack of acquisition of the current user interest, they are inferior to MSCAA.

5.3 Further Experiments

To illustrate the design effectiveness of MSCAA, we remove or replace key modules one-by-one and report results in Table 3.

Impact of Recommendation Modules. The results in (a-c) show that the missing of any type of information causes the degradation of model performance. Especially, the long-term interest is the most significant among these three for highlighting the personality of users. Moreover, (d) demonstrates the significance of contrastive loss for the recommendation. (e) suggests that unified graph modeling is pivotal for long-term interest extraction. (f) represents the GRU modules in Sect. 4.1 are replaced with "mean" operation, indicating the importance of sequential encoder of previous subsessions for recommendation.

Impact of Attribute and Conversation Policy. (g) and (h) indicate that our attribute selection policy is replaced by Max Entropy and Max User-like score rule, respectively. The results show the effectiveness of our attribute agent for selecting an appropriate attribute. And (i) means the fine-grained score information $s_{item} \oplus s_{attr}$ in conversation state is masked. From the result, this kind of information is necessary for a high-level action decision.

Table 3. Results of ablation studies on different information and model architecture.

Models	LastFM		Yelp*		Models	LastFM		Yelp*	
	SR↑	hN↑	SR↑	hN↑		SR↑	hN↑	SR↑	hN↑
(a)w/o Long-term Pref.	.255	.078	.065	.019	(f)w/ Mean	.537	.166	.262	.077
(b)w/o Short-term Pref.	.463	.143	.146	.043	(g)w/ Max Entropy	.102	.030	.032	.009
(c)w/o Current Pref.	.537	.167	.178	.052	(h)w/ Max User-like	.422	.132	.138	.041
(d)w/o InfoNCE Loss	.538	.168	.265	.078	(i)w/o $s_{item} \oplus s_{attr}$	.538	.169	.270	.079
(e)w/o RGCN	.517	.161	.261	.076	**MSCAA**	**.547**	**.170**	**.277**	**.081**

6 Conclusion

In this paper, we extend the MCR to a general CRS setting, MSMCR in which the user continues the dialogue with the system after several subsessions and might preserve no clear interest in the current subsession, and the system would proactively take actions to activate the user's dynamic interest. For this scenario, a novel framework called MSCAA is introduced to model user interests comprehensively for the recommendation, learn the flexible strategy for asking the appropriate attributes, and manage actions between asking and recommending adaptively. Our experimental results demonstrate the effectiveness of MSCAA.

Acknowledgments. The work is partially supported by the National Nature Science Foundation of China (No. 62376199, 62076184, 62076182) and Shanghai Science and Technology Plan Project (No.21DZ1204800).

References

1. Bordes, A., Usunier, N., Garcia-Duran, A., Weston, J., Yakhnenko, O.: Translating embeddings for modeling multi-relational data. In: NeurIPS (2013)
2. Chu, Z., Wang, H., Xiao, Y., Long, B., Wu, L.: Meta policy learning for cold-start conversational recommendation. In: WSDM (2023)
3. Deng, Y., Li, Y., Sun, F., Ding, B., Lam, W.: Unified conversational recommendation policy learning via graph-based reinforcement learning. In: SIGIR (2021)
4. Gao, C., Lei, W., He, X., de Rijke, M., Chua, T.S.: Advances and challenges in conversational recommender systems: a survey. In: AI Open (2021)
5. He, X., Deng, K., Wang, X., Li, Y., Zhang, Y., Wang, M.: LightGCN: simplifying and powering graph convolution network for recommendation. In: SIGIR (2020)
6. Hidasi, B., Karatzoglou, A., Baltrunas, L., Tikk, D.: Session-based recommendations with recurrent neural networks. In: ICLR (2016)
7. Hu, C., Huang, S., Zhang, Y., Liu, Y.: Learning to infer user implicit preference in conversational recommendation. In: SIGIR (2022)

8. Lei, W., et al.: Estimation-action-reflection: towards deep interaction between conversational and recommender systems. In: WSDM (2020)
9. Lei, W., et al.: Interactive path reasoning on graph for conversational recommendation. In: SIGKDD (2020)
10. Oord, A.v.d., Li, Y., Vinyals, O.: Representation learning with contrastive predictive coding. arXiv preprint arXiv:1807.03748 (2018)
11. Pang, Y., Wu, L., Shen, Q., Zhang, Y., Wei, Z., et al.: Heterogeneous global graph neural networks for personalized session-based recommendation. In: WSDM (2021)
12. Rendle, S., Freudenthaler, C., Gantner, Z., Schmidt-Thieme, L.: BPR: bayesian personalized ranking from implicit feedback. arXiv preprint arXiv:1205.2618 (2012)
13. Schlichtkrull, M., Kipf, T.N., Bloem, P., Van Den Berg, R., et al.: Modeling relational data with graph convolutional networks. In: ESWC (2018)
14. Sun, Y., Zhang, Y.: Conversational recommender system. In: SIGIR (2018)
15. Sutton, R.S., McAllester, D., Singh, S., Mansour, Y.: Policy gradient methods for reinforcement learning with function approximation. In: NeurIPS (1999)
16. Vaswani, A., et al.: Attention is all you need. In: NeurIPS (2017)
17. Xu, K., Yang, J., Xu, J., Gao, S., Guo, J., Wen, J.R.: Adapting user preference to online feedback in multi-round conversational recommendation. In: WSDM (2021)
18. Yu, F., Liu, Q., Wu, S., Wang, L., Tan, T.: A dynamic recurrent model for next basket recommendation. In: SIGIR (2016)
19. Zhang, Y., et al.: Multiple choice questions based multi-interest policy learning for conversational recommendation. In: WWW (2022)
20. Zhou, K., et al.: Leveraging historical interaction data for improving conversational recommender system. In: CIKM (2020)
21. Zhou, K., Zhou, Y., Zhao, W.X., Wang, X., Wen, J.R.: Towards topic-guided conversational recommender system. In: COLING (2020)

False Negative Sample Aware Negative Sampling for Recommendation

Liguo Chen[1], Zhigang Gong[1], Hong Xie[2(✉)], and Mingqiang Zhou[1]

[1] College of Computer Science, Chongqing University, Chongqing, China
chenliguo@stu.cqu.edu.cn, gongzhigang@cigit.ac.cn, zmqmail@cqu.edu.cn
[2] School of Computer Science and Technology, USTC, Anhui, China
xiehong2018@foxmail.com

Abstract. Negative sampling plays a key role in implicit feedback collaborative filtering. It draws high-quality negative samples from a large number of uninteracted samples. Existing methods primarily focus on hard negative samples, while overlooking the issue of sampling bias introduced by false negative samples. We first experimentally show the adverse effect of false negative samples in hard negative sampling strategies. To mitigate this adverse effect, we propose a method that dynamically identifies and eliminates false negative samples based on dynamic negative sampling (EDNS). Our method integrates a global identification module and a positives-context identification module. The former performs clustering on embeddings of all users and items and deletes uninteracted items that are in the same cluster as the corresponding user as false negative samples. The latter constructs a similarity measure for uninteracted items based on the positive sample set of the user and removes the top-k items ranked by the measure as false negative samples. Finally, we utilize the dynamic negative sampling strategy to build a sample pool from the corrected uninteracted sample set, effectively mitigating the risk of introducing false negative samples Experiments on three real-world datasets show that our approach significantly outperforms state-of-the-art negative sampling baselines.

Keywords: Negative sampling · CF · Implicit feedback

1 Introduction

Collaborative filtering (CF), a key technique in recommendation systems, learns user preferences based on users' historical behaviors and recommend items for users [7,12]. Implicit feedback has become a default scenario for CF methods in recent mainstream research [5,15]. In implicit feedback CF, each observed interaction normally represents a user preference for an item (i.e., a positive sample), while the other unobserved items are uninteracted samples. Considering that only positive samples are available, a common approach in CF to deal with the implicit feedback is to select a few instances from the uninteracted data as negative samples, also known as negative sampling [14]. Then, the CF model is optimized to assign positive samples higher scores than negative ones.

D.-N. Yang et al. (Eds.): PAKDD 2024, LNAI 14649, pp. 195–206, 2024.
https://doi.org/10.1007/978-981-97-2262-4_16

A widely used negative sampling strategy is the uniform distribution [6,14,17]. However, a uniformly sampled negative sample is very likely to be ranked correctly below the positive sample and thus the gradient of the pair is near 0, contributing little to the convergence [13]. To solve this problem, many hard negative sampling methods [8,11,13,16,19] were proposed, which oversampling the samples with high predicted scores, as it can bring more information with larger gradients for model training. Nevertheless, these hard negative sampling strategies focus on mining hard negative samples and neglect the risk of introducing false negative samples during the training process, thereby causing incorrect preference learning for the CF model. Recently, some studies on negative sampling focus on how to mitigate the effects of false negative samples [4,18,21] and have shown impressive performance in alleviating the problem of false negative samples. However, these methods only observe the difference in the prediction scores of true and false negative samples from a statistical point of view, ignoring the semantic relationships between samples.

We first experimentally show the adverse effect of false negative samples in the context of hard negative sampling strategies. Specifically, we compare the performance of two methods, i.e., original dynamic negative sampling and ideal dynamic negative sampling, that do not sample false negative samples, under different sizes of the candidate pool. We find that the performance of original dynamic negative sampling first increases and then decreases with the increase of the candidate pool size, while the performance of ideal dynamic negative sampling always increases. This demonstrates the necessity of alleviating the problem of false negative samples in hard negative sampling. To solve this problem, we develop a method called EDNS that integrates a global identification module and a positive-context identification module to dynamically identify false negative samples, and then explicitly eliminates the identified false negative samples in dynamic negative sampling. Concretely, the global identification module first performs K-means clustering on all embeddings of users and items, and then identifies uninteracted samples with the same cluster labels as users as false negative samples. Furthermore, since the semantics represented by the embeddings in the early stages of training are not reliable, we adopt a warm-start strategy, which gradually increases the acceptance rate of the instances' cluster label in the training progresses. The positives-context identification module first constructs a similarity measure for uninteracted items based on the positive sample set of the user, and then extract the top-k samples from the sorted uninteracted items based on this measure in descending order as the false negative samples. After identifying the false negative samples, we directly eliminate them from the uninteracted samples, and then use the dynamic negative sampling method to sample negative samples to improve the quality of training samples. Experimental on three real-world datasets show that our method outperforms state-of-the-art negative sampling baselines significantly.

2 Related Work

Static negative sampling samples negative samples from a fixed distribution throughout the training process. The most popular and widely used sampling strategy is random negative sampling (RNS) [14]. A heuristic strategy for the fixed distribution of negative samples is popularity-based negative sampling (PNS) [2]. This method uses the interaction frequency of items in the training data set to define the popularity, and then assigns higher negative sampling probability to items with higher popularity.

Adaptive negative sampling can adaptively sample negative samples based on the state of the current recommendation model to obtain high-quality negative instances. Many adaptive negative sampling strategies have been proposed for implicit feedback recommendation systems. Initially, researchers pay attention to the study of hard negative sampling, focusing on rank-based methods and GAN-based methods. Specifically, for rank-based method, DNS is proposed to dynamically select negative samples with the highest prediction scores. AOBPR [13] applies an adaptive and context-dependent sampling distribution to over-sample top ranked items. For GAN-based method, IRGAN [16] uses the generative network that generates items for the user and sends the items into a discriminative network to judge whether the sample is from the real data. AdvIR [11] adds perturbation to adversarial sampling, making model training more robust.

Since hard negative sampling looks for samples that are similar to positive samples in the embedding space, it inevitably causes sampling bias. Recently, researchers have focused on how to mitigate the effects of sampling bias. SRNS [4] uses the low variance of the false negative samples during the recent epochs, and over-samples items with both high prediction scores and high variances to tackle the false negative problem. GDNS [21] develops a expectation gain driven indicator to efficiently distinguish the false negative samples, and uses a group-aware optimization by constructing positive and negative groups for each user in each iteration to enhance training efficiency. GFNS [18] proposes a generalized negative sampling framework that can alleviate the impact of false negatives by introducing two robust methods: self-sampling and dynamic sub-sampling. Besides, as auxiliary knowledge, side information can alleviate the influence of sampling bias to a certain extent, such as users' connections in social networks [20], geographical locations of users [9], and additional interaction data including viewed but unclicked [3] and clicked but unpurchased [10].

3 Preliminary

Implicit feedback CF. We define the user set as $\mathbf{U}$ and the item set as $\mathbf{I}$. Denote the user-item interaction matrix as $\mathbf{X} = (x_{ui})$, where $x_{ui} = 1$ indicates that user u has interacted with item i, and $x_{ui} = 0$ indicates no interaction. Let $\mathbf{I}_u = \{i|x_{ui} = 1\}$ denote the interacted item set of user u. User preferences can be represented by predicted scores, and learning user preferences involves training a model to obtain the embedding $\mathbf{p}_u$ of user u and the embedding $\mathbf{q}_i$ of item i, which together form a score function $\hat{y}_\Theta(u, i)$ abbreviated as $\hat{y}_{ui}$. Matrix Factorization (MF) directly learns the embedding $\mathbf{p}_u$ and $\mathbf{q}_i$ of user u and item i simultaneously, and performs the inner product of the two to obtain the predicted score (i.e., $\hat{y}_{ui} = \langle \mathbf{p}_u, \mathbf{q}_i \rangle$).

The widely-adopted loss function is Bayesian Personalized Ranking (BPR): $\mathcal{L} = -\sum_{i \in \mathbf{I}_u \wedge j \in \mathbf{I} \setminus \mathbf{I}_u} \ln \sigma(\hat{y}_{ui} - \hat{y}_{uj})$, where $\sigma(x) = 1/(1+e^{-x})$ is a sigmoid function. For a user u, $\hat{y}_{ui}$ and $\hat{y}_{uj}$ are the predicted scores of item i that he has interacted with (called positive samples) and item j that he has not interacted with (called negative samples), respectively. BPR utilizes a negative sampling strategy to select negative samples from $\mathbf{I}_u^-$ as training samples, while simultaneously employing Stochastic Gradient Descent (SGD) to optimize the loss function. The gradient of BPR for a training instance pair (u, i, j) with respect to an arbitrary model parameter $\theta \in \Theta$ is:

$$\frac{\partial \mathcal{L}(u,i,j)}{\partial \theta} = -\left(1 - \sigma(\hat{y}_{ui} - \hat{y}_{uj})\right) \frac{\partial (\hat{y}_{ui} - \hat{y}_{uj})}{\partial \theta}. \tag{1}$$

Negative Sampling. To investigate the impact of false negative samples on hard negative sampling, we employ a representative approach named dynamic negative sampling (DNS) for analysis. Given a positive user-item pair (u, i), we uniformly sample M items from $\{j | j \in \mathbf{I}_u^-\}$ as the candidate pool C_u and select the negative sample with the highest predicted score. First, we denote the DNS method that may potentially sample false negative samples as DNS-Real. Then, we design an ideal method that does not sample false negative samples at all by manually excluding the false negative samples (i.e., samples in the test set) from the uninteracted samples, named DNS-Ideal. We compared the performance of DNS-Real and DNS-Ideal with different M on two public datasets MovieLens-100K and MovieLens-1M, as shown in Fig. 1. We use the same dataset processing method and optimal hyper-parameters as in Sect. 5.1, and use Recall@20 as the evaluation metrics. From the figure, we have the following findings: (1) On both datasets, the performance of the model first increases and then decreases as M increases using DNS-Real. This indicates that when M is small, the probability of the model sampling a false negative sample from C_u is small, so this sample can improve the performance of the model; and when M gradually becomes larger, the probability of the model sampling a false negative sample from C_u becomes large, so this sample may hurt the model's performance. (2) The DNS-Ideal method increases the performance of the model as M increases. Due to the fact that the probability of sampling to false negative samples in this method is 0, increasing the size of candidate pool C_u only enlarges the probability of sampling to samples with higher gradients, thus the performance of the model also gets promoted.

4 Methodology

4.1 False Negatives Identification

Global Identification. The traditional hard negative sampling strategies aim to improve the gradient of model training, ignoring the existence of false

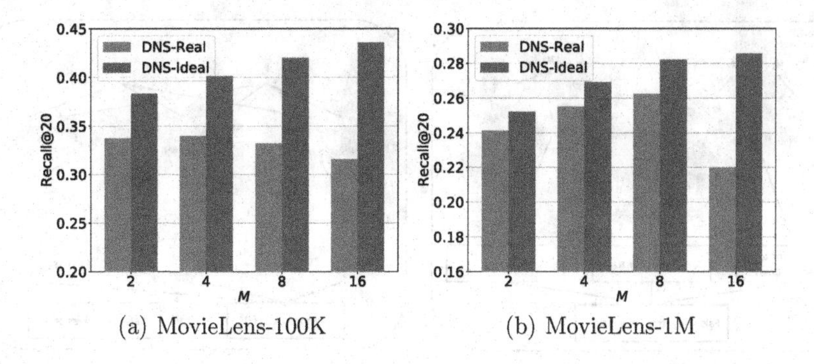

(a) MovieLens-100K (b) MovieLens-1M

Fig. 1. The impact of false negative samples on Recall@20 with different M.

negative samples. In this section, we construct a global identification module that can dynamically identify false negative samples based on the semantic relationships exhibited by the embeddings of all users and items. Specifically, we perform K-means clustering on the embeddings $\mathbf{e}_b$ of all users and items (i.e., all $\mathbf{p}_u$ and $\mathbf{q}_i$), and the embeddings are clustered into n clusters. Since a centroid of K-means clustering represent the center of a set of embeddings with the highest similarity, we can use it to represent the mined semantic relationships among instances and set the centroid embedding of k-th cluster as $\mathbf{c}_k$. Then, we calculate the distance between the instance b and n centroids separately and classify b into the cluster corresponding to the centroid with the smallest distance, that is, its cluster label is $y_b = \arg\min_k \|\mathbf{e}_b - \mathbf{c}_k\|$. During the training process, employing our proposed method of exploiting global embeddings to explore semantic relationships, an uninteracted sample j of a user u is identified as a false negative sample of u if j shares the same cluster label as u. Intuitively, the explanation is that when applying K-means clustering to the overall embeddings, the semantic relationships between instances in the recommendation system are mapped as subordination relationships between embeddings and centroids.

However, due to the unreliability of the embedding space during the early training stages of the recommendation system model, the cluster labels obtained through K-means clustering based on embeddings at that moment contain a significant amount of noise, thereby adversely affecting the identification of false negative samples. To tackle this problem, we employ a warm-start strategy, whereby in the initial stages of training, only instances with high-confidence cluster labels are identified, while the remaining instances are treated as independent individuals. Concretely, we define the confidence of a cluster label by:

$$\varphi_b = \frac{\exp\left(\frac{1}{\tau}\langle \mathbf{e}_b, \mathbf{c}_{y_b}\rangle\right)}{\sum_{k=1}^{n} \exp\left(\frac{1}{\tau}\langle \mathbf{e}_b, \mathbf{c}_k\rangle\right)}. \tag{2}$$

The elucidation of this definition posits that for a given instance b, the confidence φ_b of the cluster label will be larger when its embedding $\mathbf{e}_b$ is not only more similar to the centroid of the cluster it belongs to, but also less similar to the centroids of other clusters. Then, we introduce a dynamically changing

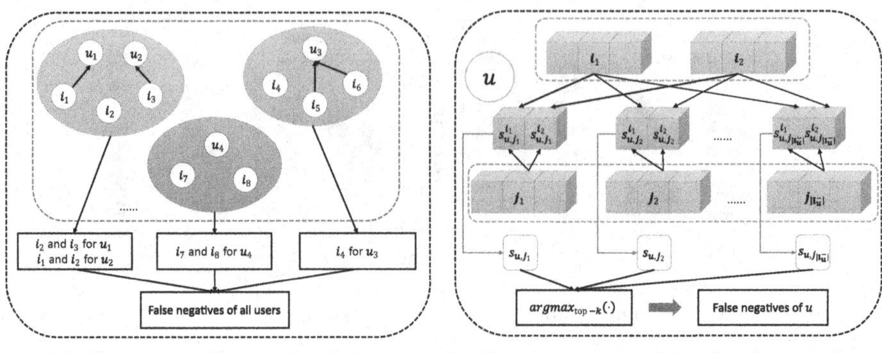

Fig. 2. The structure of the two core modules of the EDNS method.

acceptance ratio (AR) during training, which stipulates that only the cluster labels of instances belonging to the top AR proportion, sorted by the confidence of the cluster labels, will be considered reliable. Motivated by [4], we propose to linearly increase the AR_t as the epoch number t increases. Specifically, we use $AR_t = \min(t/T_0, 1)$, where T_0 denotes the threshold of stopping increase.

Positives-Dependent Identification. As mentioned above, we design a global identification module to dynamically identify false negative samples. Nevertheless, relying solely on the embeddings of all instances may not be sufficient to accurately capture the semantic relationships between users and their uninteracted samples. Given the fact that the semantics of each positive sample can only represent one aspect of user preferences, the possibility of false negative samples conforming to the semantics of these positive samples is higher than directly meeting all preferences represented by the user itself. In other words, as long as the semantics of an uninteracted sample has a high semantic similarity with the user's positive samples, it will be regarded as a false negative sample.

Therefore, we propose a positives-context identification module, only leveraging the semantic relationships between the user's positive sample set and the uninteracted sample, to adaptively detect false negative samples. First and foremost, prior to each training iteration, we undertake the computation of the similarity between each uninteracted sample j belonging to user u and all positive samples $i \in \mathbf{I}_u$. This similarity computation relies on the inner product of their respective embeddings (i.e., $s^i_{u,j} = \langle \mathbf{q}_i, \mathbf{q}_j \rangle$), resulting in a similarity vector $\mathbf{s}_{u,j} \in \mathbb{R}^{|\mathbf{I}_u|}$ for each uninteracted sample j. Then, for each $\mathbf{s}_{u,j} \in \mathbb{R}^{|\mathbf{I}_u|}$, we employ the mean aggregation strategy to aggregate them separately into the scalar $s_{u,j}$. The scalar resulting from the mean aggregation strategy can capture the shared user preferences expressed by each positive sample's semantics, serving as a measure of the likelihood of uninteracted samples being false negative samples. Lastly, we extract the top-k samples from the sorted list of $l \in \mathbf{I}^-_u$ based on their scalar $s_{u,j}$ in descending order, which will be considered as false negative samples. The value of k indicates the degree of tolerance of the algorithm to the number of false negative samples. We treat the selection of k as a data-driven problem and deem that it should be carefully adjusted in different scenarios. As discussed above, the embeddings of items are not reliable in the early stages of training. Hence, we also user a similar approach that sets $k_t = k \cdot \min(t/T_0, 1)$.

4.2 False Negatives Elimination

Given the above two identification modules, we can obtain potential set $\mathcal{F}_u^g$ by global identification and potential set $\mathcal{F}_u^{pc}$ by positives-context identification respectively, both of which contain the identified false negative samples for each user u. Our next step is to integrate $\mathcal{F}_u^g$ and $\mathcal{F}_u^{pc}$ to form $\mathcal{F}_u$. Then, we use the approach that explicitly eliminates false negative samples from $\mathbf{I}_u^-$. At last, we define the revised dynamic negative sampling distribution that replacing $\mathbf{I}_u^-$ with $\mathbf{I}_u^- \setminus \mathcal{F}_u$ as $\mathbb{P}_{elim}^{dns}$. Thus, the loss function is modified to:

$$\mathcal{L}_{elim} = -\sum\nolimits_{(u,i)\in\mathcal{R}, j\sim\mathbb{P}_{elim}^{dns}} \ln\sigma(\hat{y}_{ui} - \hat{y}_{uj}). \tag{3}$$

Algorithm 1 shows the implicit feedback CF learning framework based on EDNS.

Algorithm 1: EDNS Algorithm

Input: Training set $\mathcal{R} = \{(u, i)\}$, embedding size d, candidate pool size M, number of clusters n, value of top-k

Output: User embeddings $\{\mathbf{p}_u|u\in\mathbf{U}\} \in \mathbb{R}^d$, item embeddings $\{\mathbf{q}_i|i\in\mathbf{I}\} \in \mathbb{R}^d$

1 Initialize $\{\mathbf{p}_u|u\in\mathbf{U}\}$ and $\{\mathbf{q}_i|i \in \mathbf{I}\}$;
2 **for** $t = 1, 2, \cdots, T$ **do**
  `// Global Identification`
3   Get centroid embeddings $\mathbf{c}$ and cluster labels y via K-means($\{\mathbf{p}_u\}, \{\mathbf{q}_i\}, n$);
4   Calculate the confidence of all instances φ by (2);
5   Accept the cluster labels by a part of y with larger φ according to AR;
6   Identify the false negative samples by the cluster labels accepted and get the potential set $\mathcal{F}_u^g$ for each user u;
  `// Positives-dependent Identification`
7   Calculate the similarity vector $\mathbf{s}_{u,j} = (\langle \mathbf{q}_i, \mathbf{q}_j \rangle, i \in \mathbf{I}_u)$;
8   Use the mean aggregation on $\mathbf{s}_{u,j}$ to obtain the scalar $s_{u,j}$;
9   Identify the false negative samples by the top-k in the sorted list based on $s_{u,j}$ and get the potential set $\mathcal{F}_u^{pc}$;
  `// DNS Method`
10   Take the union of $\mathcal{F}_u^g$ and $\mathcal{F}_u^{pc}$ to form $\mathcal{F}_u$;
11   Sample a mini-batch $\mathcal{R}_{batch} \in \mathcal{R}$;
12   **for** *each* $(u, i) \in \mathcal{R}_{batch}$ **do**
13   Uniformly sample M items from $\mathbf{I}_u^- \setminus \mathcal{F}_u$ as candidate set C_u;
14   Get the negative sample j with highest score from C_u;
15   Update embeddings $\{\mathbf{p}_u\}$ and $\{\mathbf{q}_i\}$;
16   **end**
17 **end**

---

5 Experiment

5.1 Experiment Settings

Dataset. We consider three public datasets, including MovieLens-100K, MovieLens-1M, and Last.fm. We use the same disposal as [18] to process the datasets. For each user, we randomly select 80% of the items from his interaction history as the training set, and the remaining 20% as the test set. From the training set, we randomly select 10% of items as validation set to tune hyper-parameters.

Baselines. To demonstrate the effectiveness of the proposed negative sampling, we choose the following competitive methods. **RNS** [6,14,17] uses a uniform distribution to sample negative samples. **PNS** [2] adopts a fixed popularity-based distribution to sample more popular items as negative samples. **AOBPR** [13] over-samples a higher ranked uninteracted item scored by the current model with higher probability. **DNS** [19] draws a set of negative samples from a uniform distribution, and selects the one with the highest predicted score as a negative sample to update the model. **SRNS** [4] uses the low variance of the false negative samples during the recent epochs, and samples items with both high prediction scores and high variances to avoid over-sampling hard negative samples. **GDNS** [21] designs an expected gain sampler, which monitors the user's expectation of the preference gap between positive and negative samples in training process to dynamically guide negative sampling. **GFNS** [18] develops a general framework for negative sampling by introducing two robustness against false negatives.

Evaluation Metrics. Like with [1], we employ two widely-used metrics to evaluate the recommendation lists: Recall@K and NDCG@K, where K is the number of recommended items and in this paper $K = \{5, 10, 20\}$. Recall@K is the fraction of positive items in the top-K ranked list over the total number of positive items in the test set. NDCG@K measures the effectiveness of a recommendation method by considering both the relevance of the recommended items and their positions within the top-K ranked list. The final results of Recall@K and NDCG@K are the average values among all users. The implementation is similar to [1,18]. Please refer to supplementary[1] for more details.

5.2 Performance Comparison

Table 1 compares the recommendation performance for the negative sampling algorithms, where the boldface and underline are used to indicate the best and the second best respectively. From the table, we have the following key findings. The proposed EDNS consistently yields the best performance on all datasets with all evaluation metrics. The results validate that our method can effectively

[1] https://anonymous.4open.science/r/EDNS-C2E7.

identify and eliminate false negative samples in the DNS method, thus help to sample high-quality negative samples for model training. The static negative sampling strategies (i.e., RNS and PNS) perform poorly, especially PNS that uses the popularity-based sampling distribution favoring popular items. Among all adaptive sampling methods, the hard negative sampling strategies (i.e., AOBPR and DNS) place more emphasis on those samples with high predicted scores and may suffer from the existence of false negative samples, thus do harm to performance. In addition, DNS often outperforms AOBPR, and the reason is that AOBPR considers the sorting of all uninteracted samples, while DNS only needs to sort the samples in the candidate pool composed of random samples, which to some extent alleviates the problem of false negative samples. The performance of the debiased negative sampling methods (i.e., SRNS, GDNS and GFNS) in different datasets is unstable, and sometimes even worse than the performance of hard negative sampling methods. SRNS leverages prior statistical information to sample both high-scores and high-variance instances, but the simple addition combination may weaken its effectiveness. GDNS identifies true negative samples by using the expectational gain as an indicator, while ignoring samples with high predicted scores that can make a big contribution to model training. GFNS exploits the self-sampling and dynamic sub-sampling to avoid false negative samples, but the number of unknown false negative samples directly affects the robustness of this method.

5.3 Study of EDNS

Impact of Pool Sizes. We first explore the improvement of performance via eliminating false negative samples under the different sizes of candidate pool. We search the hyper-parameter M in {2, 4, 8, 16}, provide a comparison between EDNS and DNS, and give the convinced results in Fig. 3. We use MovieLens-100K and MovieLens-1M dataset to show the results and Last.fm follows the same trend, so we do not present it in the following experiments. We can observe that EDNS outperforms DNS under different sizes of candidate pool, indicating the outstanding capacity of EDNS to capture high-quality negative samples. Furthermore, we observe an interesting phenomenon: as M continues to increase, the performance gap between the two methods becomes more pronounced. One plausible explanation is that EDNS exhibits strong stability. Despite the fact that the number of false negative samples in the candidate pool will increase with the growth of M, the EDNS method effectively can identify and eliminate these false negative samples. Consequently, the quantity of false negative samples in the candidate pool is reduced.

Impact of Number of Clusters. To figure out the influence of number of clusters, we vary n in the range of {10, 50, 100, 250, 500, 750} and the experimental results are illustrated in Fig. 4. A too small number of clusters may cause a large number of samples to be divided into one cluster, resulting in an excessive number of false negative samples being determined. Conversely, A too large number of clusters can result in many samples not being correctly allocated

Table 1. Performance comparison on three datasets

Dataset	Method	Top-5		Top-10		Top-20	
		Recall	NDCG	Recall	NDCG	Recall	NDCG
100K	RNS	0.1335	0.4286	0.2162	0.4039	0.3342	0.4037
	PNS	0.1198	0.4032	0.1970	0.3784	0.3085	0.3759
	DNS	0.1391	0.4510	0.2246	0.4226	0.3410	0.4183
	AOBPR	0.1401	0.4443	0.2204	0.4123	0.3259	0.4023
	SRNS	0.1404	0.4522	0.2249	0.4219	0.3385	0.4154
	GDNS	0.1409	0.4432	0.2249	0.4165	0.3420	0.4161
	GFNS	0.1428	0.4561	0.2270	0.4242	0.3397	0.4169
	EDNS	**0.1447**	**0.4652**	**0.2296**	**0.4317**	**0.3454**	**0.4246**
1M	RNS	0.0886	0.4131	0.1474	0.3833	0.2337	0.3682
	PNS	0.0644	0.3348	0.1041	0.3008	0.1616	0.2756
	DNS	0.0991	0.4467	0.1625	0.4124	0.2551	0.3962
	AOBPR	0.1002	0.4351	0.1615	0.3957	0.2470	0.3755
	SRNS	0.1008	0.4527	0.1637	0.4170	0.2556	0.3973
	GDNS	0.0997	0.4490	0.1631	0.4123	0.2547	0.3966
	GFNS	0.1007	0.4531	0.1636	0.4159	0.2555	0.3961
	EDNS	**0.1015**	**0.4557**	**0.1642**	**0.4190**	**0.2561**	**0.3998**
Last.fm	RNS	0.0975	0.2132	0.1456	0.1729	0.2091	0.2073
	PNS	0.1127	0.2477	0.1681	0.1999	0.2381	0.2381
	DNS	0.1207	0.2638	0.1797	0.2132	0.2525	0.2531
	AOBPR	0.1202	0.2630	0.1790	0.2128	0.2512	0.2527
	SRNS	0.1219	0.2680	0.1855	0.2192	0.2617	0.2599
	GDNS	0.1222	0.2655	0.1796	0.2188	0.2620	0.2597
	GFNS	0.1226	0.2676	0.1860	0.2186	0.2612	0.2601
	EDNS	**0.1233**	**0.2700**	**0.1872**	**0.2201**	**0.2635**	**0.2609**

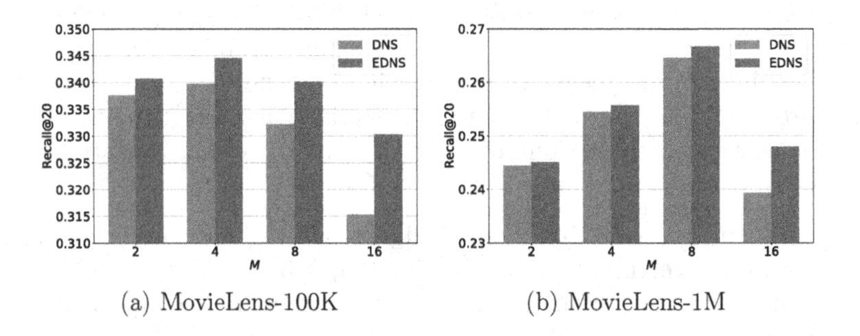

(a) MovieLens-100K (b) MovieLens-1M

Fig. 3. Impact of size of candidate pool size M.

together, rendering the semantic relationships between samples ineffective. We can observe that Recall@20 achieves its maximum value when $n = 50$ or $n = 100$, indicating that the clustering result at this setting is the best.

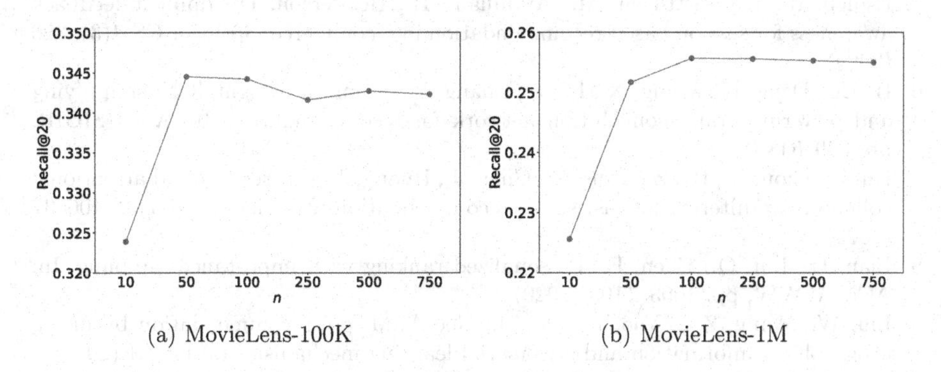

(a) MovieLens-100K (b) MovieLens-1M

Fig. 4. Impact of number of clusters.

Impact of Value of Top-k, Warm-start Strategy and Integrated Modules. Due to page limit, we present them in supplementary file.

6 Conclusion

In this paper, we propose a method EDNS that dynamically identifies and eliminates false negative samples during the training process based on dynamic negative sampling. EDNS integrates a global identification module and a positives-context identification module, which utilizes clustering on all embeddings and constructs a similarity measure based on the positive samples to identify false negative samples respectively. We also eliminate the detected samples from the sample pool to mitigate adverse effect of false negative samples. We conduct experiments on three real-world datasets to show that EDNS significantly outperforms state-of-the-art negative sampling methods.

Acknowledgment. The work of Hong Xie was supported in part by National Nature Science Foundation of China (61902042), Chongqing Natural Science Foundation (cstc2020jcyj-msxmX0652).

References

1. Chen, C., Ma, W., Zhang, M., Wang, C., Liu, Y., Ma, S.: Revisiting negative sampling vs. non-sampling in implicit recommendation. ACM TOIS **41**(1), 1–25 (2023)
2. Chen, T., Sun, Y., Shi, Y., Hong, L.: On sampling strategies for neural network-based collaborative filtering. In: ACM KDD, pp. 767–776 (2017)

3. Ding, J., Quan, Y., He, X., Li, Y., Jin, D.: Reinforced negative sampling for recommendation with exposure data. In: IJCAI, pp. 2230–2236 (2019)
4. Ding, J., Quan, Y., Yao, Q., Li, Y., Jin, D.: Simplify and robustify negative sampling for implicit collaborative filtering. In: NIPS, pp. 1094–1105 (2020)
5. Esmeli, R., Bader-El-Den, M., Abdullahi, H., Henderson, D.: Implicit feedback awareness for session based recommendation in e-commerce. Springer CS **4**(3), 320 (2023)
6. He, X., Deng, K., Wang, X., Li, Y., Zhang, Y., Wang, M.: LightGCN: simplifying and powering graph convolution network for recommendation. In: ACM SIGIR, pp. 639–648 (2020)
7. Hu, Z., Zhou, X., He, Z., Yang, Z., Chen, J., Huang, J.: Discrete limited attentional collaborative filtering for fast social recommendation. Elsevier EAAI **123**, 106437 (2023)
8. Lian, D., Liu, Q., Chen, E.: Personalized ranking with importance sampling. In: ACM WWW, pp. 1093–1103 (2020)
9. Liu, W., Wang, Z.J., Yao, B., Yin, J.: Geo-ALM: poi recommendation by fusing geographical information and adversarial learning mechanism. In: IJCAI (2019)
10. Loni, B., Pagano, R., Larson, M., Hanjalic, A.: Bayesian personalized ranking with multi-channel user feedback. In: ACM RecSys, pp. 361–364 (2016)
11. Park, D.H., Chang, Y.: Adversarial sampling and training for semi-supervised information retrieval. In: ACM WWW, pp. 1443–1453 (2019)
12. Rehman, I.u., Hanif, M.S., Ali, Z., Jan, Z., Mawuli, C.B., Ali, W.: Empowering neural collaborative filtering with contextual features for multimedia recommendation. Multimedia Syst. **29**, 2375–2388 (2023). https://doi.org/10.1007/s00530-023-01107-9
13. Rendle, S., Freudenthaler, C.: Improving pairwise learning for item recommendation from implicit feedback. In: ACM WSDM, pp. 273–282 (2014)
14. Rendle, S., Freudenthaler, C., Gantner, Z., Schmidt-Thieme, L.: BPR: bayesian personalized ranking from implicit feedback. In: UAI, pp. 452–461 (2009)
15. Tian, G., Yu, Q., Yang, H., Wang, R.: A smart contract top-n recommendation method based on implicit feedback. In: ACM RICAI, pp. 1016–1020 (2023)
16. Wang, J., et al.: IRGAN: a minimax game for unifying generative and discriminative information retrieval models. In: ACM SIGIR, pp. 515–524 (2017)
17. Wang, X., He, X., Wang, M., Feng, F., Chua, T.S.: Neural graph collaborative filtering. In: ACM SIGIR, pp. 165–174 (2019)
18. Yamanaka, Y., Sugiyama, K.: Generalized negative sampling for implicit feedback in recommendation. In: ACM WI-IAT, pp. 544–549 (2022)
19. Zhang, W., Chen, T., Wang, J., Yu, Y.: Optimizing top-n collaborative filtering via dynamic negative item sampling. In: ACM SIGIR, pp. 785–788 (2013)
20. Zhao, T., McAuley, J., King, I.: Leveraging social connections to improve personalized ranking for collaborative filtering. In: ACM CIKM, pp. 261–270 (2014)
21. Zhu, Q., Zhang, H., He, Q., Dou, Z.: A gain-tuning dynamic negative sampler for recommendation. In: ACM WWW, pp. 277–285 (2022)

Multi-sourced Integrated Ranking with Exposure Fairness

Yifan Liu[1], Weiwen Liu[2], Wei Xia[2], Jieming Zhu[2], Weinan Zhang[1], Zhenhua Dong[2], Yang Wang[3], Ruiming Tang[2], Rui Zhang[4], and Yong Yu[1(✉)]

[1] Shanghai Jiao Tong University, Shanghai, China
{sjtulyf123,wnzhang}@sjtu.edu.cn, yyu@apex.sjtu.edu.cn
[2] Huawei Noah's Ark Lab, Shenzhen, China
{liuweiwen8,xiawei24,dongzhenhua,tangruiming}@huawei.com, jiemingzhu@ieee.org
[3] East China Normal University, Shanghai, China
ywang@sei.ecnu.edu.cn
[4] ruizhang.info, Shanghai, China
rayteam@yeah.net

Abstract. Integrated ranking system is one of the critical components of industrial recommendation platforms. An integrated ranking system is expected to generate a mix of heterogeneous items from multiple upstream sources. Two main challenges need to be solved in this process, namely, (i) Utility-fairness tradeoff: an integrated ranking system is required to balance the overall platform's utility and exposure fairness among different sources; (ii) Information utilization from upstream sources: each source sequence has been carefully arranged by its provider, so how to efficiently utilize the source sequential information is important and should be carefully considered by the integrated ranking system. Existing methods generally cannot address these two challenges well. In this paper, we propose an integrated ranking model called Multi-sourced Constrained Ranking (MSCRank). It is a dual RNN-based model managing the utility-fairness tradeoff with multi-task learning, and capturing information in source sequences with a novel MA-GRU cell. We compare MSCRank with various baselines on public and industrial datasets, and MSCRank achieves the state-of-the-art performance on both utility and fairness. Online A/B test further validates the effectiveness of MSCRank.

Keywords: Integrated ranking · Exposure fairness · Recommender systems

1 Introduction

The integrated ranking system (IRS), also known as the mixed ranking system, is a rapidly emerging field driven by industrial applications in recommendation platforms [16]. Figure 1 illustrates an integrated ranking process. The IRS receives multiple ordered sequences from different upstream sources or categories

Y. Liu, W. Liu and W. Xia—Authors contributed equally to this research.

D.-N. Yang et al. (Eds.): PAKDD 2024, LNAI 14649, pp. 207–218, 2024.
https://doi.org/10.1007/978-981-97-2262-4_17

(referred to as *source sequences*), *e.g.*, articles, images, and videos. Different sources are held by different providers or third-party partners [10]. Then the system merges multiple source sequences into a single integrated list and displays it to users. The IRS has responsibilities not only to users, but also to the sequence providers [17]. The result of the IRS can significantly impact providers' income, because items' exposure opportunity directly determines the probability that users will interact with items. Every provider expects to receive a preferential portion of the exposure opportunities and competes for finite exposure positions, and thus different providers' interests may conflict. In this case, the IRS needs to *fairly* expose items from different providers to users. On the other hand, the platform also has its own utility to be improved, *i.e.*, the number of clicks or conversions. Therefore, jointly optimizing the *platform's utility* and the *exposure fairness* is the major objective for IRS with the following two challenges.

Utility-fairness Tradeoff. The platform's utility and the exposure fairness are generally conflict goals. As for the platform, to greedily maximize the platform's utility, the system tends to allocate more resources to the items of higher immediate return. Nevertheless, since items' utility or return from multiple sources may be quite different as mentioned above, purely optimizing the utility yields extremely unbalanced ranking results—items from sources with high instant returns are usually over-represented, dominating opportunities and resources; while items from newly-added or less profitable sources barely receive exposure. These unbalanced results harm source providers' fairness and revenue. Worse still, keeping recommending items from the same source may hurt user experience and thereby affect the platform's utility in the long term.

Information Utilization from Upstream Sources. The source sequences generated by corresponding providers are usually arranged by local promoting strategies. For instance, the video provider wants to promote a particular video, *e.g.*, The Peripheral, and ranks it before all other videos. Then, the source sequence's order reflects its promoting strategies, which is important for the IRS to maintain and utilize. However, due to the privacy issue, each local promoting strategy is upgraded individually, remaining unknown to the IRS. Thereby, to maintain these strategies, the IRS often tends to merge the source sequences without altering the order within each sequence [17]. We refer to it as the *order-preservation constraint* and regard it as a basic setting of our IRS. Though the relative order within each source is determined by such a constraint, how to merge different sources still relies heavily on the upstream source information, including intra-source information and inter-source information. The former reflects local promoting strategies and sequential dependencies for each source; while the latter contains the comparison of different sources' importance.

However, few studies have been devoted to addressing both challenges. Some fairness or diversity recommendation methods [2,14,20] attempt to balance the exposure fairness between multiple groups. But they mainly focus on list-level fairness rather than long-term fairness over days and do not consider the information utilization from upstream sources, which lack personalization and degrade the platform's utility. A few works consider the long-term fairness [9], but they

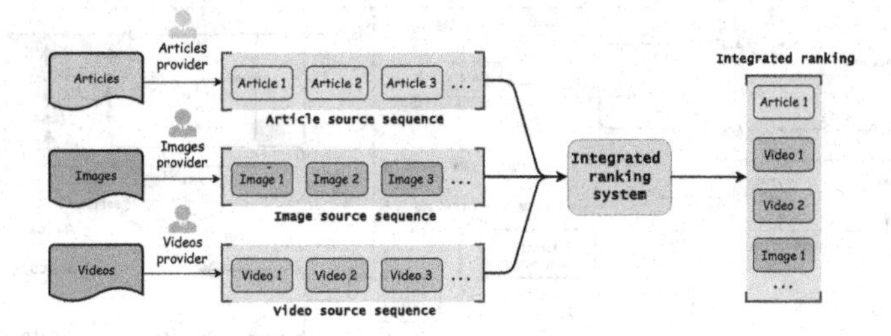

Fig. 1. An illustration of an integrated ranking process.

regard source sequences as unordered sets, ignoring the potential sequential pattern along items and breaking local promoting strategies. Some category-based recommendation methods [6,18] consider differences between different categories or sources. But they neither consider the fairness issue, nor study the critical inter-source and intra-source relationships.

Focusing on the *utility-fairness tradeoff* and the *source information utilization* challenges, we propose Multi-sourced Constrained Ranking (MSCRank) model. To manage the utility-fairness tradeoff, MSCRank builds *a utility-oriented task* and *a category prediction task* to optimize two goals, respectively. In the utility-oriented task, MSCRank predicts each item's click-through rate (CTR) [12] and calculates its expected utility, guiding the category prediction task towards a higher utility. In the category prediction task, MSCRank predicts which category should be allocated to each position with the KL loss [8] to optimize the exposure fairness. Moreover, we propose a Multi-sourced Attentive GRU (MA-GRU) structure to model the sequential knowledge from different upstream sources by considering inter-source and intra-source relationships. We pop items from source sequences to fill in categories with items to satisfy the order-preservation setting. The main contributions are concluded as follows:

- We focus on an important integrated ranking problem, and jointly analyze the utility-fairness tradeoff and the source information utilization challenges.
- We propose a novel MSCRank model with a utility-oriented task and a category prediction task to optimize utility and fairness simultaneously.
- We design a novel MA-GRU structure to capture the information and relationships from source sequences in integrated ranking.

We conduct experiments to show the effectiveness of MSCRank. It has also been deployed on the Huawei AppGallery with better online performance.

2 Problem Formulation

Formally, given c source sequences $\mathcal{S}_1, \ldots, \mathcal{S}_c$ as input. Each sequence $\mathcal{S}_i = [v_{i,1}, \ldots, v_{i,m_i}]$ contains m_i items. Integrated ranking is required to generate an integrated list Φ of M items ($M = \sum_i m_i$) to balance the overall utility (*e.g.*, number of clicks or overall revenue) and the exposure fairness while satisfying the

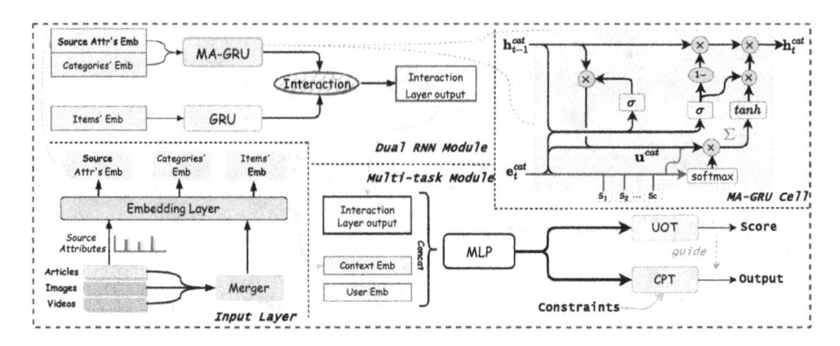

Fig. 2. The proposed MSCRank model.

order-preservation constraint. In practice, m_i is usually very large and we often focus on the top-K recommended results with $K \ll m_i$ (*e.g.*, top-10 or top-20). If $K > m_i$, the number of items may be insufficient for integrated ranking. Let $E(\Phi, i)$ be the number of exposure of source i in Φ, and $\mathbf{w} = [w_1, \ldots, w_c]$ is the desired exposure distribution. Moreover, denote the top-K items in the list Φ by $\Phi_{[:K]}$ and the position of an item v in list Φ by $\Phi(v)$. The exposure fairness and the order-preservation constraint are defined as follows.

Exposure Fairness. It is defined as the negative of the KL divergence between the category (source) distribution of the integrated list Φ and the desired distribution $\mathbf{w}$. We need to maximize the exposure fairness, *i.e.*, minimizing KL.

$$-KL([E(\Phi_{[:K]}, 1) : E(\Phi_{[:K]}, 2) : \ \ldots : E(\Phi_{[:K]}, c)] || \mathbf{w}) .$$

Order-Preservation Constraint. It is defined as follows. For example, suppose $\mathcal{S}_i$ for source i has two items $v_{i,1}$ and $v_{i,2}$, and $v_{i,1}$ is ranked before $v_{i,2}$ in the initial source sequence, then in Φ, $v_{i,1}$ should also be ranked before $v_{i,2}$.

$$\Phi_{[:K]}(v_{i,1}) < \Phi_{[:K]}(v_{i,2}) < \ldots < \Phi_{[:K]}(v_{i,m_i}) \quad \forall i = 1, \ldots, c.$$

3 Proposed Model

The structure of our proposed Multi-Sourced Constrained Ranking (MSCRank) model is presented in Fig. 2 with two main parts: a dual RNN module and a multi-task module. The dual RNN module captures item-level and category-level interactions. In the multi-task module, we build a utility-oriented task and a category prediction task to manage the utility-fairness tradeoff. In what follows, we elaborate on the details of MSCRank.

3.1 Input Layer

MSCRank takes c sequences $\mathcal{S}_1, \ldots, \mathcal{S}_c$ as input. It first merges all sequences with *a merger* into a list L_o with length M by re-scaling the ranking score provided by each source to the same range, to roughly fuse the information. It then re-ranks the merged list L_o to optimize the utility and exposure fairness.

There are generally three types of features, including user features (*e.g.*, gender, age), item features (*e.g.*, category, price), and context features (*e.g.*, time, location). Specifically for MSCRank, we split item features into category-level (item's category, tag, *etc.*) and item-level (item's name, size, *etc.*). Thus each item can be modeled from two different granularities. Then we apply embedding layers and get $\mathbf{E}^{user}$, $\mathbf{E}^{item}$, $\mathbf{E}^{cat}$ and $\mathbf{E}^{con}$, respectively. In addition, we denote the embeddings of some important attributes in source i (like items' identities, categories) as $\mathbf{E}_i^s$ (*i.e.*, Source Attr's Emb in Fig. 2).

3.2 Dual RNN Module

As discussed before, MSCRank is required to predict a suitable category for each position due to the order-preservation constraint. Therefore, category-level information is of great importance. On the other hand, category-level and item-level features along ranking lists may have different sequential patterns. Explicitly modeling item-category interactions is beneficial to address the *source information utilization* challenge. Following these ideas, we design a dual RNN module to extract item-level and category-level dependencies of the ranking list, with two RNNs and our proposed novel MA-GRU structure.

Item-Level RNN. We adopt the traditional GRU [3,4] to get sequential representations of item-level features: $\mathbf{H}^{item} = \text{GRU}\left(\mathbf{E}^{item}\right) \in \mathbb{R}^{M \times d_g}$. $\mathbf{H}^{item}$ denotes GRU hidden states, and d_g is the GRU dimension.

MA-GRU Based Category-Level RNN. In category-level RNN, we design a Multi-sourced Attentive GRU (MA-GRU) based on traditional GRU to better utilize source sequences' information, discover inter-source and intra-source relationships, and output category-level representations. It is illustrated in the upper right corner of Fig. 2 (red parts show our modifications). First, we use a weight matrix to aggregate source i information $\mathbf{S}_i$. Then we determine the percentage p_i of information that should be absorbed into the current state with a softmax function. Finally, we build hidden states of MA-GRU, which can capture the differences between categories and select the best one.

First, to aggregate the information in each source sequence i ($i = 1, \ldots, c$), we use a position-weighting method and get the aggregated representation $\mathbf{S}_i = \mathbf{W}_i^p \cdot \mathbf{E}_i^s$. In our paper, we suppose different positions are of the same importance, and regard $\mathbf{W}_i^p$ as an average weight matrix.

Then we use the softmax function to determine the percentage (attention weight) of encoded attributes representations from different sources that should remain towards the next RNN hidden state. We denote p_i as the percentage value (importance score) of source i, $\oplus$ as the concatenation, $\mathbf{e}_t^{cat}$ as the category-level embedding in position t (one row in $\mathbf{E}^{cat}$), $\mathbf{s}_{i,t}$ as the aggregated information of source i in position t (one row in $\mathbf{S}_i$). The percentage can be calculated as:

$$(p_1, \ldots, p_c) = softmax\left(\left(\mathbf{e}_t^{cat} \oplus \mathbf{s}_{1,t}\right), \ldots, \left(\mathbf{e}_t^{cat} \oplus \mathbf{s}_{c,t}\right)\right) . \quad (1)$$

Next, we use the learned percentage values to construct GRU's hidden representations in the current position t by multiplying the percentage values and corresponding sequence features. Specifically, compared with the traditional GRU, the position t's **input** for MA-GRU becomes $(\mathbf{e}_t^{cat} \oplus \mathbf{s}_{1,t} \oplus \ldots \oplus \mathbf{s}_{c,t})$, and the input for the *tanh* function changes from $\mathbf{u}^{cat}$ to $\sum_i p_i \cdot (\mathbf{u}^{cat} \oplus \mathbf{s}_{i,t})$.

Finally, applying these new input and update strategies above into MA-GRU, we get the category-level hidden states $\mathbf{H}^{item} \in \mathbb{R}^{M \times d_g}$ of all M positions.

Interaction Between Two RNNs. The output of item-level and category-level RNNs contain information at different granularities. To fuse these information, we make interactions between them to get the blended representation $\mathbf{H}^b \in \mathbb{R}^{M \times d_g}$. Concretely, we implement a element-wise interaction: $\mathbf{H}^b = \mathbf{H}^{item} \cdot \mathbf{H}^{cat}$.

3.3 Multi-task Module

To address the *utility-fairness tradeoff* challenge and optimize the utility and fairness simultaneously, we build a multi-task module. We first build a shared bottom by aggregating $\mathbf{H}^b$ with context and user embeddings, with the multi-layer perceptron (MLP): $\mathbf{H}^{fc} = MLP(\mathbf{H}^b \oplus \mathbf{E}^{con} \oplus \mathbf{E}^{user})$.

Utility-Oriented Task. To optimize the utility, we build a utility-oriented task (UOT) by predicting the CTR. The estimated CTR vector in M positions is denoted as $\mathbf{H}^{UOT} \in \mathbb{R}^M$ by applying MLP and a sigmoid function σ:

$$\mathbf{H}^{UOT} = \sigma(MLP(\mathbf{H}^{fc})) . \quad (2)$$

Combining the estimated CTR with item's price, we can calculate the expected item's utility $\mathbf{U}$ (*i.e.*,**score** in Fig. 2) at each position: $\mathbf{U} = \mathbf{H}^{UOT} * \mathbf{P}$, where $\mathbf{P} \in \mathbb{R}^M$ is the item's price. It will be used in the category prediction task.

Category Prediction Task Guided by Utility. Integrated ranking requires models to predict the suitable category for each position, and fill in items from corresponding sources. Therefore, we design a category prediction task (CPT) to predict each position's category, *i.e.*,

$$\mathbf{H}^{CPT} = softmax(MLP(\mathbf{H}^{fc})) , \quad (3)$$

where $\mathbf{H}^{CPT} \in \mathbb{R}^{M \times (c+1)}$ is an estimated click probability matrix of all categories in M positions. The first c columns represent the click probabilities of c categories, and the last column represents the probability of not being clicked.

Next we re-arrange (re-rank) rows of $\mathbf{H}^{CPT}$ according to the descending order of each element in $\mathbf{U}$ and get $\mathbf{H}_{\mathbf{x}}^{CPT}$. Rows with higher expected utility are put forward in $\mathbf{H}_{\mathbf{x}}^{CPT}$, so that MSCRank will predict categories by considering items' utility. Finally, we use the re-arranged probability matrix $\mathbf{H}_{\mathbf{x}}^{CPT}$ to optimize the exposure fairness, which will be introduced in Sec. 3.4.

3.4 Model Training

MSCRank takes click signals (0 or 1) as initial labels. Suppose $y_j \in \{0,1\}$ and $c_j \in \{1,\ldots,c\}$ are the click signal and category in position j. In the *utility-oriented task*, we use the cross-entropy loss to fit each estimated CTR h_j^{UOT} (*i.e.*, one row in $\mathbf{H}^{UOT}$) to y_j:

$$\mathcal{L}_{UOT} = -\frac{1}{M}\sum_{j=1}^{M}\left(y_j \log h_j^{UOT} + (1-y_j)\log(1-h_j^{UOT})\right) . \quad (4)$$

In the *category prediction task*, we first construct one-hot label $\mathbf{y}_j^{CPT}$ for $(c+1)$-class category classification: (i) If $y_j = 1$, then the c_j-th value is set to 1; (ii) If $y_j = 0$, then the last value is set to 1. The CPT tries to fit the predicted vector $\mathbf{h}_j^{CPT}$ (*i.e.*, one row in $\mathbf{H}^{CPT}$) to the label $\mathbf{y}_j^{CPT}$ with log loss:

$$\mathcal{L}_{CPT} = -\frac{1}{M}\sum_{j=1}^{M}\left(\mathbf{y}_j^{CPT} \cdot \log(\mathbf{h}_j^{CPT})^T\right) . \quad (5)$$

Besides, the exposure category distribution aims to be optimized towards the desired distribution $\mathbf{w}$. First, we mask the last column in the re-arranged $\mathbf{H}_{\mathbf{x}}^{CPT}$ and choose its index with the maximum probability as the predicted category. After that, we calculate the batch-level category distribution $\mathbf{w}^p$ among all categories. We then use the KL divergence [8] to optimize $\mathbf{w}^p$ towards $\mathbf{w}$ in Eq.(6). This is a batch-level loss that optimizes fairness over a period of time rather than in every list, considering users' preferences on categories.

$$\mathcal{L}_{KL} = KL\left(\mathbf{w}||\mathbf{w}^p\right) . \quad (6)$$

Combining Eq. (4), Eq.(5) and Eq.(6) with hyperparameters α and β to control the importance of tasks, the loss for MSCRank is:

$$\mathcal{L} = \mathcal{L}_{CPT} + \alpha\mathcal{L}_{KL} + \beta\mathcal{L}_{UOT} . \quad (7)$$

4 Experiments

We conduct experiments to answer three research questions:

- **RQ1**: How does MSCRank perform compared with other baselines?
- **RQ2**: How do different network structures affect MSCRank?
- **RQ3**: How does MSCRank actually perform in the real industrial platform?

4.1 Experimental Settings

Datasets. Since there are no public datasets available for integrated ranking, we construct a dataset from **PRM public dataset** [11], by remaining three most frequent categories and construct requests in original orders. We also collect an **industrial dataset** from an industrial platform for half of a month. These two datasets contain users' requests, source sequences, and users' feedbacks (*i.e.*, click signals). The detailed statistics are in Table 1. Each dataset is split into training and test sets with a proportion of 4:1.

Table 1. Statistics of two offline datasets.

	#Requests	#Users	#Category A	#Category B	#Category C
PRM dataset	7,919,659	605,668	1,411,185	291,629	311,364
Industrial dataset	194,223	184,443	9,018	4,771	–

Evaluation Metrics. According to the *utility-fairness tredeoff* challenge, we adopt commonly used *utility@K* and *fairness@K* [15,17]. *utility@K* is the average *platform's utility* per request on top-K items. *fairness@K* is defined through the KL divergence between the top-K category distribution $\mathbf{w}_K^p$ and the desired distribution $\mathbf{w}_K$ over all integrated lists. Smaller *fairness@K* indicates more balanced exposure opportunities. We also propose a novel metric α-*utility* to measure whether the utility is balanced across categories, inspired by α-*NDCG* [5]. It is defined as: α-*utility@K* $= \frac{1}{|\mathcal{R}|}\sum_{r\in\mathcal{R}}\sum_{i=1}^{K}\alpha^{N(c_{r,i})}u_{r,i}$, where $\mathcal{R}$ is the requests set, $u_{r,i}$ is the item's utility, and $\alpha = 0.5$ is a discounted factor similar to [5]. $N(c_{r,i})$ is the total counts of category $c_{r,i}$ appears in request r before position i.

Hyperparameters. We choose MSCRank's hyperparameters with grid search. The embedding size of discrete features is 64, the hidden dimension of the GRU and linear layer is 64, the maximum length of a request is 30. α and β are set to 1.0 and 1.0 by default. The optimizer is Adam; the learning rate is 10^{-5}; the batch size is 128 for the industrial dataset and 512 for the PRM dataset.

4.2 Baselines

We compare MSCRank[1] with several baselines. **MMR** [2] is a ranking algorithm that considers both diversity and relevance. **FA*IR** [20] is a fairness-based method ensuring that the proportion of protected candidates remains above a given minimum. **LinkedIn-Det** [7] proposes fairness methods to achieve the desired distribution of top-K results for some attributes. **LinkedIn-Reranker** [19] proposes a lightweight re-rank algorithm to optimize expected engagement and revenue. **DHCRS** [6] is a hierarchical RL based method to recommend items by choosing categories first.

4.3 Model Selection

Our experiments have two key metrics: *utility@K* and *fairness@K*. We hope the utility should be large while the fairness should be close to 0. However, due to the utility-fairness tradeoff challenge, optimizing one metric sometimes harms the other. Therefore, following [13], we draw Pareto Fronts in the fairness-utility plane, by adjusting the KL loss weight α for MSCRank and DHCRS, and by applying a grid search on fairness- or diversity-related hyperparameters for other baselines. To give a direct impression, we constrain utility or fairness in small ranges and then present their detailed performance.

[1] Code is available in https://github.com/sjtulyf123/MSCRank.

Table 2. Performance comparison under model selection rules on two datasets.

Methods	PRM dataset (top-10)				Industrial dataset (top-20)					
	Utility in 0.0385 ± 0.0005		Utility in 0.0355 ± 0.0005		Utility in 1.906 ± 0.002		Utility in 1.897 ± 0.002		Utility in 1.887 ± 0.002	
	Fairness	α-utility	Fairness	α-utility	Fairness	α-utility	Fairness	α-utility	Fairness	α-utility
MMR	0.3641	0.0143	–	–	–	–	0.1502	0.5867	–	–
FA*IR	–	–	–	–	0.1263	0.6084	0.0971	**0.6084**	0.0853	0.6084
LinkedIn-Det	0.3671	0.0144	0.3434	0.0132	0.1237	0.6080	0.0965	0.6081	0.0801	0.6084
LinkedIn-Reranker	–	–	–	–	0.1326	0.6083	0.0974	0.6083	0.0833	0.6083
DHCRS	0.3502	0.0094	0.3304	0.0096	0.1240	0.6032	0.1010	0.6032	0.0837	0.6030
MSCRank	**0.3205**	**0.0155**	**0.3091**	**0.0150**	**0.1095**	**0.6104**	**0.0812**	0.6067	**0.0668**	**0.6123**

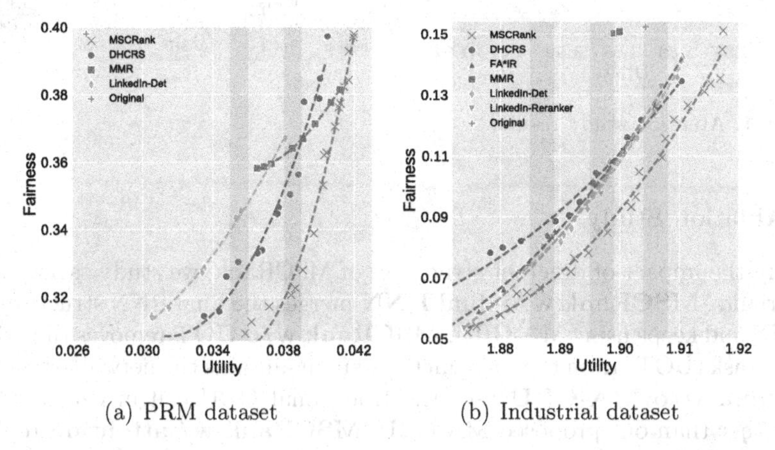

(a) PRM dataset (b) Industrial dataset

Fig. 3. Pareto Fronts of different methods on two datasets.

4.4 Performance Comparison

We draw the Pareto Fronts on both datasets in Fig. 3, where "Original" denotes the originally merged ranking lists L_o. On PRM dataset, we focus on the top-10 performance. **Note** that FA*IR and LinkedIn-Reranker are designed for two categories, so we do not compare them on PRM dataset. On industrial dataset we focus on the top-20 performance due to longer L_o, To give a detailed comparison, We select results within small utility ranges with the best *fairness@K* and calculate *α-utility@K* in Table 2. We conclude observations as follows:

Firstly, In Fig. 3, higher utility value and lower fairness value (*i.e.*, smaller KL) are better. If the results of one method are closer to the bottom-right corner than another, the utility-fairness tradeoff of this method will also be better. Therefore, our MSCRank achieves the best performance on the utility-fairness tradeoff. The fit polynomial curves also demonstrate MSCRank's superior performance. From Table 2, we can also observe the fairness improvement.

Secondly, MSCRank generally outperforms baselines in the *α-utility* metric, indicating that it can balance the utility among different categories. It is because MSCRank adopts the multi-task learning to optimize the utility while considering exposure, and uses the MA-GRU to capture the category-level information.

Lastly, Traditional methods like MMR fail to achieve diverse fairness values even if we apply a grid search on their hyperparameters, because these methods only use simple rules to achieve certain fairness or diversity goals.

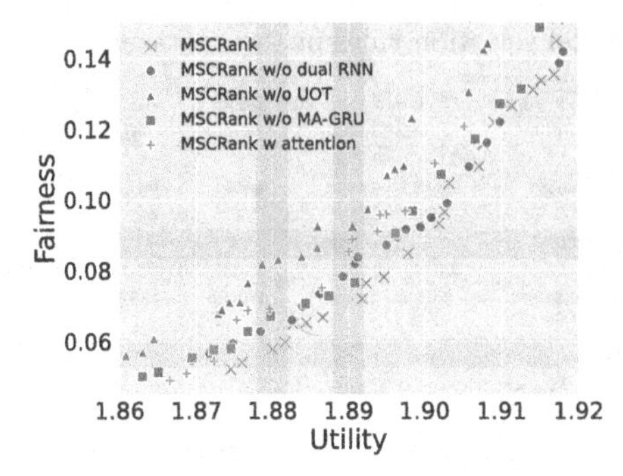

Fig. 4. Ablation study results.

Table 3. Ablation study results under different model selection rules.

top-20	Utility in 1.8800 ±0.0020	Utility in 1.8900 ±0.0020	Fairness in 0.0566 ±0.0040	Fairness in 0.0873 ±0.0040
	Fairness		Utility	
MSCRank	**0.0578**	**0.0714**	**1.8817**	**1.8996**
w/o dual RNN	0.0627	0.0782	1.8748	1.8951
w/o UOT	0.0818	0.0926	1.8645	1.8859
w/o MA-GRU	0.0671	0.0765	1.8745	1.8963
w/ attention	0.0678	0.0847	1.8736	1.8943

4.5 Ablation Study

To verify the impact of different structures of MSCRank, we study several ablation variants. **MSCRank w/o dual RNN** merges the dual RNN structure into one RNN and keeps using MA-GRU. **MSCRank w/o UOT** removes the utility-oriented task (UOT) from the loss function, and maintains the network structure. **MSCRank w/o MA-GRU** uses the traditional GRU cell in category-level RNN rather than our proposed MA-GRU. **MSCRank w/ attention** replaces the proposed MA-GRU with the classical attention layer.

We experiment on the industrial dataset and present the Pareto Fronts in Fig. 4. We also present the utility and fairness in Table 3, by restricting *utility@20* or *fairness@20* in different ranges. The performance gap between MSCRank and MSCRank w/o UOT indicates the effectiveness of our utility-oriented task. By predicting items' CTR, the UOT can guide the CPT towards higher utility. MSCRank performs better than MSCRank w/o dual RNN and w/o MA-GRU when the fairness is close to 0. It is because when we need both better fairness and higher utility, we should make good use of source sequences' information and focus more on category-level features with MA-GRU and dual RNN. When we just need high utility and do not care about the fairness (top-right corner), they become less critical. The performance of MSCRank w/ attention further indicates the necessity of MA-GRU to model relationships between sources.

4.6 Online A/B Testing

We have deployed MSCRank to Huawei AppGallery with an online A/B test. In what follows, we introduce the deployment feasibility, system, and results.

Deployment Feasibility. Before deploying the model, we first analyze the feasibility of MSCRank in terms of efficiency. MSCRank adopts dual RNN structures, and the category-level and item-level RNNs execute in parallel, so its time complexity is $O(CT)$, where T is the average sequence length and C is the cost of one GRU operation. The DHCRS baseline utilizes two RNNs to extract states, with a specifically-designed category selection rule in each position. Its

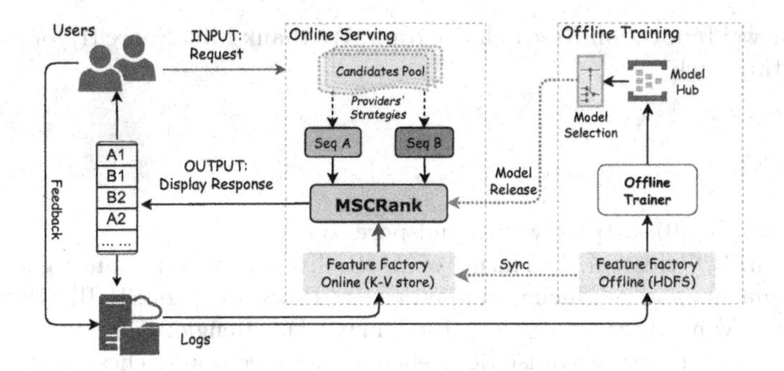

Fig. 5. An overview of the deployed IRS.

time complexity is $O(C(T+T'))$, where T' is the cost of one category selection process, which cannot be neglected. Therefore, the time complexity of MSCRank is smaller than DHCRS, and is feasible for the online deployment.

System Overview. The Huawei AppGallery has millions of daily users and tens of millions of logs. The system is illustrated in Fig. 5, including an *offline training module* and an *online serving module.* The offline trainer periodically trains and updates MSCRank. During the online serving process, MSCRank receives multiple source sequences from upstream providers, generates integrated ranking lists, and displays them to users. The online feature factory is synchronized with the offline feature factory at intervals. After receiving users' feedbacks, the system will record these logs to further improve MSCRank.

Online Results. We compare MSCRank with the method currently deployed on the Huawei AppGallery through the A/B test for consecutive five days. Both methods aim to keep the exposure fairness between sources A and B close to an ideal value and maximize the platform's overall utility. MSCRank improves the exposure fairness by **16.60%** and increases the utility by **1.52%**. Its inference time per request is around **5 ms**, satisfying the latency requirement.

5 Conclusion

In this paper, we propose a novel integrated ranking model MSCRank. To manage the utility-fairness tradeoff challenge, we adopt multi-task learning, building a utility-oriented task and a category prediction task. We also construct a dual RNN structure to further improve the utility by discovering sequential interactions on category-level and item-level features. To manage the information utilization challenge, we propose a novel MA-GRU cell, capturing inter-source and intra-source relationships. Empirical results and online A/B test results demonstrate the superior performance and efficiency of MSCRank.

Acknowledgements. The Shanghai Jiao Tong University team is partially supported by National Natural Science Foundation of China (62177033). The work is also sponsored by Huawei Innovation Research Program. We thank MindSpore [1] for the partial support of this work.

References

1. Mindspore (2020). https://www.mindspore.cn/
2. Carbonell, J., Goldstein, J.: The use of MMR, diversity-based reranking for reordering documents and producing summaries. In: Proceedings of SIGIR (1998)
3. Cho, K., Van Merriënboer, B., Bahdanau, D., Bengio, Y.: On the properties of neural machine translation: encoder-decoder approaches. arXiv preprint arXiv:1409.1259 (2014)
4. Chung, J., Gulcehre, C., Cho, K., Bengio, Y.: Empirical evaluation of gated recurrent neural networks on sequence modeling. arXiv preprint arXiv:1412.3555 (2014)
5. Clarke, C.L., et al.: Novelty and diversity in information retrieval evaluation. In: Proceedings of SIGIR (2008)
6. Fu, M., Agrawal, A., Irissappane, A.A., Zhang, J., Huang, L., Qu, H.: Deep reinforcement learning framework for category-based item recommendation. IEEE Trans. Cybern. **52**(11), 12028–12041 (2021)
7. Geyik, S.C., Ambler, S., Kenthapadi, K.: Fairness-aware ranking in search & recommendation systems with application to linkedin talent search. In: Proceedings of of KDD (2019)
8. Kullback, S.: Information theory and statistics. Courier Corporation (1997)
9. Morik, M., Singh, A., Hong, J., Joachims, T.: Controlling fairness and bias in dynamic learning-to-rank. In: Proceedings of SIGIR (2020)
10. Okura, S., Tagami, Y., Ono, S., Tajima, A.: Embedding-based news recommendation for millions of users. In: Proceedings of KDD (2017)
11. Pei, C., et al.: Personalized re-ranking for recommendation (2019)
12. Richardson, M., Dominowska, E., Ragno, R.: Predicting clicks: estimating the click-through rate for new ads. In: Proceedings of WWW (2007)
13. Sener, O., Koltun, V.: Multi-task learning as multi-objective optimization. In: Proceedings of NeurIPS (2018)
14. Sonboli, N., et al.: Librec-auto: a tool for recommender systems experimentation. In: Proceedings of CIKM (2021)
15. Wan, M., Ni, J., Misra, R., McAuley, J.: Addressing marketing bias in product recommendations. In: Proceedings of WSDM (2020)
16. Xi, Y., et al.: On-device integrated re-ranking with heterogeneous behavior modeling. In: Proceedings of KDD, pp. 5225–5236 (2023)
17. Xia, W., Liu, W., Liu, Y., Tang, R.: Balancing utility and exposure fairness for integrated ranking with reinforcement learning. In: Proceedings of CIKM (2022)
18. Xie, R., Zhang, S., Wang, R., Xia, F., Lin, L.: Hierarchical reinforcement learning for integrated recommendation. In: Proceedings of AAAI (2021)
19. Yan, J., Xu, Z., Tiwana, B., Chatterjee, S.: Ads allocation in feed via constrained optimization. In: Proceedings of KDD (2020)
20. Zehlike, M., et al.: Fa* IR: a fair top-k ranking algorithm. In: Proceedings of CIKM (2017)

Soft Contrastive Learning for Implicit Feedback Recommendations

Zhen-Hua Zhuang[1,2] and Lijun Zhang[1,2](✉)

[1] National Key Laboratory for Novel Software Technology, Nanjing University, Nanjing, China
{zhuangzh,zhanglj}@lamda.nju.edu.cn

[2] School of Artificial Intelligence, Nanjing University, Nanjing, China

Abstract. Collaborative filtering (CF) plays a crucial role in the development of recommendations. Most CF research focuses on implicit feedback due to its accessibility, but deriving user preferences from such feedback is challenging given the inherent noise in interactions. Existing works primarily employ unobserved interactions as negative samples, leading to a critical noisy-label problem. In this study, we propose SCLRec (Soft Contrastive Learning for Recommendations), a novel method to alleviate the noise issue in implicit recommendations. To this end, we first construct a similarity matrix based on user and item embeddings along with item popularity information. Subsequently, to leverage information from nearby samples, we employ entropy optimal transport to obtain the matching matrix from the similarity matrix. The matching matrix provides additional supervisory signals that uncover matching relationships of unobserved user-item interactions, thereby mitigating the noise issue. Finally, we treat the matching matrix as soft targets, and use them to train the model via contrastive learning loss. Thus, we term it soft contrastive learning, which combines the denoising capability of soft targets with the representational strength of contrastive learning to enhance implicit recommendations. Extensive experiments on three public datasets demonstrate that SCLRec achieves consistent performance improvements compared to state-of-the-art CF methods.

Keywords: Contrastive Learning · Implicit Recommendations · Collaborative Filtering

1 Introduction

In the era of information explosion, recommender system has become a crucial tool for enhancing user engagement and satisfaction by providing personalized suggestions for products [16], videos [6], among others. Collaborative filtering (CF) has been widely adopted in personalized recommendation systems [3,20,22], with the key idea that similar users tend to share similar preferences. Typically, CF models mainly rely on historical interactions to predict user interests for candidate items [26]. Most CF research [13,20] focuses on implicit

D.-N. Yang et al. (Eds.): PAKDD 2024, LNAI 14649, pp. 219–230, 2024.
https://doi.org/10.1007/978-981-97-2262-4_18

feedback which only contains user-item interactions (e.g., clicks, browsing history) because it encompasses a large volume of data and captures abundant collaborative information in a simple manner [27,28].

A persistent challenge for implicit recommendations is how to formulate the loss function based on implicit feedback [3]. In general, there are three popular types of loss functions in recommendation systems: pointwise loss [18], pairwise loss [20], and listwise loss [3]. Specifically, the contrastive learning loss [17], as a novel type of listwise loss, has been introduced to the implicit feedback recommendations due to its excellent representational capabilities [26,31]. Contrastive learning (CL) aims to learn feature representations by minimizing the distance between similar (matched) sample pairs and maximizing the distance between dissimilar (unmatched) pairs. Existing methods [26,31] assume that user-item pairs within observed interactions are matched, while user-item pairs within unobserved interactions are considered unmatched in implicit feedback. However, such hard labeling mechanism, which strictly classifies user-item pairs as either matched or unmatched, fails to account for the inherent ambiguity present in missing feedback within implicit datasets. Specifically, unobserved interactions might not indicate disinterest, but simply that the items have not been exposed to the user. Thus directly fitting implicit feedback without addressing the noise issue cannot yield optimal user representations, leading to performance degradation.

Inspired by the recent advancements in CL for noisy-label problems [4], we introduce a novel contrastive learning loss to mitigate the noise issue in implicit feedback recommendations. Our approach comprises three phases: First, we estimate a similarity matrix that captures the likeness between users and items in a batch based on their embeddings and item popularity information. Next, to utilize information from nearby samples, we leverage entropy-regularized optimal transport to obtain the matching matrix which reflects the matching degree for user-item interactions from the similarity matrix. In deviation from previous methods [26,31], our approach assigns non-zero matching values, i.e., soft targets, to unobserved user-item interactions. The soft targets provide additional supervisory signals to guide the learning of the recommendation system, uncovering unobserved user-item matching relationships and effectively alleviating the noise issue. Finally, we optimize the model using these soft targets via the contrastive learning loss, which we thus term as soft contrastive learning loss. We conduct extensive experiments on three real-world datasets and observe consistent performance improvements when optimizing a matrix factorization model using our proposed soft contrastive loss.

The main contributions of this work can be summarized as follows:

- To the best of our knowledge, SCLRec is the first work that introduces soft contrastive learning into the recommendation domain.
- SCLRec assigns soft targets to unmatched user-item pairs, providing additional supervisory signals to identify latent user-item correspondences in unobserved interactions and effectively alleviate the noise problem.
- Experiments on three public datasets show that SCLRec achieves better performance than the state-of-the-art methods.

2 Related Work

Collaborative Filtering. CF plays a crucial role in recommender systems [22]. Most CF research focuses on implicit feedback, with a prominent approach being Bayesian Personalized Ranking (BPR) [20]. BPR uniformly samples unobserved interactions as negative samples, leading to a critical noisy-label problem. Existing denoising techniques [24,27,28,30] can be divided into two categories: sample selection methods and sample re-weighting methods. Sample selection methods choose clean and information-rich samples to train the model and enhance its performance. IR [28] represents a typical sample selection approach that iteratively creates pseudo-labels based on the disparity between labels and predictions to exclude noisy samples. However, sample selection methods, while effective in gathering cleaner data, rely on the sampling distribution, potentially resulting in biased gradient estimation and degrading recommendation performance. Conversely, sample re-weighting methods differentiate between clean and noisy interactions based on the model's learning process (e.g., loss values and predictions). T-CE [27] adopts a sample re-weighting strategy, dynamically assigning lower weights to samples exhibiting high loss values under the premise that noisy samples suffer larger losses. Yet, although these methods achieve promising results, they run the risk of neglecting hard clean samples and lack of adaptivity and universality [10].

Contrastive Learning. CL is a representative self-supervised learning (SSL) method, which measures the dependency of input variables by calculating their mutual information [1]. A prominent methodology in contrastive learning is the InfoNCE loss [17], which has been extensively applied in the fields of computer vision [2,4,19] and natural language processing [9]. The InfoNCE loss aims to minimize the distance between positive sample pairs while maximizing the distance between negative pairs, thereby facilitating effective representation learning. With the growing popularity of SSL, there have been efforts [14,26,31] to incorporate contrastive loss into recommendation systems. CLRec [31] employs the InfoNCE loss to address the exposure bias in CF and enhance deep candidate generation (DCG) in terms of fairness within large-scale recommendation scenarios. DirectAU [26] explores the desired alignment and uniformity properties of CF from the perspectives of contrastive representation learning. It works to push positive pairs closer to each other and make random pairs scatter across the unit hypersphere. However, these attempts have neglected the inherent noise issue in implicit recommendations. In contrast, we assign soft targets to unmatched user-item pairs, providing additional supervisory signals to alleviate the noise issue.

3 Methodology

In this section, we first introduce some notations related to collaborative filtering and InfoNCE loss [17]. Then we demonstrate the architecture and optimization process of our proposed SCLRec model.

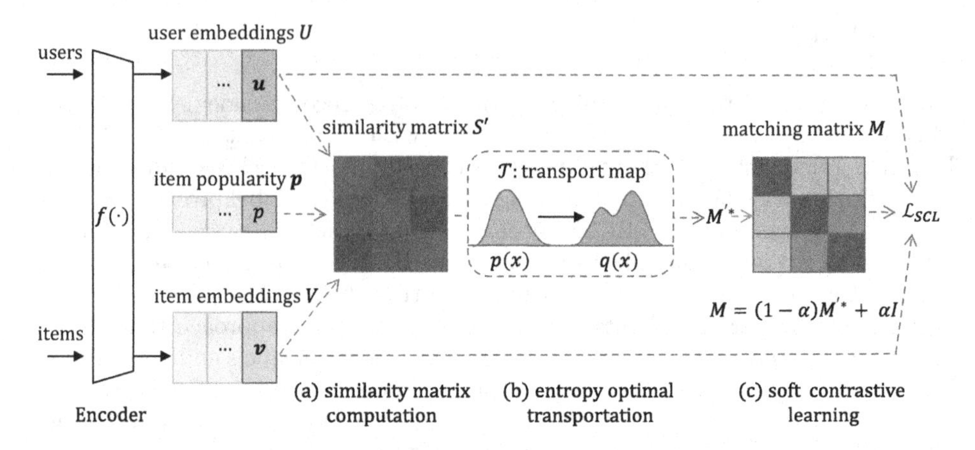

Fig. 1. Overview of the proposed SCLRec, which can mainly be divided into three parts: (a) similarity matrix computation, (b) entropy optimal transportation, and (c) soft contrastive learning.

3.1 Notations

Collaborative Filtering. Let $\mathcal{X}$ represent the set of users and $\mathcal{Y}$ denote the set of items. Given the observed user-item interactions $\mathcal{R} = \{(x, y) \mid x$ has interacted with $y\}$, the goal of CF methods is to estimate a score $s(x, y) \in \mathbb{R}$ for each unobserved interaction. The score indicates the likelihood that user x will interact with item y, and items with the highest scores for each user will be recommended. In general, most CF methods [12,20] employ an encoder network $f(\cdot)$ that maps each user and item into a low-dimensional embedding. The embeddings are further l_2-normalized to the unit hypersphere, represented as $\widetilde{f(x)}, \widetilde{f(y)} \in \mathbb{R}^d$, where d is the dimension of the latent space. We denote $\widetilde{f(x)}$ as $\mathbf{u}$ and $\widetilde{f(y)}$ as $\mathbf{v}$, then the user embeddings within the batch are denoted as U and the item embeddings are represented by V. Finally, the predicted score is defined as the similarity between the user and item representation (e.g., dot product, $s(x, y) = \mathbf{u}^T\mathbf{v}$).

InfoNCE Loss. CLRec [31] trains the user and item encoders with contrastive learning to pull the matched user-item pairs closer and push the unmatched user-item pairs farther. This is achieved by minimizing the InfoNCE loss [17], which is defined as

$$\mathcal{L}_{\text{InfoNCE}} = -\frac{1}{N}\sum_{i=1}^{N}\sum_{j=1}^{N} I_{ij} \log \frac{\exp((\mathbf{u}_i^\top \mathbf{v}_j)/\tau)}{\sum_{k=1}^{N} \exp((\mathbf{u}_i^\top \mathbf{v}_k)/\tau)}, \tag{1}$$

where $(\mathbf{u}_i^\top \mathbf{v}_j)$ is the cosine similarity between two ℓ_2-normalized embedding vectors. τ represents a temperature parameter, while N denotes the batch size, i.e., the number of user-item pairs. I_{ij} is the element of an identity matrix I with $I_{ii} = 1, \forall i$ and $I_{ij} = 0, \forall i \neq j$. Note that $\mathbf{u}_i$ and $\mathbf{v}_j$ are on the unit hypersphere.

3.2 The SCLRec Framework

As mentioned above, there have been efforts [26,31] to incorporate contrastive learning loss into recommendations. However, these efforts have overlooked the inherent noise issue in implicit recommendations. Inspired by recent breakthroughs in CL for the noisy-label issue [4], we propose SCLRec, a novel method to alleviate the noise problem in implicit recommendations.

An overview of SCLRec is illustrated in Fig. 1, specifically, we first use the encoder to generate user and item embeddings. Then we construct a similarity matrix for unmatched users and items based on those embeddings and item popularity information. Subsequently, to utilize information from nearby samples, we employ entropy optimal transport to obtain the matching matrix for unmatched user-item pairs. After incorporating information from the matched pairs, we obtain the final matching matrix which serves as soft targets to provide additional signals for enhancing the recommendations. Finally, we adopt a modified contrastive learning loss as optimization objective, transforming the one-hot hard target I from Eq. (1) into a soft target M. Hence, it is denoted as the soft contrastive learning loss. In summary, our method primarily consists of three parts: similarity matrix computation, entropy optimal transportation and soft contrastive learning, detailed in the following sections.

Similarity Matrix Computation. Naturally, we consider harnessing embedding information to calculate similarity. To obtain reliable embeddings, we incorporate the Exponential Moving Average (EMA) method [23] to stabilize the encoder. This involves constructing a teacher encoder with the same model structure as the original encoder but with parameter updates following the EMA principle, i.e., $\tilde{\theta} \leftarrow m\tilde{\theta} + (1-m)\theta$, where θ and $\tilde{\theta}$ represent the weights of the original encoder and the teacher encoder, respectively, and m is momentum parameter set to 0.9. The user and item embeddings are generated by this teacher encoder. Then we utilize these embeddings and item popularity information to compute the similarity matrix as

$$S' = \gamma_u U^T U + \gamma_v V^T V + \gamma_p \mathbf{1}^T \text{softmax}(\boldsymbol{p}) + U^T V - \eta I. \tag{2}$$

For users, the term $U^T U$ computes cosine similarity between user embeddings. Intuitively, it assumes that similar users might favor items liked by their counterparts. For items, $V^T V$ can be used to measure item similarity. Note that user and item embeddings could be regarded as representations of latent factor vectors across the dimensions of user preferences and item attributes, thus their similarities are inherently additive. Also, previous research [3] suggests that the popularity-based negative sampler usually exceeds the random negative sampler in implicit recommendations. Inspired by this, we utilize item popularity for computing the similarity. We obtain the item popularity distribution with $\boldsymbol{p} = [p_1, p_2, \cdots, p_N]$, where p_i denotes the frequency of item i for all users. Generally, more popular items are viewed as positive samples since they're more likely to be recommended. We then arrive at the formula

$\mathbf{1}^T\text{softmax}(\boldsymbol{p})$, using softmax$(\cdot)$ for normalizing the distribution. Specifically, $[\text{softmax}(\boldsymbol{p})]_i = \exp(p_i)/\sum_j \exp(p_j)$, converting the vector elements into a probability distribution. For user-item relationships, $U^T V$ captures the affinity between user and item embeddings [21]. Additionally, the term $-\eta I$ with $\eta \to \infty$ ensures diagonal elements of S' are infinitely small. We calculate the similarity matrix in Eq. (2) by linearly weighting all the aforementioned terms.

Entropy Optimal Transportation. Next, we focus on generating the matching matrix from the similarity matrix S'. A naïve approach might be to directly use the similarity values as matching degrees. However, this is an oversimplified perspective because the similarity matrix is heuristically designed and might deviate from the actual scenario. Drawing inspiration from the application of optimal transport in noisy label scenarios [4,8], we utilize information from nearby samples to estimate accurate matching values. Specifically, the matching matrix is obtained by solving the following problem:

$$M'^* = \underset{M'}{\arg\max}\langle M', S' \rangle_F + \lambda H(M'). \tag{3}$$

Here, $\langle M', S' \rangle_F$ represents the Frobenius inner product of the matching matrix M' and the similarity matrix S', and it aims to establish a direct relationship between the similarity degree and the matching strength in user-item pairs. Specifically, a higher similarity degree yields a greater matching score. $H(M') = -\sum_{ij} M'_{ij} \log M'_{ij}$ serves as an entropy regularization term to enhance robustness. Moreover, the solution to Eq. (3) has a closed-form formulation:

$$M'^* = \text{Diag}(\mathbf{r}) \exp(S'/\lambda)\text{Diag}(\mathbf{c}), \tag{4}$$

where vectors $\mathbf{r}$ and $\mathbf{c}$ can be computed using the iterative Sinkhorn-Knopp algorithm [7]. Additionally, the algorithm ensures that the sum of each row and column in M'^* equals 1. As depicted in Eq. (4), the term $\exp(S'/\lambda)$ highlights the significance of entropy regularization. Similar to the temperature parameter in CL, increasing λ will make the distribution of M'^* more dispersed, while decreasing it will yield the opposite effect.

Soft Contrastive Learning. After generating the matching matrix M'^* for unmatched user-item pairs, we can then obtain the overall matching matrix M, which serves as the soft target. To incorporate the matching information of matched user-item pairs, we define M as a linear combination of the identity matrix I and M'^*:

$$M = \alpha I + (1 - \alpha)M'^*. \tag{5}$$

We define it for two reasons: First, $\alpha \in [0, 1]$ can represent the prior matching probability (degree) between user $\mathbf{u}_i$ and item $\mathbf{v}_i$. Second, the formula can ensure that the sum of each user's matching probability (degree) with all items within a batch remains 1, and likewise for each item with all users. Consequently, in the

InfoNCE loss, the one-hot hard target I can be substituted with the soft target M, resulting in a soft contrastive learning loss:

$$\mathcal{L}_{\text{SCL}} = -\frac{1}{N}\sum_{i=1}^{N}\sum_{j=1}^{N}(\alpha I_{ij} + (1-\alpha)\, M'^{*}_{ij}) \log \frac{\exp((\mathbf{u}_i^\top \mathbf{v}_j)/\tau)}{\sum_{k=1}^{N}\exp((\mathbf{u}_i^\top \mathbf{v}_k)/\tau)}. \quad (6)$$

The formulation assigns soft targets to unmatched user-item pairs, providing additional supervision signals to better guide the learning process of recommendation systems.

4 Experiments

In this section, we evaluate our proposed model through comprehensive experiments and compare its results with current leading models on three public datasets. Our experiments are guided by the following research questions (RQs):

- **RQ1:** Does SCLRec outperform existing baselines in recommendation performance?
- **RQ2:** How do hyperparameters and components within SCLRec affect its performance?
- **RQ3:** Can SCLRec achieve denoising objectives effectively?

4.1 Experimental Settings

Datasets. We use three real-world datasets as outlined in Table 1: **MovieLens-10M** [11]: A well-known dataset containing movie ratings. **Gowalla** [5]: A dataset recording user check-in data from the Gowalla platform. **TmallBuy** [25]: An e-commerce dataset containing user purchase records on the Tmall platform. To construct implicit feedback, each entry is marked as 0/1 indicating whether the user rates the item. During preprocessing, we further ensure every user and item has at least 5 associated interactions.

Baselines. We compare the performance of SCLRec with various state-of-the-art CF methods:

General Methods: (1) **BPRMF** [20]: a typical approach optimizing matrix factorization via pairwise ranking loss; (2) **BUIR** [15]: a CF method that learns user and item embeddings only from positive interactions; (3) **LightGCN** [12]: a simplified graph convolution technique for CF.

CL-based Methods: (1) **CLRec** [31]: a method that uses the InfoNCE loss to mitigate the issue of exposure bias in CF; (2) **DirectAU** [26]: an approach that optimizes the properties of alignment and uniformity, inspired by contrastive representation learning.

Table 1. Statistics of datasets.

Dataset	#user ($\|\mathcal{X}\|$)	#item ($\|\mathcal{Y}\|$)	#inter. ($\|\mathcal{R}\|$)	avg. inter. per user	density
MovieLens	69.9k	10.2k	9998.9k	143.1	1.42%
Gowalla	29.9k	41.0k	1027.4k	34.4	0.08%
TmallBuy	413.1k	221.9k	4985.6k	12.1	0.02%

Denoising Methods: (1) **IR** [28]: a sample selection method iteratively relabels ambiguous samples to address noisy interactions; (2) **T-CE** [27]: a sample reweighting method that employs the Truncated BCE loss to assign zero weights to examples with high losses beyond a dynamic threshold.

Evaluation Protocols. We partition the datasets into training, validation, and testing sets with an 8:1:1 ratio. The evaluation metrics are Recall@N and Normalized Discounted Cumulative Gain (NDCG)@N for N = 10, 20, 50.

Implementation Details. We use Adam as the default optimizer and early stop is adopted if NDCG@20 on the validation dataset continues to drop for 10 epochs. We set the embedding size to 64 and the learning rate to 10^{-3} for all the methods. The training batch size is set to 1024 and the weight decay is tuned among [0, 10^{-8}, 10^{-6}]. The default encoder in SCLRec is a simple embedding table that maps user/item IDs to embeddings.

4.2 Overall Performance (RQ1)

The overall recommendation performance of SCLRec and various baselines are presented in Table 2. From the results, we can draw several key findings: First, we observe consistent performance improvements when comparing the proposed SCLRec with recent baselines on the MovieLens, Gowalla, and TmallBuy datasets. Furthermore, our results show that the T-CE method outperforms most CL-based and general methods on the MovieLens and TmallBuy datasets, and the IR method excels on the Gowalla dataset in a similar manner. These findings demonstrate that denoising methods perform better than most other methods, particularly in sparse datasets that are susceptible to noise. This can be attributed to the inherent noise in implicit feedback, making effective denoising techniques especially beneficial. In addition, our results indicate that the best baseline is significantly influenced by the specific characteristics of individual datasets. For example, T-CE performs notably on the MovieLens and Tmallbuy datasets, while IR is more suited for user check-in data like Gowalla. However, only SCLRec, which combines contrastive learning representations with soft target denoising ability, consistently delivers superior results across all datasets.

Table 2. Top-K recommendation performance on three datasets. The best results are in boldface, and the best baselines are underlined.

Setting		Baseline Methods							Ours
Dataset	Metric	BPRMF	BUIR	LightGCN	CLRec	DirectAU	IR	T-CE	SCLRec
MovieLens-10M	Recall@10	0.1734	0.1885	0.1946	0.2071	0.2023	0.2112	0.2108	**0.2160***
	Recall@20	0.2606	0.2725	0.2856	0.2901	0.2937	0.2985	0.3026	**0.3148***
	Recall@50	0.4081	0.4073	0.4352	0.4370	0.4379	0.4401	0.4413	**0.4669***
	NDCG@10	0.2061	0.2322	0.2427	0.2402	0.2392	0.2408	0.2417	**0.2453***
	NDCG@20	0.2256	0.2467	0.2590	0.2595	0.2585	0.2587	0.2598	**0.2616***
	NDCG@50	0.2685	0.2831	0.3003	0.3019	0.2982	0.3016	0.3021	**0.3071***
Gowalla	Recall@10	0.0866	0.0798	0.1289	0.1215	0.1394	0.1388	0.1382	**0.1420***
	Recall@20	0.1263	0.1164	0.1871	0.1755	0.2014	0.2008	0.2011	**0.2078***
	Recall@50	0.2040	0.1917	0.2934	0.2813	0.3127	0.3140	0.3134	**0.3230***
	NDCG@10	0.0622	0.0570	0.0930	0.0868	0.0991	0.0980	0.0989	**0.1008***
	NDCG@20	0.0736	0.0676	0.1097	0.1022	0.1170	0.1196	0.1184	**0.1215***
	NDCG@50	0.0926	0.0858	0.1356	0.1281	0.1442	0.1448	0.1437	**0.1489***
TmallBuy	Recall@10	0.0366	0.0385	0.0455	0.0695	0.0696	0.0709	0.0701	**0.0730***
	Recall@20	0.0470	0.0571	0.0620	0.0958	0.0952	0.0953	0.0970	**0.1011***
	Recall@50	0.0668	0.0917	0.0937	0.1390	0.1368	0.1372	0.1388	**0.1476***
	NDCG@10	0.0268	0.0220	0.0299	0.0416	0.0422	0.0418	0.0428	**0.0440***
	NDCG@20	0.0296	0.0269	0.0342	0.0486	0.0490	0.0485	0.0497	**0.0514***
	NDCG@50	0.0337	0.0341	0.0409	0.0577	0.0577	0.0562	0.0579	**0.0612***

4.3 Ablation Study (RQ2)

As shown in Table 3, we analyze the impact of various hyperparameters and components on the performance of SCLRec. Evaluations are conducted using the Recall@50 and NDCG@50 metrics on the MovieLens test dataset.

Confidence in the Implicit Datasets: As outlined in Sect. 3.2, we define α as the matching probability between the corresponding user and item in observed interactions. This indicates the confidence or the noise level of matched pairs. Through testing values of 0.80, 0.90, and 0.99 for α, we find that both low confidence (0.80) and over-confidence (0.99) compromise performance.

Coefficients of the Similarity Matrix: The computation of the similarity matrix involves user similarity, item similarity, and item popularity. In order to verify their role, we set their coefficients $\gamma_u, \gamma_v, \gamma_p$ to zero respectively. Our results show that all components contribute positively. Notably, item popularity appears to be more crucial than the others, as omitting it results in a more pronounced decline in performance.

Implications of Optimal Transport: A key aspect of SCLRec is its use of entropy optimal transport to obtain the matching matrix based on the similarity matrix. The question arises: is entropy optimal transport truly necessary? We verify the requirement by setting the number of Sinkhorn iterations to 0 and 6. The experiments indicate that using 0 iterations results in lower performance, thus highlighting the effectiveness of entropy optimal transport in mitigating

Table 3. Ablation study. SCLRec evaluated on MovieLens test set.

	α	γ_u	γ_v	γ_p	EMA	λ	#iter	Recall@50	NDCG@50
SCLRec	0.9	1.0	1.0	0.1	✓	0.1	6	**0.4669**	**0.3071**
α	0.80	1.0	1.0	0.1	✓	0.1	6	0.4283 (↓ 8.3%)	0.2848 (↓ 7.3%)
	0.99	1.0	1.0	0.1	✓	0.1	6	0.4368 (↓ 6.4%)	0.3014 (↓ 1.9%)
similarity matrix	0.9	0.0	1.0	0.1	✓	0.1	6	0.4496 (↓ 3.7%)	0.2945 (↓ 4.1%)
	0.9	1.0	0.0	0.1	✓	0.1	6	0.4487 (↓ 3.9%)	0.2927 (↓ 4.7%)
	0.9	1.0	1.0	0.0	✓	0.1	6	0.4380 (↓ 5.2%)	0.2921 (↓ 4.9%)
Sinkhorn	0.9	1.0	1.0	0.1	✓	0.05	6	0.4586 (↓ 1.8%)	0.2985 (↓ 2.8%)
	0.9	1.0	1.0	0.1	✓	0.2	6	0.4424 (↓ 5.2%)	0.2923 (↓ 4.8%)
	0.9	1.0	1.0	0.1	✓	0.1	0	0.4440 (↓ 4.9%)	0.2908 (↓ 5.3%)

data noise. We also investigate the impact of entropy regularization, dictated by λ in Eq. (4), and find that overly "hard" (0.05) or "soft" (0.2) target distribution both harm performance.

4.4 Robustness to Interaction Noises (RQ3)

To evaluate SCLRec's robustness to interaction noise, following the experimental settings of recent work [29], we incorporate specified ratios of unobserved interactions (i.e., 5%, 10%, 15%, and 20%) into the training set and test on an untouched test set. The results on MovieLens and Gowalla datasets are shown in Fig. 2. From the results, we can draw several key findings: Obviously, as the amount of added noise increases, the performance of all models declines. This is because CF models rely on user-item interactions for enhanced representations. Besides, on both datasets, SCLRec exhibits less performance degradation compared to other

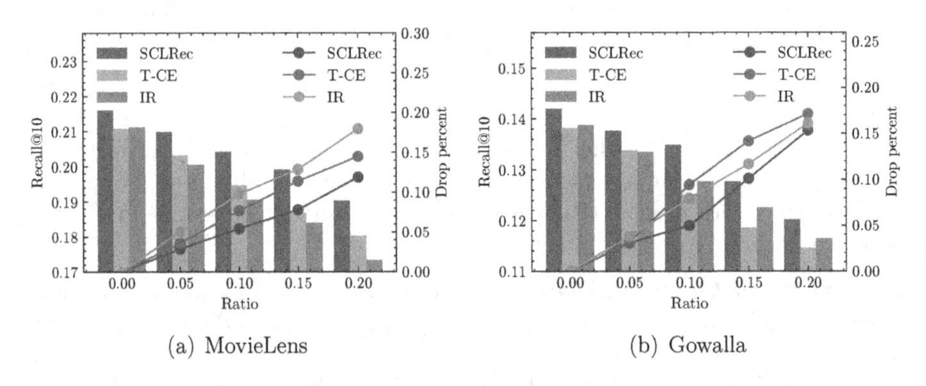

(a) MovieLens (b) Gowalla

Fig. 2. Model performance w.r.t. noise ratio. The bar represents Recall@10, and the line shows the percentage of performance degradation.

models. Interestingly, the performance gap increases as noise levels rise on the MovieLens dataset. These observations indicate that our method, leveraging entropy optimal transport to determine soft user-item matches in contrastive loss, is effective in mitigating interaction noise. Moreover, it is worth noting that SCLRec shows stronger robustness on Movielens, which is consistent with MovieLens having denser interactions than Gowalla according to Table 1.

5 Conclusion

In this study, we introduce a novel soft contrastive learning method that improves implicit feedback recommendations. Specifically, we utilize entropy optimal transport to find soft user-item matches as labels for contrastive learning. Our proposed method provides additional supervisory signals to better guide the learning process of the recommendations. Furthermore, extensive experiments on three public evaluation datasets demonstrate that SCLRec achieves better performance compared to state-of-the-art CF methods.

Acknowledgements. This work was partially supported by NSFC (62122037) and the Collaborative Innovation Center of Novel Software Technology and Industrialization. We thank Zi-Hao Qiu for the helpful discussions.

References

1. Cao, J., Cong, X., Sheng, J., Liu, T., Wang, B.: Contrastive cross-domain sequential recommendation. In: CIKM, pp. 138–147 (2022)
2. Chen, T., Kornblith, S., Norouzi, M., Hinton, G.: A simple framework for contrastive learning of visual representations. In: ICML, pp. 1597–1607 (2020)
3. Cheng, M., et al.: Learning recommender systems with implicit feedback via soft target enhancement. In: SIGIR, pp. 575–584 (2021)
4. Cheng, R., Wu, B., Zhang, P., Vajda, P., Gonzalez, J.E.: Data-efficient language-supervised zero-shot learning with self-distillation. In: CVPR, pp. 3119–3124 (2021)
5. Cho, E., Myers, S.A., Leskovec, J.: Friendship and mobility: user movement in location-based social networks. In: KDD, pp. 1082–1090 (2011)
6. Covington, P., Adams, J., Sargin, E.: Deep neural networks for youtube recommendations. In: RecSys, pp. 191–198 (2016)
7. Cuturi, M.: Sinkhorn distances: lightspeed computation of optimal transport. In: NIPS, pp. 2292–2300 (2013)
8. Damodaran, B.B., Flamary, R., Seguy, V., Courty, N.: An entropic optimal transport loss for learning deep neural networks under label noise in remote sensing images. Comput. Vis. Image Understand. **191**(C), 1–12 (2020)
9. Gao, T., Yao, X., Chen, D.: SimCSE: Simple contrastive learning of sentence embeddings. In: EMNLP, pp. 6894–6910 (2021)
10. Gao, Y., et al.: Self-guided learning to denoise for robust recommendation. In: SIGIR, pp. 1412–1422 (2022)
11. Harper, F.M., Konstan, J.A.: The movielens datasets: history and context. ACM Trans. Interact. Intell. Syst. **5**(4), 1–19 (2015)

12. He, X., Deng, K., Wang, X., Li, Y., Zhang, Y., Wang, M.: Lightgcn: simplifying and powering graph convolution network for recommendation. In: SIGIR, pp. 639–648 (2020)
13. He, X., Zhang, H., Kan, M.Y., Chua, T.S.: Fast matrix factorization for online recommendation with implicit feedback. In: SIGIR, pp. 549–558 (2016)
14. Jiang, Y., Huang, C., Huang, L.: Adaptive graph contrastive learning for recommendation. In: KDD, pp. 4252–4261 (2023)
15. Lee, D., Kang, S., Ju, H., Park, C., Yu, H.: Bootstrapping user and item representations for one-class collaborative filtering. In: SIGIR, pp. 317–326 (2021)
16. McAuley, J., Targett, C., Shi, Q., van den Hengel, A.: Image-based recommendations on styles and substitutes. In: SIGIR, pp. 43–52 (2015)
17. Oord, A.v.d., Li, Y., Vinyals, O.: Representation learning with contrastive predictive coding. arXiv preprint arXiv:1807.03748 (2018)
18. Pan, R., et al.: One-class collaborative filtering. In: ICDM, pp. 502–511 (2008)
19. Qiu, Z.H., Hu, Q., Yuan, Z., Zhou, D., Zhang, L., Yang, T.: Not all semantics are created equal: contrastive self-supervised learning with automatic temperature individualization. In: ICML, pp. 28389–28421 (2023)
20. Rendle, S., Freudenthaler, C., Gantner, Z., Schmidt-Thieme, L.: BPR: bayesian personalized ranking from implicit feedback. In: UAI, pp. 452–461 (2009)
21. Sankar, A., Wang, J., Krishnan, A., Sundaram, H.: Beyond localized graph neural networks: an attributed motif regularization framework. In: ICDM, pp. 472–481 (2020)
22. Su, X., Khoshgoftaar, T.M.: A survey of collaborative filtering techniques. In: Advances in Artificial Intelligence, pp. 1–19 (2009)
23. Tarvainen, A., Valpola, H.: Mean teachers are better role models: weight-averaged consistency targets improve semi-supervised deep learning results. In: NIPS, pp. 1195–1204 (2017)
24. Tian, C., Xie, Y., Li, Y., Yang, N., Zhao, W.X.: Learning to denoise unreliable interactions for graph collaborative filtering. In: SIGIR, pp. 122–132 (2022)
25. Tianchi: IJCAI-16 brick-and-mortar store recommendation dataset (2018). https://tianchi.aliyun.com/dataset/dataDetail?dataId=53
26. Wang, C., et al.: Towards representation alignment and uniformity in collaborative filtering. In: KDD, pp. 1816–1825 (2022)
27. Wang, W., Feng, F., He, X., Nie, L., Chua, T.S.: Denoising implicit feedback for recommendation. In: WSDM, pp. 373–381 (2021)
28. Wang, Z., Xu, Q., Yang, Z., Cao, X., Huang, Q.: Implicit feedbacks are not always favorable: iterative relabeled one-class collaborative filtering against noisy interactions. In: MM, pp. 3070–3078 (2021)
29. Wu, J., et al.: Self-supervised graph learning for recommendation. In: SIGIR, pp. 726–735 (2021)
30. Yu, W., Qin, Z.: Sampler design for implicit feedback data by noisy-label robust learning. In: SIGIR, pp. 861–870 (2020)
31. Zhou, C., Ma, J., Zhang, J., Zhou, J., Yang, H.: Contrastive learning for debiased candidate generation in large-scale recommender systems. In: KDD, pp. 3985–3995 (2021)

Dual-Graph Convolutional Network and Dual-View Fusion for Group Recommendation

Chenyang Zhou[1], Guobing Zou[1(✉)], Shengxiang Hu[1], Hehe Lv[1], Liangrui Wu[1], and Bofeng Zhang[2]

[1] School of Computer Engineering and Science, Shanghai University, Shanghai, China
{chenyangzhou,gbzou,shengxianghu,hhlv,lr_w}@shu.edu.cn

[2] School of Computer and Information Engineering, Shanghai Polytechnic University, Shanghai, China
bfzhang@sspu.edu.cn

Abstract. Group recommendation constitutes a burgeoning research focus in recommendation systems. Despite a multitude of approaches achieving satisfactory outcomes, they still fail to address two major challenges: 1) these methods confine themselves to capturing user preferences exclusively within groups, neglecting to consider user collaborative signals beyond groups, which reveal users' potential interests; 2) they do not sufficiently take into account the impact of multiple factors on group decision-making, such as individual expertise and influence, and the group's general preferences. To tackle these challenges, we propose a new model named DDGR (**D**ual-Graph Convolutional Network and **D**ual-View Fusion for **G**roup **R**ecommendation), designed to capture representations addressing two aspects: member preferences and group preferences. DDGR consists of two components: 1) a dual-graph convolutional network that combines the benefits of both hypergraphs and graphs to fully explore member potential interests and collaborative signals; 2) a dual-view fusion strategy that accurately simulates the group negotiation process to model the impact of multiple factors from member and group view, which can obtain semantically rich group representations. Thorough validation on two real-world datasets indicates that our model significantly surpasses state-of-the-art methods.

Keywords: Group Recommendation · Hypergraph Learning · Graph Convolution Networks · Attention Mechanism

1 Introduction

The popularity of social media has led to a rise in online group activities. Recommending the related item to a group is a critical task in the information system. Unlike user recommendation, group recommendation involves group decision-making which is a complex process. Each group member has their preferences, which will affect the final decision. The more complicated aspect is that the

D.-N. Yang et al. (Eds.): PAKDD 2024, LNAI 14649, pp. 231–243, 2024.
https://doi.org/10.1007/978-981-97-2262-4_19

influence of group members also changes dynamically when faced with different choices. In the group decision-making process, it is necessary to minimize conflicts among members and improve the common acceptance of members.

In group preference capture, the early methods mainly adopted predefined and fixed aggregation strategies, such as average, popularity [3], and PIT [10] etc. However, these methods cannot model the dynamic changes of the group in the face of different decisions.

Considering that different group members will have different influences on group preference, the models [1,8,14] are generated to solve it by assigning corresponding weights to each group member. However, these models tend to prioritize pairwise connections between users, overlooking high-order interactions both within and outside the group. Due to the success of hypergraphs in modeling high-order feature relationships, several group recommendation models [5,9,17] proposed hypergraph convolutional network to capture user and group-level group preferences.

Despite achieving impressive results, the aforementioned methods still have some limitations to be better explored. 1) When it comes to preference aggregation strategies, these methods only take into account collaborative signals between group members. They overlooked the optimization of individual preferences when users are outside of a group, which is problematic considering that groups are composed of individual users. This leads to inaccuracies in aggregating group preference information. 2) the final decision in group decision-making is often influenced by multiple factors, such as individual expertise, the influence of group members, and the general preferences of the group. Most models consider only one aspect without taking into account different perspectives. Oversimplification of such complex factors can lead to a biased understanding of the group's decision-making process, posing limitations to the accuracy of the recommendation system.

To solve the problems, we put forward the model named DDGR(**D**ual-Graph Convolutional Network and **D**ual-View Fusion for **G**roup **R**ecommendation). Firstly, one key consideration is that as users increasingly purchase identical products, it often signals a greater convergence in their preferences. So we model the interest-similarity graph according to the interaction data. To have a more comprehensive understanding of user preferences, we present a dual graph convolutional network that integrates graphs and hypergraphs to capture collaborative information from both within and outside user groups. Third, we design a dual-view attention mechanism fusion strategy that takes into account the impact of group members, their expertise, and the group's overall preferences on the final decision. Considering that each factor has different weights in different situations, we design an adaptive weight fusion strategy to indicate each factor's weight. We adopt a joint training strategy that combines group-item and user-item recommendations during training. The following are the key contributions of our work:

- We establish interest-similarity graphs for individual users and hypergraphs for groups based on interaction data. On this basis, we propose a dual-graph convolution network that uses hypergraphs and graphs to extract users' collaborative information within and outside the group and capture latent interests.

- We design a dual-view attention mechanism fusion strategy to model multiple factors in the group decision-making process. Meanwhile, we design an adaptive weight fusion strategy to measure the weight of different factors and obtain semantically rich group representations.
- Our proposed method is subject to extensive experimentation, incorporating two real-world datasets. The results clearly demonstrate that the method significantly exceeds most methods for group recommendation tasks.

2 Problem Formulation

In this section, we begin by concentrating on the definition of the group recommendation task. Subsequently, we provide a definition of a hypergraph.

Definition 1 (Group Recommendation). We define the set of users as $\boldsymbol{\mathcal{U}} = \{u_1, u_2, \ldots, u_{|\mathcal{U}|}\}$, the set of items as $\boldsymbol{\mathcal{I}} = \{i_1, i_2, \ldots, i_{|\mathcal{I}|}\}$, and the set of groups as $\boldsymbol{\mathcal{G}} = \{g_1, g_2, \ldots, g_{|\mathcal{G}|}\}$. The t-th group $g_t \in \mathcal{G}$ is a collection of users $\mathcal{G}_t = \{u_1, u_2, \ldots, u_i \ldots, u_{|\mathcal{G}_t|}\}$, where $u_i \in \mathcal{U}$, $|\mathcal{G}_t|$ is the size of $\mathcal{G}_t$. Let $\mathbf{R} \in \mathbb{R}^{|\mathcal{U}| \times |\mathcal{I}|}$ denote the user-item interaction matrix, where $r_{ui} = 1$ if the user u interacted with item, otherwise $r_{ui} = 0$. Let $\mathbf{S} \in \mathbb{S}^{|\mathcal{G}| \times |\mathcal{I}|}$ denote the group-item interaction matrix, where $s_{gi} = 1$ if the group g interacted with item i, otherwise $s_{gi} = 0$. In group recommendation, the goal is to suggest a list of items that a target group is likely to be interested in. Formally, this involves developing a function, f_g, which assigns a real-valued score to each item, indicating the probability of the target group g_t interacting with that item: f_g: $\boldsymbol{\mathcal{I}} \rightarrow \mathbb{R}$.

In hypergraphs, each hyperedge can connect multiple nodes, allowing for the representation of various relationships, which can support modeling multidimensional relationships in groups.

Definition 2 (Hypergraph). The hypergraph G is formally defined as $(\mathcal{V}, \mathcal{E})$, where $\mathcal{V}$ exhibits a collection of M distinct vertices, and $\mathcal{E}$ represents the set of hyperedges containing N edges. And each hyperedge $\epsilon \in \mathcal{E}$ can contain multiple vertices. The incidence matrix $\mathbf{H} \in \mathbb{R}^{M \times N}$ can represent hypergraph, where $h_{v\epsilon} = 1$ if the hyperedge ϵ contains the vertex $v \in \mathcal{V}$, otherwise $h_{v\epsilon} = 0$. The diagonal matrix $\mathbf{W} \in \mathbb{R}^{N \times N}$ is utilized to represent the weight of the hyperedges. The degree of each vertex is represented by the diagonal matrix $\mathbf{D}$, where the vertex degree D_v can be calculated as $D_v = \sum_{\epsilon=1}^{M} W_\epsilon H_{v\epsilon}$. Similarly, the diagonal matrix $\mathbf{B}$ is used to denote the degree of each hyperedge, where the hyperedge degree B_ϵ can be determined as $B_\epsilon = \sum_{\epsilon=1}^{N} H_{i\epsilon}$.

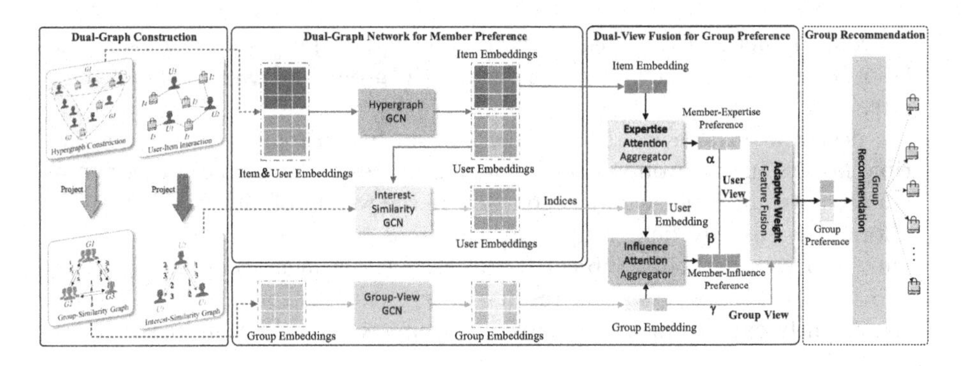

Fig. 1. The overview architecture of our proposed DDGR model, consisting of the member preference learning and group preference learning.

3 Approach

In this section, we introduce the proposed DDGR in detail, and it is composed of three vital parts: Dual-Graph Construction, Dual-graph Network for Member Preference, and Dual-view Fusion for Group Preference. Figure 1 illustrates the overall architecture of DDGR.

3.1 Dual-Graph Construction

Effective group recommendations require establishing suitable connections. We construct a hypergraph to model higher-order relationships between members and items and construct a member interest-similarity graph exploring latent interests when the user is outside the group.

Hypergraph Construction. The transformation from group interaction data to hypergraph $G_h = (\mathcal{V}_h, \mathcal{E}_h)$ is shown in Fig. 2. Group members and items form hyperedge $\mathcal{E}_H = \{u_1, u_2, \ldots, u_{|\mathcal{U}|}, i_1, \ldots, i_{|\mathcal{I}|}\}$. Unlike graphs, the members and items are explicitly connected by hyperedge in the hypergraph. It can extract many-to-many high-order relations from graphs.

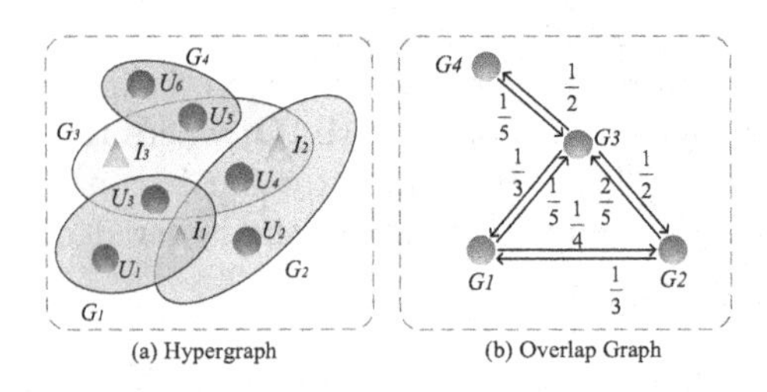

Fig. 2. Example of a hypergraph and overlap graph.

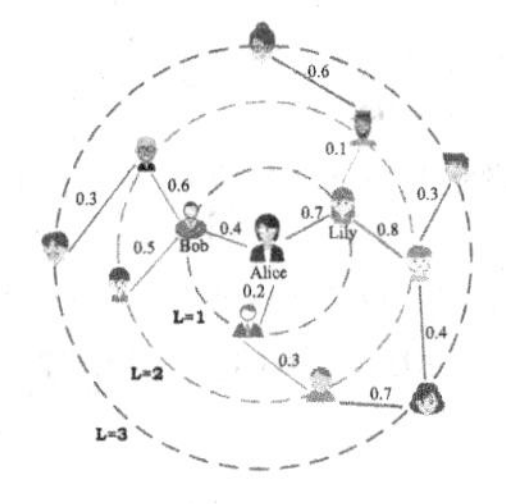

Fig. 3. Example of a member interest-similarity graph.

Member Interest-Similarity Graph Construction. The work [7] indicates that an increased level of interaction with shared products among users often signifies a higher probability of common interests, we design an interest-similarity graph $G_b = (\mathcal{V}_b, \mathcal{E}_b)$ to shows the similarity of interests between users which is shown in Fig. 3. The user i and user j are connected if they interact with common items, and we give edge weight to represent the similarity of user preferences:

$$S(I_i, I_j) = \frac{|I_i \cap I_j|}{|I_i|} \tag{1}$$

where $|I_i|$ represents the total number of items which user i interacted with, $|I_i \cap I_j|$ is the number of items which user i and j interacted together. The weight signifies the proximity between the two users, with an unequal influence on both sides.

3.2 Dual-Graph Network for Member Preference

The dual-graph network was proposed to learn collaborative high-order representation of users and items based on the hypegraph, and capture cross-user collaborative information of users based on the member interest-similarity graph.

We introduce the hypergraph convolution operation to exploit high-order interactions to learn the user's and item's dynamic representation. Let $\mathbf{X_h}^{(0)} = [\mathbf{U}; \mathbf{I}]$ be the input of hypergraph convolutional network, which is the concatenation of user embeddings $\mathbf{U} \in \mathbb{R}^{|U| \times d}$ and item embeddings $\mathbf{I} \in \mathbb{R}^{|I| \times d}$. Building upon the spectral hypergraph convolution proposed in [4], the work defines its hypergraph convolution as:

$$\mathbf{X_h}^{(l+1)} = \mathbf{D}^{-1}\mathbf{H}\mathbf{W}\mathbf{B}^{-1}\mathbf{H}^T\mathbf{X_h}^{(l)}\mathbf{\Theta_h}^{(l)} \tag{2}$$

where $\mathbf{D}$ denotes the vertex degree matrix of the hypergraph, and $\mathbf{B}$ denotes the hyperedge degree matrix. $\mathbf{W}$ is initialized as an identity matrix, indicating the assignment of equal weights for all hyperedges. $\mathbf{H}$ is an incidence matrix to delineate the relationship between nodes and hyperedges. $\mathbf{\Theta_h}^{(l)}$ is the learnable weight matrix. After passing $\mathbf{X}_0$ through $\mathbf{L}$ hypergraph convolutional layers, we get the final embeddings $\mathbf{X}_h$ by averaging embeddings obtained at each layer, where $\mathbf{X}_h = \frac{1}{L+1}\sum_{l=0}^{L}\mathbf{X}_h^{(l)}$. By leveraging the node-edge-node transformation, the hypergraph convolutional network can efficiently extract high-order correlations on the hypergraph.

The hypergraph network is employed to capture users' high-dimensional relation within the group; however, it does not acquire collaborative information from users outside the group. The member Interest-similarity graph depicts the similarity of interest and contains cross-user information. Based on this, we aim to mine the mutual influence between users and discover users' potential interests. So we apply the graph convolution operation to capture communication features. Let user embeddings $\mathbf{X_b} \in \mathbb{R}^{|U| \times d}$, which is the output of the hypergraph convolution layer, be the input of the interest-similarity graph convolutional network. The graph convolution operation is defined as:

$$\mathbf{X_b}^{(l+1)} = \mathbf{D_b}^{-1/2}\mathbf{A_b}\mathbf{D_b}^{-1/2}\mathbf{X_b}^{(l)}\mathbf{\Theta_b}^{(l)} \tag{3}$$

where $\mathbf{A_b} \in \mathbb{R}^{|U|\times|U|}$ defines An incidence matrix, $|U|$ is the number of users and $\mathbf{A_{p,q}} = \mathbf{W_{p,q}}$ according to definition of Eq. (1). $\mathbf{D_b} \in \mathbb{R}^{|U|\times|U|}$ is a diagonal degree matrix where $\mathbf{D}_{p,q} = \sum_{q=1}^{|U|} \mathbf{A}_{p,q}$.

3.3 Dual-View Fusion for Group Preference

Dual-graph network helps count for group member preference but has a relatively minor contribution toward group decision-making. Making decisions is a complex process for groups, it will be affected by many factors, such as individual expertise, the influence of group members, and the general preferences of the group. So we propose a dual-view fusion strategy to capture these factors from *member* and *group* views and get the most reasonable group preference possible.

From the members' view, we consider two factors: *Member-expertise Preference* and *Member-influence Preference*. For Member-expertise Preference, imagine a scenario where a group of friends is selecting products. If a member possesses expertise in assessing electronic products, his influence on the group's decision-making process is likely to be more significant. For Member-influence Preference, a company group discusses where to travel, and the leader may have greater decision-making power to make decisions on behalf of the group even though most members have different choices. So the varying identities and positions of group members can lead to fluctuations in the levels of influence they exert in group decision-making. We define $\alpha^E(i,t)$ denotes the weight of user u_i in the group's decision-making with respect to item i_t and $\alpha^P(i,l)$ represents the score of the user u_i's influence in the group g_l. User embedding $\mathbf{u}_i$, item embedding $\mathbf{i}_t$ and group embedding $\mathbf{g}_l$ is the input of neural attention network, which is defined as:

$$o(i,t) = \mathbf{h}^T \operatorname{ReLU}\left(\mathbf{P}_u \mathbf{u}_i + \mathbf{P}_i \mathbf{i}_t + \mathbf{b}\right), \quad \alpha^E(i,t) = \frac{\exp o(i,t)}{\sum_{i' \in \mathcal{G}_l} \exp o\left(i',t\right)} \tag{4}$$

$$p(i,l) = \mathbf{h}^T \operatorname{ReLU}\left(\mathbf{P}_u \mathbf{u}_i + \mathbf{P}_g \mathbf{g}_l + \mathbf{b}\right), \quad \alpha^P(i,l) = \frac{\exp p(i,l)}{\sum_{i' \in \mathcal{G}_l} \exp p\left(i',l\right)} \tag{5}$$

In the attention network, the activation function utilized for the hidden layer is ReLU, and we use a weight vector $\mathbf{h}$ to project the score $o(t,j)$. Finally, we calculate the group's member-expertise representation $\mathbf{g}_l^E$ and the group's member-influence representation $\mathbf{g}_l^P$ using a weighted sum operation:

$$\mathbf{g}_l^E = \sum_{\mathbf{u}_i \in \mathcal{G}_l} \alpha^E(i,t)\mathbf{u}_i, \quad \mathbf{g}_l^P = \sum_{\mathbf{u}_i \in \mathcal{G}_l} \alpha^P(i,l)\mathbf{u}_i. \tag{6}$$

By incorporating an attention mechanism, each member's contribution to group is learned from interaction data and varies dynamically based on different scenes.

In addition to aggregating the embeddings of group members, we employ dedicated group embeddings to represent groups from the group view. The intention is to take intrinsic group-level preferences into account and capture collaborative information between groups. Group decisions may not always align with individual preferences as they aim to satisfy the collective interests of the group. When a family group discusses about restaurant choices, individual preferences may differ and have their preferance. Finally, the group's choice of a restaurant aims to satisfy everyone's taste rather than selecting the one that's liked the most by everyone individually. Groups often have common members or shared purchases, leading to interaction between groups. The similarity between two groups will be higher if they share more common members or items. So we employ an overlap graph to map out inter-group connections and hidden interests, with Fig. 2 detailing the transformation and showcasing the interactions and proximity among groups. Group embeddings $\mathbf{G} \in \mathbb{R}^{|G| \times d}$ is the input of graph convolutional operation:

$$\mathbf{G}^{(l+1)} = \mathbf{D_g}^{-1/2}\mathbf{A_g}\mathbf{D_g}^{-1/2}\mathbf{G}^{(l)}\mathbf{\Theta_g}^{(l)} \tag{7}$$

where $\mathbf{D_g} \in \mathbb{R}^{|G| \times |G|}$ is a diagonal degree matrix and $\mathbf{D}_{p,q} = \sum_{q=1}^{|G|} \mathbf{A}_{p,q}$, $|G|$ is the number of groups. The incidence matrix is defined as $\mathbf{A_g} \in \mathbb{R}^{|G| \times |G|}$ and $\mathbf{A_{p,q}} = \mathbf{W_{p,q}}$. We set the matrix $\mathbf{A_g} = \mathbf{A_g} + \mathbf{I}$, where $\mathbf{I}$ is an identity matrix. Finally, we calculate the group-view representation $\mathbf{g}_l^G$.

The complexity of group decision-making lies not only in being influenced by multiple factors but also in considering the weight of each factor. So we propose an adaptive weight fusion strategy to capture the weight of each factor, resulting in a more reasonable group preference representation $\mathbf{g}_l$. We compute the influence score α of member-expertise factor:

$$\alpha = \mathbf{w}^T \operatorname{ReLU}\left(\mathbf{W}_f\left[\mathbf{g}_l^E \odot \mathbf{i}_h; \mathbf{g}_l^E; \mathbf{i}_h\right] + \mathbf{b}_f\right) \tag{8}$$

Similarly, for member-influence and group-level preference, we can get scores β and γ, respectively. By weighted addition, we get the group representation $\mathbf{g}_l$:

$$\mathbf{g}_l = \alpha \mathbf{g}_l^E + \beta \mathbf{g}_l^P + \gamma \mathbf{g}_l^G \tag{9}$$

We utilize the adaptive weight fusion network to extract key features from the three hidden group representations and blend them together seamlessly.

3.4 Group Recommendation and Model Training

After utilizing the Dual-graph Network and Dual-view Fusion, we can effectively capture cooperative signals between groups, discern the potential preferences of group members, identify group decision factors, and acquire a semantically rich group representation. Subsequently, we can proceed to match items of interest to the group.

Given that our objective is to rank items, we utilize the regression-based pairwise loss that has been motivated by the work [11].The group training set $\mathcal{O}_G$ is defined as a collection of triplets (i, j, j'), each triplet represents a scenario in which group g_i interacted with item i_j but has no prior interaction with i'_j:

$$\mathcal{L}_{\text{group}} = \sum_{(i,j,j') \in \mathcal{O}_G} \left(\hat{y}_{ij} - \hat{y}_{ij'} - 1\right)^2 \tag{10}$$

Considering the data sparsity of group interactions, we adopt a joint learning strategy that simultaneously combines the group-item interaction and user-item interaction data to enhance recommendation tasks. Similarly, we define $\mathcal{L}_{\text{user}}$ the same pairwise loss function as Eq. (10) to optimize user recommendation. As a consequence, the overall loss function is a combination of the losses incurred by both the group and user pairs:

$$\mathcal{L} = \lambda \mathcal{L}_{\text{group}} + (1 - \lambda)\mathcal{L}_{\text{user}} \tag{11}$$

where hyper-parameter λ plays a role in balancing the weights between the group and user losses. Notably, the primary focus of our task is group recommendation.

4 Experiments

This section covers the details of our experiments, including the datasets, baseline methods, and evaluation metrics. To investigate the following research questions, we carried out rigorous experiments on two openly accessible datasets.

4.1 Experimental Dataset and Setup

Datasets. To assess the performance of our proposed model, we evaluated it on two real-world datasets, namely MaFengWo and CAMRa2011. The MaFengWo dataset is a collection of user-generated travel experiences and group journeys. The CAMRa2011 dataset consists of real-world movie ratings from both individuals and groups. Detailed statistics for both datasets are presented in Table 1.

Table 1. The statistics of two datasets. U-I and G-I indicate user and item interactions, and group and item interactions, respectively.

Datasets	Users	Groups	Items	U-I	G-I	Avg. Group size
MaFengWo	5,275	995	1513	39,761	3,595	7.19
CAMRa2011	602	290	7,710	116,344	145,068	2.08

Competing Methods. To validate the performance of our group recommendation model, we conducted a comparative analysis with the following baseline: **Popularity** [3], **PIT** [10], and **COM** [16] models are mainstream probabilistic model. Baseline also includes deep learning models, which can be divided into the following: the neural network-based model(**NCF** [7]), the attention-based model(**AGREE** [1], **SIGR** [14], **GAME** [8], **GroupSA** [6], **SoAGREE** [2]), and the Hypergprah-based model(**HCR** [9], **ConsRec** [12]).

Evaluation Metrics. For the purpose of assessing group recommendation performance, we utilized commonly-used metrics called Hits Ratio (HR) and

Normalized Discounted Cumulative Gain (NDCG) at top-K recommendation list [15].

$$HR@K = \frac{\#hit@K}{|\mathcal{D}_{\text{test}}|}, \quad NDCG@K = \frac{1}{|\mathcal{D}_{\text{test}}|}\sum_{i=1}^{|\mathcal{D}_{\text{test}}|}\frac{1}{\log_2(p_{i@K}+1)} \tag{12}$$

Consequently, we divided the data into two separate sets: training set($\mathcal{D}_{train}$) and testing set($\mathcal{D}_{test}$). We performed a random sampling of the missing data to generate negative instances for each positive instance in the testing set.

4.2 Experimental Results and Analysis

In this section, we present a comparison of the recommendation performance of our proposed model to that of several baseline models. The group and user performance for the CAMRa2011 and MaFengWo, as shown in Tables 2 and 3, respectively. From the result, we can observe that:

Table 2. Performance comparison of top-K **group** recommendation on datasets.

Datasets	MaFengWo				CAMRa2011			
Metric	HR@5	NDCG@5	HR@10	NDCG@10	HR@5	NDCG@5	HR@10	NDCG@10
Popularity	0.3115	0.2169	0.4251	0.2537	0.4324	0.2825	0.5793	0.3302
PIT	0.4159	0.2965	0.5012	0.3382	0.5632	0.3741	0.7523	0.4492
COM	0.4432	0.3325	0.5528	0.3812	0.5798	0.3785	0.7695	0.4385
NCF	0.4701	0.3657	0.6269	0.4141	0.5803	0.3896	0.7693	0.4448
AGREE	0.4729	0.3694	0.6321	0.4203	0.5879	0.3933	0.7789	0.4530
SIGR	0.5041	0.3955	0.6569	0.4573	0.6172	0.4473	0.8158	0.4870
GroupSA	0.4876	0.3871	0.6409	0.4351	0.5906	0.4163	0.7800	0.4667
GAME	0.4759	0.3956	0.6346	0.4482	0.5953	0.4356	0.7957	0.4713
SoAGREE	0.4898	0.3807	0.6481	0.4301	0.5883	0.3955	0.7807	0.4575
HCR	0.7759	0.6611	0.8503	0.6852	<u>0.6772</u>	<u>0.6115</u>	0.8193	<u>0.6576</u>
ConsRec	<u>0.8844</u>	<u>0.7692</u>	<u>0.9156</u>	<u>0.7794</u>	0.6407	0.4358	<u>0.8248</u>	0.4945
DDGR	**0.9126**	**0.8517**	**0.9317**	**0.8595**	**0.7648**	**0.7548**	**0.8386**	**0.7782**

Table 3. Performance comparison of top-K **user** recommendation on datasets.

Datasets	MaFengWo				CAMRa2011			
Metric	HR@5	NDCG@5	HR@10	NDCG@10	HR@5	NDCG@5	HR@10	NDCG@10
Popularity	0.4047	0.2876	0.4971	0.3172	0.4624	0.3104	0.6026	0.3560
NCF	0.6363	0.5432	0.7417	0.5733	0.6119	0.4018	0.7894	0.4535
AGREE	0.6357	0.5481	0.7403	0.5738	0.6196	0.4098	0.7897	0.4627
SoAGREE	0.6510	0.5612	0.7610	0.5775	0.6223	0.4118	0.7967	0.4687
HCR	<u>0.7571</u>	0.6703	0.8290	0.6937	0.6731	<u>0.4608</u>	**0.8595**	<u>0.5219</u>
ConsRec	0.7725	<u>0.6884</u>	<u>0.8404</u>	<u>0.7107</u>	<u>0.6774</u>	0.4568	<u>0.8412</u>	0.5104
DDGR	**0.7930**	**0.7285**	**0.8518**	**0.7445**	**0.7475**	**0.7255**	0.8375	**0.7540**

DDGR consistently outperforms other baseline models, achieving the highest performance on both group recommendation datasets. The great improvements in performance provide further evidence of the effectiveness of our model for capturing group representation. DDGR shows a more significant improvement over the second-best model at NDCG@K compared to HR@K. It highlights DDGR's ability to prioritize items that align with the interests of user groups, resulting in more precise recommendations. It also showcases DDGR's capacity to dynamically model the decision-making process within groups and accurately capture their preferences.

In the user recommendation task, DDGR achieved excellent results in all metrics, essentially reaching optimal precision. This observation highlights that our proposed method has the ability to capture users' latent interests, thereby optimizing their preference representations.

To further explore the importance of Dual-Graph Networks and Dual-View Fusion Strategy, we conducted several ablation studies. Figure 4 shows the results of DDGR and the two variants, DDGR-G denotes the ablated model "DDGR with Dual-Graph Network only" and DDGR-V denotes "DDGR with dual-view fusion only". In both benchmark datasets, it is consistently observed that DDGR outperforms DDGR-G and DDGR-V across all evaluation metrics, indicating that Dual-Graph Network and Dual-View Fusion Strategy can facilitate each other, and combining them enables a more comprehensive capture of group interests. DDGR-V performs better than DDGR-G on both datasets. It indicates that Dual-view fusion strategy has a larger impact than the Dual-Graph Network on group representation of learning.

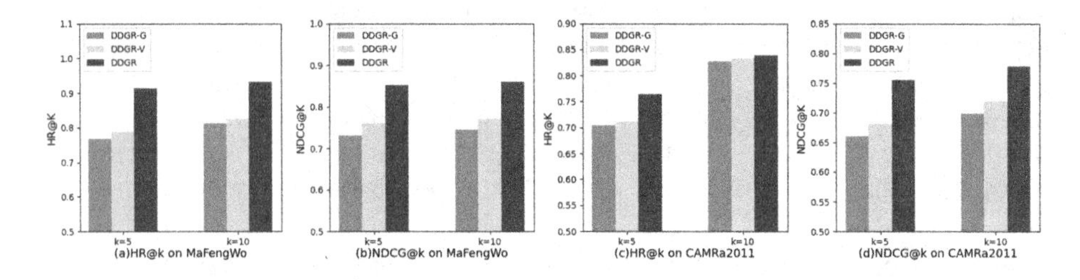

Fig. 4. The performance comparison between DDGR and its variants.

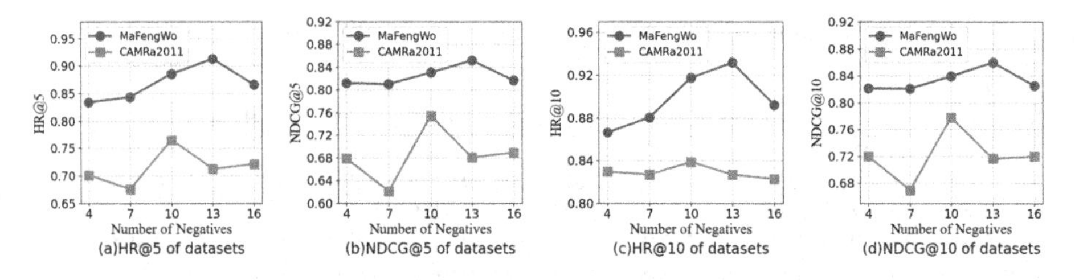

Fig. 5. The trend of HR and NDCG w.r.t. the number of negative samples.

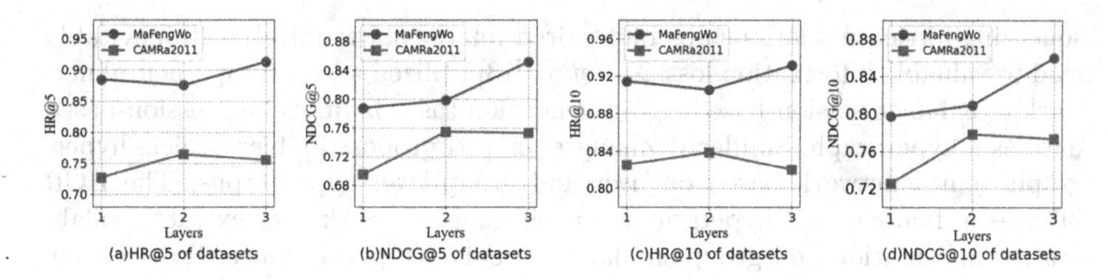

Fig. 6. The trend of HR and NDCG w.r.t. the number of convolutional layers.

4.3 Parameter Sensitivity

To investigate the impact of negative sampling and the number of convolutional layers on the performance of our model, we conducted a series of experiments. The impact of negative sampling on DDGR is presented in Fig. 5. We can observe that the performance of DDGR will not consistently increase as more negative samples are added. If there are too many negative samples in the sampling process, it's possible that the model will select items that the group would potentially be interested in, which leads to the deviation of group representation learning. So it is necessary to control them within a reasonable range. In Fig. 6, as the number of layers increases, DDGR does not consistently demonstrate improved performance on CAMRa2011. One potential explanation is that as the number of layers increases, nodes in higher levels may acquire the issue of over-smoothing. The increased complexity introduced by additional layers may not always align with the underlying data, causing difficulties in distinguishing groups and leading to adverse effects in performance gains.

5 Related Works

Group Recommendation. Early works on group recommendation used two approaches: *score aggregation* and *preference aggregation*. Score aggregation methods, such as average, least misery, and maximum satisfaction, employ a fixed strategy to calculate group representations. However, these methods couldn't capture the dynamic group decision process and neglected changes in group. Recently, with the successful development of the deep neural network, model-based approaches have achieved significant advances. AGREE [1], GroupSA [6] incorporate attention mechanisms to automatically learn and assign the user's corresponding weight to optimize the preference aggregation strategy. GAME [8] utilizes the heterogeneous information network to generate multi-view embeddings for nodes and members' weights. Although these methods offer a dynamic aggregation process of group preference, they all fail to capture complex and higher-order user interactions.

Hypergraph-Based Recommendation. In real-world scenarios, relationships among objects are more complicated than simple pairwise connections;

squeezing intricate relationships into paired relationships naively will inevitably lead to valuable information loss. Xia *et al.* [13] utilized a hypergraph neural network to enhance session-based recommendation tasks by modeling session-based data as a hypergraph. Similarly, Zhang *et al.* [17] proposed a hierarchical hypergraph neural network based on user and group-level hypergraphs. The HCR [9] uses a dual-channel hypergraph convolutional network that extracts collaborative information and group similarity to enhance performance. HyperGroup [5] proposes connecting groups as an overlapping set network to capture the similarity of groups and learn accurate group representations from group-item interactions.

6 Conclusion and Future Work

In this work, we introduced a novel model called DDGR to tackle the key challenges in group recommendation tasks: 1) how to obtain comprehensive group members' preferences from interaction data, 2) how to get a semantically rich group representation by emulating the decision-making processes. For group members' preferences, we construct the dual-graph network that combines the advantages of both hypergraphs and graphs to capture high-order and pairwise relationships between users. For group representation learning, we propose a dual-view fusion strategy that considers the impact of multiple factors in decision-making from two views, allowing us to accurately simulate the decision-making process. Comprehensive experiments on datasets demonstrate that DDGR surpasses other approaches in group recommendation tasks. In the future, we will combine group recommendations with large language models.

Acknowledgments. This work was supported by National Natural Science Foundation of China (No.62272290, 62172088) and Shanghai Natural Science Foundation(No.21ZR1400400).

References

1. Cao, D., He, X., Miao, L., An, Y., Yang, C., Hong, R.: Attentive group recommendation. In: The 41st International ACM SIGIR Conference on Research and Development in Information Retrieval, pp. 645–654 (2018)
2. Cao, D., He, X., Miao, L., Xiao, G., Chen, H., Xu, J.: Social-enhanced attentive group recommendation. IEEE Trans. Knowl. Data Eng. **33**(3), 1195–1209 (2019)
3. Cremonesi, P., Koren, Y., Turrin, R.: Performance of recommender algorithms on top-N recommendation tasks. In: ACM Conference on Recommender Systems, pp. 39–46 (2010)
4. Feng, Y., You, H., Zhang, Z., Ji, R., Gao, Y.: Hypergraph neural networks. In: Proceedings of the AAAI Conference on Artificial Intelligence, vol. 33, pp. 3558–3565 (2019)
5. Guo, L., Yin, H., Chen, T., Zhang, X., Zheng, K.: Hierarchical hyperedge embedding-based representation learning for group recommendation. ACM Trans. Inf. Syst. **40**(1), 1–27 (2021)

6. Guo, L., Yin, H., Wang, Q., Cui, B., Huang, Z., Cui, L.: Group recommendation with latent voting mechanism. In: IEEE International Conference on Data Engineering, pp. 121–132. IEEE (2020)
7. He, X., Liao, L., Zhang, H., Nie, L., Hu, X., Chua, T.S.: Neural collaborative filtering. In: International Conference on World Wide Web, pp. 173–182 (2017)
8. He, Z., Chow, C.Y., Zhang, J.D.: GAME: learning graphical and attentive multi-view embeddings for occasional group recommendation. In: International ACM SIGIR Conference on Research and Development in Information Retrieval (2020)
9. Jia, R., Zhou, X., Dong, L., Pan, S.: Hypergraph convolutional network for group recommendation. In: IEEE International Conference on Data Mining, pp. 260–269. IEEE (2021)
10. Liu, X., Tian, Y., Ye, M., Lee, W.C.: Exploring personal impact for group recommendation. In: ACM International Conference on Information and Knowledge Management, pp. 674–683 (2012)
11. Wang, X., He, X., Nie, L., Chua, T.S.: Item silk road: recommending items from information domains to social users. In: International ACM SIGIR Conference on Research and Development in Information Retrieval, pp. 185–194 (2017)
12. Wu, X., et al.: ConsRec: learning consensus behind interactions for group recommendation. In: Proceedings of the ACM Web Conference 2023, pp. 240–250 (2023)
13. Xia, X., Yin, H., Yu, J., Wang, Q., Cui, L., Zhang, X.: Self-supervised hypergraph convolutional networks for session-based recommendation. In: Proceedings of the AAAI Conference on Artificial Intelligence, vol. 35, pp. 4503–4511 (2021)
14. Yin, H., Wang, Q., Zheng, K., Li, Z., Yang, J., Zhou, X.: Social influence-based group representation learning for group recommendation. In: International Conference on Data Engineering, pp. 566–577. IEEE (2019)
15. Yin, H., Wang, Q., Zheng, K., Li, Z., Zhou, X.: Overcoming data sparsity in group recommendation. IEEE Trans. Knowl. Data Eng. **34**, 3447–3460 (2020)
16. Yuan, Q., Cong, G., Lin, C.Y.: COM: a generative model for group recommendation. In: ACM SIGKDD International Conference on Knowledge Discovery and Data Mining, pp. 163–172 (2014)
17. Zhang, J., Gao, M., Yu, J., Guo, L., Li, J., Yin, H.: Double-scale self-supervised hypergraph learning for group recommendation. In: ACM International Conference on Information and Knowledge Management, pp. 2557–2567 (2021)

TripleS: A Subsidy-Supported Storage for Electricity with Self-financing Management System

Jia-Hao Syu[1], Rafal Cupek[2], Chao-Chun Chen[3], and Jerry Chun-Wei Lin[2](✉)

[1] National Taiwan University, Taipei, Taiwan
f08922011@ntu.edu.tw
[2] Silesian University of Technology, Gliwice, Poland
Rafal.Cupek@polsl.pl, jerry.chun-wei.lin@polsl.pl
[3] National Cheng Kung University, Tainan, Taiwan
chaochun@mail.ncku.edu.tw

Abstract. In this paper, we propose a Subsidy-Supported Storage (also called TripleS) to assist grid management. Q-learning algorithms first determine the origin subsidies, and the proposed self-financing mechanism then balances the expected costs and gains, and generates the final subsidies. During market equilibrium, energy storage is fully charged when there is excess electricity and discharged when there is insufficient electricity. The electricity market then calculates the cash flow of the subsidies, and the remaining cash is used to make up for the self-discharge loss of the storage units. Experimental results demonstrate the effectiveness of the proposed TripleS in maintaining grid stability.

Keywords: Electricity grid · Management · Storage · Self-Financing

1 Introduction

Climate change is accelerated by the extensive use of fossil fuels in human activities, and the most critical aspect is to develop clean and renewable energy. However, integrating renewable energy can be challenging due to lower price competitiveness and higher production volatility. Therefore, a sustainable grid is essential to incentivize renewable energy and ensure grid stability.

Electricity subsidies are a common economic scheme that provides financial incentives for renewable energy [4]. Positive subsidies lower costs, encourage production, stimulate investment, and promote growth. Negative subsidies, on the other hand, increase costs, discourage production, inhibit investment, and reduce growth. The determination of electricity subsidies is mainly influenced by government policies and objectives, taking into account environmental goals (reducing greenhouse gas emissions) and economic development (energy independence, stability, and growth). Artificial Intelligence (AI) research on electricity has the potential to provide valuable insights and decision-making.

D.-N. Yang et al. (Eds.): PAKDD 2024, LNAI 14649, pp. 244–255, 2024.
https://doi.org/10.1007/978-981-97-2262-4_20

To enhance the stability of the electricity grid, the concept of operating reserve is developed, which refers to the extra power capacity to ensure reliability and stability [5]. Relevant research on operating reserves mainly focuses on renewable energy, and lacks a macroscopic perspective on the entire electricity grid. In addition, electricity storage is also effective in ensuring grid stability, the purpose of which is to store excess electricity and release it when needed [1]. Electricity storage can reduce the risk of blackouts and increase stability; however, there is an inevitable phenomenon of self-discharge, which means that stored energy is gradually lost over time, even when not in use.

In this paper, we propose Subsidy-Supported Storage (TripleS) to assist grid management. TripleS starts with Q-learning algorithms to determine origin subsidies for different types of electricity supplies. The self-financing mechanism then balances the expected costs and gains, and generates the final subsidies. During reaching the market equilibrium, energy storage is fully charged (buy) when there is excess electricity (supply is greater than demand), and discharged (sell) when electricity is insufficient (demand is greater than supply). After market equilibrium is reached, the system will calculate the cash flow of subsidies and electricity storage. To support the operation of storage, the remaining cash of subsidy is used to make up for the self-discharge loss of storage units.

Experimental results demonstrate the effectiveness of the proposed TripleS in maintaining grid stability with self-financing management. To demonstrate the robustness, we execute TripleS under various attack scenarios, and TripleS achieves zero failure rate in most experiments (8 out of 10), and obtains extremely low failure rates of 0.01% in exceptions (2 out of 10), which outperforms the state-of-the-art models significantly. We further investigate the influence of the self-discharge rate, and the proposed TripleS is less affected by the rate, but the traditional mechanism has a tendency to increase the failure probability. Experimental results demonstrate that the traditional storage mechanism is not suitable for uncertain environments such as attacks and high self-discharge rates. On the contrary, the proposed TripleS is robust to various uncertain environments, and significantly improves the stability of electricity grids with self-financing management.

2 Literature Review

2.1 Electricity Subsidy and Operating Reserve

Electricity subsidy is an economic scheme that provides financial incentives to various electricity suppliers [4]. Positive subsidies reduce costs, and stimulate investment; negative subsidies increase costs, and inhibit investment. By determining the subsidy for each supplier, the electricity supply distribution can be shaped into the desired distribution. Electricity subsidies are mainly determined by government policies [4]. Yang et al. [12] studied the effectiveness of subsidies for renewable energy investment between two parties. Gustian et al. [2] adopted classification and optimization algorithms to determine the electricity subsidies.

The research on electricity subsidies is still in the early stages, and less attention has been paid to the equilibrium of supply and demand in the electricity market.

Operating reserve refers to extra electricity capacity to ensure the stability of the grid, and can compensate for real-time shortages while reducing energy efficiency (generating electricity but not connecting to the grid) [5]. Therefore, grid management systems must carefully balance stability and efficiency. To measure reserve capacity, the operating reserve rate (ORR) is defined as excess (unused) electricity capacity divided by supply capacity. If the ORR becomes negative, it means the demand for electricity is greater than the supply, which will lead to a blackout of the entire grid. Most research on operating reserve focuses on renewable energy. Holttinen et al. [3] adopted the wind integration analysis to set the operating reserve. Stiphout et al. [7] studied the impact of operating reserve on the investment of renewable energy. Relevant research mainly focuses on renewable energy, lacking a macroscopic perspective on the electricity grid.

2.2 Electricity Management System

The first research focusing on both subsidies and ORR belongs to CAEMS [9], which is a call-auction-based energy management system that defines standard environments to simulate the electricity supply, demand, and equilibrium. In addition, CAEMS [9] proposed the adoptive subsidies (SUB) and the dynamic operating reserve multiplier ($DMOR$) to shape supply distribution and maintain grid stability, respectively. SUB and $DMOR$ are determined by close-form solutions and updated daily to manage the electricity grid.

However, the close-form solutions are not flexible and not explainable for grid management; therefore, Q-learning-based energy management systems [8], QEMS, are developed to manage the electricity grid through clear states and intuitive actions. Through the interaction between the Q-tables and the environment of the electricity grid, the Q-learning algorithms become more adept at decision making. Consequently, QEMS achieves better management performance in both supply distribution shaping and grid stability maintenance (ORR).

Several studies have started to consider cyber-physic attacks in the electricity grid, such as malicious supplier attacks (MS) [10,11]. MS attacks indicate malicious suppliers that provide wrong supply information to the management system. Related studies [10,11] proposed attention-based deep learning models to classify suppliers (normal or abnormal) and predict the true supply ratio. The literature focuses on adjusting market equilibrium and reserving extra electricity to deal with attacks, and lacks discussions of electricity storage that is effective and efficient in insecure environments.

2.3 Electricity Storage

Electricity storage aims to store excess electricity and release it when needed, which is an essential component of grid management [1] to balance electricity supply and demand in real-time. For example, the traditional energy storage mechanism is fully charging storage when the ORR is positive (supply exceeds demand, excess capacity), and is discharged when the ORR is negative (demand exceeds supply, requiring capacity).

However, there is an inevitable phenomenon of self-discharge, which refers to the gradual loss of energy from the electricity storage device even when it is not in use [6]. Self-discharge is affected by various factors such as temperature, humidity, and energy conversion rate, and has a significant impact on the energy efficiency and lifetime of the device. Therefore, the control of self-discharge is an important aspect of the research of electricity storage technologies.

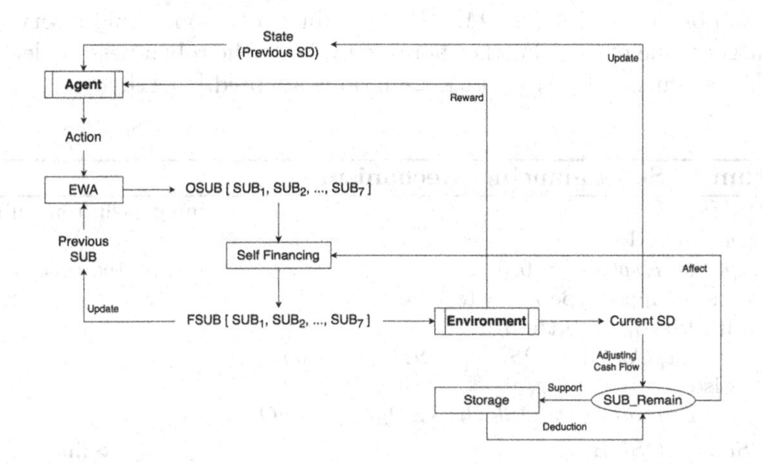

Fig. 1. Flowchart of the proposed TripleS

3 Problem Definition and Simulation Environment

For a fair comparison, we inherit the simulation environment of the electricity grid from [9]. There are 7 types of electricity suppliers with initial supply distribution $SD = [0.29, 0.03, 0.36, 0.08, 0.09, 0.02, 0.13]$. Each supplier provides the supply plan for the following day with a minimum willing price to sell and the planned supply quantity. The call-auction system [9] will collect the information, adjust the price by the subsidy SUB, and create the aggregate supply curve.

The call-auction system also collects historical information to generate the aggregated demand curve. By the supply and demand curves, the system derives the equilibrium price and quantity ($EquP$ and $EquQ$), which apply to all transactions. The electricity supply with a willing price smaller (larger) than $EquP$ will (not) be traded, and will generate profit (loss) and increase (deduct) the capital. If capital becomes negative, the supplier is bankrupted and does not supply electricity. Note that all characteristics of suppliers (capital, willing price, supply quantity, and so on) are randomly sampled by definition in [9].

However, there would be an unexpected shortage of supply (real quantities are smaller than planned quantities) and a variance in demand in the electricity market. [9] proposed a dynamic factor $DMOR$ to reserve capacities as backup, and multiplied the quantities of the demand curve by $1 + DMOR$, specifically. The daily operating reserve rate ORR_d is defined as the unused electricity capacity divided by the supplied electricity capacity.

In this paper, we adopt the Q-learning algorithm [8] to determine the daily $DMOR$ and SUB through clear states (the current value is greater than, close to, or less than the target value) and intuitive actions. In summary, the research question is determining daily subsidies for 7 types of electricity suppliers ($SUB_i, i = 1, \ldots, 7$ to shape the electricity supply distribution to the target distribution of $\overline{\text{SD}}$) and daily $DMOR$ (to maintain the operating reserver rate at the target value of $\overline{\text{SD}}$). Furthermore, to evaluate the robustness of electricity supply management, the MS attack scenario is adopted [10,11].

Algorithm 1. Self-Financing Mechanism

```
1: SUB_Remain = 0                                        ▷ remaining cash from subsidies
2: for each day d do
3:     ExpCost, ExpGain = 0, 0                            ▷ expected cost and gain
4:     for each supply type i ← 1 to 7 do
5:         if OSUB_{i,d} > 0 then
6:             ExpCost + = OSUB_{i,d} · SD_{i,d-1} · EquQ_{d-1}
7:         else
8:             ExpGain − = OSUB_{i,d} · SD_{i,d-1} · EquQ_{d-1}
9:     FSUB = OSUB                                        ▷ final subsidies
10:    if ExpCost > ExpGain+ SUB_Remain then
11:        Factor = (SUB_Remain + ExpGain) / ExpCost
12:        for each supply type i ← 1 to 7 and FSUB_{i,d} > 0 do
13:            FSUB_i × = Factor
14:    Market Equilibrium at EquP_d, EquQ_d, and SD_d
15:    for each supply type i ← 1 to 7 do                 ▷ adjusting remaining
16:        SUB_Remain − = FSUB_{i,d} · SD_{i,d} · EquQ_d
17:    for each storage j do                              ▷ supporting storage
18:        SUB_Remain − = SC_j · δ̄ · EquP_d
```

4 Proposed TripleS

We proposed a subsidy-supported storage (TripleS) to efficiently and effectively assist grid management. Suppose the initial capacity of electricity storage is 5% (the initial electricity capacity is 50,000 megawatts), which includes multiple distributed storage units in the market. The capacities and capitals of storage units are randomly sampled from normal distributions (mean of 10 and variance of 2 megawatts; mean of 100 and variance of 20 dollars). In addition, there is an inevitable self-discharge rate δ in each storage unit, which is also sampled from a normal distribution (mean of 5% and variance of 1%). The self-discharge indicates the stored electricity capacity will decrease δ each day.

The flowchart of the proposed TripleS is illustrated in Fig. 1, and the backbone of TripleS is Q-learning algorithms referred from [8]. Note that the initialization, parameter setting, and Q-table updating mechanism of TripleS are the same as [8]. For 7 electricity supply types, there are 7 Q-tables that act as agents to determine the daily subsidy SUB_i, where $i = 1, \ldots, 7$. Each Q-table i takes the state SD_i to make the action A_i. An exponential moving average (EMA) mechanism is applied to smooth the subsidies, denoted as:

$$\begin{aligned} SUB_{i,d} &= (1-\alpha)SUB_{i,d-1} + \alpha(SUB_{i,d-1} + A_i), \\ &= SUB_{i,d-1} + \alpha A_i, \end{aligned} \tag{1}$$

where α is a factor to control the scale of EMA, and d is the index of day. The SUB_i adjusted by EMA is called original subsidies ($OSUB$). Then, a self-financing mechanism is designed to achieve sustainable subsidies and support electricity storage, introduced in Algorithm 1.

The self-financing mechanism first calculates the expected costs and gains of subsidies ($ExpCost$ and $ExpGain$, lines 3 to 8), which are the original subsidies multiplied by the expected quantity (quantity on the previous day, $SD_{i,d-1} \cdot EquQ_{d-1}$, in lines 5 and 8). If the expected cost ($ExpCost$) is greater than the sum of the expected gains and remaining cash ($ExpGain$ + SUB_Remain), the subsidies must be adjusted to achieve self-financing management, which means reducing the cost and positive subsidies by a factor defined in line 11. The factor is the sum of the expected profit and remaining cash ($ExpGain$ + SUB_Remain) divided by the expected cost ($ExpCost$). Therefore, all positive subsidies are multiplied by the factor, and expected costs are reduced to achieve self-financing management (lines 12 to 13). The adjusted subsidies are referred to as final subsidies ($FSUB$). After market equilibrium (line 14), the remaining cash (SUB_Remain) is adjusted for real costs and gains, which are the final subsidies ($FSUB$) multiplied by the real supply quantity of the day ($SD_{i,d} \cdot EquQ_d$, line 16).

The electricity storage mechanism in TripleS is simply designed to fully charge when $ORR_d > 0$ and discharge when $ORR_d < 0$ to maintain the stability of the electricity grid. Note that both charging and discharging electricity are traded at the equilibrium price $EquP$. Unlike traditional storage mechanisms, the charging and discharging sequences of TripleS are based on the ranking of δ, i.e., the lower δ, the higher the ranking of charging and discharging. Moreover, due to the inevitable self-discharge, the proposed TripleS subsidizes storage by the product of the storage capacity, the average δ, and the equilibrium price $EquP$, i.e., the average loss due to self-discharge. The mechanism is described in line 18 of Algorithm 1, where SC is the stored capacity of each storage unit and $\overline{\delta}$ is the average δ of all storage units.

With the designed electricity storage mechanism, the storage units are supported by the remaining cash of the subsidy to prevent bankruptcy, and are fairly subsidized by the equilibrium price and the average self-discharge rate.

Additionally, the self-financing mechanism is a sustainable mechanism to support undersupplied electricity type and storage units, which not only shape the electricity supply distribution but also maintain the stability of the grid.

5 Experimental Results

Six benchmark systems are adopted for comparison, including TRA [9], STA [9], CAES [9], QEMS [8], SDM [10], and CEE [11] systems. Note that the CEE and SDM systems are state-of-the-art grid management systems, and obtain deep learning modules to predict and revise the supply information. We compare the proposed TripleS with the traditional storage mechanism (TS), which charges when $ORR > 0$ and discharges when $ORR < 0$ without subsidizing.

For measurement, we use three indicators to assess the supply distribution shaping, namely ConDay, Con%, and MSD, related to [11]. ConDay and Con% are the average days to convergence and the percentage of convergence. The convergence means the mean absolute error (MAE) between the target and the (last 30-day) average supply distribution is continuously lower than 10%; MSD is the MAE between the target and the daily supply distribution ($\overline{\text{SD}}$ and SD). We also adopt two indicators to evaluate the performance of gird stability, which are MORR and Fail. MORR is the MAE between the target and daily operating reserve rate ($\overline{\text{ORR}}$ and ORR). Fail is the probability that ORR is lower than zero, causing electricity failure to the entire grid. Among all indicators, the smaller values are the better performance, except Con%.

All the following experiments are simulated for 1,825 days and repeated 10 times to obtain the average performance; the scale of the EMA, α, is set to 0.5. The experimental setups of the Q-learning algorithms follow [8], and the setups of the deep-learning models follow [11]. Moreover, the implementation of the proposed TripleS is written in Python and can be found on GitHub[1].

5.1 Performance Evaluation

In this section, we evaluate the grid management performance without an attack scenario, where the results are presented in Table 1. There are six sub-tables in Table 1, corresponding to five target supply distributions and the average performance (of the five subtables), where $\overline{\text{SD}}1$ is [0.30, 0.10, 0.10, 0.15, 0.10, 0.10, 0.15], $\overline{\text{SD}}2$ is [0.25, 0.05, 0.20, 0.10, 0.15, 0.15, 0.10], $\overline{\text{SD}}3$ is [0.20, 0.10, 0.20, 0.10, 0.15, 0.10, 0.15], $\overline{\text{SD}}4$ is [0.20, 0.05, 0.15, 0.15, 0.15, 0.15, 0.15], and $\overline{\text{SD}}5$ is [0.15, 0.05, 0.10, 0.10, 0.20, 0.20, 0.20]. Additionally, the bold values are the top-2 performance of the column (among all systems).

In Table 1, it can be seen that the benchmark systems from TRA, STA, and CAES do not converge well (extremely low Con% and high ConDay). In contrast, the QEMS, QEMS+TS, and QEMS+TripleS systems achieve excellent

---
[1] https://github.com/JiaHao-Syu/Subsidy-Supported-Storage-for-Electricity-Management-Systems.

and similar performance in shaping the supply distribution (Con%, ConDay, and MSD) and always achieve perfect Con%. For grid stability (MORR and Fail), the reference systems of TRA and STA still perform poorly, while the CAES and QEMS systems perform better. Compared to QEMS and QEMS+TS, QEMS+TS has the traditional storage mechanism, which reduces MORR by 0.04% and Fail by 0.03% on average (last subtable). Compared to QEMS and QEMS+TripleS, the proposed TripleS slightly reduces MORR by 0.06% but drastically reduces Fail by 0.14% on average. The proposed TripleS achieves the perfect Fail (0%) in 4 of 5 experiments, and has only 0.01% in the exception.

After observing the experimental results, we found that most electricity storage units of TS are bankrupt due to self-discharge. In contrast to our proposed TripleS, TripleS supports storage units by the subsidy remaining under the self-financing mechanism. In summary, TripleS performs excellently in designing supply distribution and significantly improves grid stability. The phenomenon demonstrates the effectiveness of the proposed TripleS in self-financing management, i.e., subsidizing storage without disturbing the design of supply distribution.

5.2 Performance Evaluation Under MS Attack

To evaluate the robustness, we adopt the MS attack in the electricity grid, and compare the state-of-the-art systems, QEMS [8], SDM [10], and CEE [11]. Note that the benchmark systems of TRA, STA, and CAES obtain worse performance in all measurements; therefore, we ignore them in this section. The experimental results under the MS attack are listed in Table 2.

It can be seen that almost all systems achieve perfect Con% (100%), with the exception of 90% for CEE with $\overline{\text{SD}}5$. For the ConDay measurement, the top 2 performances often belong to QEMS and CEE, which achieve average ConDays of 268 and 419, respectively. Similar results are found in the MSD measurement, and the top 2 performances also belong to QEMS and CEE with average MSDs of 8.00% and 8.10%, respectively. After embedding with storage mechanisms (both TS and the proposed TripleS), there are slightly negative effects on shaping the supply distribution under the MS attack. Take CEE as an example; the average ConDay increases slightly from 419 to 475 (CEE +TS) and 446 (CEE +TripleS), and the average MSD increases slightly from 8.10% to 8.64% (CEE +TS) and 8.28% (CEE +TripleS). We suspect that the reason is that the remaining subsidy money is used to support storage units, which slightly slows down the design of the supply distribution. Compared to TS and TripleS, the proposed TripleS could further mitigate the negative effects.

AS for MORR measurement, the top-2 performance often belongs to QEMS and CEE +TripleS, which achieve average MORRs of 4.04% and 4.64%, respectively. The experimental results show that the proposed TripleS could slightly reduce the MORR (from 4.88% to 4.78% in SDM; from 4.84% to 4.64% in CEE), while the traditional TS slightly increases the MORR (from 4.88% to 4.96% in SDM; from 4.84% to 4.90% in CEE). Encouragingly, the proposed TripleS significantly reduces the Fail, always achieving perfect Fail (0%) with CEE +TripleS

Table 1. Performance evaluation of grid management without attack scenario

	System	Con%	ConDay	MSD	MORR	Fail
$\overline{\text{SD}}1$	**TRA** [9]	30%	1,819	23.60%	172.20%	95.41%
	STA [9]	10%	1,825	24.00%	172.90%	95.71%
	CAES [9]	20%	1,775	28.80%	**3.70%**	0.19%
	QEMS [8]	**100%**	**496**	**7.30%**	4.00%	0.12%
	QEMS+TS	**100%**	590	7.80%	**3.90%**	**0.07%**
	QEMS+TripleS	**100%**	**558**	**7.50%**	**3.90%**	**0.00%**
$\overline{\text{SD}}2$	**TRA** [9]	10%	1,824	24.70%	161.80%	95.73%
	STA [9]	10%	1,824	23.30%	176.20%	95.75%
	CAES [9]	10%	1,813	25.90%	**3.70%**	0.18%
	QEMS [8]	**100%**	**491**	**7.20%**	**4.00%**	0.14%
	QEMS+TS	**100%**	**488**	**7.00%**	4.10%	**0.10%**
	QEMS+TripleS	**100%**	559	7.40%	**4.00%**	**0.00%**
$\overline{\text{SD}}3$	**TRA** [9]	0%	1,825	25.00%	171.30%	95.64%
	STA [9]	0%	1,825	25.40%	172.40%	95.80%
	CAES [9]	30%	1,781	26.80%	**3.60%**	0.14%
	QEMS [8]	**100%**	**298**	**7.50%**	4.00%	**0.12%**
	QEMS+TS	**100%**	**489**	7.70%	**3.90%**	**0.12%**
	QEMS+TripleS	**100%**	577	**7.30%**	**3.90%**	**0.00%**
$\overline{\text{SD}}4$	**TRA** [9]	0%	1,825	28.10%	181.20%	95.64%
	STA [9]	0%	1,825	29.10%	173.40%	95.73%
	CAES [9]	0%	1,825	34.70%	**3.70%**	0.20%
	QEMS [8]	**100%**	**433**	7.30%	**4.00%**	0.14%
	QEMS+TS	**100%**	462	**7.10%**	**4.00%**	**0.10%**
	QEMS+TripleS	**100%**	**217**	**7.00%**	4.10%	**0.01%**
$\overline{\text{SD}}5$	**TRA** [9]	0%	1,825	41.30%	192.10%	95.87%
	STA [9]	0%	1,825	43.20%	187.70%	95.70%
	CAES [9]	0%	1,825	43.50%	**3.70%**	**0.14%**
	QEMS [8]	**100%**	950	9.70%	4.60%	0.16%
	QEMS+TS	**100%**	**766**	**9.20%**	4.50%	0.16%
	QEMS+TripleS	**100%**	**753**	**9.50%**	**4.40%**	**0.00%**
Average	**TRA** [9]	8%	1,824	28.54%	175.72%	95.66%
	STA [9]	4%	1,825	29.00%	176.52%	95.74%
	CAES [9]	12%	1,804	31.94%	**3.68%**	0.17%
	QEMS [8]	**100%**	**534**	7.80%	4.12%	0.14%
	QEMS+TS	**100%**	559	**7.76%**	4.08%	**0.11%**
	QEMS+TripleS	**100%**	**533**	**7.74%**	**4.06%**	**0.00%**

Table 2. Performance evaluation of grid management under MS attack scenario

	System	**Con%**	**ConDay**	**MSD**	**MORR**	**Fail**
$\overline{SD}1$	**QEMS** [8]	**100%**	**105**	**7.30%**	**3.90%**	0.83%
	SDM [10]	**100%**	**186**	7.70%	4.90%	0.12%
	CEE [11]	**100%**	254	**7.60%**	4.80%	0.14%
	SDM+TS	**100%**	387	8.00%	4.70%	0.11%
	CEE+TS	**100%**	216	8.50%	5.00%	0.24%
	SDM+TripleS	**100%**	374	7.70%	**4.30%**	**0.01%**
	CEE+TripleS	**100%**	389	7.70%	4.40%	**0.00%**
$\overline{SD}2$	**QEMS** [8]	**100%**	**267**	**7.40%**	**3.90%**	0.77%
	SDM [10]	**100%**	373	8.00%	4.90%	0.25%
	CEE [11]	**100%**	**302**	**7.50%**	**4.50%**	0.19%
	SDM+TS	**100%**	463	7.80%	4.60%	0.24%
	CEE+TS	**100%**	377	8.20%	4.90%	0.27%
	SDM+TripleS	**100%**	374	8.10%	**4.50%**	**0.00%**
	CEE+TripleS	**100%**	327	7.70%	**4.50%**	**0.00%**
$\overline{SD}3$	**QEMS** [8]	**100%**	**114**	**7.60%**	**4.00%**	0.70%
	SDM [10]	**100%**	198	8.10%	4.50%	0.25%
	CEE [11]	**100%**	**170**	**7.40%**	4.60%	0.12%
	SDM+TS	**100%**	354	7.80%	**4.40%**	0.15%
	CEE+TS	**100%**	264	7.90%	4.50%	0.34%
	SDM+TripleS	**100%**	328	8.30%	4.50%	**0.00%**
	CEE+TripleS	**100%**	200	7.80%	**4.40%**	**0.00%**
$\overline{SD}4$	**QEMS** [8]	**100%**	**207**	**7.90%**	**4.10%**	1.11%
	SDM [10]	**100%**	439	8.40%	4.70%	0.24%
	CEE [11]	**100%**	406	**7.70%**	4.80%	0.12%
	SDM+TS	**100%**	673	8.60%	5.20%	0.23%
	CEE+TS	**100%**	497	8.40%	4.80%	0.13%
	SDM+TripleS	**100%**	**256**	8.40%	4.80%	**0.00%**
	CEE+TripleS	**100%**	505	8.20%	**4.60%**	**0.00%**
$\overline{SD}5$	**QEMS** [8]	**100%**	**649**	**9.80%**	**4.30%**	0.86%
	SDM [10]	**100%**	972	10.00%	5.40%	0.20%
	CEE [11]	90%	963	10.30%	5.50%	0.09%
	SDM+TS	**100%**	**593**	10.70%	5.90%	0.30%
	CEE+TS	**100%**	1,023	10.20%	**5.30%**	0.22%
	SDM+TripleS	**100%**	794	**9.70%**	5.80%	**0.00%**
	CEE+TripleS	**100%**	808	10.00%	**5.30%**	**0.00%**
Average	**QEMS** [8]	**100%**	**268**	**8.00%**	**4.04%**	0.85%
	SDM [10]	**100%**	434	8.44%	4.88%	0.21%
	CEE [11]	98%	**419**	**8.10%**	4.84%	0.13%
	SDM+TS	**100%**	494	8.58%	4.96%	0.21%
	CEE+TS	**100%**	475	8.64%	4.90%	0.24%
	SDM+TripleS	**100%**	425	8.44%	4.78%	**0.00%**
	CEE+TripleS	**100%**	446	8.28%	**4.64%**	**0.00%**

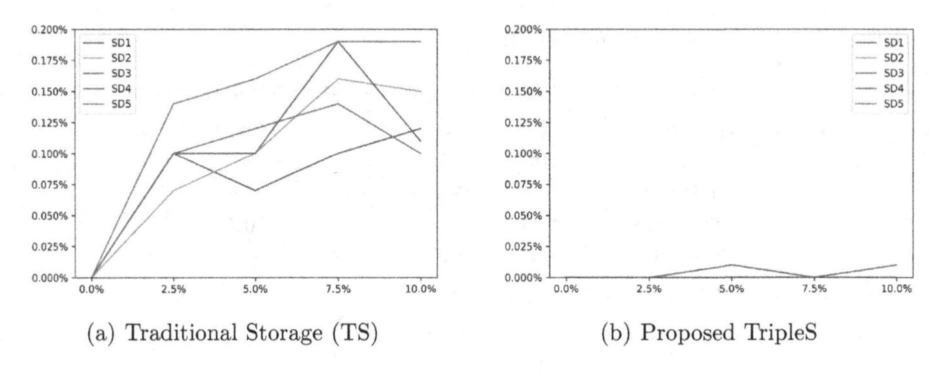

(a) Traditional Storage (TS) (b) Proposed TripleS

Fig. 2. Fail (y-axis) under different self-discharge factors δ (x-axis)

and often (4 out of 5 experiments) achieving 0% Fail with SDM+TripleS. The only exception is SDM+TripleS with $\overline{\text{SD}}1$, and only achieves an extremely low Fail of 0.01%. In contrast to traditional storage, TS not only slightly increases the average MORR, but also significantly increases the average Fail of the CEE model (from 0.13% to 0.24%). The experiments show that TS is not always applicable under the MS attack. In summary, the proposed TripleS maintains excellent performance in grid stability under the MS attack, achieving the perfect Fail (0%) in almost all experiments. The experimental results demonstrate the effectiveness and robustness of the proposed TripleS under the MS attack.

5.3 Influence of Self-discharge

Since the drawback of storage mechanisms is self-discharge, we study the influence of self-discharge as a sensitivity analysis. We simulate with different means of δ (0.0%, 2.5%, 5.0%, 7.5%, and 10.0%) without an attack scenario. The Fail of the traditional storage mechanism TS and the proposed TripleS under different δ are exhibited in Fig. 2, where the x- and y-axes are the means of δ and Fail.

In Fig. 2(a), it can be found that the Fail of TS gradually increases as the increasing δ. Under δ of 0.0%, TS can also achieve perfect Fail (0%), in all experiments; contrary, when δ is 10.0%, TS obtains high Fails of 0.1% to 0.2%. The phenomenon shows that the traditional storage mechanism is only applicable under an ideal environment, which has zero δ and no attack (observation from Sect. 5.1). In Fig. 2(b), the proposed TripleS often achieves perfect Fail (0%), and has only an extremely low Fail of 0.01% in $\overline{\text{SD}}4$ with δ of 5.0% and 10.0%. Experimental results show that the proposed TripleS is less affected by δ, and has robustness under uncertain environments (different δ and attacks).

6 Conclusion

We propose a Subsidy-Supported Storage (TripleS) to assist grid management. Experimental results demonstrate the effectiveness of the proposed TripleS in maintaining grid stability with self-financing management. Under various attack scenarios, TripleS outperforms the state-of-the-art models significantly. The proposed TripleS is robust against various uncertain environments, and significantly improves the stability of electricity grids with self-financing management. In the future, we will trade storage capacities to profit from price fluctuations.

References

1. Fuchs, G., Lunz, B., Leuthold, M., Sauer, D.U.: Technology overview on electricity storage. ISEA, Aachen, Juni **26** (2012)
2. Gustian, D., Sembiring, F., Amelia, R., Nurhasanah, E., Waelah, S., Anggraeni, N.: Comparison data mining based on optimization algorithms in receiving electricity subsidies. In: 2020 6th International Conference on Computing Engineering and Design, pp. 1–6 (2020)
3. Holttinen, H., et al.: Methodologies to determine operating reserves due to increased wind power. IEEE Trans. Sustain. Energy **3**(4), 713–723 (2012)
4. Newell, R.G., Pizer, W.A., Raimi, D.: Us federal government subsidies for clean energy: design choices and implications. Energy Econ. **80**, 831–841 (2019)
5. Rashidi-Nejad, M., Song, Y., Javidi-Dasht-Bayaz, M.: Operating reserve provision in deregulated power markets. In: IEEE Power Engineering Society Winter Meeting. Conference Proceedings, vol. 2, pp. 1305–1310 (2002)
6. Ryu, H., Ahn, H., Kim, K., Ahn, J., Cho, K., Nam, T.: Self-discharge characteristics of lithium/sulfur batteries using TEGDME liquid electrolyte. Electrochim. Acta **52**(4), 1563–1566 (2006)
7. Stiphout, V.A., De Vos, K., Deconinck, G.: The impact of operating reserves on investment planning of renewable power systems. IEEE Trans. Power Syst. **32**(1), 378–388 (2016)
8. Syu, J.H., Lin, J.C.W., Fojcik, M., Cupek, R.: Q-learning based energy management system on operating reserve and supply distribution. Sustain. Energy Technol. Assess. **57**, 103264 (2023)
9. Syu, J.H., Lin, J.C.W., Srivastava, G.: Call auction-based energy management system with adaptive subsidy and dynamic operating reserve. Sustain. Comput.: Inform. Syst. **36**, 100786 (2022)
10. Syu, J.H., Lin, J.C.W., Yu, P.S.: Anomaly detection networks and fuzzy control modules for energy grid management with Q-learning-based decision making. In: Proceedings of the 2023 SIAM International Conference on Data Mining, pp. 397–405 (2023)
11. Syu, J.H., Srivastava, G., Fojcik, M., Cupek, R., Lin, J.C.W.: Energy grid management system with anomaly detection and Q-learning decision modules. Comput. Electr. Eng. **107**, 108639 (2023)
12. Yang, Y.C., Nie, P.Y., Liu, H.T., Shen, M.H.: On the welfare effects of subsidy game for renewable energy investment: toward a dynamic equilibrium model. Renew. Energy **121**, 420–428 (2018)

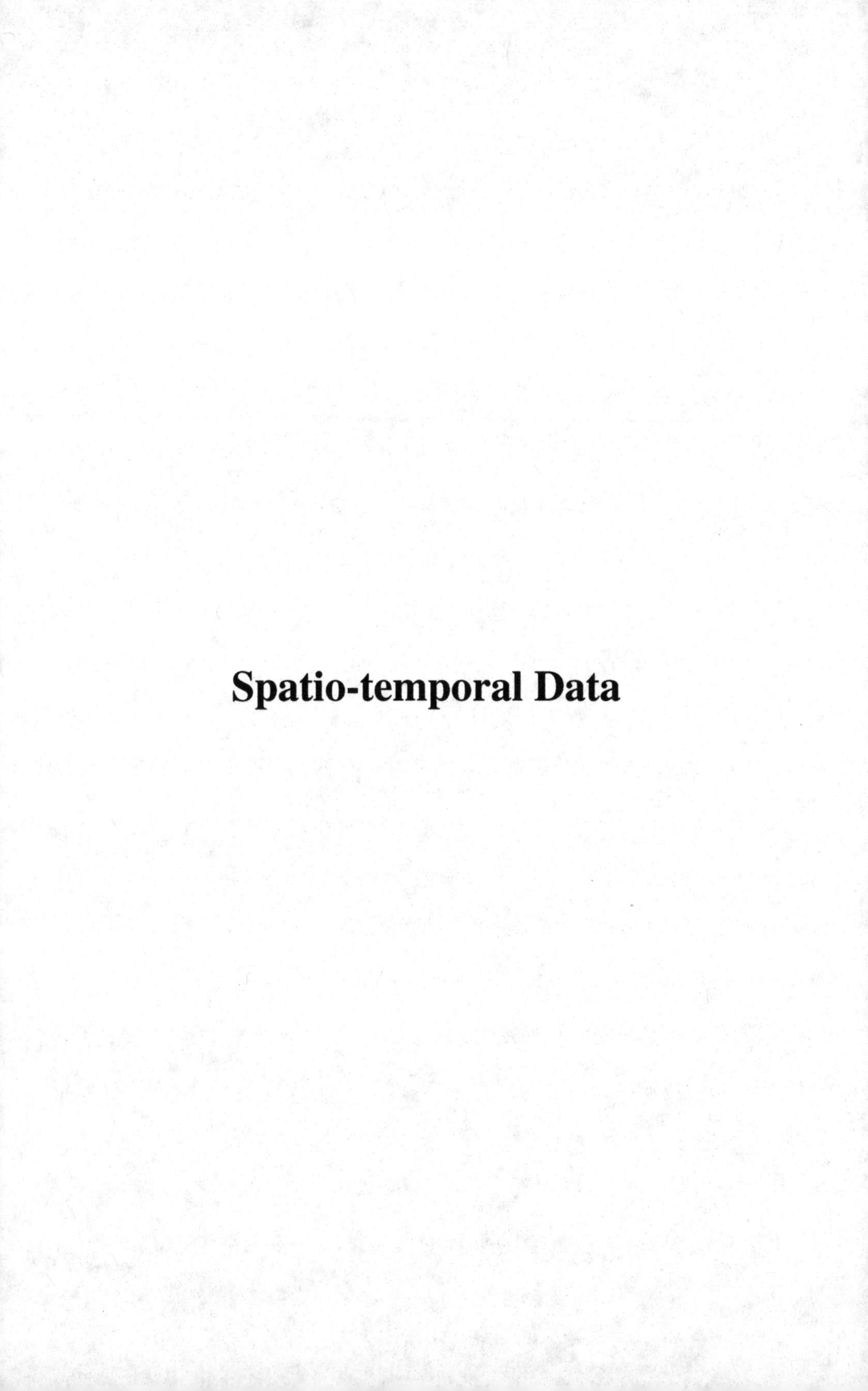

Spatio-temporal Data

Mask Adaptive Spatial-Temporal Recurrent Neural Network for Traffic Forecasting

Xingbang Hu[1], Shuo Zhang[1], Wenbo Zhang[1], and Hejiao Huang[1,2](✉)

[1] School of Computer Science and Technology, Harbin Institute of Technology, Shenzhen, China
{22s151067,21s151165}@stu.hit.edu.cn, huanghejiao@hit.edu.cn
[2] Guangdong Provincial Key Laboratory of Novel Security Intelligence Technologies, Shenzhen, China

Abstract. How to model the spatial-temporal graph is a crucial problem for the accuracy of traffic forecasting. Existing GNN-based work mostly captures spatial dependencies by using a pre-defined graph for close nodes and a self-adaptive graph for distant nodes. However, the predefined graphs cannot accurately represent the genuine spatial dependency due to the complexity of traffic conditions. Furthermore, existing methods cannot effectively capture the spatial heterogeneity and temporal periodicity in traffic data. Additionally, small errors in each time step will greatly amplify in the long sequence prediction for a sequence-to-sequence model. To address these issues, we propose a novel framework, MASTRNN, for traffic forecasting. Firstly, a novel mask-adaptive matrix is proposed to enhance the pre-defined graph, which is learned through node embedding. Secondly, we assign identity embeddings to each node and each time step in order to capture the spatial heterogeneity and temporal periodicity, respectively. Thirdly, a multi-head attention layer is employed between the encoder and decoder to alleviate the problem of error propagation. Experimental results on three real-world traffic network datasets demonstrate that MASTRNN outperforms the state-of-the-art baselines.

Keywords: Traffic forecasting · Spatial-Temporal graph data · Graph convolution

1 Introduction

Recently, many countries have made a strong commitment to developing the Intelligent Transportation System (ITS). As an integral part of ITS, traffic prediction aims to forecast future traffic conditions (e.g., volume, speed) in road networks. This prediction is based on historical observations and is of significant value in optimizing the allocation of transportation resources. For example,

D.-N. Yang et al. (Eds.): PAKDD 2024, LNAI 14649, pp. 259–270, 2024.
https://doi.org/10.1007/978-981-97-2262-4_21

transportation agencies can reduce traffic congestion by relying on accurate traffic prediction [1].

Nowadays, graph neural networks (GNNs) [3] have been shown to be suited for modeling the spatial dependencies of traffic data. GNNs have achieved great success in handling graph data because they can represent spatial-temporal dependencies using graphs and sequences, respectively. Despite their effectiveness, GNN-based models still have three major limitations in traffic prediction.

Firstly, the predefined graph structure of data is based on the distance between two nodes, which means that it cannot accurately reflect the genuine spatial dependencies among nodes due to the complexity of traffic conditions. Although many researchers now propose to use adaptive graphs [4] to solve this problem, they still use the unsuitable predefined graphs as the primary and the adaptive graph as the secondary for graph convolution. However, if the adaptive graph is employed alone, it will lead to performance degradation due to the difficulty of training.

Secondly, although these methods can capture the spatial and temporal dependencies, they ignore the spatial heterogeneity and temporal periodicity of traffic time series and cannot effectively capture them.

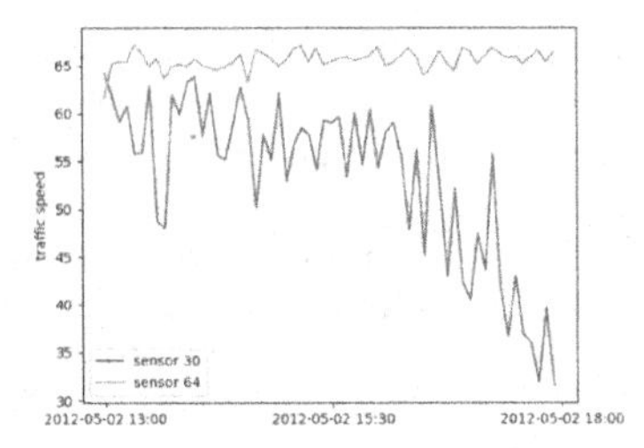

(a) **Traffic speed on sensor 29 and 301** in METR-LA dataset.

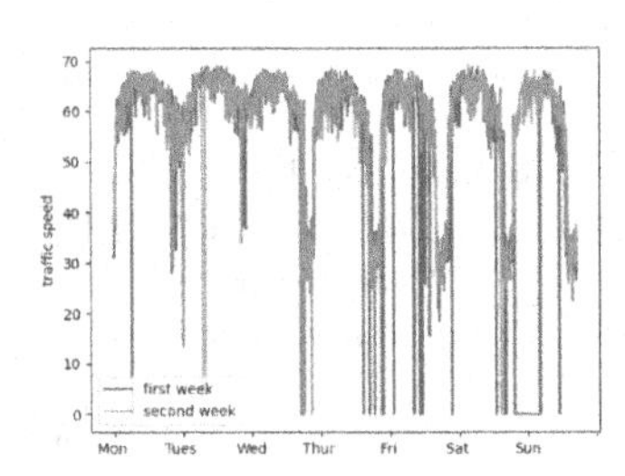

(b) **Two successive weeks traffic** speed of sensor 6 in METR-LA dataset.

Fig. 1. Visualization of the spatial heterogeneity and the temporal periodicity.

- Spatial heterogeneity. Li *et al.* [5] proved that GCN is actually a special form of Laplacian smoothing. This finding highlights the issue of oversmoothing, where the representation of each node is forced to be close to its neighbors. However, due to the specific static attributes (e.g., Point of Interest, road width, speed limit) of different spatial regions, two adjacent regions may exhibit dissimilar patterns during certain periods. As shown in Fig. 1(a), although sensor 30 and 64 are two neighboring regions in the METR-LA dataset, they exhibit completely different changes after 4:00 p.m. We refer to such observation as spatial heterogeneity. However, we cannot get more information about the specific static attributes of spatial regions, which is a problem for us to capture spatial heterogeneity.

- Temporal periodicity. Due to the periodic nature of human behavior, traffic data also exhibits clear periodic patterns. For example, an employee goes to work at 9:00 a.m. and finishes at 5:00 p.m. each weekday. On weekends, he or she is at home. This leads to variability and periodicity in traffic flow at each time of day. As shown in Fig. 1(b), we selected the traffic speed data of two successive weeks on sensor 6 from the METR-LA dataset. We can observe that the data largely overlap during the two-week period. We refer to such observation as temporal periodicity. However, human activities in specific period are very complex, which is a problem for us to capture temporal periodicity.

Thirdly, due to the vanishing gradient problem, traditional RNN architectures struggle to capture and model these long-term dependencies effectively [6]. Furthermore, in a sequence-to-sequence model, small errors in each time step will amplify greatly in the long sequence prediction [1]. It is a challenging problem that how to solve error propagation.

To address the aforementioned three challenges, we propose a Mask Adaptive Spatial-Temporal Recurrent Neural Network (MASTRNN) for the traffic forcasting. The contributions of this work are summarized as follows:

- We construct a mask-adaptive matrix to enhance the pre-defined graph. Our proposed mask-adaptive matrix is capable of adjusting the weights of neighboring nodes in a predefined graph without any guidance of prior knowledge. It means that future researchers will no longer have to struggle to find the right way to build pre-defined graph.
- We design a Spatial-Temporal Graph Convolutional Recurrent Unit (STGCRU) that enables the simultaneous capture of spatial and temporal dependencies. Moreover, the inclusion of spatial and temporal embedding addresses the spatial heterogeneity and temporal periodicity, respectively.
- The attention mechanism is proposed to alleviate the problem of error propagation. Considering the aforementioned components, we propose the MASTRNN model, which is based on the encoder-decoder architecture, for accurate traffic prediction.
- We evaluate our MASTRNN on three real-world traffic datasets. The results show that our proposed MASTGNN significantly outperforms the state-of-the-art traffic forecasting methods.

2 Related Work

With the availability of large-scale traffic data, deep learning models have been employed to solve traffic prediction problems. Previous deep learning models simply formulated the problem as a simple time series prediction task [7–13]. Nowadays, researchers are combining graph convolution with deep learning methods to propose spatial-temporal graph neural networks (STGNNs). These networks have achieved significant success in addressing spatial and temporal dependencies. The majority of STGNNs can be divided into the following three categories: RNN-based, CNN-based, and attention-based approaches. RNN-based STGNNs inject a graph convolution operation into the recurrent cell (e.g. GRU), filtering the historical information and hidden states passed to itself. For example, DCRNN [2] proposed a new diffusion convolution for traffic and injected it into a recurrent cell. CNN-based STGNNs employ CNNs to capture temporal dependencies, resulting in faster processing compared to RNN-based models due to their ability to leverage parallel computing. STGCN [14] combines graph convolution with a standard 1D convolution. However, limited by the kernel size, it is also difficult for 1D CNN to capture long-term temporal correlations. Graph WaveNet [4] employed dilated causal convolution to capture long sequence dependencies. Attention-based STGNNs, such as GMAN [1] proposed spatial and temporal attention mechanisms with gated fusion to model complex spatial-temporal correlations. ASTGNN [15] designed a trend-aware multi-head attention mechanism specialized for time series prediction tasks, which captures the local context in time series. Besides, more and more researchers now combine the transformer with graph convolution and achieve a promising result. PDFormer [16] design a spatial self-attention module to capture the dynamic spatial dependencies.

3 Model Architecture

3.1 Problem Definition

A traffic network is represented by a graph $G = (V, E, A)$, where V is a set of nodes $|V| = N$, E is the set of edges $|E| = M$ and $A \in \mathbb{R}^{N \times N}$ is a weighted adjacency matrix that expresses the reachability between nodes. Given the historical traffic signals $X_{t-T_h:t} = [X_{t-T_h+1}, \cdots, X_{t-1}, X_t] \in \mathbb{R}^{T_h \times N \times C}$, where T_h, C denote the number of passed time steps and features of each node(e.g., speed, flow) respectively. Let $X_{t+1:t+T_f} = [X_{t+1}, X_{t+2}, \cdots, X_{t+T_f}] \in \mathbb{R}^{T_f \times N \times C}$ represent the traffic signals for the next T_f steps. The problem of traffic forecasting is to learn a function f mapping $X_{t-T_h:t}$ and $X_{t+1:t+T_f}$, as shown below:

$$[X_{t-T_h:t}; G] \xrightarrow{f} X_{t+1:t+T_f} \tag{1}$$

3.2 Mask-Adaptive Matrix

Graph convolution plays an important role in extracting node's features given its spatial dependency. Li *et al.* [2] proposed modeling the traffic flow as a diffusion process, which is defined as:

$$X \star_{\mathcal{G}} \Theta = \sum_{k=0}^{K} P_f^k X W_{k1} + P_b^k X W_{k2}, \tag{2}$$

where $P^k \in \mathbb{R}^{N\times N}$ denote a k order transition matrix. $A \in \mathbb{R}^{N\times N}$ denote the road network graph adjacency matrix. There are two directions of information diffusion: the forward directions transition matrix $P_f = A/rowsum(A)$ and backward directions transition matrix $P_b = A^T/rowsum(A^T)$.

Pre-defined graph adjacency matrix A take a leading role in the performance of graph convolution. However, the existing pre-defined graph adjacency matrix measure the similarity between pairs of nodes by a distance metric and it cannot reflect the accurate spatial dependency between nodes due to complex traffic conditions. Furthermore, $P_f = A/rowsum(A)$ may not be suitable for the diffusion progress of traffic. In our work, we propose a mask-adaptive matrix that enhances the diffusion process, which is defined as follows:

$$\begin{cases} P_f = tanh(E_1 E_2^T) \odot A \\ P_b = (tanh(E_1 E_2^T))^T \odot A^T \end{cases} \tag{3}$$

where $E_1, E_2 \in \mathbb{R}^{N\times d}$ denote the two learnable parameter embeddings. We use the tanh activation function to generate negative values for the traffic outflow from the node. $A \in \mathbb{R}^{N\times N}$ denote the predefined sparse adjacency matrix and be leveraged as a mask to avoid overfitting.

In order to model spatial dependency between distant nodes, we combine the mask-adaptive matrix with the self-adaptive matrix and propose a new graph convolution as follows:

$$X \star_{\mathcal{G}} \Theta = P_f X W_1 + P_b X W_2 + A_{adp} X W_3, \tag{4}$$

$$A_{adp} = softmax(relu(E_3 E_4^T)), \tag{5}$$

where $E_3, E_4 \in \mathbb{R}^{N\times d}$,$W_{1,2,3} \in \mathbb{R}^{C\times M}$ denote the learnable project matrix.

3.3 Spatial Temporal Identity Embedding

In order to model the spatial heterogeneity, we assign each node an identity embedding $E^{Nid} \in \mathbb{R}^{N\times d}$ to learn the specific static attributes of different spatial regions. In order to capture the temporal periodicity, we divide a day into N_t time steps. We assign each time step two identity embeddings: $E^{Did} \in \mathbb{R}^{N_d\times d}$, $E^{Tid} \in \mathbb{R}^{N_t\times d}$ to indicate which day of week, and which time of day. N is the number of nodes and $N_d = 7$ is the number of days in a week. Note that all identity embeddings are trainable.

In our work, we inject the above node, day, and time identity embedding into the Gated Recurrent Unit (GRU) [8], which leads to our proposed Spatial Temporal Graph Convolutional Recurrent Unit (STGCRU).

$$\begin{cases} r_t = \sigma(([X_t, H_{t-1}] \star_{\mathcal{G}} \Theta_r) + [E^{Nid}, E_t^{Did}, E_t^{Tid}]W_r + b_r) \\ u_t = \sigma(([X_t, H_{t-1}] \star_{\mathcal{G}} \Theta_u) + [E^{Nid}, E_t^{Did}, E_t^{Tid}]W_u + b_u) \\ C_t = tanh(([X_t, (r_t \odot H_{t-1})] \star_{\mathcal{G}} \Theta_C) + [E^{Nid}, E_t^{Did}, E_t^{Tid}]W_C + b_C) \\ H_t = u_t \odot H_{t-1} + (1 - u_t) \odot C_t \end{cases} \tag{6}$$

where $X_T \in \mathbb{R}^{N \times C}, H_t \in \mathbb{R}^{N \times h}$ denote the input and output at time t respectively. $\star_{\mathcal{G}}$ denote the above graph convolution in Equation(4). r_t, u_t, c_t represent the reset gate, update gate, candidate hidden state, respectively. $E^{Nid} \in \mathbb{R}^{N \times d}$ denote node embedding, $E_t^{Did}, E_t^{Tid} \in \mathbb{R}^d$ denote day, and time identity embedding at time step t, respectively. $\Theta_{r,u,C}$ are the kernel parameters of the corresponding filters, and $W_{r,u,C} \in \mathbb{R}^{3d \times h}$ denotes the learnable projection matrices.

3.4 Multi-head Attention Layer

We leverage the sequence-to-sequence architecture to model temporal dependencies. Both the encoder and the decoder are composed of a stack of STGCRU. Motivated by the attention mechanism [1,19,20]. We add a multi-head attention layer to convert the hidden states of encoder to the decoder. The attention function of node v_i can be expressed as:

$$Q_s^t = H_{t \in dec}^{v_i} W_s^Q, K_s = H_{enc}^{v_i} W_s^K, V_s = H_{enc}^{v_i} W_s^V \tag{7}$$

$$head_s^t = Attention_s(Q_s^t, K_s, V_s) = softmax(\frac{Q_s^t (K_s)^T}{\sqrt{d}}) V_s \tag{8}$$

$$\hat{H}_{t \in dec}^{v_i} = Concat(head_1, \cdots, head_s) W^O \tag{9}$$

where $H_{t \in dec}^{v_i} \in \mathbb{R}^{1 \times h}$ denote the decoder hidden state of the node v_i at time step t and $H_{enc}^{v_i} \in \mathbb{R}^{T_h \times h}$ denote the encoder hidden state of the node v_i at all time step. Given attention head s, $Q_s^t \in \mathbb{R}^{1 \times h}, K_s, V_s \in \mathbb{R}^{T_h \times h}$ represent the query, key, value matrices respectively. W_s^Q, W_s^K, W_s^V, W^O are learnable parameters. $\hat{H}_{t \in dec}^{v_i} \in \mathbb{R}^h$ is the output of multi-head attention layer. To take the position into account, we choose fixed positional encoding [11] for encoder hidden states.

In the end, we design a gated fusion mechanism to adaptively fuse the $\hat{H}_{t \in dec}$ and $H_{t \in dec}$ representations, which is defined as:

$$\tilde{H}_{t \in dec} = sigmoid(\alpha) \hat{H}_{t \in dec} + (1 - sigmoid(\alpha)) H_{t \in dec} \tag{10}$$

where $\alpha \in \mathbb{R}^1$ is a learnable parameter and $\tilde{H}_{t \in dec}$ is the hidden state input of decoder at time step t.

3.5 Framework of MASTRNN

As shown in Fig. 2, MASTRNN follows the sequence-to-sequence architecture, where the encoder encodes the historical traffic signals and the decoder predicts the output sequence. Both the encoder and decoder are composed of STGCRU, where graph convolution operations (Sect. 3.2) and spatial-temporal identity embedding (Sect. 3.3) are injected. A multi-head attention layer (Sect. 3.4) is employed between the encoder and decoder to alleviate the problem of error propagation. This layer connects the hidden states of the encoder with the decoder.

We choose to employ Mean Absolute Error (MAE) to optimize our model:

$$\mathcal{L}(\hat{X}_{t+1:t+T_f}, X_{t+1:t+T_f}; \Theta) = \frac{1}{T_f N C_{out}} \sum_{i=1}^{T_f} \sum_{j=1}^{N} \sum_{k=1}^{C_{out}} |\hat{X}_{ijk} - X_{ijk}| \tag{11}$$

Motivated by the research [17,21], curriculum learning is employed to train our model, which is an effective strategy for RNN models.

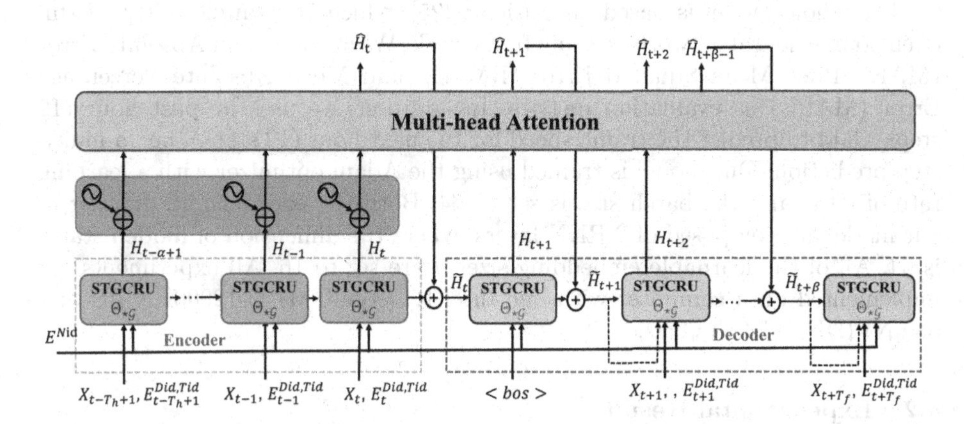

Fig. 2. Framework of MASTRNN.

4 Experiments

4.1 Experimental Settings

We verify our model on three standard benchmark datasets. METR-LA and PEMS-BAY were released by Li [2]. PEMSD7M were released by Yu [14]. METR-LA covers 207 traffic sensors lasting from 2012/5/1 to 2012/6/30 with a 5-minute sampling rate. PEMS-BAY covers 325 traffic sensors lasting from 2017/1/1 to 2017/5/31 with a 5-minute sampling rate. PEMSD7M covers 228 traffic sensors lasting from 2012/5/1 to 2012/6/30 with a 5-minute sampling rate. In all of those datasets, we apply Z-score normalization for input. We split all datasets with ratio 7:1:2 into training sets, validation sets, and test sets. For the adjacency matrix of nodes, we follow the procedure of DCRNN [2]. It is constructed by road network distance using thresholded Gaussian kernel [22].

We compare MASTRNN with the following eight baselines: 1) HA: Historical Average, which uses averaged values from historical days; 2) FC-LSTM [23]: a sequence-to-sequence model with fully-connected LSTM layers; 3) DCRNN [2]: Diffusion convolutional recurrent neural network, which proposes a new diffusion graph convolution for traffic and injects it into recurrent cell; 4) Graph Wavenet [4]: It combines diffusion graph convolutions with gated 1D dilated convolutions and proposes self-adaptive adjacency matrix; 5) GMAN [1]: An attention-based model, which stacks spatial, temporal and transform attentions; 6) DGCRN [17]: Dynamic Graph Convolutional Recurrent Network, which leverages and extracts dynamic characteristics from node attributes in a seq2seq architecture; 7) STID [24]: A simple yet effective baseline for MTS forecasting by attaching spatial and temporal identity information based on simple MLP; 8) MegaCRN [18]: Meta-Graph Convolutional Recurrent Network, which proposes a novel spatial-temporal graph structure learning mechanism for traffic forecasting.

The whole model is based on BasicTS [25], which is a simple but powerful open-source neural network training framework. We choose Mean Absolute Error (MAE), Root Mean Squared Error (RMSE), and Mean Absolute Percentage Error (MAPE) as evaluation metrics. In addition, we use the past hour (12 steps) data to predict the traffic speed for the next hour (12 steps), i.e., a multi-step prediction. Our model is trained using the Adam optimizer with a learning rate of 0.01, and the batch size is set to 64. Both the encoder and decoder of our model are composed of 2 RNN layers, where the dimension of hidden states is 64. All of the learnable embedding sizes d are set to 16. All experiments are implemented on a computer with one Intel(R) Core(TM) i9-12900F CPU and one NVIDIA 3090Ti GPU.

4.2 Experimental Result

As shown in Table 1, our model has demonstrated to achieve state-of-the-art performance on three traffic speed datasets. Besides, we also observe that: (1) Compared to traditional traffic forecasting models, deep learning methods perform better, which indicates the effectiveness of deep neural networks in traffic forecasting. Non-deep learning methods such as HA perform worse due to their strong assumptions (e.g., stationarity or linearity) about the data. (2) Compared to FC-LSTM, GNN-based methods achieve better results, demonstrating that incorporating the information of the road network is essential for traffic forecasting. (3) The attention mechanism excels in capturing long-term dependencies, such as GMAN and our model, which means that attention mechanism make our model more suitable for long-term prediction tasks. (4) Due to the self-adaptive matrix of Graph WaveNet, its performance remains very promising, even when compared to newer work such as STID. This demonstrates the necessity of adaptive graphs learned from data. Furthermore, our model not only employ the self-adaptive matrix but also enhance the pre-defined graph, which achieve a more promising performance. (5) Because DGCRN and MegaCRN captures the dynamic characteristics of the spatial topology, it achieves the best performance compared to other baselines. However, although our MASTRNN applies a static graph, the temporal identity embedding enables our model to capture temporal changes and achieve the better performance.

Table 1. Main results.

Methods	METR-LA			PEMS-BAY			PEMSD7M		
	MAE	RMSE	MAPE	MAE	RMSE	MAPE	MAE	RMSE	MAPE
HA	4.14	7.77	12.90%	2.88	5.58	6.76%	4.28	7.68	11.5%
FC-LSTM	3.75	7.10	9.92%	2.15	4.45	5.05%	3.23	6.27	8.20%
DCRNN	3.02	6.23	8.31%	1.59	3.69	3.58%	2.85	5.67	7.17%
Graph WaveNet	3.03	6.14	8.28%	1.59	3.68	3.59%	2.73	5.33	6.66%
GMAN	3.08	6.46	8.67%	1.58	3.66	3.53%	2.77	5.73	7.26%
DGCRN	2.95	6.05	7.88%	1.57	3.62	3.48%	2.81	5.75	7.25%
STID	3.11	6.48	9.11%	1.56	3.63	3.52%	2.69	5.46	6.90%
MegaCRN	2.95	6.02	8.11%	1.54	3.53	3.45%	2.61	5.23	6.46%
MASTRNN	**2.84**	**5.86**	**7.74%**	**1.51**	**3.48**	**3.35%**	**2.56**	**5.20**	**6.33%**

4.3 Ablation Study

To validate the effectiveness of our work, we will conduct ablation studies on the METR-LA dataset. Specifically, we will analyze the impact of three components: the mask-adaptive graph, the spatial-temporal identity embedding, and the multi-head attention layer. As shown in Table 2, we have named the variants of MASTRNN as follows: 1) w/o mask-ada: We apply the diffusion matrix to replace the mask-adaptive matrix; 2) w/o stid: We remove the $E^{Nid}, E^{Did}, E^{Tid}$ in STGCRU; 3) w/o attention: we remove the multi-head attention layer; 4) Additionally, GWNet mask-ada represents the variants that apply the mask-adaptive graph to replace the diffusion graph in GWNet.

Effect of Mask-Adaptive Matrix. The results of w/o mask-ada and GWNet mask-ada suggest that the mask-adaptive graph is important for discovering hidden spatial dependencies. As shown in Fig. 3, compared to the original adjacency matrix, the learned forward mask-adaptive adjacency matrix eliminates the influence of most edges and retaining only a few important edges. It demon-

Table 2. Ablation study.

Variants	@Horizon3			@Horizon6			@Horizon12		
	MAE	RMSE	MAPE	MAE	RMSE	MAPE	MAE	RMSE	MAPE
MASTRNN	**2.54**	**4.92**	**6.46%**	**2.89**	**5.96**	**7.89%**	**3.32**	**7.14**	**9.66%**
w/o mask-ada	2.60	5.05	6.67%	2.96	6.08	8.06%	3.37	7.21	9.79%
w/o stid	2.60	5.01	6.60%	3.01	6.12	8.16%	3.50	7.37	10.1%
w/o attention	2.54	4.93	6.47%	2.92	6.01	7.94%	3.34	7.15	9.71%
GWNet	2.69	5.15	6.90%	3.07	6.22	8.37%	3.53	7.37	10.01%
GWNet mask-ada	2.64	5.06	6.69%	3.02	6.13	8.07%	3.48	7.30	9.77%

strates that although two nodes are geographically close, the association is weak. Our proposed methods is capable of adjusting the weights of pre-defined graph.

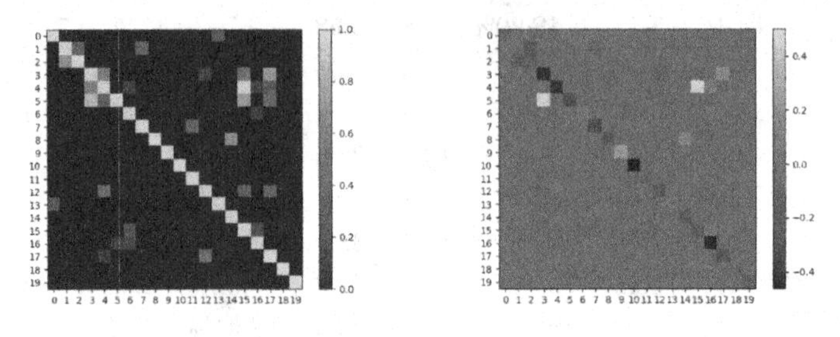

(a) **Original matrix for first 20 nodes.** (b) Learned forward mask-adaptive matrix for first 20 nodes.

Fig. 3. The heatmap of original and learned forward mask-adaptive adjacency matrix.

Effect of ST Identity Embedding. The results of w/o stid suggest that the spatial and temporal identity embeddings are crucial for capturing spatial heterogeneity and temporal periodicity. In Fig. 4, we set the size d of the spatial and temporal identity embeddings to 2 and then trained the MASTRNN to obtain a more accurate visualization. Firstly, Fig. 4(a) shows that despite the spatial proximity of sensors 3 and 4, their node embeddings exhibit very different characteristics, which demonstrates the spatial heterogeneity. Secondly, Fig. 4(b) demonstrates that the day identity embedding of weekdays is similar, while that of weekends is very different. Thirdly, the time identity embedding exhibits cyclic variation in Fig. 4(c), which represents the daily periodicity.

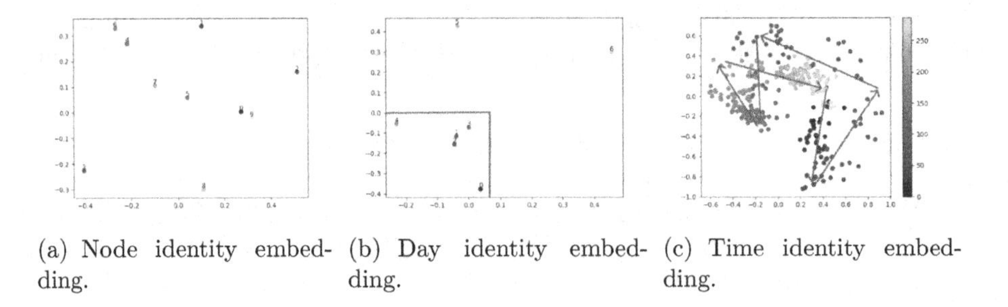

(a) Node identity embedding. (b) Day identity embedding. (c) Time identity embedding.

Fig. 4. Visualization of learned spatial and temporal identity embeddings.

5 Conclusion

In this paper, we propose a novel model to predict traffic conditions. Our model can not only capture spatial-temporal dependencies but also discover spatial heterogeneity and temporal periodicity. Specifically, we proposed an effective method to enhance the predefined graph, which can discover the hidden spatial dependencies. On three public traffic datasets, MASTRNN achieves state-of-the-art results.

References

1. Zheng, C., Fan, X., Wang, C., Qi, J.: GMAN: a graph multi-attention network for traffic prediction. In: AAAI, pp. 1234–1241 (2020)
2. Li, Y., Yu, R., Shahabi, C., Liu, Y.: Diffusion convolutional recurrent neural network: data-driven traffic forecasting. In: ICLR (2018)
3. Wu, Z., Pan, S., Chen, F., Long, G., Zhang, C., Philip, S.Y.: A comprehensive survey on graph neural networks: theoretical basis and empirical results. J. Transp. Eng. **32**(1), 4–24 (2020)
4. Wu, Z., Pan, S., Long, G., Jiang, J., Zhang, C.: Graph WaveNet for deep spatial-temporal graph modeling. In: IJCAI (2019)
5. Li, Q., Han, Z., Wu, X.M.: Deeper insights into graph convolutional networks for semi-supervised learning. In: AAAI (2018)
6. Hochreiter, S., et al.: Gradient flow in recurrent nets: the difficulty of learning long-term dependencies (2001)
7. Graves, A.: Long short-term memory. Supervised sequence labelling with recurrent neural networks, pp. 37–45 (2012)
8. Chung, J., Gulcehre, C., Cho, K., Bengio, Y.: Empirical evaluation of gated recurrent neural networks on sequence modeling. In: NIPS (2014)
9. Yu, F., Koltun, V.: Multi-scale context aggregation by dilated convolutions. arXiv preprint: arXiv:1511.07122 (2015)
10. Van Den Oord, A., et al.: WaveNet: a generative model for raw audio. In: 9th ISCA Speech Synthesis Workshop, pp. 125–125 (2016)
11. Vaswani, A., et al.: Attention is all you need. In: Advances in Neural Information Processing Systems, vol. 30 (2017)
12. Zhou, H., et al.: Informer: beyond efficient transformer for long sequence time-series forecasting. In: AAAI, pp. 1106–11115 (2021)
13. Zhou, T., Ma, Z., Wen, Q., Wang, X., Sun, L., Jin, R.: FedFormer: frequency enhanced decomposed transformer for long-term series forecasting. In: ICML, pp. 7268–27286 (2022)
14. Yu, B., Yin, H., Zhu, Z.: FedFormer: Spatio-temporal graph convolutional networks: a deep learning framework for traffic forecasting. In: IJCAI, pp. 634–3640 (2018)
15. Guo, S., Lin, Y., Wan, H., Li, X., Cong, G.: Learning dynamics and heterogeneity of spatial-temporal graph data for traffic forecasting. IEEE Trans. Knowl. Data Eng. **34**(11), 5415–5428 (2021)
16. Kipf, T.N., Welling, M.: Semi-supervised classification with graph convolutional networks. In: International Conference on Learning Representations (2018)
17. Li, F., et al.: Dynamic graph convolutional recurrent network for traffic prediction: benchmark and solution. ACM Trans. Knowl. Discov. Data **17**(1), 1–21 (2023)

18. Jiang, R., et al.: Spatio-temporal meta-graph learning for traffic forecasting. In: AAAI, pp. 8078–8086 (2023)
19. Wang, X., et al.: Traffic flow prediction via spatial temporal graph neural network. In: Proceedings of the Web Conference 2020, pp. 1082–1092 (2020)
20. Shao, Z., et al.: Decoupled dynamic spatial-temporal graph neural network for traffic forecasting. Proc. VLDB Endowment **15**(11), 2733–2746 (2022)
21. Wu, Z., Pan, S., Long, G., Jiang, J., Chang, X., Zhang, C.: Connecting the dots: multivariate time series forecasting with graph neural networks. In: SIGKDD, pp. 753–763 (2020)
22. Shuman, D.I., Narang, S.K., Frossard, P., Ortega, A., Vandergheynst, P.: The emerging field of signal processing on graphs: extending high-dimensional data analysis to networks and other irregular domains. IEEE Sign. Process. Mag. **30**(3), 83–98 (2013)
23. Sutskever, I., Vinyals, O., Le, Q.V.: Sequence to sequence learning with neural networks. In: Advances in Neural Information Processing Systems, vol. 27 (2014)
24. Shao, Z., Zhang, Z., Wang, F., Wei, W., Xu, Y.: Spatial-temporal identity: a simple yet effective baseline for multivariate time series forecasting. In: CIKM, pp. 4454–4458 (2022)
25. Liang, Y., Shao, Z., Wang, F., Zhang, Z., Sun, T., Xu, Y.: BasicTS: an open source fair multivariate time series prediction benchmark. In: International Symposium on Benchmarking, Measuring and Optimization, pp. 87–101 (2022)

Distributional Kernel: An Effective and Efficient Means for Trajectory Retrieval

Yuanyi Shang[1,2], Kai Ming Ting[1,2], Zijing Wang[1,2(✉)], and Yufan Wang[1,2]

[1] National Key Laboratory for Novel Software Technology, Nanjing University, Nanjing, China
{shangyy,tingkm,wangyf,wangzj}@lamda.nju.edu.cn

[2] School of Artificial Intelligence, Nanjing University, Nanjing, China

Abstract. In this paper, we propose a new and powerful way to represent trajectories and measure the distance between them using a distributional kernel. Our method has two unique properties: (i) the identity property which ensures that dissimilar trajectories have no short distances, and (ii) a runtime orders of magnitude faster than that of existing distance measures. An extensive evaluation on several large real-world trajectory datasets confirms that our method is more effective and efficient in trajectory retrieval tasks than traditional and deep learning-based distance measures.

Keywords: Trajectory Representation and Measure · Trajectory Retrieval · Distributional Kernel

1 Introduction

Trajectory retrieval is one of the most important tasks in trajectory data mining which aims at finding the most similar trajectory in the dataset given the query trajectory. A trajectory is often represented by a sequence of location points representing the route of a moving object.

An effective distance measure is essential for finding similar trajectories. However, current trajectory distance measures have issues with both effectiveness and efficiency. Many distance measures fail to retrieve similar trajectories because of the lack of the identity property. In other words, these measures might assign a shorter distance to two dissimilar trajectories than that to two similar trajectories. Moreover, the commonly used distance measures, such as Dynamic Time Warping (DTW) distance [21] and Hausdorff distance [13], have high time complexities of at least $O(n^2)$, making them difficult to be applied to large datasets.

Efforts over the last 50 years have mainly focused on improving the efficiency issue in the field of trajectory retrieval [12]. Despite these efforts, the problem of effectiveness remains unsolved. In this paper, we reveal for the first time that the main issue with existing methods is that they lack an essential property of

D.-N. Yang et al. (Eds.): PAKDD 2024, LNAI 14649, pp. 271–283, 2024.
https://doi.org/10.1007/978-981-97-2262-4_22

a metric called the identity property. This leads to dissimilar trajectories being ranked higher than similar trajectories in trajectory retrieval tasks.

Our contributions are:

1. Proposing to use a distributional kernel for trajectory embedding and distance measurement, and verifying its effectiveness and efficiency on several large real-world trajectory datasets.
2. Analysing the importance of the identity property for trajectory data. We find that some existing trajectory measures do not have the identity property that the proposed distribution-based method has, which is the reason why it performs better on trajectory retrieval tasks than existing methods.
3. Introducing a simple yet efficient algorithm to solve the Similar Subtrajectory Search problem. This no-learning algorithm is orders of magnitude faster than two state-of-the-art algorithms while producing similar accuracy.

2 Related Work

We survey existing trajectory distance measures and similar subtrajectory search methods here.

2.1 Distance Measures for Trajectories

We classify current trajectory distance measures into three categories: point-based methods, distribution-based methods, and deep learning methods.

For **point-based methods**, Hausdorff distance [13] and Fréchet distance [1] use the distance of two specific points from each trajectory to represent the distance of two trajectories. Dynamic time warping (DTW) distance [21] and its variants calculate the distance by summing the Euclidean distance of the points along the warping path of the two trajectories. All the above measures have a high time complexity of $O(n^2)$, where n is the length of trajectories. Even the fastest known DTW has a time complexity of $O(n^2/\log\log n)$ [5].

Distribution-based methods treat a trajectory as a set of independent and identical distributed (i.i.d.) points sampled from an unknown distribution and use the distance between two distributions to measure the distance between two trajectories. For example, Earth Mover Distance (EMD) measures the distance between two probability distributions estimated from spatiotemporal trajectories [18]. NeuroSeqRet [6] measures the similarity between time series by comparing the generative distribution between the query-corpus sequence pairs. Temporal Kernel SQFD (tkSQFD) [14] proposes a set kernel based on signature quadratic form distance [2] to measure the distance between two sets of feature vectors of time series, where the set of feature vectors is treated as a distribution.

Deep learning methods can be divided into two sub-types. The first type involves learning the embedding of trajectories. For example, t2vec and play2vec map a trajectory to a point in a new feature space and use Euclidean distance to measure the distance between them. The second type of method uses deep learning to approximate traditional distance metrics and accelerate calculations.

A recent deep learning method called NeuTraj [19] aims to learn a distance $\Theta'(\cdot,\cdot)$ which approximates a given distance metric, such as DTW distance or Hausdorff distance, reducing the time complexity from $O(n^2)$ to $O(n)$. TrajGAT [20] is an improvement over NeuTraj.

However, these methods could not solve the issue of effectiveness because none of these methods, including deep learning-based methods, have provided evidence that they have the identity property. From a different perspective, we suggest using a distributional kernel to measure the similarity between trajectories in Sect. 3 and analyze the importance of the identity property for a trajectory distance measure in Sect. 4.

2.2 Similar Subtrajectory Search

Similar subtrajectory search (SimSub) problem [17] aims to find a portion of a given trajectory that is most similar to a query trajectory. RLS-Skip [17] is a deep reinforcement learning method that identifies an action trajectory by maximizing a reward function. GBAF [4] is a state-of-the-art method that transforms the subtrajectory search problem into a reading comprehension problem in Natural Language Processing. In other words, it searches for a paragraph from an article (representing a subtrajectory from a trajectory) that is consistent with the meaning expressed in the query sentence.

We show that subtrajectory search can be achieved effectively and efficiently with the proposed distributional kernel without learning in Sect. 5.

3 Distributional Kernel for Trajectory Similarity Measure

Unlike Hausdorff and Fréchet distances that treat each trajectory as a set of points, we assume that the points in a trajectory are i.i.d. samples from an unknown probability density function (pdf) in $\mathbb{R}^d$ domain, which is the same assumption made in EMD [18].

With the i.i.d. assumption, we propose to use Isolation Distributional Kernel (IDK) [15], a data-dependent distributional kernel, to measure the similarity between two trajectories. IDK is based on kernel mean embedding (KME) [9] that maps a distribution $\mathcal{P}$ into a reproducing kernel Hilbert space (RKHS) via its feature map Φ. For two distributions $\mathcal{P}_X$ and $\mathcal{P}_Y$, IDK is given as:

$$\mathcal{K}(\mathcal{P}_X, \mathcal{P}_Y) = \frac{1}{|X||Y|} \sum_{\mathbf{x}\in X, \mathbf{y}\in Y} \kappa_I(\mathbf{x}, \mathbf{y}) = \frac{1}{t}\langle \Phi(\mathcal{P}_X), \Phi(\mathcal{P}_Y)\rangle \quad (1)$$

where Isolation Kernel (IK) $\kappa_I(\mathbf{x}, \mathbf{y}) = \frac{1}{t}\langle \phi(\mathbf{x}), \phi(\mathbf{y})\rangle$ is a data-dependent kernel which has a finite-dimensional feature map $\phi(\cdot)$ [16]; and $\Phi(\mathcal{P}_X) = \frac{1}{|X|}\sum_{\mathbf{x}\in X} \phi(\mathbf{x})$ is the kernel mean map of $\mathcal{K}$. κ_I could be replaced by Gaussian kernel κ_G to produce GDK, which is commonly used in KME. We propose to use IK because it produces a finite-dimensional feature map with $O(n)$ time complexity.

The proposed $\mathcal{K}$-based distance measure is given as:

$$\Theta_{\mathcal{K}}(\mathcal{P}_X, \mathcal{P}_Y) = 1 - \frac{1}{t}\langle \Phi(\mathcal{P}_X), \Phi(\mathcal{P}_Y) \rangle \tag{2}$$

To make the distance more amenable to an indexing method, we propose two variants of the distance based on $\Theta_{\mathcal{K}}$:

$$\Theta_{MMD}(\mathcal{P}_X, \mathcal{P}_Y) = \|\Phi(\mathcal{P}_X) - \Phi(\mathcal{P}_Y)\|^2 \tag{3}$$

$$\Theta_{\mathcal{K} \circ \mathcal{K}_I}(\mathcal{P}_X, \mathcal{P}_Y) = 1 - \frac{1}{t}\langle \phi\left(\Phi\left(\mathcal{P}_X\right)\right), \phi\left(\Phi\left(\mathcal{P}_Y\right)\right) \rangle \tag{4}$$

Table 1. Example trajectories and distance results of different measures.

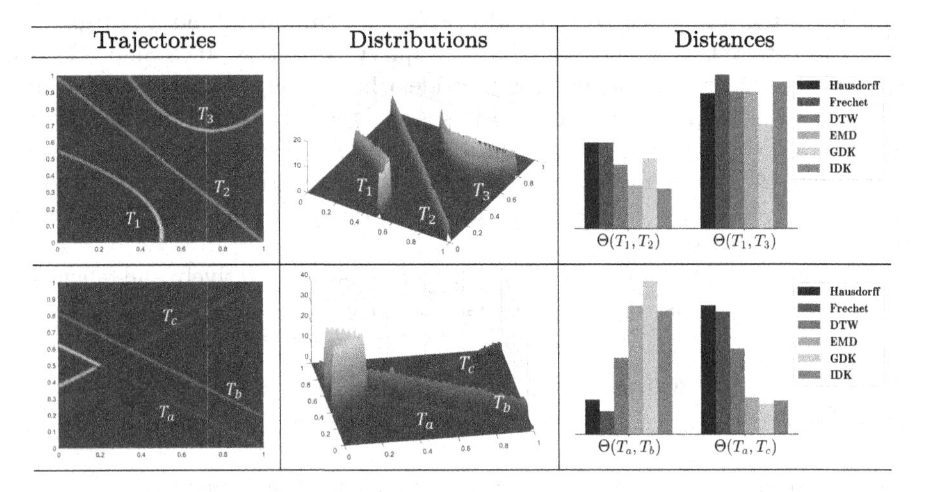

Here we illustrate the features of our proposed distributional kernel-based measure through two examples. Firstly, we show that the results obtained from different measures are consistent with human intuition. As shown in the first row of Table 1, all distance measures provide similar results, indicating that two trajectories that are spatially closer have a shorter distance than those that are far apart, which aligns with our expectations.

Secondly, two trajectories with the same shape may have different distributions. For example, a moving object may start with a slow speed in one section and then accelerate in another section of the same trajectory. As shown in the second row of Table 1, T_a and T_b have similar shapes but different point distributions, while sample points of T_a and T_c are concentrated in the same area. The results obtained from different measures indicate that distribution-based methods (EMD, GDK, and IDK) are more sensitive to the distribution of points in a trajectory than to its shape as they consider T_a to be more similar to T_c than T_b. This property is particularly important when we are interested in the specific behavior of the moving object, such as stopping or accelerating, rather than just the shape of the trajectory.

For the dataset used in this paper, considering only the spatial information of the trajectory sampling points is enough to achieve fair results. We discuss the effect of temporal information on trajectories in Sect. 7.

4 Identity Property

The identity property is one of the four axioms of a distance metric, i.e., $\Theta(x, y) = 0$ if and only if $x = y$. However, all point-based trajectory distance measures mentioned in Sect. 2 do not satisfy this property. In other words, under some conditions, their measured distance between two similar trajectories is longer than that between two dissimilar trajectories.

The identity property of a point-based trajectory distance measure is hard to ascertain because a trajectory is represented by a series of sampled points. As a result, even if two trajectories are from an identical path, they may have completely different sets of points. Yet, we want a measure to have zero or very short distance between two trajectories from the identical path.

In this section, we show that the $\mathcal{K}$-based distance measures, proposed in the last section, have the identity property that is not present in other measures. This property is the key that enables the proposed measures to produce a better retrieval accuracy than other measures.

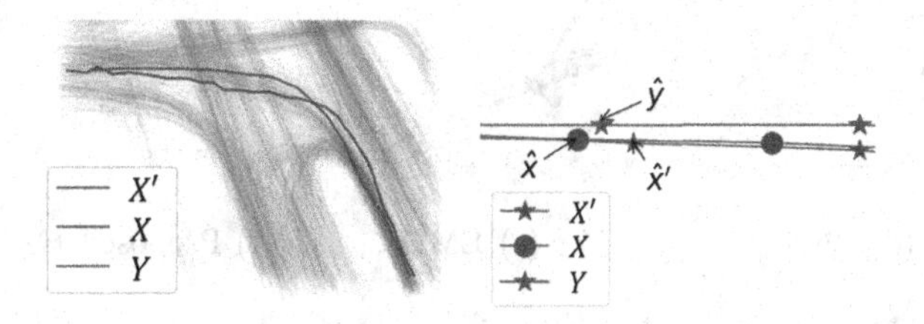

Fig. 1. An example illustrating that Hausdorff distance does not have the identity property. The left subfigure has three trajectories with specific sampled points shown in the right subfigure. The effective point-pair $[\![\hat{\mathbf{x}}, \hat{\mathbf{y}}]\!]$ has shorter distance than that of $[\![\hat{\mathbf{x}}, \hat{\mathbf{x}}']\!]$. Therefore, Hausdorff distance measured X to be more similar to Y than to X'.

In Fig. 1, X is a trajectory more similar to X' than Y because X and X' are sampled points from alternative points on the same original trajectory, but Y is one from a different path close by.

Hausdorff distance and **Fréchet distance** do not have the identity property because they use the distance between two effective points $\hat{\mathbf{x}}, \hat{\mathbf{y}}$ to determine the distance between two trajectories X, Y, i.e., $\Theta(X, Y) := \|\hat{\mathbf{x}}, \hat{\mathbf{y}}\|$. Suppose the effective points between X and X' are $\hat{\mathbf{x}}_\imath \in X, \hat{\mathbf{x}}' \in X'$, and that between X and Y are $\hat{\mathbf{x}}_\jmath \in X, \hat{\mathbf{y}}' \in Y'$. The identity property is not satisfied when $\|\hat{\mathbf{x}}_\imath, \hat{\mathbf{x}}'\| \geq \|\hat{\mathbf{x}}_\jmath, \hat{\mathbf{y}}\|$, which is the case, shown in Fig. 1.

DTW distance has a similar issue as it uses the summation of Euclidean distance of a series of point-pairs from two aligned trajectories: $\Theta_W(X,Y) := \sum_{[\![\hat{\mathbf{x}},\hat{\mathbf{y}}]\!] \in E} \|\hat{\mathbf{x}} - \hat{\mathbf{y}}\|$, where E is the set of effective point pairs after aligning. Under some conditions, the sum of the point-to-point distance between similar trajectories $\Theta_W(X,X')$ could be larger than that between dissimilar trajectories $\Theta_W(X,Y)$, violating the identity property.

In contrast, the identity property holds for **distributional kernel** $\mathcal{K}$. Let trajectory X be similar to X', but different from Y, which means $\mathcal{P}_X = \mathcal{P}_{X'}$ and $\mathcal{P}_X \neq \mathcal{P}_Y$. The feature map Φ of $\mathcal{K}$ has been shown to be injective [10,15], where $\|\Phi(\mathcal{P}_X) - \Phi(\mathcal{P}_{X'})\| = 0$ if and only if $\mathcal{P}_X = \mathcal{P}_{X'}$. This is equivalent to the identity property of a distance measure, where $\Theta_\mathcal{K}(\mathcal{P}_X,\mathcal{P}_{X'}) < \Theta_\mathcal{K}(\mathcal{P}_X,\mathcal{P}_Y)$.

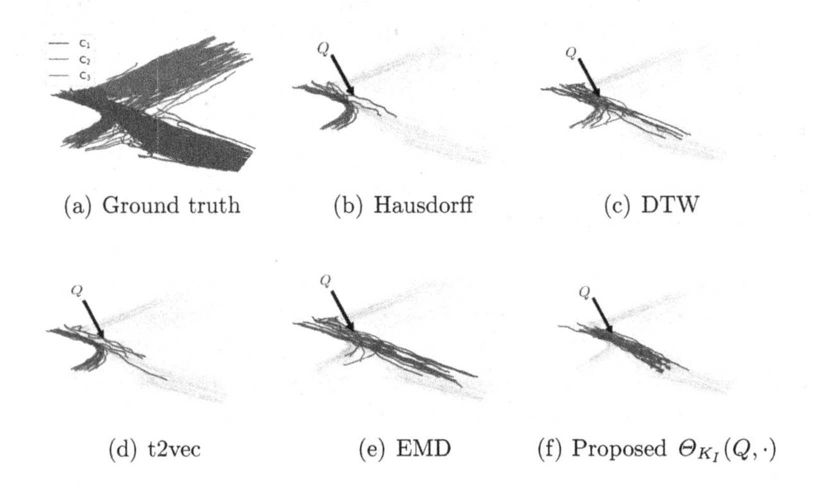

(a) Ground truth (b) Hausdorff (c) DTW

(d) t2vec (e) EMD (f) Proposed $\Theta_{\mathcal{K}_I}(Q,\cdot)$

Fig. 2. The 20-nearest-neighbours from the same query trajectory Q found by five different distance measures.

The lack of the identity property for distance measures has a significant impact on trajectory retrieval tasks. We conduct trajectory retrieval on the VRU dataset [3], which consists of three classes, as shown in Fig. 2(a). For a query trajectory selected from class C_1, all of its nearest neighbors shall also belong to the same class C_1. However, a significant portion of the 20 nearest neighbors computed by Hausdorff distance, DTW distance, and t2vec belong to another class C_2, as shown in Fig. 2(b), 2(c) and 2(d), respectively. Earth Mover Distance outperforms the other three distances due to the use of distribution information. Our proposed distributional kernel, which is the only measure with the identity property, successfully identified all 20 trajectories in class C_1, as presented in Fig. 2(f).

5 Similar Subtrajectory Search

We have thus far considered the similarity between trajectories by taking a trajectory as a whole to retrieve similar trajectories in a database. In another scenario, one may wish to find the most similar subtrajectories in the same database.

A formal definition of *similar subtrajectory search (SimSub)* problem has been provided previously [17]. Instead of searching in the whole database, it focuses on finding the subtrajectories of one data trajectory:

Definition 1. *Similar subtrajectory search (SimSub).* *Given a data trajectory* $X = \ulcorner \mathbf{x}_1, \ldots, \mathbf{x}_i, \ldots, \mathbf{x}_n \urcorner$ *and a query subtrajectory* $Q_s = \ulcorner \mathbf{q}_1, \ldots, \mathbf{q}_i, \ldots, \mathbf{q}_m \urcorner$*, where* $m < n$*, SimSub aims to find subtrajectory* $X_{[a^*,b^*]} = \ulcorner \mathbf{x}_{a^*}, \ldots, \mathbf{x}_{b^*} \urcorner, 1 \leq a^* \leq b^* \leq n$ *of* X *which is most similar to* Q_s *according to a trajectory similarity measurement* $\Theta(\cdot,\cdot)$*, i.e.,* $[a^*, b^*] = \arg\min_{1 \leq a \leq b \leq n} \Theta(X_{[a,b]}, Q_s)$.

RLS-Skip [17] and GBAF [4] are two deep learning methods targeted to SimSub. However, we find that SimSub is not a challenging problem. Once we have identified the data trajectory X that is most similar to the query subtrajectory Q_s, finding $X_{[a^*,b^*]}$ of X (denoted as $X_s \prec X$) can be achieved effectively by creating a bounding box around the area covering Q_s; and then extract $X_s \prec X$ that lies within the box. The procedure is shown in Algorithm 1.

Algorithm 1. Most Similar Subtrajectory Search

Input: Subtrajectory Query $Q_s = \ulcorner \mathbf{q}_1, \ldots, \mathbf{q}_m \urcorner$; trajectory dataset $\mathcal{D}$
Output: Subtrajectory $X_s \prec X \in \mathcal{D}$ which is most similar to Q_s

1: $\hat{X} = \arg\min_{X \in \mathcal{D}} \Theta(X, Q_s)$.
2: In each dimension $i \in [1, d]$, $l^{(i)} = \min_{j \in [1,m]} q_j^{(i)}$; $h^{(i)} = \max_{j \in [1,m]} q_j^{(i)}$.
3: Build a bounding box $B \subset \mathbb{R}^d$ where $l^{(i)} - \epsilon \leq B^{(i)} \leq h^{(i)} + \epsilon$ in each dimension i.
4: Return subtrajectory $\hat{X}_s \prec \hat{X}$ that is in B.

6 Empirical Evaluation

6.1 Experimental Design and Settings

The experiments are designed to answer the following questions:

1. Is the proposed distributional measure effective and robust in computing the distance between trajectories in trajectory retrieval?
2. Is Algorithm 1 effective for the similar subtrajectory search problem compared with the SOTA SimSub methods?

To answer the first question, we build three IK-based distance measures $\mathcal{K}_I, MMD_I$ and $\mathcal{K} \circ \kappa_I$ based on Eqs. 2, 3, and 4 in Sect. 3, and replace the Isolation Kernel to Gaussian kernel to build two variants $\mathcal{K}_G$ and MMD_G. We compare these five distributional measures with existing trajectory distance measures: (i) point-based measures: DTW [21] and Hausdorff Distance [13]; (ii) deep learning methods: t2vec [8], NeuTraj [19] and TrajGAT [20]; (iii) distribution-based measures: DBF [18], NeuroSeqRet (NSEQ) [6] and tkSQFD [14]. We examine the performance of these measures on trajectory retrieval tasks, including 1-Nearest Neighbor search and k-Nearest Neighbor search. We also investigate the impact of noise and down-sampling on these tasks to assess the robustness of different distance measures. To answer the second question, we compare Algorithm 1 with two state-of-the-art deep learning methods: GBAF [4] and RLS-Skip [17] on the SimSub problem.

We adopt a straightforward approach to generate the ground truth for the retrieval tasks. For each trajectory X in the dataset $\mathcal{D}$, we split it into two trajectories Q and Q' by alternately sampling the points on the trajectory, thus generating two datasets $\mathcal{B}$ and $\mathcal{B}'$.

For 1-Nearest Neighbor (1NN) search, each trajectory $Q \in \mathcal{B}$ is used as the query trajectory to retrieve its nearest neighbor from dataset $\mathcal{B}'$. We use the rank of Q' in the returned rank list to evaluate the performance of each distance measure. The smaller the average rank (close to 1) the better the retrieval result.

For K-Nearest Neighbor (KNN) search, we firstly identify the set of KNNs of trajectory Q in $\mathcal{B}$ with a target distance measure and use the corresponding set in $\mathcal{B}'$ as ground truth. We then search the KNNs of Q in $\mathcal{B}'$ using the target distance measure and evaluate the performance based on accuracy.

For SimSub problem, we randomly select two points from the trajectory Q as the beginning and end to generate subtrajectories Q_s. Next, we use $\mathcal{K}_I$ to find the most similar trajectory of Q_s in dataset $\mathcal{B}'$ as the data trajectory. Different SimSub methods are applied to find the similar subtrajectory $\hat{Q}'_s$. We use Jaccard distance [7] as the evaluation measure to compute the similarity between the query subtrajectory Q_s and the result subtrajectory $\hat{Q}'_s$. A good method for finding the most similar subtrajectory has J-distance close to 0.

The data characteristics of the real-world datasets used are given in Table 2. For the Geolife dataset, we use trajectories that have at least 8 points.

Table 2. Real-world datasets. min – max $|X|$ indicates the minimum and maximum numbers of points of individual trajectories in a dataset.

Dataset	#Points	min – max $\lvert X \rvert$	#trajectories
Proto	14,698,842	91–3,836	100,000
Geolife	24,876,978	8–92,645	17,521
Detrac	445,052	11–2,120	5,356
Casia	143,383	16–612	1,500

6.2 Experimental Results

1-Nearest Neighbor search: Table 3 shows the average rank of 1NN search on the two largest datasets: Proto and Geolife. We have the following observations:

1. $\mathcal{K} \circ \kappa_I$ is the best method in every experiment without exception. It also shows better robustness to noise.
2. $\mathcal{K}_I$ (with Isolation Kernel) is always better than $\mathcal{K}_G$ (with Gaussian Kernel); and MMD_I is almost always better than MMD_G with only one exception.
3. All five proposed no-learning distributional methods performed better than other methods in every experiment, followed by DBF and t2vec.
4. NeuTraj and TrajGAT approximate Hausdorff and DTW distance through deep learning, but they sacrifice accuracy for efficiency, leading to poorer search outcomes than the original distance measures.

Table 3. Average rank result of 1NN search. NR is % Gaussian noise added.

	NR	Distribution Based							Point-to-point distance based							
		Proposed					Existing		tkSQFD	t2vec	NeuTraj		TrajGAT		Original	
		$\mathcal{K}_I$	MMD_I	$\mathcal{K} \circ \kappa_I$	$\mathcal{K}_G$	MMD_G	DBF	NSEQ	κ_G	ℓ_2	DTW	HD	DTW	HD	DTW	HD
Proto	0	1.2	1.3	**1.1**	1.4	1.5	2.9	10.1	5.2	3.3	45.5	40.8	41.6	39.5	38.6	38.2
	20	2.4	2.8	**1.7**	2.8	2.7	5.1	20.3	9.6	6.2	62.7	66.8	59.7	64.9	58.9	62.1
	40	5.6	5.7	**5.5**	7.8	7.7	10.7	65.2	20.8	14.7	192.8	140.9	184.1	129.2	180.8	122.4
	60	29.7	39.1	**20.7**	32.9	44.7	74.5	89.4	82.7	61.0	387.6	343.8	380.1	340.1	375.4	332.7
Geolife	0	**1.1**	**1.1**	**1.1**	3.1	3.1	3.5	7.1	5.3	3.7	5.3	16.8	4.9	16.3	4.2	15.8
	20	1.5	1.4	**1.2**	6.2	6.1	7.1	34.6	25.8	7.2	52.6	45.7	51.7	42.7	49.7	40.4
	40	2.0	2.8	**1.9**	15.0	15.9	25.1	41.9	38.7	26.0	68.5	58.4	66.7	57.5	62.7	55.2
	60	7.2	8.6	**6.8**	22.2	25.1	30.2	64.4	50.2	30.8	102.4	85.0	99.8	81.8	96.4	79.8

K-Nearest Neighbor search: Table 4 shows the 5-nearest neighbor search results on 4 trajectory datasets. The proposed distributional measures are more accurate in retrieval and less affected by high rates of down-sampling as compared to the existing measures. This result is consistent with the outcome of the 1NN search.

SimSub problem: Table 5 displays the J-distance result and runtime cost of three SimSub methods. Algorithm 1 outperforms the other two methods in terms of the J-distance. Additionally, our method is at least ten times faster than the other methods, despite their use of GPU acceleration.

Table 4. 5-nearest neighbor search results on the trajectory datasets.

	Proto	Geolife	Detrac	Casia
$\mathcal{P}$ measures	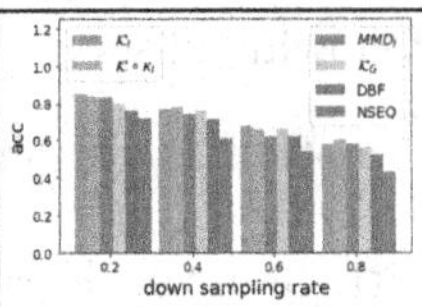 acc / down sampling rate ($\mathcal{K}_I$, $\mathcal{K}^* \circ \mathcal{K}_I$, MMD_I, $\mathcal{K}_G$, DBF, NSEQ)	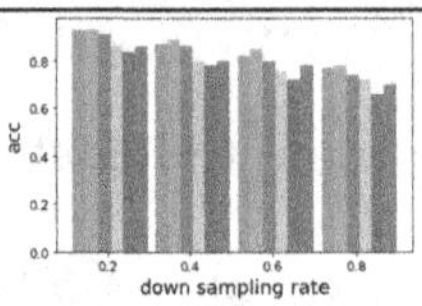 acc / down sampling rate	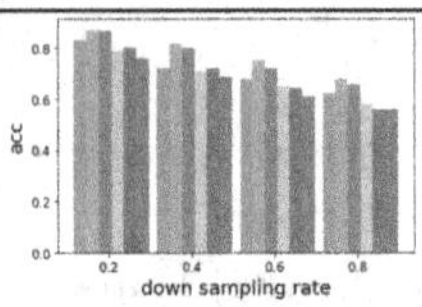 acc / down sampling rate	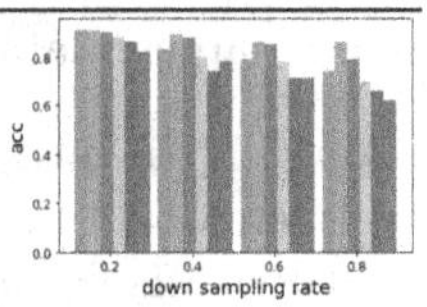 acc / down sampling rate
$\mathcal{K}_I$ vs others	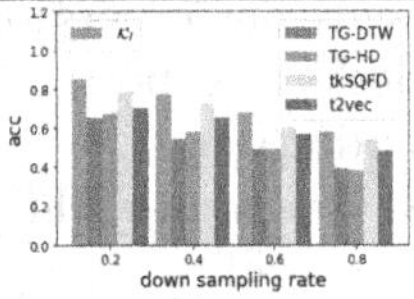 acc / down sampling rate ($\mathcal{K}_I$, TG-DTW, TG-HD, tkSQFD, t2vec)	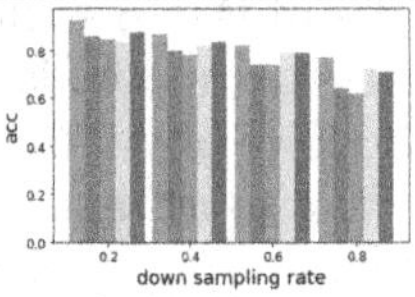 acc / down sampling rate	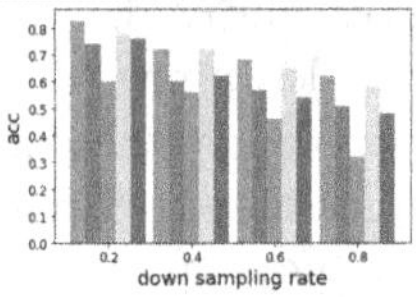 acc / down sampling rate	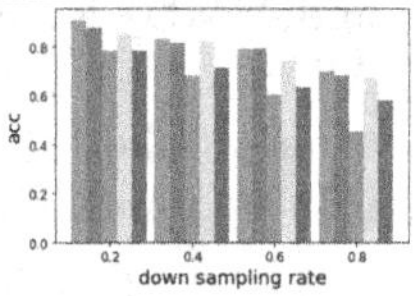 acc / down sampling rate

7 Discussion

7.1 How Important Is the Temporal Information?

The findings presented in Table 3 demonstrate that t2vec and tkSQFD, which both consider temporal information, have better results than point-based methods such as Hausdorff and Fréchet distance. However, DBF with EMD and our proposed distributional kernel methods outperform all other methods, indicating that distributional information is more important than temporal information in trajectory retrieval tasks.

Table 5. J-distance scores and runtimes for the most similar subtrajectory search. Our algorithm ran on CPU while the other two methods ran on GPU.

		Proto	Geolife	Detrac	Casia
J-distance	GBAF	.44	.32	.35	.24
	RLS	.44	.32	.37	.26
	Ours	.42	.32	.33	.19
Train (hours)	GBAF	.22	.20	.15	.04
	RLS	41.80	38.30	15.40	6.30
	Ours	.02	.02	.01	.00
Search (seconds)	GBAF	.21	.52	.12	.12
	RLS	.32	.52	.12	.12
	Ours	.02	.03	.02	.01

However, it is worth noting that some trajectories may be similar in shape but differ in direction. In cases where the order of points is important, any of the distribution-based methods can include this information by simply adding a dimension of order. This extra dimension can be used to differentiate between two trajectories that share a similar shape but have opposite orders.

7.2 Runtime Efficiency and Indexed Search

Table 6. Brute force search vs indexed search on the Geolife dataset. The result is based on 20NN search. All deep learning methods use GPU; others use CPU. The one-off index tree building times for all methods are in the same order.

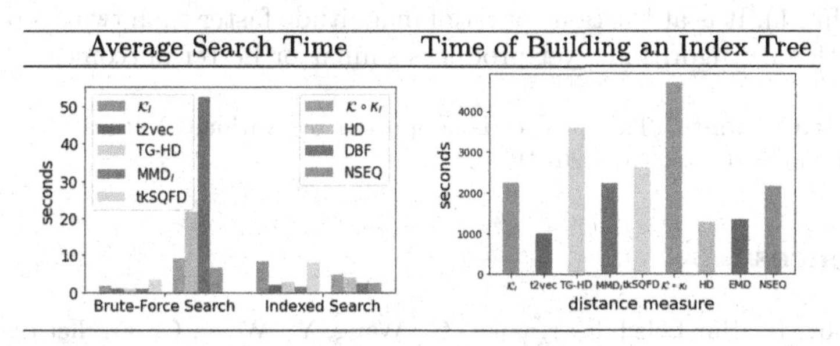

Here we investigate the relative runtime efficiency and the ability to use an index scheme. The ball tree index [11] is used to perform the indexed search for the distributional kernel-based measures ($\mathcal{K}_I$, MMD$_I$ and $\mathcal{K} \circ \kappa_I$) and tkSQFD; DBF [18] is used to accelerate EMD and LSH [6] is used for NSEQ. The result shown in Table 6 can be summarized as follows:

1. MMD$_I$ is the fastest method for both brute-force and ball-tree searches.
2. DBF using the Earth Mover Distance, is the slowest in brute-force search. It benefits the most from an index, but still runs slower than the proposed MMD$_I$ without indexing.
3. Index is counterproductive for five measures: $\mathcal{K}_I$, t2vec, TG-HD, MMD$_I$ and tkSQFD. However, brute-force searches using any of these measures (except tkSQFD) run faster than any measures using an index.
4. $\mathcal{K} \circ \kappa_I$ is the only distributional measure that benefits from an index.

Generally, we recommend MMD$_I$ without an index for fast runtime and $\mathcal{K} \circ \kappa_I$ with an index for high retrieval accuracy.

8 Conclusion

Treating a trajectory of points as a set of i.i.d. points generated from an unknown distribution, distribution-based measures perform better than point-based methods and deep learning methods in both 1NN and KNN searches. This indicates the importance of distribution information for trajectory data.

We are the first to point out that the identity property is crucial for trajectory distance measurement. Our analyses show that traditional distance measures, including Hausdorff, Fréchet and DTW distances, do not have the identity property, while our proposed $\mathcal{K}$-based measures do, which is the key factor that makes $\mathcal{K}$ outperforms all other methods in the experiments.

Our empirical evaluation suggests that MMD_I is the most efficient measure which is faster than any of the deep learning methods. For tasks that require high retrieval accuracy, we recommend using $\mathcal{K} \circ \kappa_I$, which has the highest retrieval accuracy among all measures.

Using a distributional kernel, we have demonstrated that the SimSub problem can be solved simply and efficiently through the proposed no-learning method (Algorithm 1). It is at least one order of magnitude faster than two deep learning state-of-the-art algorithms, yet produces similar or better outcomes.

Acknowledgements. This project is supported by National Natural Science Foundation of China (Grant No. 62076120).

References

1. Aronov, B., Har-Peled, S., Knauer, C., Wang, Y., Wenk, C.: Fréchet distance for curves, revisited. In: Azar, Y., Erlebach, T. (eds.) Algorithms - ESA 2006. Lecture Notes in Computer Science, vol. 4168, pp. 52–63. Springer, Berlin (2006). https://doi.org/10.1007/11841036_8
2. Beecks, C., Uysal, M.S., Seidl, T.: Signature quadratic form distance. In: Proceedings of the ACM International Conference on Image and Video Retrieval, pp. 438–445 (2010)
3. Chen, H., Liu, Y., Hu, C., Zhang, X.: Vulnerable road user trajectory prediction for autonomous driving using a data-driven integrated approach. IEEE Trans. Intell. Transp. Syst. **24**(7), 7306–7317 (2023)
4. Deng, L., Sun, H., Sun, R., Zhao, Y., Su, H.: Efficient and effective similar subtrajectory search: a spatial-aware comprehension approach. ACM Trans. Intell. Syst. Technol. **13**(3), 1–22 (2022)
5. Gold, O., Sharir, M.: Dynamic time warping and geometric edit distance: breaking the quadratic barrier. ACM Trans. Algorithms **14**(4), 1–17 (2018)
6. Gupta, V., Bedathur, S., De, A.: Learning temporal point processes for efficient retrieval of continuous time event sequences. In: Proceedings of the AAAI Conference on Artificial Intelligence,
7. Levandowsky, M., Winter, D.: Distance between sets. Nature **234**(5323), 34–35 (1971)
8. Li, X., Zhao, K., Cong, G., Jensen, C.S., Wei, W.: Deep representation learning for trajectory similarity computation. In: Proceedings of the 34th International Conference on Data Engineering, pp. 617–628. IEEE (2018)
9. Muandet, K., et al.: Kernel mean embedding of distributions: a review and beyond. Found. Trends® Mach. Learn. **10**(1-2), 1–141 (2017)
10. Muandet, K., Fukumizu, K., Sriperumbudur, B., Schölkopf, B.: Kernel mean embedding of distributions: a review and beyond. Found. Trends Mach. Learn. **10**(1–2), 1–141 (2017)
11. Omohundro, S.M.: Five Balltree Construction Algorithms. International Computer Science Institute Berkeley, Berkeley (1989)
12. Su, H., Liu, S., Zheng, B., Zhou, X., Zheng, K.: A survey of trajectory distance measures and performance evaluation. VLDB J. **29**, 3–32 (2020)
13. Taha, A.A., Hanbury, A.: An efficient algorithm for calculating the exact Hausdorff distance. IEEE Trans. Pattern Anal. Mach. Intell. **37**(11), 2153–2163 (2015)

14. Tavenard, R., Malinowski, S., Chapel, L., Bailly, A., Sanchez, H., Bustos, B.: Efficient temporal kernels between feature sets for time series classification. In: Ceci, M., Hollmen, J., Todorovski, L., Vens, C., Dzeroski, S. (eds.) Machine Learning and Knowledge Discovery in Databases. Lecture Notes in Computer Science(), vol. 10535, pp. 528–543. Springer, Cham (2017). https://doi.org/10.1007/978-3-319-71246-8_32
15. Ting, K.M., Xu, B.C., Washio, T., Zhou, Z.H.: Isolation distributional kernel: a new tool for kernel based anomaly detection. In: Proceedings of the ACM SIGKDD Conference on Knowledge Discovery & Data Mining, pp. 198–206 (2020)
16. Ting, K.M., Zhu, Y., Zhou, Z.H.: Isolation Kernel and its effect on SVM. In: Proceedings of the ACM SIGKDD International Conference on Knowledge Discovery & Data Mining, pp. 2329–2337 (2018)
17. Wang, Z., Long, C., Cong, G., Liu, Y.: Efficient and effective similar subtrajectory search with deep reinforcement learning. Proc. VLDB Endowment **13**(12), 2312–2325 (2020)
18. Yang, W., Wang, S., Sun, Y., Peng, Z.: Fast dataset search with earth mover's distance. Proc. VLDB Endowment **15**(11), 2517–2529 (2022)
19. Yao, D., Cong, G., Zhang, C., Bi, J.: Computing trajectory similarity in linear time: a generic seed-guided neural metric learning approach. In: IEEE International Conference on Data Engineering, pp. 1358–1369. IEEE (2019)
20. Yao, D., Hu, H., Du, L., Cong, G., Han, S., Bi, J.: TrajGAT: a graph-based long-term dependency modeling approach for trajectory similarity computation. In: Proceedings of the 28th ACM SIGKDD Conference on Knowledge Discovery & Data Mining, pp. 2275–2285 (2022)
21. Yi, B.K., Jagadish, H.V., Faloutsos, C.: Efficient retrieval of similar time sequences under time warping. In: Proceedings of the 14th International Conference on Data Engineering, pp. 201–208. IEEE (1998)

Multi-agent Reinforcement Learning for Online Placement of Mobile EV Charging Stations

Lo Pang-Yun Ting, Chi-Chun Lin, Shih-Hsun Lin, Yu-Lin Chu, and Kun-Ta Chuang(✉)

Department of Computer Science and Information Engineering, National Cheng Kung University, Tainan, Taiwan
{lpyting,megaa,shlin}@netdb.csie.ncku.edu.tw, f74091019@gs.ncku.edu.tw, ktchuang@mail.ncku.edu.tw

Abstract. As global interest shifts toward sustainable transportation with the proliferation of electric vehicles (EVs), the demand for an efficient, real-time, and robust charging infrastructure becomes increasingly pronounced. This paper introduces an approach to address the imbalance between the surging EV demand and the existing charging infrastructure: the concept of Mobile Charging Stations (MCSs). The research develops an algorithm for the dynamic placement of MCSs to significantly reduce the waiting time for EV owners. The core of this research is the Two-stage Placement and Management with Multi-Agent Reinforcement Learning (2PM-MARL) for a dynamic balancing of charging demand and supply. The complexity of the problem is elaborated by showing the NP-hard nature of the MCS placement issue through a relation to the Uncapacitated Facility Location Problem (UFLP), underscoring the computational challenges and emphasizing the need for intelligent real-time solutions. Our framework is validated through comprehensive experiments using real-world charging session data. The results exhibit significant reductions in the waiting time, suggesting the potential practicality and efficiency of our proposed model.

1 Introduction

In this decade, as the climate crisis and environmental protection are of utmost importance, global businesses have been rapidly adapting to sustainable practices. A standout in this paradigm shift is the automotive sector, which has seen a dramatic growth in electric vehicles (EVs) - the leading edge of environmentally friendly transportation for the future [7]. This marks a major shift from traditional gasoline-powered vehicles and is the beginning of a new era in the history of automobiles. The growing popularity of electric vehicles has made it indispensable to have a charging infrastructure that can fulfill all demands. A comparative analysis of the infrastructure across regions like the United States manifests a striking imbalance that the ratio of EVs to charging ports ranges from 19.27 (Florida) to 41.34 (New Jersey) [1]. This discrepancy underscores the intertwined complexities of time, capital, and logistics inherent in establishing new charging units. Nevertheless, the complexities of this dilemma are multifaceted. Temporal fluctuations in charging demand add another layer of complexity. The increase in the

D.-N. Yang et al. (Eds.): PAKDD 2024, LNAI 14649, pp. 284–296, 2024.
https://doi.org/10.1007/978-981-97-2262-4_23

need for charging during rush hours usually leads to longer waiting times at charging stations. On the contrary, resources are often not used to their full potential during off-peak hours, indicating that there are problems with their distribution and management in terms of ineffectiveness.

A potential solution involves an online Electric Vehicle Charging Scheduling (EVCS) system that uses real-time data to schedule EV charging requests. Previous research has explored various methods for EVCS, including mixed-integer programming, demand response, dynamic pricing, and reinforcement learning [4,6,12,17]. However, the effectiveness of EVCS depends on EV owners' willingness to adjust their charging times, which can be challenging in areas with limited charging infrastructure. Furthermore, understanding the changing behaviors and expectations of EV owners becomes essential. With the increase in EV adoption, owners have developed a profound expectation for quick charging [5], intensifying the need for a widely accessible, fast-charging infrastructure.

In response, we provide a solution from the perspective of Mobile Charging Stations (MCSs). As a new form of charging facility, MCSs have the potential to charge EVs at any location and time, without dependence on the permanent infrastructure of fixed charging stations (FCSs). A lot of new businesses have implemented a charging-as-a-service model based on MCS[1,2] but so far these services have only been utilized for on-demand request of emergency purpose. Motivated by their flexibility, we design an online algorithm for dynamic placement of MCSs, aiming to reduce the time cost for EV owners waiting to charge. Our method is designed with EV owners' perspectives in mind. When an EV user makes a charging request, the system provides the owner with real-time information about the location and usage of all available charging stations. After the owner reserves a station and starts charging at the scheduled time, the system updates the usage information for that station. On the other hand, a Mobile Charging Station Management System determines the optimal placement of MCSs each hour, initially using historical data to compute the point-based optimal placement, followed by a region-based approach that plans the dynamic placement of MCSs with the sliding-window update of best-utilized location. In conclusion, this study introduces the **2PM-MARL** (standing for **Two**-stage **P**lacement and **M**anagement with **M**ulti-**A**gent **R**einforcement **L**earning) framework, contributing to ongoing discussions and developments around the EV charging infrastructure. Key contributions of this study include: (a) Exploring a new application of multi-agent reinforcement learning in the EV charging infrastructure domain, which could help balance charging demand and supply with a dynamic, online, and adaptable solution. (b) Conducting a detailed analysis that suggests the MCS placement problem's NP-hard nature by mapping it to the Uncapacitated Facility Location Problem (UFLP), underscoring the computational complexity involved and the potential for real-time solutions as the proposed 2PM-MARL framework. (c) Performing extensive validation of the 2PM-MARL framework using real-world charging session data, demonstrating reductions in waiting times, which could highlight the practicality and efficiency of the proposed model.

[1] LIGHTNING MOBILE, https://lightningemotors.com/.

[2] SparkCharge, https://www.sparkcharge.io/.

2 Preliminaries

2.1 Related Works

EV Charging Station Scheduling. A Lyapunov optimization method-based user-oriented EV control scheme was proposed to manage a dynamic population of EVs [12]. This approach was advanced by a heuristic fuzzy inference system-based algorithm (FISA) [6]. For considering user behaviors, [17] proposed an online charging registration mechanism incorporating an incentive-based framework to boost user adherence to charging schedules. The work proposed in [13] utilized a Markov decision process to devise a model-free reinforcement learning approach for charging station coordination, while [14] proposed a smart reservation system based on multicriteria decision-making techniques, managing parking slot availability and optimizing user acceptance ratios. Furthermore, the importance of dynamic traffic conditions and uncertain future requests was emphasized through a deep reinforcement learning method [8]. However, these studies focus solely on fixed charging stations, limiting their flexibility and showing constraints in addressing dynamic and uncertain factors in scheduling.

Fixed Charging Station Placement Optimization. Our work intersects with several studies in the domains of charging station placement and optimization strategies. [18] uses deep reinforcement learning to optimize the location of charging stations in urban areas. [11] extended the placement problem to road networks, proposing the Social-Aware Optimal EV Charger Deployment (SOCD) that also factored in multiple types of charging plugs at each station. Our work builds upon these concepts by implementing a more comprehensive and efficient optimization algorithm that factors in not just location but also the timing of charging demand.

Dynamic Vehicle Management. We explore two key aspects of dynamic vehicle management: (i) Mobile Charging Station (MCS) management and (ii) Fleet management. Regarding MCS management. [19] introduced a system that uses reinforcement learning to determine the number and routes of Mobile Energy Disseminators (MEDs). [2] developed an optimal scheduling framework to minimize users' costs and charging times by strategically placing charging stations. [10] investigated the deployment of idle MCSs using federated learning. However, these studies primarily focus on optimizing global performance metrics like average displacement error or waiting times, without ensuring benefits for each charging station. For example, [19] employed a reinforcement learning-based method considering the dynamic information of each MED. In the realm of dynamic fleet management, multi-agent techniques are commonly used to enhance taxi services. [9] implemented explicit coordination among agents for large-scale fleet management but did not address the competitive aspects crucial in our approach. Advancing demand and supply optimization, [20] and [16] introduced innovative strategies. [20] proposed the SOUP method, which focuses on spatial-temporal demand forecasting and competitive supply using the ST-GCSL model for request predictions and a DROP algorithm for agent routing. However, the competitive supply aspect was not fully considered. Conversely, [16] presented the V1D3 framework for order dispatching and vehicle repositioning, prioritizing efficiency over system fairness.

In conclusion, while multi-agent environments aid in dynamic fleet management, applying these techniques to MCS management remains challenging. Our focus extends beyond operational efficiency and user experience in the EV charging network to ensure balanced utilization among MCSs, guaranteeing benefits for each charging station. Therefore, our work leverages the beneficial aspects of these studies while offering a unique perspective through the 2PM-MARL framework.

2.2 Problem Definition

In this subsection, we provide a series of detailed preliminary definitions.

Def. 1 Road Network: We denote $G = (V, E)$ as a road network, representing the geographical framework within which EVs operate. V denotes the set of vertices (or nodes, indicating intersections or endpoints in the city), and E represents the set of edges (or road segments connecting these nodes).

Def. 2 Charging Station: Stations are represented as $\mathbb{S} = \mathbb{F} \cup \mathbb{M}$, where $\mathbb{F}$ is the existing charging infrastructure of FCSs (fixed charging stations). Each station $i \in \mathbb{F}$ is associated with information (v_i, c_i, a_i^t), where v_i denotes the location of station i, c_i is the number of chargers at i, and a_i^t is the available time of i at time period t. The set of MCSs (mobile charging stations) is denoted as $\mathbb{M}$. Each station $j \in \mathbb{M}$ is associated with (v_j^t, c_j, a_j^t) which respectively denotes the location of station j at time period t, the number of chargers at j, and the available time of j at time period t.

Def. 3 MCS Placement Plan: This represents the strategy for the placement of MCSs at each time period t, which allows us to adapt to the fluctuating charging demand across the city. Our goal is to optimize this plan, expressed as $\mathcal{P}_t = \{\ v_j^t \mid \forall j \in \mathbb{M}\ \}$, to effectively meet the demand.

Def. 4 Charging Demand: D_W represents all the charging demands within time window W. Each charging demand $d = (v, p, t) \in D_w$ includes v as the location of charging demand, t as the time period when this demand occurs, and p denotes the requested power of charging demand.

Def. 5 Indicator: The indicator function, represented as $I : D_W \rightarrow \mathbb{S}$, tracks the relationship between demand and stations. For each charging demand d, a station $s \in \mathbb{S}$ is chosen to charge the EV, denoted as $I(d) = s$. The inverse $I^{-1}(s) = \{d \mid \forall d \in D_W,\ I(d) = s\}$ gives the set of charging demands chosen to charge at station s.

Def. 6 Inconvenience Score: This encapsulates the time cost borne by EV owners, including time spent in transit to stations and waiting for charging:

$$IS(s) = \sum_{d \in I^{-1}(s)} TravelingTime(d) + WaitingTime(d). \tag{1}$$

Def. 7 Perfect Plan (Objection function): Our objective is to optimize the placement of MCSs such that the total inconvenience score, summed across all stations and times, is minimized within a given time window. The objective function is expressed as:

$$Perfect\ Plan : \mathcal{P} = arg\ min_{\mathcal{P}} \sum_{s \in \mathbb{S}} IS(s). \tag{2}$$

2.3 Problem Complexity and NP-Hardness

Here, we dissect the complexity of our problem by establishing it as NP-hard, which sets the direction for our approach in developing suitable algorithms to tackle it. This lies in a connection to a classic computational optimization problem - the *Uncapacitated Facility Location Problem (UFLP)* [3]. This NP-hard problem features a set of potential facility locations F, and a set of customers C. Each facility $f \in F$ incurs cost $c(f)$ upon opening, while each customer $c \in C$ has allocation cost $a(c, f)$ to a facility f. The objective is to open a set of facilities and allocate each customer to their nearest facility such that the total cost, comprising both opening and allocation costs, is minimized:

$$\min\Big(\sum_{f \in F} c(f) * x_f + \sum_{c \in C, f \in F} a(c, f) * y_{cf}\Big), \tag{3}$$

where x_f is a binary variable that equals 1 if f is opened and 0 otherwise, and y_{cf} is another binary variable that equals 1 if c is assigned to f and 0 otherwise. Now, we create a modified version of UFLP that suits our context.

Set of Customers and Facilities. Each intersection in our road network $G = (V, E)$ is seen as both a customer and a potential facility location. This gives us $C = F = V$.

Opening Cost of Facilities. In our context, the cost of opening a charging station at any location isn't a concern. Hence, the opening cost of each facility is zero, i.e., $c(f) = 0$.

Allocation Cost. The allocation cost signifies the inconvenience score that a vehicle would incur if it needs to be charged. This is the sum of the traveling time to and the waiting time at the charging station: $a(c, f)$ is defined as $a(d, v) = TravelingTime(d) + WaitingTime(d)$.

Our problem's cost function thus becomes:

$$\min\Big(\sum_{d \in D, v \in V} a(d, v) * x_v\Big), \sum_{v \in V} x_v = |\mathbb{S}|, \tag{3'}$$

where $a(d, v)$ is the cost of demand d if it is charged at the station located at v. x_v is a binary variable that equals 1 if a charging station is placed at v and 0 otherwise.

In mapping our problem onto the UFLP and introducing necessary modifications, we have effectively created a parallel to UFLP that mirrors our problem's constraints and objectives. Considering that UFLP is an established NP-hard problem, it follows that our problem is also NP-hard.

3 Methodology

3.1 Static Placement: Shortest Traveling Heuristic Algorithm

The proposed 2PM-MARL methodology operates in two stages - static placement and dynamic management. The preparatory static placement stage begins at the start of each day, leveraging historical charging request data to calculate optimal charging station locations. This stage is vital for initiating a strategic layout for the MCSs. Once the initial configuration is finished, we move on to the dynamic management phase.

Here, we employ the multi-agent reinforcement learning-based Value Decomposition Networks (VDN) algorithm to manage the movement of MCSs in real time, routing them to suitable locations based on evolving demand. We begin by detailing the static placement stage. Our algorithm assumes that each charging demand will be met at the nearest charging station, i.e., the station entailing the shortest traveling time. Given this, we redefine our indicator function as $D_W \xrightarrow{I} \mathbb{S} \equiv D_W \xrightarrow{loc} V \xrightarrow{I'} \mathbb{S}$. Consequently, we can compute the inconvenience score with respect to location $v \in V$ rather than the specific charging demand $d \in D_W$.

- First, we define a location function *loc*, $loc : D_W \to V$ to indicate the location v of each charging demand *d*, i.e., $loc(d) = \{v\}$. The inverse $loc^{-1}(v) = \{d \mid \forall d \in D_W, loc(d) = v\}$ provides the set of charging demands located at location v.
- The modified Indicator I' is defined as $I' : V \to \mathbb{S}$. For each charging demand at location v, the closest station s is chosen to meet the demand. This selection is governed by $I'(v) = s = \arg\min_{s' \in \mathbb{S}} Distance(v, s')$. The inverse $I'^{-1}(s) = \{v \mid \forall v \in V, I'(v) = s\}$, on the other hand, provides the set of locations that are chosen to be charged at the station with id equal to s.
- The Inconvenience Score constituting the time cost before charging starts is computed by:

$$IS(s) = \sum_{v \in I^{-1}(s)} \sum_{d \in loc^{-1}(v)} TravelingTime(d) + WaitingTime(d). \quad (1')$$

- The Perfect Plan $\mathcal{P}$ in Time Window W, which forms the objective function for our static placement stage, is thus expressed as:

$$Perfect\ Plan: \mathcal{P} = \arg\min_{\mathcal{P}} \sum_{s \in \mathbb{S}} IS(s). \quad (2')$$

Having established these definitions and assumptions, we proceed with a greedy algorithm (see Algorithm 1) to determine the optimal static placement plan. The process begins by calculating the overall inconvenience score for each node when it's hypothetically included in the current plan. From these nodes, we identify the node v with the lowest overall inconvenience score and add it to the plan $\mathcal{P}$. This process is performed iteratively for M rounds in which each iteration improves the plan. After M rounds, we obtain a plan for the initial placement of MCSs that significantly reduces the inconvenience score.

3.2 Dynamic Management: Two-Phase Management Algorithm

Following the static placement phase, the 2PM-MARL methodology advances into the dynamic management phase, which comes into action in real-time to manage the movement of MCSs based on the evolving demand. At its core, the dynamic phase employs a *multi-agent reinforcement learning-based Value Decomposition Networks (VDN)* algorithm [15]. Here, we elaborate on this phase by detailing the structure and functions of

Algorithm 1: Shortest Traveling Placement Heuristic Algorithm

Input : Road network $G = (V, E)$, Set of FCSs $\mathbb{F}$, Set of MCSs $\mathbb{M}$, Set of charging demand D_W
Output: MCS placement plan $\mathcal{P}$

```
   // Initialization
1  S ← F;
2  M ← |M|;
3  P ← ∅;
   // Iterate to determine MCS location
4  for i = 1 to M do
5      for each node v' in V do
6          Location of MCS_i set to v';
7          S' = S ∪ {MCS_i};
8          Compute overall inconvenience score Σ_{s∈S'} IS(s) against D_W;
9      end
       // Pick vertex with lowest overall inconvenience Score to place MCS
10     v = arg min_{v'∈V} Σ_{s∈S'} IS(s);
11     P ← P ∪ {v};
12     S ← S ∪ {MCS_i with location v};
13 end
14 return P;
```

its integral components, such as the hexagon grid representation of the road network, the Markov Decision Process (MDP), and the reward system. We will also explain how the VDN algorithm functions within this framework.

Definitions. We adopt a *hexagonal grid* as that employed in [9] to represent the road network due to its advantages in spatial analysis. A hexagonal tessellation offers equal distance to its adjacent cells and avoids the directional bias presented in a rectangular grid. Each hexagon in the grid is identified by a unique ID that enables us to map real-world locations to the corresponding grid cell.

Markov Decision Process. We model the Mobile Charging Station Placement problem as a Markov Decision Process (MDP) $\mathcal{G}$ for N agents, which is defined by $\mathcal{G} = (N, S, A, P, R, \gamma)$ where N, S, A, P, R, γ is the number of agents (MCSs), sets of states, joint action space, transition probability functions, reward functions, and a discount factor respectively. Their definitions are given as follows.

State: The state $s_t \in S$ encapsulates the full condition of the system at time t. It is a vector that contains the state of all agents $s_t^m = \{GridInfo^K,\ Util,\ Time\}$ for $1 \leq m \leq N$:

- $GridInfo^K$ denotes information regarding the K nearest grids for a given MCS. It is a set of tuples $(Demand,\ FcsNum,\ CV_f)$ where: 1) $Demand$ refers to the number of charging requests that appeared in the last time step in each grid. 2) $FcsNum$ signifies the number of fixed charging stations (FCSs) in each grid. 3) CV_f represents the Coefficient of Variance of utilization of FCSs in each grid, defined as $CV_f = \frac{U_f^{std}}{U_f^{avg}+\epsilon}$ where U_f^{std} and U_f^{avg} are the standard deviation and average of all FCSs' utilization, respectively. ϵ is a small constant to prevent division by zero.
- $Util$ represents the utilization information of each grid. $Util = (U_m,\ U_f^{std})$ in which U_m is the utilization of the m-th agent (MCS).

– $Time$ denotes the time information of each grid. $Time = (ToD, DoW)$ where ToD and DoW stand for the time of the day and day of the week, respectively.

Action: The action for each MCS is represented by a vector $a_t^m \in \{k\}_{k=0}^{K}$. Figure 1 shows an example for $K = 18$ (two outer-ring hexagonal grids). In (a), we have $a_4^2 = 9$, which means the 2nd MCS is commanded to relocate from its current grid to the 9th nearby grid at the beginning of the 4th time period. In (b), we have $a_3^5 = 0$, indicating that the 5th MCS shall stay at the same grid in the 3rd time period.

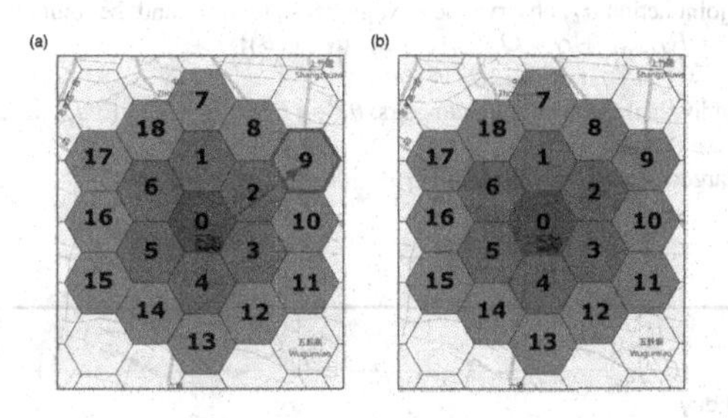

Fig. 1. The illustration of action with $K = 18$.

Reward: The reward function is constructed to reflect the performance of the environment. For each state s, it is calculated as:

$$reward(s) = \frac{1}{1 + e^{IS(s)\cdot\lambda}} \cdot utilization(s),$$

where the term $\frac{1}{1+e^{IS(s)\cdot\lambda}}$ ensures the reward is higher when the inconvenience score IS is low. The λ here is a parameter that controls the sensitivity of the reward to the inconvenience score. $utilization(s)$ is the utilization of the state, which measures how well the resources (MCSs and FCSs) are being used. Then the agent reward R_m^{t+1} for each MCS m is defined as $reward(m)$, the potential reward R_f is the sum of rewards of all FCSs $\sum_{f\in\mathbb{F}} reward(f)$, and the global reward R_{total} for all agents is $\sum_m (R_m + R_f)$.

Value Decomposition Networks. The core of VDN lies in learning a centralized Q-function Q_{total} with weights θ that can be decomposed into individual Q-functions Q_i for each agent i, represented as:

$$Q_{total}(s_t, a_t; \theta) \approx \sum_{\text{agent } i} Q_i(s_t^i, a_t^i; \theta_i).$$

The overall target, TD_{target}, is then:

$$TD_{target} = R_{total} + \gamma * Q_{total}(s_{t+1}, a_{t+1}; \theta).$$

Algorithm 2: 2PM-MARL

1 Initialize the environment and agents (MCSs);
2 Initialize the centralized Q-function Q_{total} with random weights θ;
3 $(E, T, T_u) \leftarrow$ (number of episodes, number of time periods, update period);
4 **for** *episode = 1 to E* **do**
5   Reset the environment to initial state by **Algorithm 1**;
6   **for** *t = 1 to T* **do**
7   Observe the current global state s_t;
8   **for** *each agent m* **do**
9   Observe the local state s_t^m;
10   Choose an action a_t^m based on the current Q-function, sampling by using ϵ-greedy for exploration;
11   **end**
12   Execute the joint action a_t, observe the next global state s_{t+1} and the joint reward R_{total};
13        $TD_{target} = R_{total} + \gamma * Q_{total}(s_{t+1}, a_{t+1}; \theta)$;
14        $TD_{error} = TD_{target} - Q_{total}(s_t, a_t; \theta)$;
15   Update the individual Q-function parameters: $\theta_i \leftarrow \theta_i - \alpha * \nabla_{\theta_i}(TD_{error})^2$;
16   **if** *t mod T_u == 0* **then**
17   Reset target Q-network with Q_{total};
18   **end**
19   **end**
20 **end**

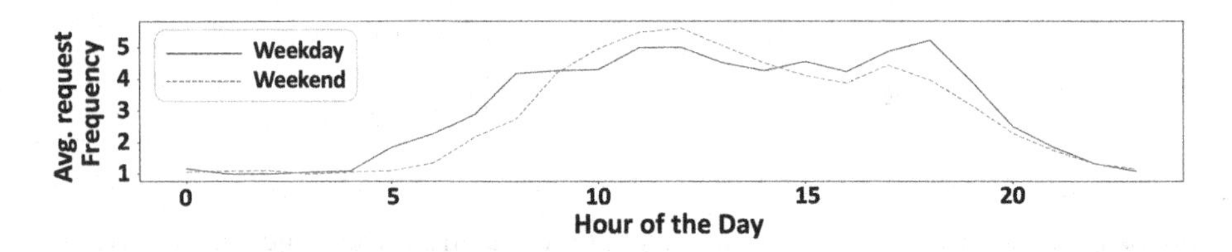

Fig. 2. The average request frequency by hour for weekdays and weekends.

And the temporal-difference error TD_{error} which the model seeks to minimize is:

$$TD_{error} = TD_{target} - Q_{total}(s_t, a_t; \theta).$$

The individual Q-network weights are updated using a gradient descent step: $\theta_i \leftarrow \theta_i - \alpha * \nabla_{\theta_i}(TD_{error})^2$, where α is the learning rate controlling the speed of learning, and γ is the decay factor discounting the future rewards. The aim is to understand that future collective behavior can be beneficial even when each agent acts independently. We now present the 2PM-MARL algorithm (see Algorithm 2).

4 Experimental Results

To validate the effectiveness of the proposed 2PM-MARL model, a series of comprehensive experiments were conducted. The process of these experiments involved the use of a real-world dataset, simulation techniques, comparative baselines, and thorough performance analysis. Details are given in the following subsections.

4.1 Dataset and Preprocessing

We used a real-world charging session dataset[3] from Palo Alto, a city in California, USA, selecting the charging records from January 2019 to December 2019. This dataset contains 35,919 records, with an EV per Charger ratio of 3.39. Each data entry includes crucial information such as the charging station id, the geographic location of the station (site), the type of connector used, the start and end times of the charging session, and the total amount of energy (kWh) consumed during the session.

In Fig. 2, we visualized the average request frequency by hour for weekdays and weekends in the dataset. Figure 2 reveals peak and off-peak periods, especially evident in the blue line representing weekdays where pronounced dual peaks occur around typical commuting hours. Such patterns indeed vary across different times, weekdays, and weekends, thereby affirming the necessity of learning time-related charging behaviors.

To simulate a scenario representative of a high-demand, low-supply charging infrastructure, we artificially enlarged the original dataset four times, resulting in a total of 143,676 records and an EV per Charger ratio of 13.57. This data expansion aims to replicate an environment where the available charging stations are insufficient to meet the charging demand for EVs. A charging behavior simulator was also developed as part of the simulation process. This simulator estimates the attractiveness of each charging station based on factors such as driving distance to the station and immediacy of charging need, assigns a corresponding score to each station, and then selects a station for the EV to charge based on a probability proportional to its score.

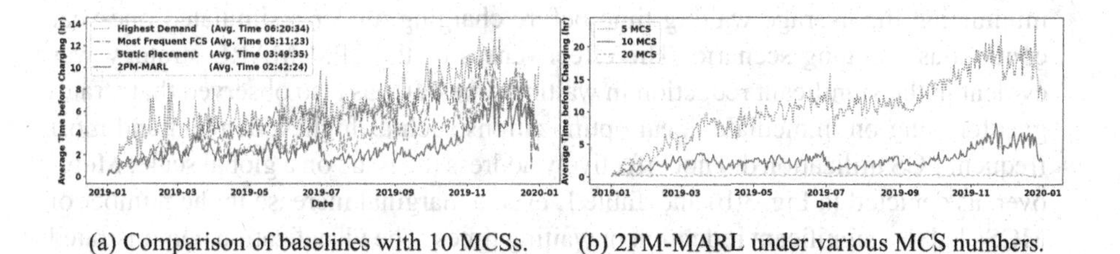

(a) Comparison of baselines with 10 MCSs. (b) 2PM-MARL under various MCS numbers.

Fig. 3. Algorithm Performance Comparison.

Table 1. 2PM-MARL under different number of MCSs.

Number of MCS	EV per Charger	Avg. Time	Avg. Utilization
5	10.4	10:40:29	63.8%
10	8.4	02:42:24	53.2%
20	6.1	00:13:47	40.0%

[3] https://data.cityofpaloalto.org/dataviews/257812/ELECT-VEHIC-CHARG-STATI-83602/.

4.2 Baselines and Experiment Settings

We established several baselines for comparison, each representing a different MCS management strategy as follows. (i) **Highest Demand:** the MCSs are relocated to the grid with the highest charging demand at each time step. (ii) **Most Frequent FCS Utilization:** the MCSs are relocated to the grid where the FCSs are most frequently utilized at each time step. (iii) **Static Placement Algorithm:** The MCSs' actions are determined at each time step only by the Static Placement algorithm (Algorithm 1).

The primary evaluation metrics in our experiments are as follows.

- **Average Time before Charging (Time Cost):** This measures the average waiting time of an EV before it can start charging. Our goal is to minimize this time cost.
- **Coefficient of Variance (CV) of Utilization:** This measures the dispersion of a probability distribution of utilization rates. The lower the CV, the more balanced the utilization across all charging stations.

The experiments were conducted in a simulated environment of the city, which was divided into 121 hexagon grids. For each set of experiments, we used three different numbers of MCSs, namely 5, 10, and 20.

4.3 Performance Comparison

- **Average Time before Charging (Time Cost):** As shown in Fig. 3(a), the 2PM-MARL strategy consistently outperformed all other baseline strategies in terms of minimizing the average waiting time before charging, under a simulated one-year continuous charging scenario. The effectiveness of the 2PM-MARL strategy was evident in the significant reduction in waiting times. It was also observed that strategies focusing on immediate, local optimizations such as highest demand and most frequent FCS utilization did not effectively address the issue on a global scale. Moreover, as depicted in Fig. 3(b) and Table 1, even a marginal increase in the number of MCSs led to a significant reduction in waiting times. The visualization demonstrated the ability of the 2PM-MARL strategy to effectively manage the MCS schedule, thereby addressing the issue of heavy loads at a few charging stations and promoting balanced utilization across all stations.
- **Coefficient of Variance (CV) of Utilization:** As demonstrated in Fig. 4 and Table 2, 2PM-MARL consistently achieved the lowest CV values in both 5 MCSs and 20 MCSs scenarios. This indicates that our method ensures a more balanced and efficient use of charging resources compared to other strategies.

In summary, results across all these analyses indicate that 2PM-MARL not only reduces the waiting time for charging but also ensures a balanced utilization of charging stations, especially in high-demand scenarios. This validates the superiority of the 2PM-MARL strategy over other common charging station management strategies.

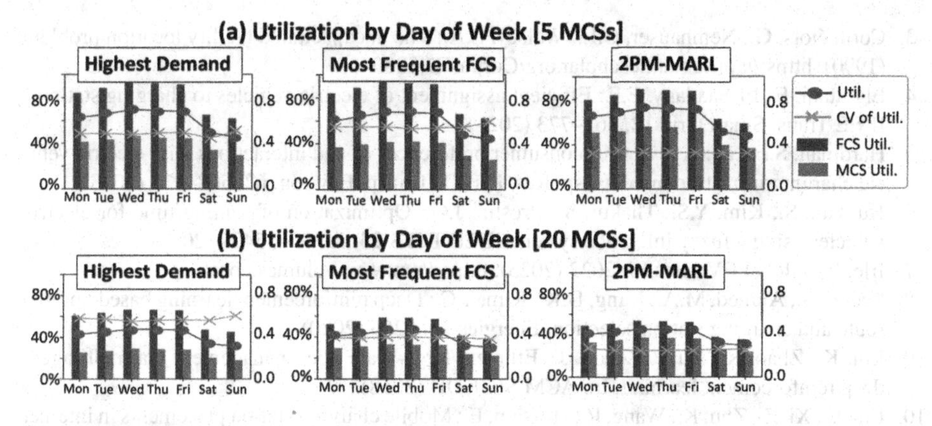

Fig. 4. The Average and Coefficient of Variance Utilization with 5 and 20 MCSs.

Table 2. Comparison with different number of MCS.

	5 MCSs (EV per Charger = 10.4)		20 MCSs (EV per Charger = 6.1)	
Algorithm	Avg. Utilization	Avg. CV	Avg. Utilization	Avg. CV
Highest Demand	61.4%	0.484	42.1%	0.572
Most Frequent FCS	59.7%	0.551	40.3%	0.392
2PM-MARL	63.8%	**0.325**	40.0%	**0.259**

5 Conclusions

This study proposes the 2PM-MARL method for optimizing the deployment and management of mobile charging stations in EV networks. By minimizing pre-charge delay and balance the utilization of high-load fixed charging stations, 2PM-MARL enhances the efficiency of charging networks and significantly improves user experience. Comprehensive experiments demonstrated that our method, which leverages adaptability as a robust solution to manage the unpredictability of EV charging demand, performs effectively across various charging behaviors and demand patterns. Such a data-driven responsive approach can enhance EV charging networks' performance, paving the way for a more efficient and user-friendly future in electric mobility.

Acknowledgement. This paper was supported in part by National Science and Technology Council (NSTC), R.O.C., under Contract 112-2221-E-006-158 and 113-2622-8-006-011-TD1.

References

1. Charging stations by state (2021). https://evadoption.com/ev-charging-stations-statistics/charging-stations-by-state/
2. Afshar, S., Disfani, V.R.: Optimal scheduling of electric vehicles in the presence of mobile charging stations. In: IEEE PESGM (2022)

3. Cornuéjols, G., Nemhauser, G.L., Wolsey, L.A.: The uncapacitated facility location problem (1990). https://api.semanticscholar.org/CorpusID:8880493
4. Elghitani, F., El-Saadany, E.F.: Efficient assignment of electric vehicles to charging stations. IEEE Trans. Smart Grid **12**, 761–773 (2021)
5. Hardman, S., et al.: A review of consumer preferences of and interactions with electric vehicle charging infrastructure. Transp. Res. Part D: Transp. Environ. **62**, 508–523 (2018)
6. Hussain, S., Kim, Y.S., Thakur, S., Breslin, J.G.: Optimization of waiting time for electric vehicles using a fuzzy inference system. IEEE T-ITS **23**, 15396–15407 (2022)
7. Irle, R.: Global EV sales for 2022 (2023). https://www.ev-volumes.com/
8. Lee, K.B., Ahmed, M.A., Kang, D.K., Kim, Y.C.: Deep reinforcement learning based optimal route and charging station selection. Energies **13**, 6255 (2020)
9. Lin, K., Zhao, R., Xu, Z., Zhou, J.: Efficient large-scale fleet management via multi-agent deep reinforcement learning. In: ACM SIGKDD (2018)
10. Liu, L., Xi, Z., Zhu, K., Wang, R., Hossain, E.: Mobile charging station placements in internet of electric vehicles: a federated learning approach. IEEE T-ITS **23**, 24561–24577 (2022)
11. Liu, Q., Zeng, Y., Chen, L., Zheng, X.: Social-aware optimal electric vehicle charger deployment on road network. In: ACM SIGSPATIAL (2019)
12. Moradipari, A., Alizadeh, M.: Pricing and routing mechanisms for differentiated services in an electric vehicle public charging station network. IEEE Trans. Smart Grid **11**, 1489–1499 (2019)
13. Sadeghianpourhamami, N., Deleu, J., Develder, C.: Definition and evaluation of model-free coordination of electrical vehicle charging with reinforcement learning. IEEE Trans. Smart Grid **11**, 203–214 (2018)
14. Sadreddini, Z., Guner, S., Erdinç, O.: Design of a decision-based multicriteria reservation system for the EV parking lot. IEEE TTE **7**, 2429–2438 (2021)
15. Sunehag, P., et al.: Value-decomposition networks for cooperative multi-agent learning (2017). arXiv:abs/1706.05296
16. Tang, X., et al.: Value function is all you need: a unified learning framework for ride hailing platforms. In: ACM SIGKDD (2021)
17. Ting, L.P.Y., Wu, P.H., Chung, H.Y., Chuang, K.T.: An incentive dispatch algorithm for utilization-perfect EV charging management. In: Gama, J., Li, T., Yu, Y., Chen, E., Zheng, Y., Teng, F. (eds.) Advances in Knowledge Discovery and Data Mining. Lecture Notes in Computer Science(), vol. 13282, pp. 132–146. Springer, Cham (2022). https://doi.org/10.1007/978-3-031-05981-0_11
18. von Wahl, L., Tempelmeier, N., Sao, A., Demidova, E.: Reinforcement learning-based placement of charging stations in urban road networks. In: ACM SIGKDD (2022)
19. Yan, L., Shen, H., Kang, L., Zhao, J., Xu, C.: Reinforcement learning based scheduling for cooperative EV-to-EV dynamic wireless charging. In: IEEE MASS (2020)
20. Zheng, B., et al.: Soup: spatial-temporal demand forecasting and competitive supply in transportation. IEEE TKDE **35**, 2034–2047 (2023)

Localization Through Deep Learning in New and Low Sampling Rate Environments

Thanh Dat Le(✉) and Yan Huang

University of North Texas, Denton, TX 76205, USA
thanhle5@my.unt.edu, yan.huang@unt.edu

Abstract. Source localization in wireless networks is essential for spectrum utilization optimization. Traditional methods often require extensive transmitter information while existing deep learning approaches perform poorly in new and low sampling rate environments. We introduce LocNet, a deep learning approach that overcomes these limitations using a compact UNet-like architecture incorporating environmental maps. Unlike other deep learning strategies, LocNet adopts loss functions designed explicitly for imbalanced data, moving beyond the conventional mean-square error loss. Our comparative analysis reveals that LocNet outperforms other deep learning models by more than a factor of two. This advancement underscores LocNet's suitability for real-world deployment across diverse operational contexts.

Keywords: Deep Learning · Source Localization · Imbalanced Data

1 Introduction

Source localization, a foundational problem with decades of research, has broad applications including forest fire tracking [16], search and rescue [1] and pinpointing radiation sources [12]. In the wireless sensor network domain, particularly for spectrum monitoring, source localization optimizes spectrum utilization by detecting and removing unauthorized transmitters [20].

Traditional transmitter localization methods include Time-Of-Arrival (TOA), Angle-Of-Arrival (AOA) and Received Signal Strength (RSS). While TOA and AOA are effective; they require information about the transmitter's structure and power as well as costly hardware to acquire the measurements [22]. On the other hand, RSS which measures the raw signal power from a sensor at a receiving location, does not rely on transmitter information.

Given the advancements in artificial intelligence, particularly in deep learning, a data-driven technique is able to leverage the RSS to localize transmitters. Prior works employed deep learning [8,10,18,20–22] to localize a randomly placed transmitter in an study area and did outperform the traditional methods [3,6,7,13]. However, there are two major challenges that were not addressed:

D.-N. Yang et al. (Eds.): PAKDD 2024, LNAI 14649, pp. 297–308, 2024.
https://doi.org/10.1007/978-981-97-2262-4_24

(1) Low sampling rate: existing methods require a large number of sensors to capture extensive RSS values and do not perform well in low sampling rate environments; and (2) Unknown environment: performance of existing methods diminish in large and unknown environments that were not covered during training [18,21]. We will address these two challenges and make the following contributions:

- **Compact Deep Learning Model**: We propose LocNet, a compact UNet based regression model, optimized for stable training and robust performance for localization. The model is orders of magnitudes smaller than many State-Of-The-Art (SOTA) deep learning models.
- **High Performance in Low Sampling Rate and Unknown Environments**: We employ a loss function that tailors to the imbalanced dataset, moving beyond the Mean-Square Error (MSE) loss approach in training. Additionally, we integrate environment maps into the framework to improve localization performance. This integration boosts the model's effectiveness in completely unknown environments that are not included in the training dataset.
- **Extensive Experiments**: To test the robustness of our model, we compare LocNet with four SOTA deep learning based models across various unseen environments with sampling rates ranging from 0.01% to 0.1%. Results show that LocNet consistently outperforms the SOTA models. We conducted an ablation study to show the effectiveness of our design choices and provide further visual analysis of the best performing and worst performing environments to understand the impact of the environments and sample distribution.

In Sect. 2, we define the problem of transmitter localization. Section 3 contains literature reviews of the prior works in this area. Section 4 defines our methodology; including the base architecture and comparison with Unet. Section 5 describes the dataset, presents the experiment design and discusses the experimental results. Section 6 provides the directions of future research.

2 Problem Definition

Consider a geographic Region Of Interest (ROI) discretized into a two-dimensional grid of $\mathrm{H} \times \mathrm{W}$; where H and W are the height and width of the grid respectively with $(\mathrm{H}, \mathrm{W} \in \mathbb{Z}^{+})$. Let S be a matrix with a size of $\mathrm{H} \times \mathrm{W}$ where S_{ij} represents a reading from a sensor if it is non-zero and no information nor buildings present. Additionally, a matrix E of $\mathrm{H} \times \mathrm{W}$ represents the environment mask where buildings are represented by -1, sensors are represented as 1 and vacancy (neither 1 nor -1) is represented as 0. The objective is to localize the transmitter location with S and the environmental mask E.

3 Related Works

Deep learning techniques [8,10,18,20–22] have achieved better localization performance than traditional methods [3,6,7,13]. These methods typically yield either the coordinate-based or the heatmap-based result for transmitter localization.

3.1 Coordinate Based Localization

The coordinate-based approach determines the (x, y) coordinates of the transmitters on a 2D map as the output of the localization process.

Zhang et al. [21] pioneered integrating deep learning in localizing a transmitter via RSS. They proposed a novel hybrid architecture that combines a Hidden Markov Model (HMM) with Multiple Layers of Perceptrons (MLP). The MLP is a feature extractor in this design, while HMM serves as a localizer to predict the transmitter's (x, y) coordinates. For data collection, Zhang et al. partitioned the ROI into multiple cells; each cell covered a small area and contained five (5) randomly placed signal sensors. The model subsequently learns from this data. The method works well in known and small-scale training environments.

Wang et al. [18] took a similar approach. Instead of randomly placing sensors in each cell, Wang et al. uniformly deployed 5 to 12 sensors across the ROI. The authors introduced a Multi-Task Gated Convolution Neural Network (MT-GCNN), combining the Gated 1D Convolution Neural Net with skip connections (GCNN) and two MLP heads. These MLP heads were responsible for finding the transmitters' containment cells and estimating the distance between the transmitters and the cells' center. A separate linear regression calculates the transmitters' location using the model's outputs.

DeepTxFinder [22] can localize transmitters with any sensor deployment across the ROI; where each is detected by training a separate deep learning model that shares a similar architecture. Each model had several Convolution Layers for feature extractions and an MLP for prediction. The system first predicts the number of transmitters in ROI and then selects the model to pinpoint them. The model performs well at high sampling rate environments.

3.2 Heatmap Based Localization

The heatmap-based deep learning technique converts sample readings into a 2D heatmap; with each pixel representing the probability of a transmitter's presence. Post-processing techniques, such as thresholding or the argmax functions, are used to identify the transmitter location.

DeepMTL [20] uses a modified object detection algorithm YOLOv3 [14] to generate a heatmap for transmitters' location detection. The authors demonstrated that YOLOv3 outperformed manual tuning via argmax or thresholding in localization. DeepMTL achieved remarkable performance in scalability and localization as compared to DeepTxFinder.

MSLocNet [8] and TL;DL [10] employ autoencoders to address the low sampling rate problem. An autoencoder comprises an encoder for extracting essential input features (latent spaces) and a decoder for generating a localized heatmap. Lin et al. implemented ResNet blocks [4] within the encoder. A ResNet block is characterized by multiple stacked convolution layers that are interspersed with skip connections that bypass selected layers. Furthermore, they innovated by introducing fusion blocks in the decoder. These blocks are a combination of bilinear upsampling and convolution layers that are specifically designed for effective heatmap generation. Mitchell et al. employed the original UNet architecture [15]. Both models used manual thresholding for localization and excelled at low sampling rates.

Among the six deep learning models above, the model proposed by Zhang et al. [21] is specifically designed for the same fixed training and testing environment while the model by Wang et al. [18] requires the sensors to be uniformly distributed. We will focus on localization with randomly distributed sensors where the sampling rate can be very low and the training and testing environments are disjoint. We compare with DeepTxFinder [22], DeepMTL [20], MSLocNet [8], TL;DL. [10]; which suffer performance degradation when dealing with such environments.

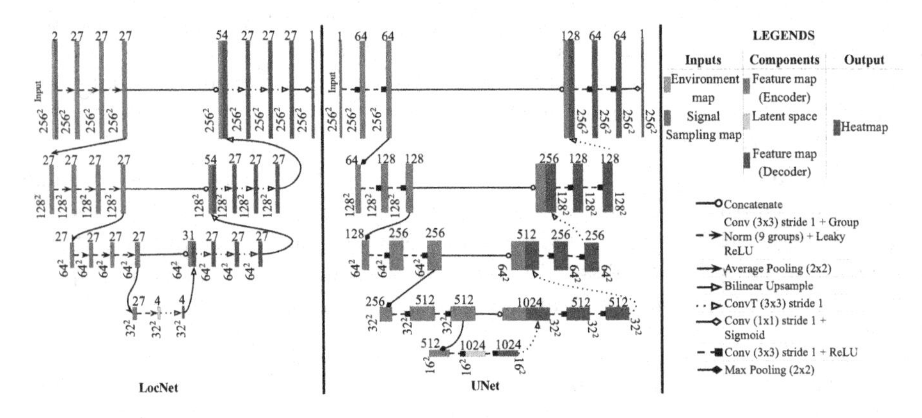

Fig. 1. Our proposed LocNet (128K parameters) in comparison with original UNet (31M parameters). Top number represents the number of channels and bottom number represents the channels's dimension. Bar width is proportional to number of channels.

4 Method

Deep Learning Autoencoder. Autoencoder is a widely-used deep learning architecture for various tasks. It comprises two primary components: Encoder and Decoder. The encoder compresses the input(s) (or downsampling high-dimension features) into lower-dimension features; which is achieved by using different deep learning layers such as convolutional layers, pooling layers or fully-connected layers. In contrast, the Decoder utilizes the encoder's output to decompress and attempt to reconstruct the original input. Depending on the training data and ground truths, the decoder can learn tasks like image restoration, generation, and more.

UNet. The UNet architecture, part of the deep learning autoencoder family, was first introduced by [15]. UNet's essential feature is the skip-connections, which link corresponding encoder-decoder layers between pooling and upsampling layers. These connections enhance information retention, facilitate the fusion of high-level features and improve gradient flow; thus aiding generalization for high-level encoders. Frost et al. [10] adopted this architecture in radio localization and designated it as TL;DL (as shown in the right of Fig. 1).

The Proposed LocNet. Drawing from studies [8,10], we hypothesize that an architecture similar to AutoEncoder is well-suited for localization tasks. Our proposed LocNet model, as shown in Fig. 1, is inspired by AutoEncoder architectures of Tenganya et al. [17] and Locke et al. [9] initially designed for generating radio propagation maps. We adapt UNet architecture as well as the features from the inspired models. UNet's skip connections are beneficial in handling highly sparse sampling maps by retaining and transferring information from encoder to decoder for better utilization. Moreover, LocNet utilizes both an environment map and a sampling map for localization to aid the model in generalizing to new environments for localizing a transmitter.

Distinct from the TL;DL, which follows the original UNet [15], LocNet is characterized by its compactness by reducing the number of channels of convolution as well as deconvolution layers. This design avoids unnecessary feature reductions in both the encoder and decoder. Additionally, we employ convolution transposes ("deconvolution layers") following bilinear upsampling to align with the decoder's purpose. This allows recovery of granularity from the latent space; serving as a "reversing" encoder. Group Normalization is integrated before the activation layer of each convolution and deconvolution layer. This technique groups the correlated features, helps convolution adjust the filters for better extraction and improves training stability. A sigmoid activation is applied at the end of the decoder to generate a probability map. To pinpoint the transmitter's (x, y) coordinate, we applied argmax function on the decoder's output. The detailed illustration of LocNet can be found in Fig. 1.

Focal Loss. Previous works [8,10,20] have employed Mean-Squre Error (MSE) as a loss function for the localization problem. However, localization is a problem with highly imbalanced data and MSE applies equal weighting to all samples' losses. There are loss functions specifically designed to handle imbalances in image processing domain including Balanced Binary Cross-Entropy (BBCE), Focal Loss (FL) and Focal Tversky loss (FT). We introduce these losses first and provide insights on why FL is chosen for LocNet for the problem of localization.

Balanced Binary Cross Entropy (BBCE): BBCE is a modified version of Binary Cross Entropy (BCE). BCE calculates the log predictions of each location for each class, as illustrated in Eq. 1. BBCE introduces a balancing coefficient, denoted as α, which adjusts the weight of the loss contributions from each class to address the class imbalance problem, as illustrated in Eq. 2.

$$BCE(p,y) = \begin{cases} -log(p) & \text{if } y = 1 \\ -log(1-p) & \text{otherwise} \end{cases} \tag{1}$$

$$BBCE(p,y) = \begin{cases} - * \alpha * log(p) & \text{if } y = 1 \\ -(1-\alpha) * log(1-p) & \text{otherwise} \end{cases} \tag{2}$$

Focal Tversky (FT): FT loss is an enhanced variant of Tversky loss, as illustrated in Eq. 3, incorporating a gamma (γ) exponent term to prioritize the learning of the minority class. Tversky loss is a function that introduces a trade-off between false positives and false negatives; thus facilitating a controlled balance (β) when assessing the similarity between predicted and actual samples.

$$Tversky(p,y) = \frac{1 + p * y}{1 + y * p + \beta * (1-y) * p + (1-\beta) * y * (1-y)} \tag{3}$$

$$FT(p,y) = (1 - Tversky(p,y))^{\gamma} \tag{4}$$

Focal Loss (FL): FL loss function addresses the imbalance between classes and directs a deep learning model to focus on the minority class; as illustrated in Eq. 5. It has two tuning parameters: γ and α. The γ parameter regulates the loss by down-weighting the "easy" or majority class samples; thus allowing the model to concentrate on the "hard" or minority class samples. Meanwhile, α provides an additional weighting for balancing the loss contributions between the two classes.

$$FL(p,y) = \begin{cases} -\alpha * log(p) * (1-p)^{\gamma} & \text{if } y = 1 \\ -(1-\alpha) * log(1-p) * p^{\gamma} & \text{otherwise} \end{cases} \tag{5}$$

All three of the above loss functions are designed to handle imbalanced data and have the ability to direct LocNet's focus toward localizing a transmitter; which is a minor class in our dataset. We choose FL as the loss function to for LocNet based on the following observations: BBCE solves the imbalanced problem but lacks a mechanism to focus on a minority region such as FL. FT may not effectively work for our task since it is primary used for segmentation tasks and addresses data imbalance indirectly by tuning false positives and false negatives based on predictions.

5 Experimental Design and Results

In the experiments, we compare our LocNet model with four other SOTA models:

- *LocNet*: Our model.
- *TL;DL*: Inspired from UNet architecture [10].
- *MSLocNet*: Adopting AutoEncoder architecture with Residual Block [8].

- *DeepMTL*: A Two-stages localizer; combination between heatmap generator model with YOLOv3 [20].
- *DeepTxFinder*: coordinate (x, y) based regression that utilizing Convolution with Fully Connected Neural Network [2].

We adopt Root Mean Square Error of euclidean distance between two points (RMSE) extensively used in the SOTA models as the metric to compare localization performance between models.

$$RMSE = \sqrt{\frac{1}{N} * \sum_{n=1}^{N} Euclidean\_Distance(p_n, t_n)^2} \tag{6}$$

where N is number of testing sampling maps, p_n is the predicted transmitter location, t_n the ground truth transmitter location for map n, and $Euclidean\_Distance(p_n, t_n)$ is the Euclidean distance between the two locations.

(a) Simple environments that LocNet has the lowest localization error (increasing RMSE in row major order).

(b) Challenging environments that LocNet has highest localization error (decreasing RMSE in row major order).

Fig. 2. Test environment visualization.

5.1 Dataset

The RadioMapSeer dataset [19] is used to create the low-sampling dataset for benchmarking our proposal against SOTA methods in outdoor localization. It comprises 701 maps, which are taken from OpenStreetMap [11] in different areas in major cities such as Ankara, Berlin, Glasgow, Ljubljana, London and Tel Aviv. Each map is a 2D binary image, where a value of 1 indicates a building is present and 0 indicates a building is not present. For each map there are 80 transmitter locations randomly distributed. Each transmitter location has a generated signal propagation map by using the Dominant Path Model (DPM) method via WinProp program [5] resulting in $701 \times 80 = 56,080$ signal maps. The transmitter, receivers, and buildings have heights of 1.5 m, 1.5 m, and 25 m [19]. The simulations are stored as 2D grayscale images of 256×256 pixels; in which 1 pixel covers an area of 1 m^2 [19]. Figure 2 shows 16 test environment maps with 8 easy and 8 most challenging environments based on our model performance.

For our benchmark, we sought to create a low-sampling dataset derived from the RadioMapSeer dataset. To this end, for each environment map and transmitter pair, we generate 5 sampling maps by using uniform sampling method with the sampling rate between 0.01% to 0.1% of the total pixels represented. Each sampling is represented by two matrices: (1) a signal map with sampling readings on sampled locations and zero on other locations; (2) a companion environment mask map with 1 representing sampled locations, 0 indicating "masked-out" areas, and −1 as building locations. This methodology yields 280,400 low-sampling maps from the 56,080 full signal maps. Based on the environments, We divide this dataset into training, validation, and testing sets (501/100/100). We ensure that the environments in all sets are entirely separate to avoid data leakage.

Implementation: We trained our model for 100 epochs with batches size of 64 maps. We employed the AdamW optimizer with a learning rate of 5e-4. In comparing between losses, we set $\alpha = 0.75$ for both BBCE and FL, $\gamma = 3$ for FL and FT, and $\beta = 0.7$ for FT loss. We reconstructed SOTA models based on the methodologies described in their research publications.

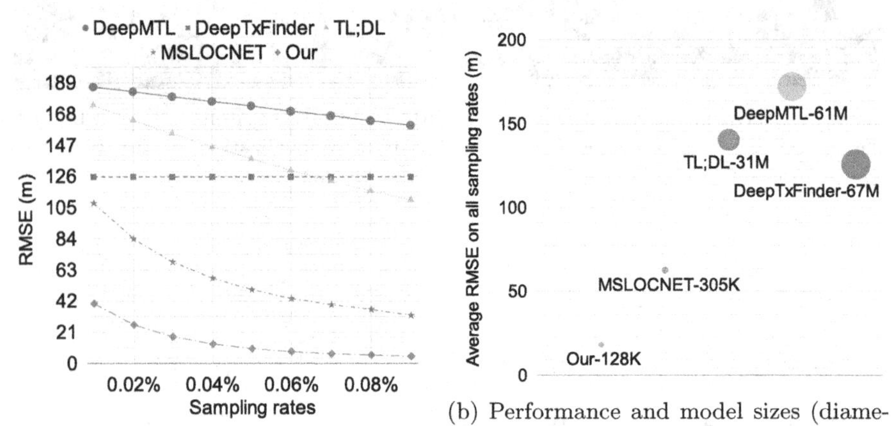

(a) Performance of our LocNet model and SOTAs.

(b) Performance and model sizes (diameter and numbers representing numbers of trainable parameters)

Fig. 3. Performance Comparison with SOTAs

5.2 Comparing with SOTAs

As shown in Fig. 3a, LocNet outperforms other SOTA models consistently across all sampling rates in the test sets where environments are not seen in training data, reducing the localization error by more than a factor of two. Moreover, all models' performances increase as the sampling rate increases, which reveals all models benefit from more samples. Among the SOTAs, MSLocNet consistently has the smallest performance gap from LocNet. Figure 3b shows that LocNet is three orders of magnitudes smaller than DeepTxFinder and less than half the size of MSLocNet in training parameters, which enables edge device deployment.

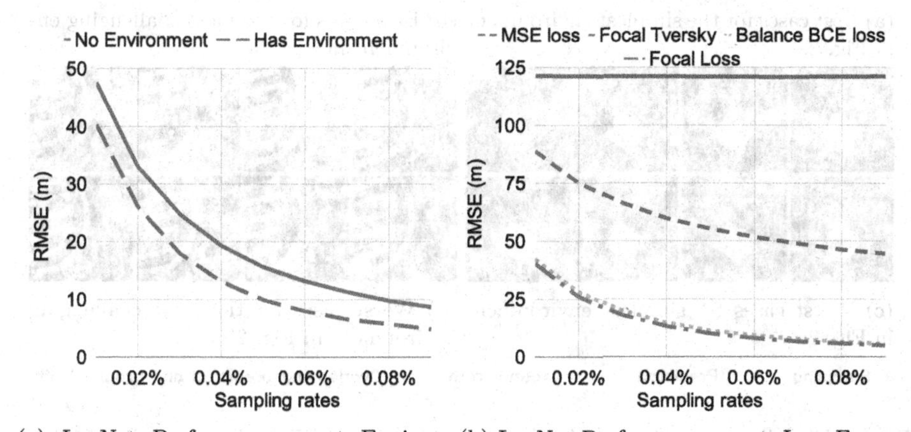

(a) LocNet Performance w.r.t Environment Map (b) LocNet Performance w.r.t. Loss Functions

Fig. 4. Ablation study on environment map and loss functions.

5.3 Ablation Study

Environment vs No Environment: As shown in Fig. 4a, fusing the environment map with the sampling map helps improve the localization performance. On average, the localization error is reduced by at least 15% for each sampling rate.

Comparing Between Losses: As shown in Fig. 4b, FT and MSE result in the poorest performance. With MSE, our model typically estimates locations to be approximately 44 m less accurate than using FL or BBCE. FL and BBCE demonstrated similar effectiveness. We believe the reason can be that both FL and BBCE employ log-form error calculation after performing a sigmoid function on predictions, which can result in significant loss values for incorrect predictions. These high loss values can lead to a high changing rate in gradients, thus enhancing the model's learning capability and likely avoiding saddle points. On the contrary, MSE and FT do not exhibit similar rates of gradient movement, which potentially causes the model's learning to hit a plateau.

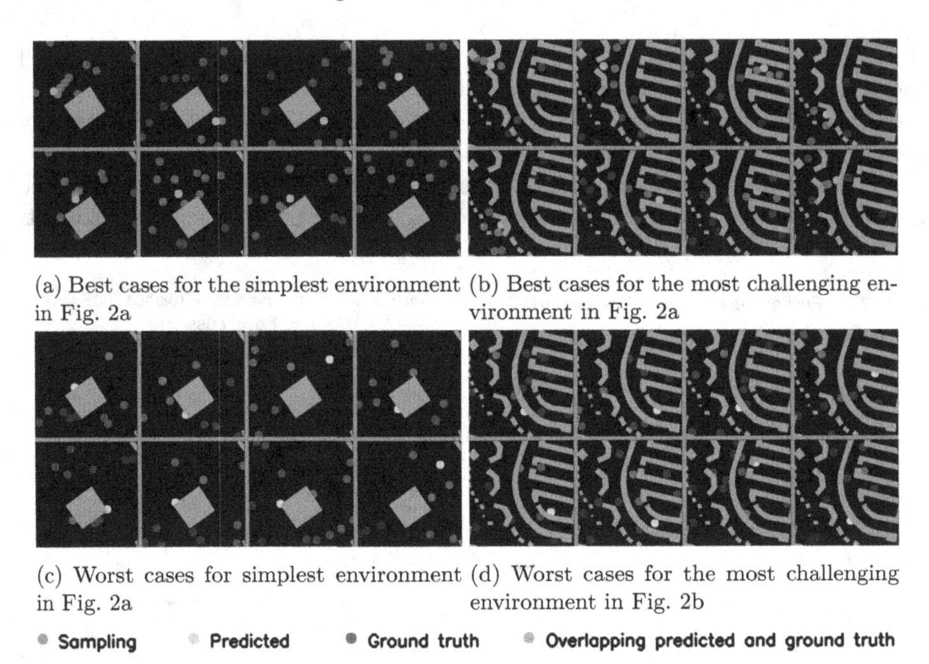

(a) Best cases for the simplest environment in Fig. 2a
(b) Best cases for the most challenging environment in Fig. 2a
(c) Worst cases for simplest environment in Fig. 2a
(d) Worst cases for the most challenging environment in Fig. 2b

Fig. 5. Visualization of best and worse cases (dot sizes enlarged for visualization purpose).

5.4 Visual Analysis

In testing, LocNet performed better in environments with fewer buildings and more open spaces, as shown in Fig. 2a, likely due to simpler signal propagation estimation in such settings. In contrast, dense urban areas in Fig. 2b cause signal attenuation, which increases the complexity of signal propagation estimation and leads to a reduction in localization performance. LocNet becomes more effective when more sensors are closer to the transmitter as shown between Fig. 5a and 5c and between Fig. 5b and 5d. Having more sensors near the transmitter simplifies signal propagation's estimation and assists the model in focusing on a specific area for localization, thus increasing the performance.

6 Conclusion and Future Work

We have presented LocNet, a specialized model for localizing a transmitter in sparse sampling maps. We employ UNet architecture, enhanced with Group Normalization, and utilize the FL function to deal with imbalanced and sparse data issues. Additionally, experimental evidence supports the notion that providing an environmental map of the area of interest further refines localization efficiency. We plan to extend our model's capabilities to localize a transmitter with different power strengths and improve our architecture to achieve greater performance in localization tasks.

Acknowledgement. The research was sponsored by the Army Research Laboratory and was accomplished under Cooperative Agreement Number W911NF-23-2-0014. The views and conclusions contained in this document are those of the authors and should not be interpreted as representing the official policies, either expressed or implied, of the Army Research Laboratory or the U.S. Government. The U.S. Government is authorized to reproduce and distribute reprints for Government purposes, not withstanding any copyright notation herein.

References

1. Atif, M., Ahmad, R., Ahmad, W., Zhao, L., Rodrigues, J.J.P.C.: UAV-assisted wireless localization for search and rescue. IEEE Syst. J. **15**(3), 3261–3272 (2021)
2. Bizon, I., Nimr, A., Schulz, P., Chafii, M., Fettweis, G.P.: Blind transmitter localization using deep learning: a scalability study. In: IEEE Wireless Communications and Networking Conference (WCNC) (2023)
3. Destino, G., Abreu, G.: On the maximum likelihood approach for source and network localization. IEEE Trans. Signal Process. **59**(10), 4954–4970 (2011)
4. He, K., Zhang, X., Ren, S., Sun, J.: Deep residual learning for image recognition. In: 2016 IEEE Conference on Computer Vision and Pattern Recognition (CVPR) (2016)
5. Hoppe, R., Wölfle, G., Jakobus, U.: Wave propagation and radio network planning software winprop added to the electromagnetic solver package FEKO. In: International Applied Computational Electromagnetics Society Symposium - Italy (ACES), pp. 1–2 (2017)
6. Khaledi, M., et al.: Simultaneous power-based localization of transmitters for crowdsourced spectrum monitoring. In: Proceedings of the 23rd Annual International Conference on Mobile Computing and Networking, pp. 235–247 (2017)
7. Lin, L., So, H., Chan, Y.: Accurate and simple source localization using differential received signal strength. Digit. Signal Process. **23**(3), 736–743 (2013)
8. Lin, M., Huang, Y., Li, B., Huang, Z., Zhang, Z., Zhao, W.: Deep learning-based multiple co-channel sources localization using bernoulli heatmap. Electronics **11**(10) (2022)
9. Locke IV, W.A.: Deep learning approaches to radio map estimation. Master thesis. UNT Digital Library, University of North Texas (2023)
10. Mitchell, F., Baset, A., Patwari, N., Kasera, S.K., Bhaskara, A.: Deep learning-based localization in limited data regimes. In: Proceedings of the ACM Workshop on Wireless Security and Machine Learning, pp. 15–20 (2022)
11. OpenStreetMap (2023). https://www.openstreetmap.org. Accessed 10 Oct 2023
12. Pinto, L.R., et al.: Radiological scouting, monitoring and inspection using drones. Sensors **21**(9) (2021)
13. Rahman, M.Z., Habibi, D., Ahmad, I.: Source localisation in wireless sensor networks based on optimised maximum likelihood. In: Australasian Telecommunication Networks and Applications Conference (2008)
14. Redmon, J., Farhadi, A.: Yolov3: an incremental improvement. arXiv preprint arXiv:1804.02767 (2018)
15. Ronneberger, O., Fischer, P., Brox, T.: U-net: convolutional networks for biomedical image segmentation. In: Navab, N., Hornegger, J., Wells, W., Frangi, A. (eds.) MICCAI 2015. LNCS, vol. 9351, pp. 234–241. Springer, Cham (2015). https://doi.org/10.1007/978-3-319-24574-4_28

16. Sharma, A., Singh, P.K., Kumar, Y.: An integrated fire detection system using IoT and image processing technique for smart cities. Sustain. Urban Areas **61**, 102332 (2020)
17. Teganya, Y., Romero, D.: Deep completion autoencoders for radio map estimation. IEEE Trans. Wireless Commun. **21**(3), 1710–1724 (2022)
18. Wang, W., Zhu, L., Huang, Z., Li, B., Yu, L., Cheng, K.: MT-GCNN: multi-task learning with gated convolution for multiple transmitters localization in urban scenarios. Sensors **22**(22) (2022)
19. Yapar, Ç., Levie, R., Kutyniok, G., Caire, G.: Dataset of pathloss and ToA radio maps with localization application. arXiv preprint arXiv:2212.11777 (2022)
20. Zhan, C., Ghaderibaneh, M., Sahu, P., Gupta, H.: Deepmtl: deep learning based multiple transmitter localization. In: IEEE 22nd International Symposium on a World of Wireless, Mobile and Multimedia Networks (WoWMoM) (2021)
21. Zhang, W., Liu, K., Zhang, W., Zhang, Y., Gu, J.: Deep neural networks for wireless localization in indoor and outdoor environments. Neurocomputing **194**, 279–287 (2016)
22. Zubow, A., Bayhan, S., Gawłowicz, P., Dressler, F.: Deeptxfinder: multiple transmitter localization by deep learning in crowdsourced spectrum sensing. In: 2020 29th International Conference on Computer Communications and Networks (ICCCN) (2020)

MPRG: A Method for Parallel Road Generation Based on Trajectories of Multiple Types of Vehicles

Bingru Han[1,2], Juanjuan Zhao[1(✉)], Xitong Gao[1], Kejiang Ye[1], and Fan Zhang[1]

[1] Shenzhen Institute of Advanced Technology, Chinese Academy of Sciences, Shenzhen, China
jj.zhao@siat.ac.cn

[2] University of Chinese Academy of Sciences, Beijing, China

Abstract. Accurate and up-to-date digital road maps are the foundation of many applications, such as navigation and autonomous driving. Recently, the ubiquity of GPS devices in vehicular systems has led to an unprecedented amount of vehicle sensing data for map inference. Existing trajectory-based map generation methods are difficult to accurately generate parallel roads where the GPS positioning errors are large, and the sampling frequency is low. In this paper, we propose a novel method MPRG to discover parallel roads based on the differences between free and fixed trajectories from different types of vehicles. This method can serve as a plugin for any existing map generation method. MPRG extracts highly discriminative features by utilizing the spatial distribution and regional correlation information of trajectories from different vehicle types. Then, the multidimensional features are fed into an SVM classification model suitable for small sample to identify and generate the parallel roads. We apply MPRG to three advanced road generation methods using GPS data from Shenzhen. The results show that we can significantly improve the performance of parallel road generation.

Keywords: map generation · spatio-temporal data · GPS trajectories

1 Introduction

Accurate and timely updates of road network maps are fundamental for many location-based applications. Manual creation and updating of maps are costly, time-consuming, and labor-intensive, which cannot meet the increasing demand for timely and accurate vehicle sharing services and navigation services. In recent years, GPS sensors have been widely deployed on mobile platforms such as cars and smartphones, providing new opportunities for generating and updating underlying road networks. However, due to complex factors such as satellite fluctuations and sampling errors [1], the collected GPS data is sparse and biased, making road generation based on GPS trajectories a challenging problem and a major research direction in intelligent transportation in recent years.

D.-N. Yang et al. (Eds.): PAKDD 2024, LNAI 14649, pp. 309–321, 2024.
https://doi.org/10.1007/978-981-97-2262-4_25

In scenarios with limited data scale, low sampling rate, and high noise, directly connecting two consecutive points of the same trajectory may not accurately reflect the actual road shape. Existing trajectory generation methods mainly rely on heuristic approaches that model road sections from local to global. These methods first extract local features of roads based on aggregated trajectory points or segments in neighboring areas, and then connect these local regions using spatial proximity to form the overall road topology.

These methods perform well in generating simple road networks with obvious shape differences, long distances between roads, and small GPS errors. However, they face difficulties in accurately identifying parallel roads in scenarios with high road density, hierarchical heterogeneity (e.g., elevated bridges, underground passages), and big GPS errors. These methods often mistakenly identify two nearby parallel roads as one road. For example, Fig. 1(a) shows a map of a parallel road area in Shenzhen, (b) is a GPS point cloud based on taxi trajectory data, and (c) is a road network map of the parallel road region generated using existing road generation methods. As can be observed, generating maps using taxi point clouds fails to distinguish closely spaced parallel roads. This situation worsens with lower sampling rates.

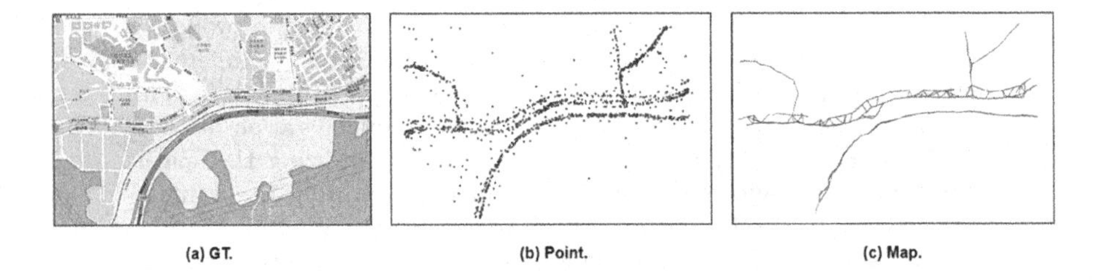

(a) GT. (b) Point. (c) Map.

Fig. 1. Motivation Examples.

In order to tackle the aforementioned issues, this study devises a parallel road discrimination plugin based on diverse trajectory types, allowing for seamless integration with existing methods to detect parallel road structures and improve the quality of generated maps. The basic idea is that the trajectories of vehicles operating in the city can be divided into two types: fixed trajectories and free trajectories. These trajectory trajectories exhibit two types of differences in distribution: (i) Inter-class differences: Fig. 2(b) shows the point cloud map of two closely located roads. One road has both taxi and bus trajectories, while the other road does not have bus routes. (ii) Intra-class differences: On the two parallel roads in Fig. 2(c), buses are present, but the distribution of bus routes is different on different roads. Therefore, we can identify and generate parallel roads by analyzing the differences and complementarity of the GPS distributions of buses and taxis in the same region.

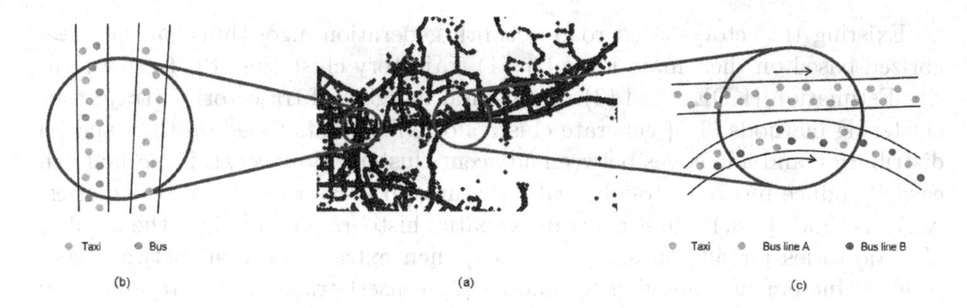

Fig. 2. Vehicle type distribution on underlying unknown roads.

In general, the main contributions of this study are as follows:

- We propose for the first time the idea of using vehicle types to assist in detecting parallel road in the road network generation problem based on trajectories. It can serve as a plugin for any existing road map generation method, detecting and distinguishing parallel road structures to improve the effectiveness of road structure inference.
- We propose a multi-modal parallel road structure generation method MPRG based on multiple types of vehicles. We extract features of multi-modal trajectories from different perspectives of vehicle types, and utilize their spatial distribution and regional association information to extract highly discriminative features. Then, we fuse the multi-dimensional features and use an SVM classification model suitable for small sample to identify the existence of parallel road and clustering the trajectories according to the features, improving the algorithm for road generation on parallel roads.
- We apply MPRG to real GPS trajectory datasets of buses and taxis in Shenzhen, and evaluate it on three advanced GPS road generation methods, namely Biagioni [2] and Kharita [3] and Karagiorgou [8]. The results demonstrate that we can significantly improve the quality of road network generation, especially the effectiveness of identifying parallel road structures.

2 Related Work

Existing map generation methods can be categorized into two categories: those based on aerial imagery and those based on trajectory data.

Aerial imagery has been widely used for automatic road network inference, and various methods have been proposed to extract digital maps from aerial images. Mnih et al. [16] used feedforward neural networks to detect roads. Máttyus et al. [17] applied CNN segmentation and extracted road network graphs through post-processing pipelines. However, there are many cases where roads are difficult to identify due to occlusions from tall buildings, overpasses, trees, etc. Additionally, the lack of real-time aerial data makes it challenging to update the latest road networks.

Existing trajectory-based road network generation algorithms can be categorized based on their main ideas [2]: (1) trajectory clustering, (2) Kernel Density Estimation (KDE), and (3) incremental merging of trajectories. Trajectory clustering methods [3–5] generate clusters of GPS points based on their spatial distribution and add edges between adjacent clusters. However, this method can easily produce incorrect topological structures in areas with dense trajectories. KDE methods [2,6,10] first generate a spatial histogram and weight the number of trajectories passing through each cell, then extract the road network centerlines. Incremental merging techniques [7,8] insert trajectories into an empty map, and the merging process of this method only operates on consecutive observations each time. However, when the trajectory sampling rate is low, it can lead to missing road segments.

Recently, researchers have proposed methods that combine or extend these techniques. Chen et al. [9] proposed utilizing prior knowledge of the real road network to learn complex topological structures. Wang et al. [11] constructed maps based on the local density field of GPS points, but this method is affected by GPS noise. Additionally, it cannot handle non-planar road structures. Katsikouli et al. [12] generated optimal intermediate trajectory paths based on beamformed trajectories, but this technique also relies on high-frequency trajectories with less noise to generate maps. Although these methods have improved upon previous approaches, they easily merge parallel roads incorrectly in complex situations.

3 Method

3.1 Framework

Given noisy trajectory points collected on parallel roads, the most straightforward approach is to utilize the difference in trajectory density for identification. However, setting the threshold is difficult when the positioning error in trajectories is significant. In this section, we extract some complementary features from multiple types of vehicles.

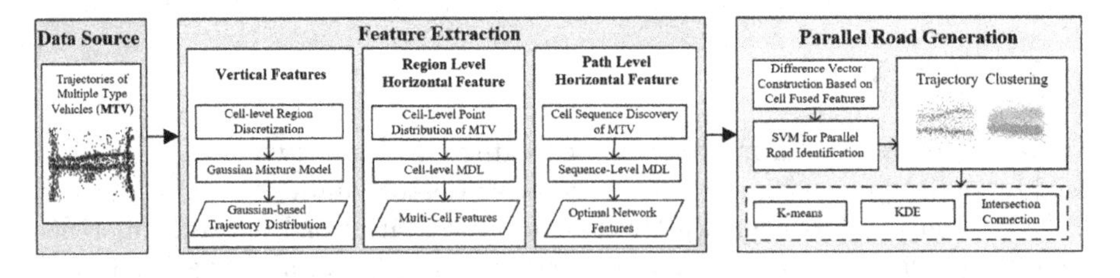

Fig. 3. The Framework of MPRG.

As shown in Fig. 3, specifically, MPRG for generating parallel road based on multiple types of vehicles consists of two modules: (1) Feature extraction and (2) Parallel road generation model. Firstly we extracts features in three dimensions, including vertical features (to discover the Gaussian distribution characteristics of trajectories), as well as regional level horizontal features (to discover the inter-class and intra-class differences in trajectory distribution between cells) and path level horizontal features (to discover the sequential order of trajectories passing through continuous cells). In the parallel road generation model, the extracted features are used to calculate the vector of multiple differences is input into an SVM model to classify the parallel road regions. Based on this, trajectories within the parallel road regions are clustered, improving the algorithm for road generation.

3.2 Feature Extraction

Vertical Features (F_V). Previous road generation works are mostly based on a priori knowledge: the road centerline is the line with the highest GPS density [2,3,14]. As shown in Fig. 4, the GPS density decreases with the increasing distance from the road centerline. We assume that each road contributes to a Gaussian distribution. This paper employs the Constrained Gaussian Mixture Model (CGMM) to model the floating vehicle trajectories of each road.

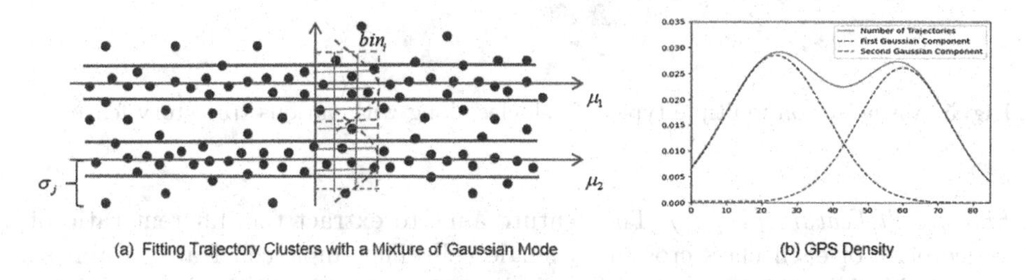

(a) Fitting Trajectory Clusters with a Mixture of Gaussian Mode (b) GPS Density

Fig. 4. The Gaussian distribution of trajectories on roads.

Figure 4(a) is a schematic representation of the point distribution of trajectories on parallel road. The target area is initially discretized into equally sized region (1 m * 1 m), in this paper called *cell*, the number of times each cell traversed is then calculated. When the number of roads, k, is known, the parameters of the Gaussian Mixture Model (GMM) can be predicted using the Expectation-Maximization (EM) algorithm. However, in practical calculations, the existence and number of parallel roads are unknown. A common approach is to exhaustively consider all possible values of k and choose the model parameters that yield the best fit. The formula is as follows:

$$k = max\left\{-\frac{1}{B_n}\sum_{i=1}^{B_n} logp(bin_i, \theta_k) + \lambda R(Dw, \theta_k)\right\}$$

$$L(bin_i, p(bin_i, \theta_k)) = -log(p(bin_i, \theta_k))$$

In the above text, k represents the number of Gaussian components; B_n is the total number of cells in the sampling region; Dw represents the road width.

Through model fitting, EM prediction, and calculation result correction, we can obtain k, which represents the number of parallel roads. $L(bin_i, p(bin_i, \theta_k))$ is the empirical risk used to evaluate the suitability of the corresponding model. Based on these, we generate F_v follows:

$$F_V = \{k, L(bin_i, p(bin_i, \theta_k))\}$$

Regional Level Horizontal Features (F_R). We first extract the distribution characteristics of vehicles in each individual single-cell, and then based on the idea that adjacent cells on the same parallel road have similar trajectory class distributions, we extract multi-cell characteristics formed by adjacent cells.

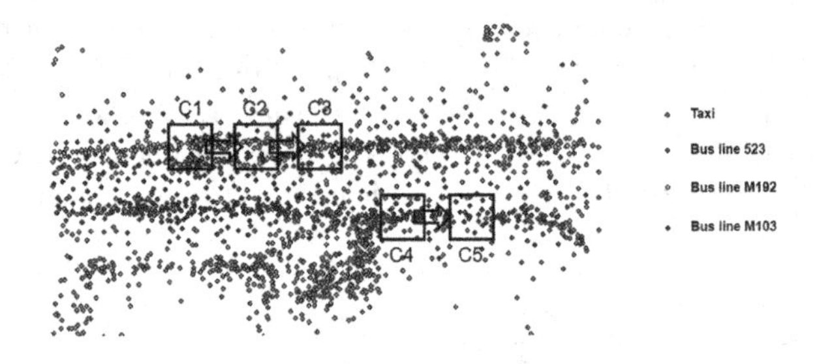

Fig. 5. Visualisation multiple types of vehicles along with various trajectory features.

Single-Cell Feature (F_R^{CiX}). This feature aims to extract the different ratio of trajectories of each class crossing a single cell. The single-cell features aim to extract the class of trajectories passing through each cell and their proportions, and to discover the differences in trajectory distribution between classes and within classes.

For example, Fig. 5 shows the extracted features from the trajectory data, which include N trajectories in total, with N_t trajectories from taxis and N_b trajectories from buses. $C1 - C5$ represent the sub-cells divided in the road region. There are $N1$ trajectories passing through $C1$, among them there are $N1_t$ trajectories from taxis and $N1_b$ trajectories from buses. Within the fixed trajectories, there are $N1_b^A$ trajectories from class A. Based on this information, we generate a single-cell feature as follows:

$$F_R^{C1A} : \{C1\} \Rightarrow \{classA\}, sup(F_R^{C1A}) = \frac{N1_b^A}{N_b}, conf(F_R^{C1A}) = \frac{N1_b^A}{N1_b}$$

The support is defined as the proportion of trajectories passing through $C1$. The confidence is the ratio of the number of class A trajectories passing through $C1$ to the total number of fixed trajectories passing through $C1$.

Multi-cell Feature (F_R^Y). The multi-cell feature aggregates adjacent cells with similar single-cell features and discovers the distribution of different trajectory classes within the cells.

The multi-cell feature emphasizes cells where the preponderance of trajectories originates from a common class. We adopt a bottom-up spatial aggregation approach to unveil multi-cell features. This method involves merging two smaller cells into a larger region using the Minimum Description Length (MDL) gain [15]. Then, we discover best pair C_i and C_j from CS such that their merged region, denoted as R, leads to the highest positive MDL gain. The MDL gain when merging C_i and C_j into region R is given as follows:

$$log(\frac{|CS|}{|CS-1|}+|R|log(|R|)-|C_i|log(|C_i|)-|C_j|log(|C_j|)-\frac{|C_i||C_j|Dist(C_i,C_j)^2}{(2\sigma\sigma^2 ln2)(|C_i|+|C_j|)}$$

Where $Dist(C_i, C_j)$is a Euclidian distance between centroid of C_i and C_j, and σ is a constant estimating the spatial distribution of sampling points.

In the scenario where comparatively more trajectories traverse the aggregated generated R, the ensuing procedures unfold as follows: Initially, elementary regional level horizontal features are formulated for the aggregated generated R. Then, a series of region features are systematically generated through a depth-first search process. This iterative process persists until the point at which the Minimum Description Length (MDL) gain turns negative. Subsequently, post-processing is applied to the region features by a substantial number of region features, aiming to eliminate redundant features. Ultimately, the result is the retrieval of the regional level horizontal features set.

Path Level Horizontal Features (F_P). The path level horizontal features based on multiple types of vehicles is a sequence of continuous cells that are traversed by trajectories, signifying feature at path level.

For example, considering the three $C1$, $C2$, and $C3$ in Fig. 5, among the total of $N\{1,2,3\}$ trajectories, where $N\{1,2,3\}_b$ represents free trajectories, $N\{1,2,3\}_b^A$ represents trajectories from class A, and trajectories from class B, pass through a series of cells simultaneously. Using this information, the following two path features can be generated:

$$F_P^{\{1,2,3\}A} : \{C1, C2, C3\} \Rightarrow \{classA\},$$

$$sup(F_P^{\{1,2,3\}A}) = \frac{N\{1,2,3\}_b^A}{N_b}, conf(F_P^{\{1,2,3\}A}) = \frac{N\{1,2,3\}_b^A}{N\{1,2,3\}_b}$$

We consolidate the path level horizontal features into a path feature network with MDL-based approach. The resulting F_P network, denoted as $N =< C, E >$, forms a directed multi-graph where each C represents a cell, and each edge E from C_m to C_n signifies the movement from C_m to C_n. Employing an MDL-based approach, we derive the optimal network from given trajectory datasets. In essence, this method identifies cells and subsequently establishes edges between these cells.

We clarify the identification of path level horizontal features using the path feature network N. The path feature network N is a directed multi-graph, permitting multiple edges between two cells. A directed simple graph N_0 is derived from the directed multi-graph N, so we perform a graph traversal method to extract path rules. Edges in N_0 with the same source and target cells are merged to simplify the initial directed multi-graph by eliminating duplicate edges.

3.3 Parallel Road Generation Model

Parallel Road Region Identification Model. In this section, the cells of the road regions are optimally divided based on the three-dimensional features extracted in the previous section. The optimal segmentation method is extended to identify homogeneous cells. Then a difference vector between regions is input into an SVM model to identify parallel road regions.

For each cell, through the above feature extraction, we can obtain the sub-regions based on multi-cell feature homogeneous aggregation and path homogeneous aggregation that it belongs to. So for each cell, we construct a sub region belonging vector. For example, if the number of sub-regions is N, then we construct an N-dimensional vector, where the sub-region to which the cell belongs has a value of 1 and the other values are 0. For vertical features, we use their Gaussian components to generate a similar vector for the cell. All vectors are connected to obtain the final feature vector for each cell.

Then, using a commonly used vector dissimilarity calculation method based on Euclidean distance, Fisher's optimal segmentation is applied to generate inter-class dissimilarity distribution values. After normalizing the road sub-regions and features, we can apply existing classification techniques to classify target regions. In this study, support vector machines (SVM) are used for region classification. SVM assumes a training sample set that contains training samples belonging to two classes. It separates the samples into two classes using a separating hyperplane, and the optimal separating hyperplane has the maximum margin from the data on both sides. Gaussian radial basis function (RBF) is selected as the kernel function of SVM classifier.

$$k(X,Y) = exp\frac{|X-Y|^2}{2\sigma^2}$$

Where $|X - Y|$ denotes the distance between two vectors, and σ is constant.

Parallel Road Trajectory Clustering Model. After finding the road regions where parallel roads are located, we construct a trajectory clustering model for parallel roads to separate trajectories on two different parallel roads. This model applies feature $F = \{F_V, F_R, F_P\}$ to the original trajectory dataset D and computes a score for each trajectory by applying the selected feature. Using the transformed score, k-means clustering is applied to generate sets of trajectories on different roads. We use k-means clustering for free trajectories to discover different underlying roads. For fixed trajectories, we treat bus trajectories on the same route as the same cluster, the average link Agglomerative hierarchical clustering method [18] was then used to merge the clusters from bottom to top and then iterating the process until two clusters are left. Our method can incorporate the connection part of road points or segments from any road generation method, without connecting road points or segments from different trajectory clusters, to reduce incorrect connections and redundant roads of parallel roads.

4 Evaluation

4.1 Data Set and Ground Truth

As mentioned earlier in this paper, our map generation process utilizes data generated by Shenzhen floating vehicles equipped with BeiDou GPS devices. Our GPS dataset covers 10 administrative districts and 626 different buslines in Shenzhen, from which we selected 100 regions of approximately 1.5 km x 2 km with parallel roads, as well as 100 regions without parallel roads. On average, each region contains 428 trajectories. For all datasets, we conducted stratified holdout evaluation experiments, using half of the data for training and the other half for validation.

In this paper, trajectories include two types of vehicle tracks: taxi tr_t and bus tr_b. A taxi GPS point is represented as a quadruple (id, t, lon, lat). A bus GPS point is represented as a quintuple $(id, buslineid, t, lon, lat)$. Both types of GPS point sequences are organized in chronological order.

Parallel roads are the distances between the centerlines of two or more roads in a road network, which are maintained within a short distance (set to 20 m–70 m in our work)and sustained for a certain length (set to over 1 km in our work).

To validate the accuracy of the generated maps, we used existing road maps, specifically OpenStreetMap, as the ground truth.

4.2 Metrics

For the classification results of the model, we report the accuracy (ACC), recall (Recall), and F1 score (F1-Score), which are commonly used metrics in classification problems. The F1-score is defined as the harmonic mean of ACC and Recall and ranges from 0 to 1, calculated as:

$$F1 - score = \frac{2 * ACC * Recall}{ACC + Recall}$$

For the generation results of the model, we apply the evaluation method from [2] to evaluate the inference of mapping topology and geometric structure. Given the generated map G and the ground truth map GT from a snapshot of the real world in the OpenStreetMap (OSM), we randomly select n starting points and perform DFS on both G and GT, calculating Precision, Recall, and F1-score.

4.3 Baselines

Our approach can be used as a plugin for any GPS trajectory-based road network generation method. In this paper, we apply our proposed method to three representative map generation methods and compare their results.

- Kharita [16] is a point-based clustering algorithm that initially selects initial cluster centroids. Then, Kharita executes an edge generation step, where MPRG is incorporated. The output of the point clustering phase is a set of cluster centroids. In the clustering output of the MPRG method, edges are added between centroids if they belong to the same trajectory cluster. Subsequently, a refinement step is executed.
- Biagioni [2] involves smoothing the map using a kernel density estimator and generating a centerline through a skeletonization process. It further performs post-processing to eliminate redundant roads. Based on this, we integrate MPRG into the post-processing phase. We transform trajectory clusters into cell clusters, remove edges between different cell clusters, reducing spurious road segments between parallel road sections.
- Karagiorgou [8] identifies locations where vehicles make turns in trajectories and clusters them together firstly. In the second step, connectivity between intersections is derived from trajectories, and the intersections are connected. We add edges only between intersections within the same cluster cell in the MPRG output. The process then proceeds to the third step, where redundant links between points are cleaned and merged to produce the final output.

4.4 Evaluation Results

Classification Results. We conducted tests in 100 road areas in Shenzhen to evaluate the existence of parallel roads using MPRG. In order to better demonstrate the performance of the method, we compared the classification results of MPRG with a method that solely uses trajectory density peaks. As shown in Table 1, the results showed that except for the case where neither of the two roads included bus trajectories, the method based on multiple types of vehicles significantly improved the identification of parallel roads. MPRG performed best in scenarios where both roads had bus routes, as the different fixed trajectories on the two roads provided highly discriminative features. In contrast, in scenarios where only one road had fixed trajectories, the features that could be provided by the road with only free trajectories were limited. In addition, the method that solely uses density peaks performed poorly in scenarios where only one road had fixed trajectories. This is because in this scenario, a portion of the fixed trajectories occurred on the main road and the auxiliary road. In this case, the distance between the two parallel roads is much smaller than in other scenarios, making it difficult to distinguish between them.

Quantitative Comparison. We report three baseline solutions and the F1-score of our proposed plugin in Fig. 6. To demonstrate the robustness of our work,

Table 1. The performance of MPRG on Kharita in three scenarios with different trajectory distributions.

Classification Accuracy	Metric	No bus on two Parallel Roads	Bus on one Parallel Roads	Bus on two Parallel Roads	All
Peak Density	ACC	0.833	0.646	0.750	0.710
	Recall	0.667	0.583	0.700	0.640
	F1-score	0.741	0.613	0.724	0.673
MPRG	ACC	0.833	0.917	0.975	0.930
	Recall	0.667	0.875	0.950	0.880
	F1-score	0.741	0.896	0.962	0.904

we reduced the sampling rate of the trajectory data for evaluation. The results show that our plugin consistently outperforms the baselines in all three types of baseline methods. At higher sampling rates, our plugin achieves improvements of 0.095, 0.038, and 0.052 compared to the best baseline algorithms, respectively. As the sampling rate decreases, the existing methods fail to handle the parallel road structure effectively, while our plugin maintains its performance due to the original baselines.

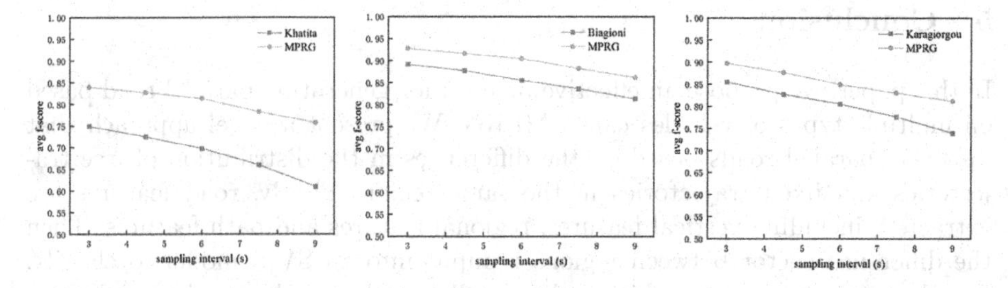

Fig. 6. F1-scores for different map inference solutions. MPRG produces significantly more accurate results.

Visual Comparison. Figure 7 presents the outputs of three mapping algorithms on a region in Shenzhen that contains parallel roads. This region includes two parallel roads in close proximity due to the presence of an elevated bridge, making it one of the most challenging cases for differentiating parallel roads. The corresponding trajectories and ground truth map of the baseline methods are depicted in the figure. The results indicate that the Kharita produces a significant number of redundant edges between parallel roads, explaining its poor scoring performance. Biagioni exhibits the least performance degradation with decreasing sampling rates, but it tends to merge two parallel roads when dealing with parallel road scenarios. Karagiorgou identifies the two parallel roads as one.

MPRG successfully discovers and distinguishes parallel roads on all three baseline methods, without generating excessive redundant edges, which determines the higher quality of the generated map compared to the baselines.

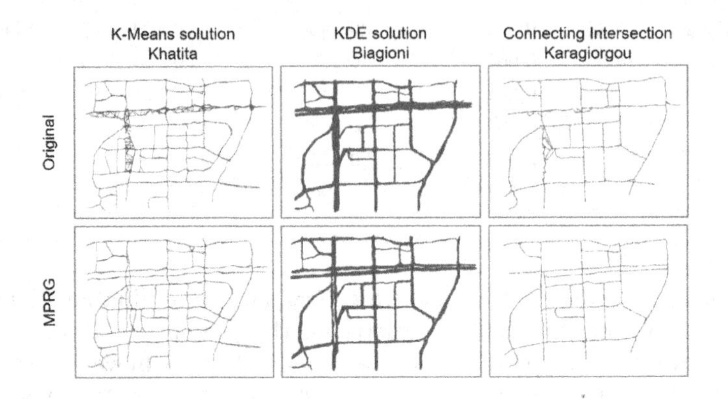

Fig. 7. MPRG promotes the performance of map generation.

5 Conclusion

In this paper, we propose an effective method for generating parallel road based on multiple types of vehicles called MPRG. We present a novel approach that discovers parallel roads based on the differences in the distribution of free trajectories and fixed trajectories in the same region. Firstly, road features are extracted, including vertical features, regional features and path features. Then the difference vector between regions is input into an SVM model to classify. Based on this, trajectories within the parallel roads are clustered according to the aforementioned features to improve the road generation algorithm on parallel roads. Extensive experiments are conducted on real-world GPS data, and MPRG is evaluated with three road generation methods. We demonstrate that MPRG consistently outperforms the baselines.

Acknowledgement. This study was funded by the National Natural Science Foundation of China (No. 62372443, No. 62376263), Shenzhen Industrial Application Projects of undertaking the National key R & D Program of China (No. CJGJZD20210408091600002).

References

1. Biagioni, J., Eriksson, J.: Inferring road maps from global positioning system traces: survey and comparative evaluation. TRR-JTRB **2291**(1), 61–71 (2012)

2. Chao, P., Hua, W., Mao, R., et al.: A survey and quantitative study on map inference algorithms from GPS trajectories. IEEE Trans. Knowl. Data Eng. **34**(1), 15–28 (2020). https://doi.org/10.1109/TKDE.2020.2977034
3. Edelkamp, S., Schrödl, S.: Route planning and map inference with global positioning traces. In: Klein, R., Six, H.W., Wegner, L. (eds.) Computer Science in Perspective. LNCS, vol. 2598, pp. 128–151. Springer, Heidelberg (2003). https://doi.org/10.1007/3-540-36477-3_10
4. Li, J., et al.: An automatic extraction method of coach operation information from historical trajectory data. J. Adv. Transp. (2019)
5. Guo, Y., Li, B., Lu, Z., Zhou, J.: A novel method for road network mining from floating car data. Geo-Spat. Inf. Sci. **25**, 197–211 (2022)
6. Jiang, Y., Li, X., Li, X., Sun, J.: Geometrical characteristics extraction and accuracy analysis of road network based on vehicle trajectory data. J. Geo-inf. Sci. **14**(2), 165–170 (2012)
7. Ahmed, M., Karagiorgou, S., Pfoser, D., Wenk, C.: A comparison and evaluation of map construction algorithms using vehicle tracking data. GeoInformatica **19**(3), 601–632 (2015)
8. Karagiorgou, S., Pfoser, D., Skoutas, D.: Segmentationbased road network construction. In: Proceedings of the 21st ACM SIGSPATIAL International Conference on Advances in Geographic Information Systems, pp. 460–463. ACM (2013)
9. Chen, C., Lu, C., Huang, Q., Yang, Q., Gunopulos, D., Guibas, L.: City-scale map creation and updating using GPS collections. In: Proceedings of the 22nd ACM SIGKDD International Conference on Knowledge Discovery and Data Mining, pp. 1465–1474. ACM (2016)
10. Chen, C., Cheng, Y.: Roads digital map generation with multi-track GPS data. In: International Workshop on Geoscience and Remote Sensing (2008)
11. Wang, Y., et al.: Regularity and conformity: Location prediction using heterogeneous mobility data. In: KDD 2015, pp. 1275–1284. ACM (2015)
12. Katsikouli, P., Sarkar, R., Gao, J.: Persistence based online signal and trajectory simplication for mobile devices. In: Proceedings of the 22nd ACM SIGSPATIAL International Conference on Advances in Geographic Information Systems, pp. 371–380. ACM (2014)
13. Davies, J.J., Beresford, A.R., Hopper, A.: Scalable, distributed, real-time map generation. IEEE Pervasive Comput. **5**(4), 47–54 (2006)
14. Goodman, N.R.: Statistical analysis based on a certain multivariate complex gaussian distribution (an introduction). Ann. Math. Stat. **34**(1), 152–177 (1963)
15. Stanojevic, R., Abbar, S., Thirumuruganathan, S., Chawla, S., Filali, F., Aleimat, A.: Robust road map inference through network alignment of trajectories. In: ICDM, pp. 135–143. SIAM (2018)
16. Mnih, V., Hinton, G.E.: Learning to detect roads in high-resolution aerial images. In: Daniilidis, K., Maragos, P., Paragios, N. (eds.) ECCV 2010. LNCS, vol. 6316, pp. 210–223. Springer, Heidelberg (2010). https://doi.org/10.1007/978-3-642-15567-3_16
17. Máttyus, G., Luo, W., Urtasun, R.: Deeproadmapper: extracting road topology from aerial images. In: International Conference on Computer Vision, vol. 2 (2017)
18. Miller, H.J., Han, J.: Geographic Data Mining and Knowledge Discovery, 2nd edn. Taylor & Francis Group, London (2009)

GSPM: An Early Detection Approach to Sudden Abnormal Large Outflow in a Metro System

Li Sun[1,2], Juanjuan Zhao[1(✉)], Fan Zhang[1], and Kejiang Ye[1]

[1] Shenzhen Institute of Advanced Technology, Chinese Academy of Sciences, Shenzhen, China
jj.zhao@siat.ac.cn

[2] University of Chinese Academy of Sciences, Beijing, China

Abstract. Early detection of Sudden Abnormal Large Outflow (SALO) aims to determine abnormal large outflows and locate the station where real-time outflow significantly exceeds expectations. SALO serves as a crucial indicator for city administration to identify emerging crowd gathering events as early as possible. Existing solutions can't work well for SALO prediction due to the lack of modeling the dynamic gathering trend of passenger flows in SALO instances, characterized by strong randomness and low probability. In this paper, we propose a novel Gathering Score based Prediction Method, called GSPM, for SALO prediction. GSPM introduces a gathering score to quantify the dynamic gathering trend of abnormal online flows, limits the SALO location to a few candidate stations, and locates it using a utility-theory-based model. This method is built on key data-driven insights, such as obvious increases in online flows before SALO occurrences, and passengers are more inclined to gather near stations. We evaluate GSPM with extensive experiments based on smart card data collected by Automatic Fare Collection system over two years. The results demonstrate that GSPM surpasses the results of state-of-the-art baselines.

Keywords: Crowd Gathering · Metro Network · Flow Prediction

1 Introduction

Large crowd gatherings result in a significantly higher number of people arriving at specific confined areas in a short time interval. Common causes include major social events like concerts, celebrations, and exhibitions. The huge amount of crowd in a short time interval may bring great challenges to public security. A recent example is the Stampede in Itaewon, South Korea, on the night of October 29, 2022, where at least 159 people were killed and 196 others were injured. Detecting crowd gatherings early is crucial for city administration and transportation agencies to design effective transportation schedules and ensure better crowd organization. Modern metros have become crucial components of

D.-N. Yang et al. (Eds.): PAKDD 2024, LNAI 14649, pp. 322–335, 2024.
https://doi.org/10.1007/978-981-97-2262-4_26

urban traffic. The variation of passenger demand can reflect that of whole city to a large extent. Automatic Fare Collection (AFC) systems, widely adopted in modern metros, not only serve the original purpose of ticketing but also generate a large volume of transaction data with high penetration rates, which can be utilized for crowd gathering forecasts in the metro system.

This paper focuses on forecasting large crowd gatherings by early detection of SALO in a metro system. Here, SALO refers to the situation where the real-time observed outflow of a metro station in a short time interval far exceeds the expected value (as shown in Fig. 1). SALO serves as the initial and significant indication of a large crowd gathering, but it is highly unpredictable due to the complex gathering trend in passenger flows gathering. Generally, the expected outflows are calculated in two ways [11,12]: (1) The average of historical outflows at the corresponding time interval in a day. (2) The outflow from the most recent time interval. However, predicting SALO instances, which exhibit distinctive variation patterns (as shown in Fig. 1) and typically occur at only a few stations during a target time slot [7], is challenging. While traffic flow prediction has been extensively studied [11,12,25], existing approaches, which primarily address general cases with clear periodicity and temporal correlation, are not suitable for SALO prediction. Recent work on abnormal traffic flow warnings [21,22] provides a warning signal only when there is a significant abnormal increase in outflow volume that reaches a predefined threshold. Additionally, Zhou *et al.* [28] locate crowd gathering using a gathering directed acyclic graph built by taxi GPS, but this approach is not applicable to our sparse observed trajectories, which only consist of the start and end information of each trip.

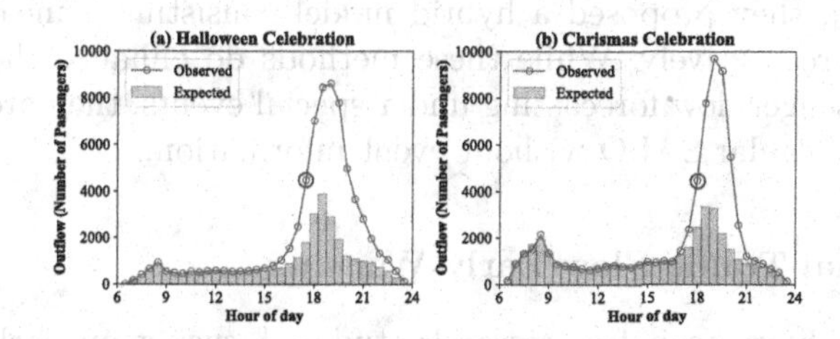

Fig. 1. Some cases of SALO. The SALO is labeled by a blue circle.

Overall, by addressing the lack of modeling the complex dynamic gathering trend of online passengers, we aim to enhance the accuracy of SALO detection. To overcome limitations in prior studies, we conduct a comprehensive analysis of long-term smart card transaction data, obtaining key insights: (1) Significant deviation between observed online flows and expected values occurs before SALO. (2) Passengers exhibit a preference for nearby stations compared to normal instances. (3) SALO occurrence probability is linked to station features such as remoteness and historical SALO frequency. Based on these findings, we propose a SALO detection method, called GSPM. The primary contributions of this work include: (1) A two-step prediction approach surpassing state-of-the-art baselines; (2) An effective gathering score measurement for quantifying the complex dynamic abnormal online flow gathering trend; (3) A utility-theory-based model for locating SALO station using specific features of each station.

2 Related Work

In this section, we briefly discuss two research areas closely related to this work: traffic flow prediction and early-warning model for abnormal traffic flow.

2.1 Traffic Flow Prediction

Numerous effective models have been proposed for short-term traffic flow prediction in prior research. Such as ARIMA [20], Support Vector Machine [8], Kalman Filtering [4], Artificial Neural Network [14], Gradient Boosted Machine [26]. With the rapid development of deep learning recently, recurrent neural network [27], long short-term memory (LSTM) [5,13], Graph Convolutional Network [1] and Graph Neural Network [18] have been applied to railway passenger flow forecasting.

In summary, current approaches consider strong spatial-temporal correlations in urban traffic flow, relying on a consistency model assuming that all past traffic flows influence future flows at all stations. Consequently, their fail to capture the dynamic gathering trend of abnormal online passengers, therefore, cannot predict SALO.

In recent years, some studies have focused on passenger flow prediction under events [2,23,24], they proposed a hybrid model consisting a mean part and a volatility part, respectively. While these methods do enhance the accuracy of short-term passenger flow forecasting under special events, they are not suitable for predicting irregular SALO without event information.

2.2 Abnormal Traffic Flow Early-Warning

Recently, there have been few research studies focusing on early warning of abnormal traffic flow. Wang *et al.* [22] used an ESWD algorithm to detect large crowd gatherings. Huang *et al.* [7] leveraged a concept of anomalous mobility networks to identify crowd gatherings. Li *et al.* [11] proposed a multi-scale radial basis function network to predict irregular alighting ridership before a concert. Huang *et al.* [6] proposed an optimized SAX algorithm to identify historical sudden passenger flows under large-scale events. Overall, these methods are designed for abnormal flow detection rather than forecasting the location.

3 Model Overview

In this section, we introduce definitions and the problem formulation, provide data-driven insights, and illustrate the method framework.

3.1 Definitions and Problem Formulation

Definition 1 (*Trip*). A trip tr, is related to four attributes s_i, s_o, t_i, t_o, representing entrance station, exit station, entrance time and exit time.

Definition 2 (*Time Interval*). The daily operation time of a metro system is divided into a fixed number of time intervals $\{T_1, T_2, \dots T_{|T|}\}$ with equal length of τ minutes. T_t contains a time range of $((t-1)\cdot\tau, t\cdot\tau]$.

Definition 3 (*Inflow/Outflow*). We use inflow $I_{i,t}$ (outflow $O_{i,t}$) to denote the amount of passengers who entering (exiting) metro system from station l_i at the time interval T_t. They are calculated by the Eq. 1 and Eq. 2, where $|.|$ is used to calculate the amount of items in the corresponding collection.

$$I_{i,t} = |\{tr|tr.s_i = l_i, tr.t_i \in T_t\}|. \tag{1}$$

$$O_{i,t} = |\{tr|tr.s_o = l_i, tr.t_o \in T_t\}|. \tag{2}$$

Definition 4 (*OD flow*). We use OD flow $C_{i,j,t',t}$ to represent the amount of passengers who enter metro system from station l_i at $T_{t'}$, and exit at station l_j at T_t (where $(t \geq t')$. N is the maximum number of time intervals taken between any two stations. $C_{i,j,t',t}$ is calculated by Eq. 3.

$$C_{i,j,t',t}{=}|\{tr|tr.s_i{=}l_i, tr.t_i{\in}T_{t'}, tr.s_o{=}l_j, tr.t_o{\in}T_t\}|. \tag{3}$$

Definition 5 (*Online flow*). Given $T_{t'}$ and T_t with $t \geq t'$ in a day, we use online flow $A_{i,t',t}$, calculated by Eq. 4, to represent the amount of passengers who have entered from station l_i at $T_{t'}$ and have not exit before T_t.

$$A_{i,t',t} = I_{i,t'} - \sum_{m=t'}^{t-1}\sum_{j=1}^{|S|} C_{i,j,t',m}. \tag{4}$$

Definition 6 (*SALO*). Given two predefined thresholds Ω_1 and Ω_2, we label the $O_{i,t}$ as SALO, if it satisfies: i) the difference of the $O_{i,t}$ with corresponding history average at same time interval $(\hat{O}_{j,t})$ is bigger than Ω_1, that is $\Delta_v = O_{j,t} - \hat{O}_{j,t} > \Omega_1$, and ii) the difference of the $O_{i,t}$ with that of previous time interval is bigger than Ω_2, that is $\Delta_p = O_{j,t} - O_{j,t-1} > \Omega_2$.

Problem Formulation: Given $I_{i,t}$, $O_{i,t}$, $C_{i,j,t',t}$,$A_{i,t,n}$, where $t' \leq t \leq n-1$, we aims to detect SALO at T_n and then locate the SALO station.

3.2 Data-Driven Insights

We first divide all outflow instances into two categories, SALO set O_o and normal set O_n, then understand SALO as follow:

Difference of Passenger Flow: (1) *Past outflows of SALO station*: In general, outflows have had a small increase relative to expected values before SALO as shown in Fig. 1. In the following, we dynamically extract the outflow sequence of a station l_i before time interval T_n, denoted as $\widetilde{O}_{i,n}$, where the item $O_{i,n-k}$ satisfy the condition that the deviation $O_{i,n-k} - \hat{O}_{i,n-k} > 0$. And we calculate the cumulative outflow deviation, denoted as $VO_{i,n}$, as a feature for SALO localization. (2) *Recent Online flow*: Fig. 2 shows the deviation of online flows before SALO instances. We can observe the obvious increases (the positive deviation) of online flows originating from most stations.

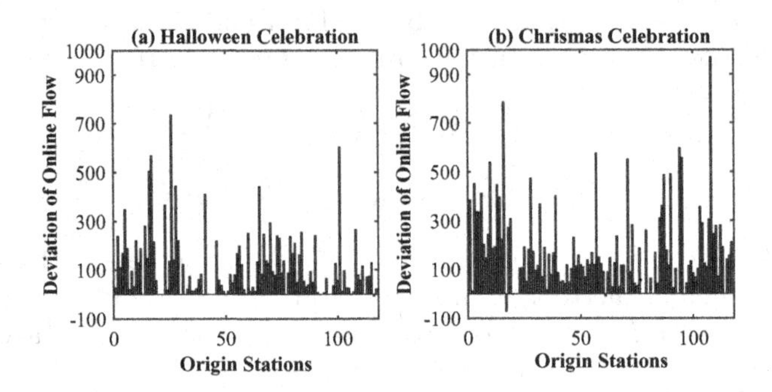

Fig. 2. The origin distribution of online flow deviations for two cases of SALO.

Gathering Trend: (1) *OD flow distribution*: The correlation coefficient of OD flow deviations (i.e., the difference of observed OD flows involved in SALO with baselines) and baseline flows is 0.560. This is consistent with the common sense that passengers are more inclined to gather at familiar locations. (2) *Time cost*: We calculate the average travel time of trips involved in SALO and that of normal instances, resulting in 29.60 min and 22.00 min, respectively. The average travel time for SALO is less than that for normal instances, illustrating that passengers are more inclined to gather at nearby stations. That is easy to understand according to the Tobler's first law of geography [15]: everything is related to everything else, but near things are more related to each other. In the following, we use $C_{i,j}$ to denote the average travel time between l_i and l_j.

Characteristic of SALOstation: (1) *SALO remoteness*: The SALO frequency of all stations in metro network are shown in Fig. 3(a), which are larger on central areas than suburbs. This result motivates us to extract the regional characteristic of each station [19], e.g., the remoteness of each station, denoted as R_i, calculated by Eq. 5. The corresponding geographic distribution of remoteness of all stations is represented in Fig. 3(b). We can observe the obvious agreement between the geographic distribution of remoteness and the SALO frequency. (2) *SALO frequency*: The SALO instances are unevenly distributed in geography.

Generally, a higher frequency of past SALO occurrences (denoted as F_i) correlates with an increased likelihood of SALO events in the future.

$$R_i = \frac{\sum_{tr \in TR, tr.s_o \neq s_j, tr.s_d = s_i} (tr.t_d - tr.t_o)}{|\{tr | tr \in Tr_{all}, tr.s_o \neq s_i, tr.s_d = s_i\}|}. \tag{5}$$

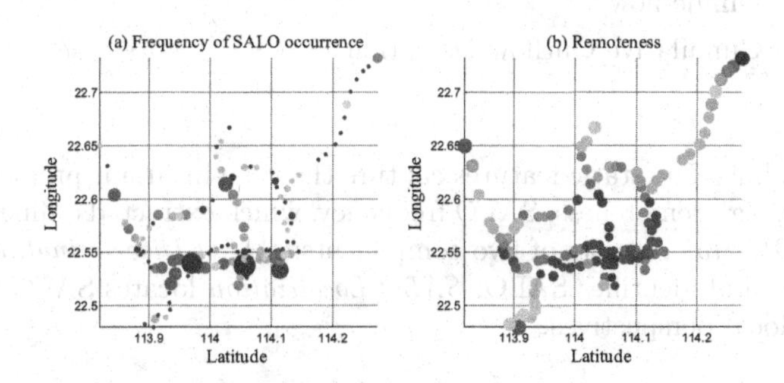

Fig. 3. (a) Geographic distribution of SALO frequency F_i. (b) Geographic distribution of station remoteness R_i.

3.3 Framework

Based on the above analysis, our solution for SALO forecast is that: the abnormal large-scale online flow deviation can be considered as the source of SALO, which is defined as critical flows. The characteristic of gathering trend in normal cases can be used to qualify the gathering trend of critical flows, and limit the gathering destinations. Finally, the SALO-related characteristic of stations can be used to further locate the SALO. Therefore, we build a method, called GSPM, as shown in Fig. 4, which consists of *Feature Extractor* and *SALO Detector*.

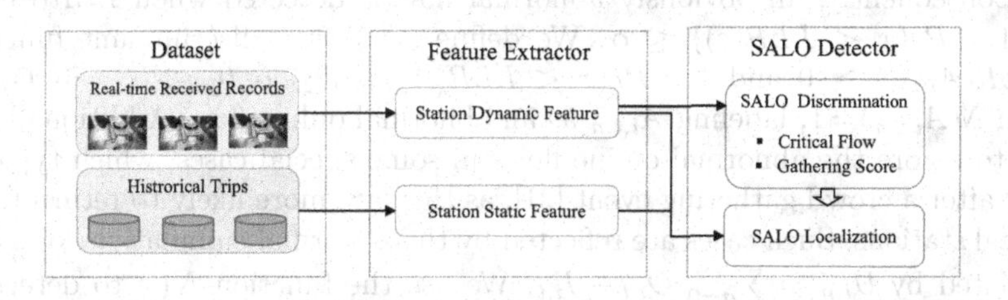

Fig. 4. Framework of the predictive model.

Feature Extractor extracts two categories of SALO related features: *Dynamic features* and *Static features*. The dynamic features capture the dynamics of traffic flows at different stations over time, e.g., inflow, outflow, OD flow, which

Table 1. SALO-related Features of each Station

Dynamic Features	Static Features
Inflow	Average Travel Time
Outflow	Remoteness
OD flow	SALO Frequency
Online-flow	
Cumulative Outflow Deviation	

extracted online. The static features capture the time-invariant profiles of different stations, e.g., remoteness, SALO frequency, which extracted offline (Table 1).

SALO Detector consists of two components: *SALO Discrimination* extracts critical flows and identifies SALO. *SALO Localization* locates SALO from both global and local perspectives.

4 SALODetector

4.1 SALO Discrimination

Critical Flow Extraction: This section aims to extract critical flow based on the abnormal large-scale observed online flows.

Since online flows can be well approximated by a Poisson distribution [16,25, 28] with a mean of $\hat{A}_{i,t,n}$, we use Eq. 6 to calculate the abnormal online flows:

$$LLR(e) = \begin{cases} C_e \cdot \log(\frac{C_e}{B_e}) + (B_e - C_e) & \text{if } C_e \geq B_e, \\ 0 & otherwise, \end{cases} \tag{6}$$

where C_e is the observed flow, and B_e is the baseline flow at the same time interval. A higher $LLR(e)$ indicates a greater level of abnormality. Therefore, we use the chi-square test to analyze the significance level of the anomaly [17, 28]. Consequently, an obviously abnormal flow is detected when $LLR(e) > 0$ and $1 - Pr(x < LLR(e)) \leq \alpha$. We define $N(.)$ as a discriminant function. If $LLR(A_{i,t,n}) > 0$ and $1 - Pr(x < LLR(A_{i,t,n})) \leq \alpha$ (e.g., $\alpha = 0.05$), we assign $N(A_{i,t,n})=1$, labeling $A_{i,t,n}$ as an abnormal online flow. Additionally, we need to ignore the abnormal online flows in some special cases, which typically occur after a crowd gathering event [21], as they are more likely to return to the original stations. Such cases are reflected by the abnormal cumulative flows $U_{i,n}$, calculated by $U_{i,n} = \sum_{t=0}^{n-1} O_{i,t} - I_{i,t}$. We use the function $N(.)$ to determine the abnormality of the cumulative flow, as it can also be well approximated by a Poisson distribution. Therefore, given $R_{i,t,n}$, we determine whether it is a critical flow by checking that $N(A_{i,t,n})=1$ and $N(U_{i,n})=0$. *Definition 8* (*Critical flow*). The deviation flow between observed online flow $A_{i,t,n}$ and corresponding baseline flow $\hat{A}_{i,t,n}$, calculated by $R_{i,t,n} = (A_{i,t,n} - \hat{A}_{i,t,n})$, is a critical flow only if $N(A_{i,t,n})=1$ and $N(U_{i,n})=0$.

Gathering Score Calculation: In this section, we aim to quantify the gathering trend of all critical flows to each destination station. We use a metric $\phi_{j,n}$, called *gathering score* to measure the gathering trend of all critical flows to each destination station l_j during time interval T_n.

$$\phi_{j,n} = \sum_{t=n-N}^{t=n-1} \sum_{s_i \in S \cap s_i \neq s_j} \phi^j_{i,t,n}, \tag{7}$$

where $\phi^j_{i,t,n}$ is the gathering score of station l_j gathered from the critical flows across all stations l_i entering at time T_t. The $\phi^j_{i,t,n}$ is calculated by:

$$\phi^j_{i,t,n} = N(R_{i,t,n}) \cdot R_{i,t,n} \cdot f(\gamma^j_{i,t,n}, C_{i,j}), \tag{8}$$

$$f(\gamma^j_{i,t,n}, C_{i,j}) = \gamma^j_{i,t,n} \cdot (1 + e^{-C_{i,j} \cdot \mu}). \tag{9}$$

The function $f(.)$ is employed to model the relationship between the gathering score and the critical flow, combining OD flow distribution and average travel cost between stations. The $\gamma^j_{i,t,n}$ represents the average proportion of online passengers $A_{i,t,n}$ gathering to the station s_j during the predicted time interval T_n. Equation 8 is utilized to capture the influence of travel time. The parameter μ is employed to control the extent of the decreasing gathering trend as travel cost increases. We set μ to a value that minimizes the average rank of SALO stations based on gathering scores in descending order.

SALO Identification: To identify a SALO, we initially calculate the relative increase, $\Delta\phi_{j,n}$, for each station, calculated as $\Delta\phi_{j,n} = \phi_{j,n} - \phi_{j,n-1}$. Subsequently, we assess the SALO occurrence probability of each destination station by comparing the $\Delta\phi_{j,n}$ with a predefined threshold, Φ. If no station satisfies $\Delta\phi_{j,n} > \Phi$, there is no SALO in the target time interval. Otherwise, a SALO is detected. The primary challenge in this step lies in determining the value of Φ.

To calculate Φ, we extract its practical value range based on two sets: SALO set O_o and normal set O_n. The distributions of $\Delta\phi$ in the two datasets are then calculated. Generally, the average gathering scores in O_o are greater than those in O_n. We determine the range of Φ that maximizes the coverage rate of SALO instances compared to normal instances. Subsequently, we select a value based on *Recall*, *Precision*, and F1-score (details in Sect. 4.3).

4.2 SALO Localization

In this section, we select a station with the highest occurrence probability of SALO at the predicted time interval based on the gathering scores and the station features, considering both global and local perspectives.

From Global Perspective: We select the top K stations with the maximum gathering scores as the candidate locations, denoted as $L_c = \{l_{c1}, l_{c2}, \ldots, l_{cK}\}$. The value of K is chosen to meet the condition: $1 - Pr(X < K) < \alpha$, where $\alpha = 0.05$, for instance.

From Local Perspective: We select a station with the highest SALO occurrence probability from the set L_c. This task can be considered as a classification problem, characterized by diverse input features and a dynamic output class. To address the SALO localization issue, we chose a utility-theorem-based model. Consequently, a station's SALO occurrence probability is intricately connected to its utility. The utility value U_{ck} for a given station l_{ck} with feature F_c is determined by the function $U_{ck} = f_U(F_{ck})$, and the SALO occurrence probability is calculated using Eq. (10).

$$Pr(l_{ck}) = \frac{e^{U_{ck}}}{\sum_{l_{ci} \in L_c} e^{U_{ci}}}. \tag{10}$$

The multinomial logit model is commonly used for modeling discrete choice problem. However, it proves insufficient for locating SALO due to its neglect of the nonlinear relationship between features. Instead, we choose a nonlinear utility function represented by a neural network and embedded into the classical logit function: $U_j = uti(F_j; \Theta)$, where Θ represents the training parameters. It's worth noting that the input features in the score function uti vary across different stations, while the parameters Θ are shared among all stations in L_c. Given the limited number of SALO instances, we apply a two-layer convolutional neural networks to the input features, providing each station with the same utility functions. Then, a softmax layer ensures that the output satisfies $\sum Pr(l_c) = 1$. The station with the highest probability is considered the SALO station.

4.3 Experiment Setting

Dataset: The dataset we use covers a two-year period, from January 1, 2014, to December 31, 2015, comprising over 29.5 billion transaction records. In the ShenZhen metro system, daily operations run from 06:30 AM to 23:59 PM. We divide the operation time into 30-minute intervals, resulting in 35 intervals. For baseline passenger flows, we average corresponding flows from the previous month at the same time interval in a day, separately calculating for weekends and weekdays due to different traffic patterns. We set Ω_1 and Ω_2 to 1500 and 1000 based on Shenzhen Metro experiments. Consequently, we categorize all outflows into two sets: the SALO set $O+$ and the normal set $O-$. In our dataset, 36 days have missing data, resulting in a collection of 2,866,220 outflow samples from 118 stations. We identified a total of 76 SALO instances. We partition the data into non-overlapped training set and test data set by ratio of 1:1. To avoid to get biased results, we use the data of the first and third weeks of every month as training set, the other data as test data set to guarantee the two data sets distributed in all months of years.

Evaluation Metrics: We predict SALO in a metro system during the next time interval. We use *Recall*, *Precision* and F1-score to measure the prediction results.

4.4 Parameter Effect

In the SALO Discrimination, we utilize two parameters: μ and Φ. μ controls the penalty factors of travel time, while Φ serves as the threshold for identifying SALO and determining its location. We conducted multiple experiments with different values and selected the optimal configuration.

Parameter μ: We uses the function in Eq. 8 to capture the influence of the travel time on passengers' gathering trend, the result is presented in Fig. 5(a). Consequently, we set μ to 4, achieving the best effect where the average ranking of SALO stations is lower than five. Furthermore, the improved results with $\mu = 4$ compared to $\mu = 0$ confirm the insight that users tend to gather at nearby stations in SALO instances compared to normal ones.

Parameter Φ: In our data, the gathering scores of SALO instances range from 90 to 500. We evaluate the effects of Φ on SALO prediction by comparing the *Recall*, *Precision* and F1-score under different values of Φ. In our application, a false negative instance is far more harmful than a false positive instance. Therefore, we select the Φ value that yields a high F1-score under the condition that *Recall* is greater than an acceptable value (e.g., 0.6). The results are represented in Fig. 5(b). We can observe that when $\Phi = 170$, we get the best F1-Score 0.718 with *Precision* 0.642 and *Recall* 0.814.

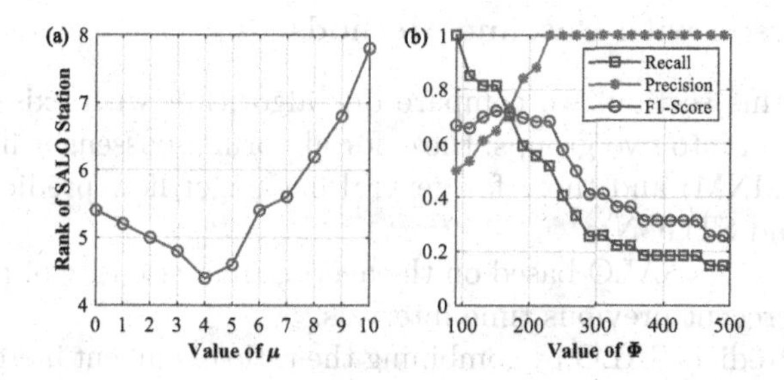

Fig. 5. (a) Effects of parameter μ. (b) Effects of parameter Φ.

4.5 Case Study

In this section, we use two SALO cases to illustrate the process of our method.

Cases 1: On October 31, 2014 at 6:00~9:00 PM, there was a big celebration activity for Halloween held in the Window of World & Happy Valley of Shenzhen, China. There are more than 50,000 passengers exiting from the Window of the World station.

Cases 2: On December 24, 2014 at 6:00~8:00 PM, there was a big activity held in Shenzhen North station for celebrating Christmas. There are more than 30,000 passengers participating the snow watching party.

We calculate the gathering scores of these two stations during the identified SALO interval, as illustrated in Fig. 6. Although SALO stations generally exhibit higher gathering scores, they don't always have the maximum values; some other stations may have higher scores. To address this, we select the top 5 stations (covering 95% of SALO) with the largest gathering scores and employ utility theorem-based models for further SALO localization.

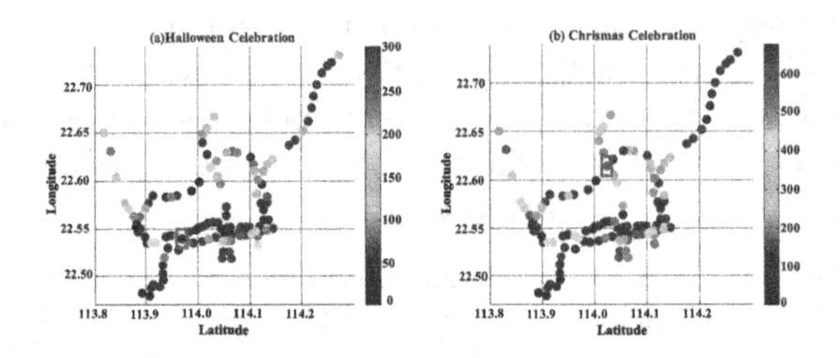

Fig. 6. Gathering scores of two cases of SALO. The gathering scores of all stations are represented by graduated colour, from green (smallest) to red (largest). (Color figure online)

4.6 Comparison with Baseline Methods

Baseline: In this section, we compare our algorithm with existing methods, categorizing them into two groups: those for abnormal passenger flow prediction (PAOM and AMNM) and those for general passenger flow prediction (ARMA, GBM, GRU, and STGCN).

PAOM: Predicts SALO based on the average increase rate of past abnormal outflow during recent previous time intervals.

AMNM: Predicts SALO by combining the ratio of current increase in inflows at origin stations in a anomalous mobility network [7] to that of recent previous time interval, with the increase in outflow at recent previous time interval.

ARMA (Auto-Regression-Moving-Average): Predicts the outflow of a station during a time interval based on the outflows of recent previous time intervals for that station.

GBM: Gradient boosted machine is gradient boosting decision tree based regression method implemented using Light-GBM [9].

GRU: Gated Recurrent Unit network [3] is a simplified version of LSTM for mining patterns in time series but is less prone to overfitting.

STGCN: A spatial-temporal graph convolution model based on the spatial method [10], successfully generalizes CNN to graph-structured data.

Experiment Result: In Table 2, our method exhibits the highest recall (0.814) and F1-score (0.718) with relatively high precision (0.642) compared to other methods. Specifically, methods in the first category outperform those in the second category for SALO prediction. Within the first category, while PAOM is relatively good, it only covers limited SALO instances. AMNM is better than PAOM, considering both outflows and the current inflows, although it doesn't model real-time traffic flow gathering trends. In the second category, ARIMA performs the worst due to its inability to extract complex non-linearity in metro data. On the other hand, GBM, GRU, and STGCN achieve much better performances, indicating their ability to capture non-linear correlations between current passenger flow and its historical observations. GRU outperforms GBM, highlighting a long temporal dependency in traffic flows. STGCN, focusing on modeling spatio-temporal dependencies, achieves the best performance, demonstrating the spatio-temporal dependencies of traffic flows in the metro system.

Table 2. Comparison with baseline methods

Model	Recall	Precision	F1-score
PAOM	0.412	0.668	0.510
AMNM	0.441	0.684	0.536
ARIMA	0.214	0.571	0.311
GBM	0.242	0.593	0.344
GRU	0.325	0.611	0.424
STGCN	0.364	0.609	0.456
GSPM	**0.814**	0.642	**0.718**

5 Conclusion

For public safety and transportation management, predicting SALO is crucial. However, it is a challenging problem due to its strong randomness, low probability based on limitation of available data. In this paper, we propose a two-step method driven by data insights, called GSPM. GSPM first extracts the dynamic critical flows as the initial source of SALO, quantifies their gathering trend to each destination station, and limits the gathering destination to a few stations. Finally, station characteristics (e.g., previous outflow, remoteness, SALO frequency) are combined to locate the SALO based on a utility theory-based model. We validate the effectiveness of GSPM by conducting extensive experiments based on smart card transaction data collected by AFC system over two years in Shenzhen, China. The evaluation results demonstrate that GSPM surpasses the results of state-of-the-art baselines.

Acknowledgement. This study was funded by the National Key R&D Program of China (No. 2023YFC3321600), National Natural Science Foundation of China (No. 62372443, No. 62376263), Shenzhen Industrial Application Projects (No. CJGJZD20210408091600002).

References

1. Ali, A., et al.: Exploiting dynamic spatio-temporal graph convolutional neural networks for citywide traffic flows prediction. Neural Netw. **145**, 233–247 (2022)
2. Chen, E., et al.: Subway passenger flow prediction for special events using smart card data. IEEE Trans. Intell. Transp. **21**(3), 1109–1120 (2019)
3. Chung, J., et al.: Empirical evaluation of gated recurrent neural networks on sequence modeling. In: NIPS 2014
4. Diao, Z., et al.: A hybrid model for short-term traffic volume prediction in massive transportation systems. IEEE Trans. Intell. Transp. **20**(3), 935–946 (2018)
5. Fu, X., et al.: Short-term prediction of metro passenger flow with multi-source data: a neural network model fusing spatial and temporal features. Tunn. Undergr. SP Technol. **124**, 104486 (2022)
6. Huang, H., et al.: Identifying subway passenger flow under large-scale events using symbolic aggregate approximation algorithm. Transp. Res. Rec. **2676**(2), 800–810 (2022)
7. Huang, Z., et al.: A mobility network approach to identify and anticipate large crowd gatherings. Transp. Res. B-Methodol. **114**, 147–170 (2018)
8. Jeong, Y.S., et al.: Supervised weighting-online learning algorithm for short-term traffic flow prediction (2013)
9. Ke, J., et al.: Short-term forecasting of passenger demand under on-demand ride services: a spatio-temporal deep learning approach. Transp. Res. C-Emerg. Technol. **85**, 591–608 (2017)
10. Li, C., et al.: Spatio-temporal graph convolution for skeleton based action recognition. In: AAAI 2018 (2018)
11. Li, Y., et al.: Forecasting short-term subway passenger flow under special events scenarios using multiscale radial basis function networks. Transp. Res. C-Emerg. Technol. **77**, 306–328 (2017)
12. Liu, L., et al.: A novel passenger flow prediction model using deep learning methods. Transp. Res. C-Emerg. Technol. **84**, 74–91 (2017)
13. Liu, Y., et al.: Deeppf: a deep learning based architecture for metro passenger flow prediction. Transport Res C-Emerg. Technol. **101**, 18–34 (2019)
14. Lv, Y., et al.: Traffic flow prediction with big data: a deep learning approach. IEEE Trans. Intell. Transp. Syst. **16**(2), 865–873 (2015)
15. Miller, H.J.: Tobler's first law and spatial analysis. Ann. Assoc. Am. Geogr. **94**(2), 284–289 (2004)
16. Murthy, A.S.N., Mohle, H.: Application of poisson distribution. American Society of Civil Engineers (2015)
17. Neill, D.B.: Expectation-based scan statistics for monitoring spatial time series data. Int. J. Forecast. **25**(3), 498–517 (2009)
18. Ou, J., et al.: STP-TrellisNets: spatial-temporal parallel trellisnets for metro station passenger flow prediction. In: CIKM (2020)
19. Toto, E., et al.: Pulse: a real time system for crowd flow prediction at metropolitan subway stations. In: ECML PKDD (2016)

20. Vanajakshi, L.: Short-term traffic flow prediction using seasonal arima model with limited input data. Eur. Transp. Res. Rev. **7**, 1–9 (2015)
21. Wang, H., et al.: Early warning of burst passenger flow in public transportation system. Transp. Res. C-Emerg. Technol. **105**, 580–598 (2019)
22. Wang, H., et al.: Online detection of abnormal passenger out-flow in urban metro system. Neurocomputing **359**, 327–340 (2019)
23. Wen, K., et al.: A decomposition-based forecasting method with transfer learning for railway short-term passenger flow in holidays. Expert Syst. Appl. **189**, 116102 (2022)
24. Xue, G., et al.: Forecasting the subway passenger flow under event occurrences with multivariate disturbances. Expert Syst. Appl. **188**, 116057 (2022)
25. Zhang, J., et al.: A real-time passenger flow estimation and prediction method for urban bus transit systems. IEEE Trans. Intell. Transp. Syst. **18**(11), 3168–3178 (2017)
26. Zhang, Y., Haghani, A.: A gradient boosting method to improve travel time prediction. Transp. Res. C-Emerg. Technol. **58**, 308–324 (2015)
27. Zhou, F., et al.: Reinforced spatiotemporal attentive graph neural networks for traffic forecasting. IEEE Internet Things J. **7**(7), 6414–6428 (2020)
28. Zhou, X., et al.: A traffic flow approach to early detection of gathering events. In: ACM SIGSPATIAL (2016)

FMSYS: Fine-Grained Passenger Flow Monitoring in a Large-Scale Metro System Based on AFC Smart Card Data

Li Sun[1,2], Juanjuan Zhao[1](✉), Fan Zhang[1], Rui Zhang[3], and Kejiang Ye[1]

[1] Shenzhen Institute of Advanced Technology, Chinese Academy of Sciences, Shenzhen, China
jj.zhao@siat.ac.cn
[2] University of Chinese Academy of Sciences, Beijing, China
[3] Shenzhen Institute of Beidou Applied Technology Co., Ltd., Shenzhen, China

Abstract. In this paper, we investigate the real-time fine-grained passenger flows in a complex metro system. Our primary focus is on addressing crucial questions, such as determining the number of passengers on a moving train and in specific station areas (e.g., access channel, transfer channel, platform). These insights are essential for effective traffic management and ensuring public safety. Existing visual analysis methods face limitations in achieving comprehensive network coverage due to deployment costs. To overcome this challenge, we introduce FMSYS, a cloud-based analysis system leveraging smart card data for efficient and reliable real-time passenger flow predictions. FMSYS identifies each passenger's travel patterns and classifies passengers into two groups: regular (D-group) and stochastic (ND-group). It models stochastic movement of passengers using a state transition process at the group level and employs a combined approach of KNN and Gaussian Process Regression for dynamic state transition prediction. Empirical analysis, based on six months of smart card transactions in Shenzhen, China, validates the effectiveness of FMSYS.

Keywords: Fine-grained passenger flows · smart card

1 Introduction

Nowadays, metro systems have become a preferred mode of transportation in numerous cities due to their efficiency and punctuality. Rapid urbanization has further increased passenger flow, leading to the issues of overcrowding not only on trains but also in platform, access and transfer tunnel of stations. Understanding the fine-grained patterns of passenger flows is crucial for optimizing train scheduling, improving passenger evacuation strategies, and providing passengers with more options for journey planning.

While devices like laser scanners or video monitors can collect passenger flow data, they face limitations in covering all metro areas due to technical or cost constraints. Their accuracy may also decrease in crowded conditions. Automatic Fare

D.-N. Yang et al. (Eds.): PAKDD 2024, LNAI 14649, pp. 336–349, 2024.
https://doi.org/10.1007/978-981-97-2262-4_27

Collection (AFC) system offers a unique opportunity to understand passenger travel patterns [8,10,11]. However, leveraging this data to enhance the efficiency and reliability of the metro transportation system is still in its early stages.

Considering a passenger in a metro system who does not tap out, we lack details about passengers' movements, such as their exact location and states at any specific timestamp during the journey. The AFC system, designed for fare purposes and not mobility modeling, only records when and where a user enters and exits the metro system. Understanding the real-time locations of passengers within a metro system is crucial for determining the number of passengers in specific areas, posing our first challenge. Additionally, during peak hours, when numerous passengers engage in activities like entering, exiting, transferring, and boarding trains simultaneously, efficiently updating all passenger flows is the second challenge. In this paper, we develop an online analysis system, named FMSYS, designed to infer and predict fine-grained metro passenger flows. The primary contributions of this work include:

(1) We propose an approach for analyzing passengers' spatio-temporal travel patterns using smart card transaction data. These patterns enable us to accurately infer a regular passenger's destination and route choice, given the entrance metro station and time.
(2) We model the stochastic movement of passengers in the system over time as a state transition process. To efficiently and robustly predict the real-time state transition of passengers in ND-group, we design a dynamic KNN and GPR combined prediction approach at the group level, running on a cloud system.
(3) We evaluate FMSYS using real-world smart card transaction data collected in Shenzhen, China over a six-month period. The results, compared with baselines, validate the effectiveness of our system.

2 Problem and Solution Overview

2.1 Preliminaries

Definition 1. (*Metro system*) A metro system M is associated with lines $L = \{l_1, l_2, \cdots, l_n\}$ and stations $S = \{s_1, s_2, \cdots, s_n\}$. Stations are categorized as either general stations, served by a single line, or transfer stations for line transfers. We use $P_{i,m}$ to represent the station platform of line l_m at station s_i.

Definition 2. (*Trip*) A trip is associated with origin station s_o, destination station s_d, starting time t_o and end time t_d.

Definition 3. (*Passenger states*) A trip could be divided into four or five types of segments [7,9]: walking-in time (ETT), waiting time (WTT), on-train time (OTT), transferring time (TFT), and walking-out time (EXT). We define passenger states as *Walking-in*, *Waiting*, *On-train*, *Transferring*, and *Walking-out*. Each passenger state κ_i associated with a beginning time $\kappa_i.b$, ending time $\kappa_i.e$ and a passenger state $\kappa_i.s$. The transitions between states are visualized in Fig. 1 according to the entire process of a trip.

Definition 4. (*Walking Time*) A trip involves three types of walking time: Walking time from the entry gate of s_j to platform $P_{j,m}$ ($ETT_{j,m}$), the transfer time from $P_{j,m}$ to $P_{j,n}$ ($TFT_{j,m,n}$), and the walking time from platform $P_{j,m}$ to exit gate of station s_j ($EXT_{j,m}$), respectively. While the walking time constitutes a small portion of a trip, we do not consider differences in user walking speeds.

Problem Statement: Based on real-time received smart card transaction data and train operation records, this paper aims to estimate the population of fine-grained passenger flows in five types of states:

$I_{j,m}$: Walking from the entry gate of station s_j to the platform $P_{j,m}$.

$W_{j,m}$: Waiting for the train on platform $P_{j,m}$.

$T_{j,m,q}$: On train q of line l_m in the section between station s_j and next adjacent one.

$R_{j,m,n}$: Walking between two platforms, $P_{j,m}$ and $P_{j,n}$, for a line change.

$O_{j,m}$: Walking from platform $P_{j,m}$ to the exit gate of station s_j.

2.2 Motivation

Prior research [7,9] revealed that a significant portion of metro passengers exhibit regular and consistent travel behavior, such as commuting to and from work during morning and evening peaks. Their route choices are also stable. These travel patterns serve as a valuable basis for predicting real-time mobility and location. Consequently, we can infer the location and state of such passengers in real-time by combining the train schedule with other factors, such as walking time, transfer time and walking-out time.

For stochastic passengers, inferring their destination and route directly from previous trips is challenging. A straightforward approach to estimate real-time location would involve initially measuring all destination and route probabilities for each Origin-Destination (OD) pair and then aggregating this information to estimate the current location. For instance, to estimate the probability of a passenger waiting on platform $P_{i,m}$ based on the tap-in time t and station s, we need to calculate the probability for choosing each line l by aggregating all probabilities of reachable destinations and routes from line l. However, predicting all passengers at same time would result in a computationally intensive workload.

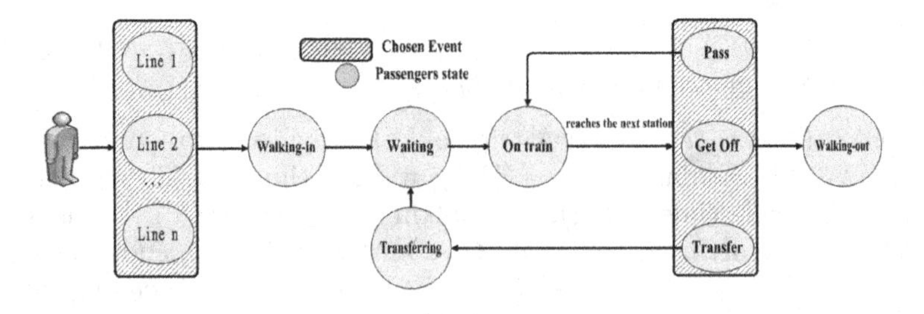

Fig. 1. Passenger state transition and random events

To enhance efficiency and scalability, we predict the detailed flows of stochastic passengers at a group level. Rather than employing two separate prediction stages for the destination and path of each passenger, we depict it as a unified state transition process based on the collective movement of passengers in the metro system. The trip process, based solely on tap-in records, is illustrated in Fig. 1. Our problem is formulated by two random events: 1) Selection of the line by a passenger upon entering a metro station s with multiple physical lines (abbreviated as the line selection event); 2) Actions such as getting off, transferring, or passing through when a train arrives at a station s (abbreviated as the action selection event). Knowing the probabilities of these two events allows us to estimate fine-grained passenger flows by taking into account the train timetable and other relevant factors.

2.3 FMSYSOverview

We build a cloud-based system named FMSYS, illustrated in Fig. 2, for estimating fine-grained metro passenger flow in real-time. The system first extracts individual passenger Origin-Destination (OD) and route choice travel patterns from long-term smart card transaction records in offline mode. Subsequently, passengers are classified into two groups based on their destination predictability: the D-group and the ND-group. Leveraging data-driven learning to model the travel state transitions of ND-group passengers, we introduce an online algorithm executed on the Storm platform for analyzing passenger movement in metro systems.

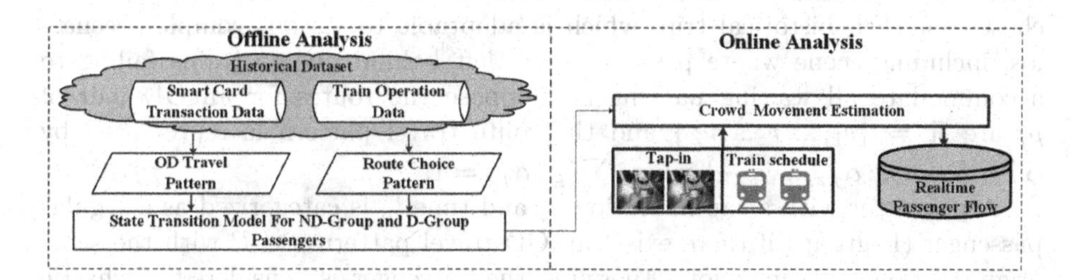

Fig. 2. Overview of the FMSYS framework.

3 Passenger State Transition Estimation

3.1 Individual Travel Pattern Analysis

OD Travel Pattern Extraction: An OD travel pattern for a passenger signifies frequent trips between specific stations within a defined timeframe, and the trip proportion surpassing a specified threshold η (e.g., 60%). Various passengers may exhibit varying numbers of OD travel patterns. Given the trip sequence of a passenger, we first discretize the travel time. The central idea is to divide time into sequential and overlapped times slots. None-overlapped time slots cannot expressively denote a duration: for example, it is challenging to analyze a trip from 08:30~09:29 using two consecutive time slots such as 08:00~08:59 or 09:00~09:59. Instead, we use overlapped time slot division, such as 08:00~10:59, 09:00~11:59.

Initially, for each passenger, we capture the travel status for every hour on each day based on all his/her history trips, denoted by a matrix H. Each row corresponds to a weekday, and each column represents one hour (e.g., H_1 for 00:00~00:59, H_2 for 1:00~1:59, $\cdots$). Each element of H indicates whether the card is active during the corresponding hour and day. Subsequently, an aggregation is performed. We use F_i to represent the number of trips in time slot i for all days and calculated it as $F_i = \sum_{j=1}^{D_{num}} (H_{j,i} + H_{j,i+1} + H_{j,i+2})$. Here D_{num} is the number of active days. A time slot covers 3 consecutive hours, considering that metro trips in our data seldom last more than 3 h.

Next, a kernel density-based clustering algorithm is employed to extract temporal travel patterns, classifying all trips into groups based on their time points within a day. For each cluster, we check if there is a subset of trips sharing similar OD pairs. If the proportion of such trips compared to the total days exceeds the threshold η, a travel pattern is established. Through this process, we extract all OD travel patterns for each passenger, denoted as $P = \{p_1, p_2, ..., p_{|P|}\}$. A travel pattern p includes the origin station s_o, destination station s_d, start time t_o, end time t_d, and proportion r.

Route Choice Pattern Extraction: For each OD travel pattern $p_j \in P$ of a passenger, we extract the likelihood of choosing each route, referred to as the route travel patterns. We apply a method introduced in [9] to extract the route chosen for each historical trip, which is adaptable to various complex scenarios, including scene where passengers are left behind due to trains failing to accommodate all waiting passengers. Suppose the routes for an OD pair of p_j are $R = \{r_1, ..., r_z ..., r_Z\}$ and the route travel pattern is represented by $\boldsymbol{\alpha}_j = \{\alpha_{j,1}, ..., \alpha_{j,z}, ..., \alpha_{j,Z}\}$ where $\sum_{z=1}^{Z} \alpha_{j,z} = 1$.

A passenger with a tap-in station s_o and time t_o is categorized as a regular passenger (D-group) if there exists an OD travel pattern $p \in P$ with the same origin station and time slot; otherwise, the passenger is considered stochastic (ND-group). For regular passengers, we predict the destination and route based on the OD and route travel patterns. The real-time location can be estimated by combining train schedule, along with other factors, i.e., walking time.

3.2 State Transition Estimation for ND-Group

As outlined in Sect. 2.2, we forecast the probabilities of two events for stochastic passengers (ND-group) at the group level. We divide a day into fixed time slots (e.g., half-hour interval) and predict the probability for each time slot.

Passengers' dynamic line (action) selection requires reliable predictions from the most updated model. To address this, we propose a lazy learning approach that integrates *KNN* (K-nearest Neighbor) and *GPR* (Gaussian Process Regression). While *GPR* is effective in handling data relationships, its cubic learning computation and quadratic space requirement pose scalability challenges. In this paper, we propose a *KNN* and *GPR* combined method, utilizing *KNN* to select the most informative data for Gaussian Process Regression.

In the rest of this section, we first obtain the ground truth for passengers' line or action selection along with relevant features based on historical trips. We then provide a detailed explanation of the combined *KNN* and *GPR* method.

Features and Ground Truth Extraction

(1) Feature Extraction (Model Input)

- *Time features:* We use $F_t = (dy, Ts)$ to represent the day of the week and the time slot of the day, which are significant in passengers' trip prediction due to the transport periodicity.
- *Traffic flow features:* Passenger line selection at a station s_i during a time slot T_k is influenced by their behavior in the preceding time slots. We compute the proportions of completed trips that entered the metro system at station s_i and selected each line during the most recent N time slots, $T_{k-1} \sim T_{k-N}$.

  Similarly, when a train on line l_m arrives at a station s_i during a time slot T_k, passengers' action selection is connected to their behavior in past time slots. We extract three types of passenger flows that reflect the three actions: i) Passengers exiting the metro system from station s_i; ii) Passengers exiting the metro system from the following stations of line l_m; iii) Passengers transferring to other lines $l_n \in L^i$ at station s_i. We user $F_l = (f_{os}, f_{ol}, f_{tl})$ to represent these three type of passenger flows.

(2) State Transition Extraction (Model Output)

- *Line selection proportion:* As mentioned in Definition 3, each historical trip can be divided into four or five types of segments. Utilizing these segments, for passengers entering the metro system at station s_i during a time slot T_k on any given day, the proportion choosing a line $l_m \in L^i$ and switching to state $I_{i,m}$ is calculated by Eq. 1. $\#L_k(I_{i,m})$ represents the number of passengers in state $I_{i,m}$ after entering metro system at station s_i during time slot T_k, calculated by Eq. 2 based on the segments $tr.\kappa_k$ of all observed trips TR throughout the day. The notation $|.|$ calculates the number of items in the given collection.

$$Pr(I_{i,m}|T_k) = \frac{\#L_k(I_{i,m})}{\sum\limits_{m \in L^i} \#L_k(I_{i,m})}. \tag{1}$$

$$\#L_k(I_{i,m}) = |\{tr|tr \in TR, tr.\kappa_k.b \in T_k, tr.\kappa_k.s = I_{i,m}\}| \tag{2}$$

- *Action selection proportion:* In a given day, within the time interval T_k, the proportion of passengers converting from the state $T_{i,m,q}$ to a state s in $S = \{O_{j,m}, T_{j,m,q}\} \cup \{R_{j,m,n} | l_n \in L^j \& l_n \neq l_m\}$ is calculated by Eq. 3.

Here, Q_k represents trains of line l_m arriving at station s_j (the adjacent station of station s_i) during time slot T_k. The $\#C_k(s,q,j)$ signifies the number of passengers in state $T_{i,m,q}$ converting to state s after the train q arrives at station s_j, calculated by Eq. 4.

$$Pr(T_{i,m,q} \rightarrow s|T_k) = \frac{\sum_{q\in Q_k} \#C_k(s,q,j)}{\sum_{s'\in S}\sum_{q\in Q_k} \#C_k(s',q,j)} \tag{3}$$

$$\#C_k(s,q,j) = \left|\{tr|tr \in TR, tr.\kappa_k.s = T_{i,m,q}, tr.\kappa_{(k+1)}.s = s, tr.\kappa_{(k+1)}.b \in T_k\}\right|. \tag{4}$$

K-Nearest Neighbor. *KNN* is used to find the data that are close to the current condition. Efficient prediction depends on our ability to capture the spatio-temporal correlation from historical data. The condition is represented as a feature vector x, which includes two types of features: time and traffic flow, denoted as $x = \{F_t, F_l\}$. The target value y is the line or action selection proportion.

Since traffic exhibits obvious periodicity, including a 24-hour and 7-day periodicity, and strong correlation between near time slots, we only search in the historical data with the same day indicator and adjacent time slots.

For each pair of historical data y and target data y', we calculate their distance using Eq. 5. Here, f and f' represent the features of the two data, $C(f,f')$ is the correlation coefficient between the time features, and $|f - f'|$ is the Manhattan distance of the traffic flow features.

$$dist(y,y') = \left(\prod_{f\in F_t} C(f,f')\right) \times \left(\sum_{f\in F_f} |f - f'|\right) \tag{5}$$

Gaussian Process Regression. The K neighbors selected by the *KNN* algorithm are used as input for the computation of *GPR* to estimate line or action selection proportions. Let $H \equiv \{X = (x_i, \cdots, x_k), Y = (y_i, \cdots, y_k)\}$ represent the set of K-nearest neighbors, where x_i is the input feature vector, and y_i denotes the target value (line or action selection). The objective of *GPR* is to identify a function f that describes the relationship between x_i and y_i. This function can be characterized by a mean function $m(x)$ and a covariance function $k(x,x')$, given as $f = GP(m(x), k(x,x'))$, where x and x' denote two input feature vectors. The crucial aspect of *GPR* lies in choosing a suitable kernel function k. In this study, we adopt one of the most commonly used function, the squared exponential kernel function, i.e., $k(x,x') = \sigma^2 \exp\left(-\frac{(x-x')^2}{2\ell^2}\right)$. Given input features x_c, the corresponding target is predicted using *GPR* as follows.

$$y_c|x_c, X, Y \sim GP(k(x_c,X)[k(X,X)]^{-1}Y, k(x_c,x_c) - k(x_c,X)[k(X,X)]^{-1}k(X,x_c)) \tag{6}$$

4 Online Analysis

Online analysis is running on the *Storm* platform, a free distributed real-time computation system. The process involves a topology represented as a network of *spouts* and *bolts*. A *spout* allows users to generate source stream, while a *bolt* receives a number of input streams, processes them, and possibly emits new streams. The topology of our system is shown in Fig. 3, containing one spout and two bolts.

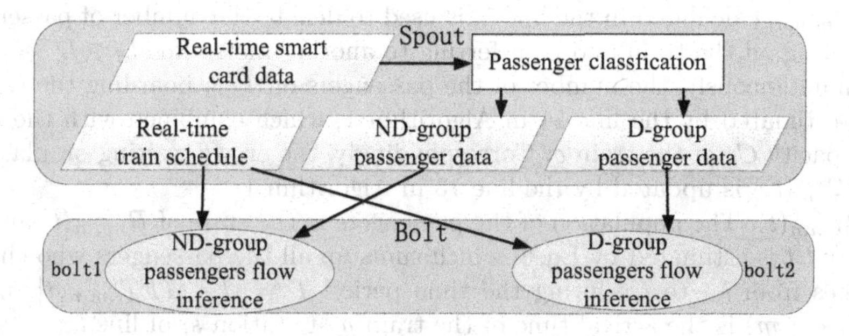

Fig. 3. The processing topology of our system.

The spout receives two types of data. One type is tap-in records, which are classified into two groups: D-group and ND-group. For the passengers of D-group, the tap-in record, together with the inferred destination, is sent to the first bolt. The tap-in record of ND-group is directly sent to the second bolt. Another type of received data is the train schedule, which includes arrival and departure times at each station. This information is directly forwarded to the two bolts, leading to the real-time update of fine-grained passenger flows. The final fine-grained passenger flow results from aggregating that of the two groups.

In the first bolt, the destinations and routes of the passengers in the D-group are determined. The real-time locations can be directly updated along with train operation time and walking time; therefore, the fine-grained D-group passenger flows can be accurately calculated. Our primary focus is on bolt1 for the ND-group.

The line and action proportions for ND-group passengers are predicted in real-time per time slot using a combined *KNN* and *GPR* approach. The population of the fine-grained ND-group passenger flows across five states $\mathcal{I}_{i,m}(t)$, $\mathcal{R}_{i,m,n}(t)$, $\mathcal{W}_{i,m}(t)$, $\mathcal{O}_{i,m}(t)$, and $\mathcal{T}_{i,m,q}$ is calculated or updated as follows. We establish a queue Q_i for each station s_i to manage tap-in passengers. Upon receiving a tap-in record, the passenger is inserted into the corresponding queue.

(1) $\mathcal{I}_{i,m}(t)$: The number of passengers in state $I_{i,m}$ at time t is estimated by Eq. 7, where $t' = (t - ETT_{i,m})$, and $Q_i(t', t)$ is the number of tap-in passengers at station s_i during the time period $(t', t]$ in queue Q_i, and $I(t)$ denotes the time slot to which the time point t belongs.

$$\mathcal{I}_{i,m}(t) = Q_i(t', t) \times Pr(I_{i,m}|I(t)) \tag{7}$$

(2) $\mathcal{T}_{i,m}(t)$: The number of passengers on a train changes when the train arrives and departs from a station. For instance, when a train q of line l_m arrives at station s_i at time t, the number of passengers on the train is updated by subtracting the passengers getting off the train and adding passengers boarding the train, as shown in the line 12 in Algorithm 1, where $s_{i'}$ is the preceding station of s_i. Intermediate variables are explained as follows. $U_{i,m,q}$ estimated by the line 2 in Algorithm 1, is the number of passengers who get off the train q and exit the metro system. If s_i is a transfer station, $C_{i,m,n,q}$, calculated in the line 5, is used to denote the number of passengers getting off the train and transferring to another metro line $l_n \in L^i$. Simultaneously, the number of the passengers on $P_{i,m}$ boarding the train q is estimated by the line 11 in Algorithm 1, which combined with the train capacity C_q of the train q. Correspondingly, the crowd waiting on platform ($\mathcal{W}_{i,m}(t)$) is updated by the line 13 in Algorithm 1.

(3) $\mathcal{R}_{i,m,n}(t)$: The population of the passengers in the state of $R_{i,m,n}(t)$ at time point t is estimated by Eq. 8, which sums up all the passengers who change lines from l_m to l_n during the time period $T = (t - TFT_{i,m,n}, t]$, where $T_a(q, i, m)$ is the arrival time of the train q at station s_i of line l_m.

$$\mathcal{R}_{i,m,n}(t) = \sum\nolimits_{T_a(q) \subset T} \mathcal{C}_{i,m,n,q}, \tag{8}$$

(4) $\mathcal{W}_{i,m}(t)$: The population of the passengers in the state of $W_{i,m}$ at the time point t is estimated by the Eq. 9, where t' is the time when the former train left station s_i, we use T to represent the time period $T = (t' - TFT_{i,m}, t - TFT_{i,m}]$.

$$\mathcal{W}_{i,m}(t) = \mathcal{W}_{i,m}(t') + \sum_{k \subset (t' - ETT_{i,m}, t - ETT_{i,m}]} \mathcal{I}_{i,m}(k) + \sum_{l_n \in L^i} \sum_{T_a(i,m,q) \subset T} C_{i,n,m,q}, \tag{9}$$

(5) $O_{i,m}(t)$: The population of the passengers in the state of $O_{i,m}(t)$ at time point t is predicted by Eq. 10, which sums up the value of $U_{i,m,q}$ for the trains whose arrival time at the station s_i are during time interval $T = (t - ETT_{s_i, l_m}\ t]$. $T_{arrv}(i, m, q)$ represents the arrival time of train q at the station s_i.

$$\mathcal{O}_{i,\ m}(t) = \sum\nolimits_{T_a(i,m,q) \subset T} U_{i,m,q}, \tag{10}$$

The final fine-grained passenger flow results from aggregating the corresponding flows of the two groups.

Algorithm 1. PassUpdateWhenTrainArrive

1: **procedure** STATEUPDATE
2: $U_{i,m,q} \leftarrow T_{i',m,q} \times Pr(T_{i',m,q} \rightarrow O_{i,m} | I(t))$
3: **if** the type of s_i is *Trs* **then**
4: **for** l_n *in* L^i **do**
5: $C_{i,m,n,q} = T_{i',m,q} \times Pr(T_{i',m,q} \rightarrow R_{i,m,n} | I(t))$
6: **end for**
7: **end if**
8: $N_f = U_{i,m,q} + \sum_{n \in F} C_{i,m,n,q}$
9: $\alpha = \mathcal{W}_{i,m} \div (C_q - (T_{i',m,q} - N_{gf}))$
10: $\alpha = \alpha \geq 1?1 : \alpha$
11: $N_o = \alpha \times (C_q - (T_{i',m,q} - N_f))$
12: $T_{i,m,q} = T_{i,m,q} - N_f + N_o$
13: $\mathcal{W}_{i,m}(t) = \mathcal{W}_{i,m}(t) - N_o$
14: **end procedure**

5 Experimental Analysis

5.1 Experimental Setting

Dataset: We utilize the smart card data and the complete train schedules obtained from the metro system in Shenzhen, China. The dataset is associated with 4 million smart card identifiers and have more than 300 million transaction records spanning six month, from January 1, 2018 to June 30, 2018.

Baselines: The problem addressed in most prior works is predicting the OD flow, which constitutes only one stage of this paper. The fine-grained passenger flows predicted in this paper need to consider not only the OD flow but also the path, train operation time, and other factors. Here, we extend the OD flow prediction methods by incorporating path choice as the historical average.

EMP: It is based on naive empirical knowledge, utilizing historical average OD matrices at the same time slot.

FCLSTM [1]: It is an artificial neural network (RNN) architecture well-suited for processing and making predictions based on time series data.

MRSTN [5]: It uses a multi-resolution spatio-temporal deep Learning approach to predict the short-term OD matrices in metro system.

5.2 Experiment Result

The passengers in the four states (I, W, R, O) may change at every time point. For simplicity, we compare the predicted result with the real values at 5-minute slots. The number of passengers in the state of T only changes during the interval between the train arriving at and leaving a station, so we evaluate it at the time the train leaves each station.

Two metrics, Mean Absolute Percentage Error (MAPE) and Mean Absolute Error (MAE), are employed to evaluate the model's performance and all baselines. To ensure method robustness and result stability, we conducted the experiments five times for all methods. The final outcomes are summarized in Table 1. Notably, our model demonstrates superior performance across all metrics. Among the three baselines, Emp exhibits the weakest performance. Its reliance solely on the periodicity characteristics of traffic demand limits its ability to model dynamicity. Although FC-LSTM and MRSTN can capture flow dynamicity, their lack of consideration for differences between regular and occasional trips, as well as the dynamic nature of path selection, hinders their direct application to our problem.

$$MAPE = \frac{1}{N}\sum_{i=1}^{N}(\frac{|x_i^{'} - x_i|}{x_i}) \tag{11}$$

$$MAE = \frac{1}{N}\sum_{i=1}^{N}|x_i^{'} - x_i| \tag{12}$$

Table 1. Comparison with baselines.

Methods	I		W		T		R		O	
	MAPE	MAE	MAPE	MAE	MAPE	MAE	MAPE	MAE	MAPE	MAE
EMP	0.27	34.22	0.27	59.32	0.26	140.08	0.27	88.74	0.29	45.34
FCLSTM	0.20	25.83	0.20	45.28	0.20	104.54	0.20	67.21	0.20	32.71
MRSTN	0.16	23.24	0.19	41.06	0.16	90.21	0.17	59.29	0.21	33.02
FMSYS	**0.11**	**17.20**	**0.10**	**11.08**	**0.10**	**68.607**	**0.11**	**34.69**	**0.11**	**16.47**

Different Time Period. We compare prediction results across various daily time periods, specifically distinguishing between weekdays and weekends. We divide a day into peak hours (07:00~09:00, 17:00~19:00) and off-peak hours (other times) for comparison, and the results are detailed in Table 2. Notably, the overall MAPE is consistently less than 12%, with an average below 10.5%, affirming the effectiveness and accuracy of our approach for the five fine-grained passenger flows. Further analysis reveals that predictions during off-peak hours on weekdays exhibit slightly inferior performance compared to peak hours. This discrepancy is attributed to the higher volume of regular passengers, especially during AM peek hours. Similarly, the average RME during weekends is marginally worse than that on weekdays, for similar reasons outlined in Table 2.

Table 2. Mean Absolute Percentage Error (MAPE) of FMSYS during different time slots

	I	W	T	R	O
Weekday-Peek	0.086	0.084	0.094	0.097	0.099
Weekday-Low	0.103	0.101	0.105	0.107	0.110
Weekend-Peek	0.121	0.124	0.115	0.106	0.125
Weekend-Low	0.122	0.124	0.126	0.107	0.127

The Fig. 4(a) illustrate the travel pattern ratio where the specified threshold $\eta > 80\%$. Figure 4(b) show the inference of destinations on a day. We can observed that during AM and PM peak hours, the destination of regular passenger is significantly accurate. The finding agrees with reality as well, i.e., most users regularly and stably commute back and forth to work during peak hours on weekdays.

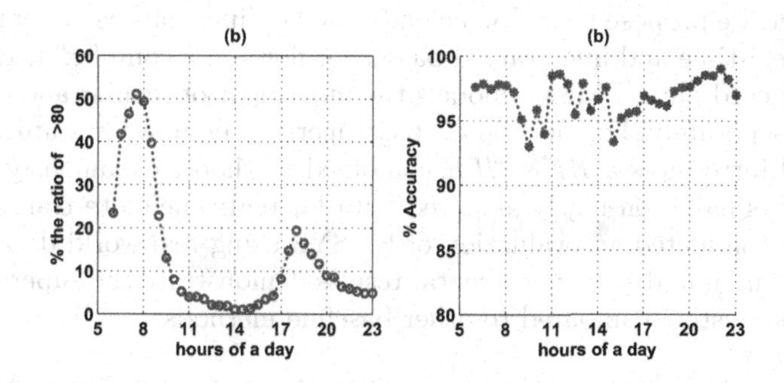

Fig. 4. Destination inference.

6 Related Work

Visual analysis-based approach: Crowd counting is an effective tool for situational awareness in public places. Recent research on crowd counting demonstrates the efficacy of convolution neural networks (CNNs) [3,6] due to their strong capability of automatic feature extraction. However, despite using very deep models and complex architectures, the accuracy gains over dense and large datasets are reasonably low. Crowd counting in the subway station scene faces most of the challenges in the general crowd counting scene, such as occlusion, scale change, distortion of perspective, rotation, illumination change, complex background, uneven crowd distribution, etc. Due to the limited coverage of equipment, the real-time accurate computing ability of the whole network state is limited.

AFC data-based method: AFC data has been widely employed for estimating passenger flow in subway systems. Hao *et al.* [2] proposed a sequence-to-sequence learning with attention mechanism. Sajanraj *et al.* [1] utilized LSTM for OD flow prediction, and [5] employed multi-resolution spatio-temporal deep learning for Short-term OD matrix prediction. These studies focus more on the prediction of travel demand. However, effective and timely metro passenger organization requires more detailed crowd data of stations and trains. Over the years, researchers have explored historical passenger flow assignment. For instance, Lee *et al.* [4] proposed a robust approach to recognize route-use patterns of transit users by incorporating the number of routes as an unknown parameter into a Bayesian framework. Zhao *et al.* [9] split each history trip into some segments with states based on time matching of AFC data with train schedule data. While this method is useful for estimating historical crowd conditions, it faces challenges in meeting the needs of real-time prediction due to delayed tap-out records.

7 Conclusion

In this paper, we propose FMSYS, a cloud-based online analysis system designed to predict real-time and fine-grained passenger flows in a complex metro system using smart card data. FMSYS models the movement of regular and occasional passengers separately. For occasional passengers, it employs a state transition process and introduces a *KNN-GPR* combined method, continuously updating its predictions based on newly acquired data for real-time state transition forecasting. We conducted an evaluation of FMSYS using real-world datasets from Shenzhen, China, and the experimental results demonstrate the superior performance of our system compared to other baseline methods.

Acknowledgement. This study was funded by the National Key R&D Program of China (No. 2023YFC3321600), National Natural Science Foundation of China (No. 62372443, No. 62376263), Shenzhen Industrial Application Projects (No. CJGJZD2021 0408091600002).

References

1. Chu, K.F., Lam, A.Y., Li, V.O.: Deep multi-scale convolutional LSTM network for travel demand and origin-destination predictions. IEEE Trans. Intell. Transp. Syst. **21**(8), 3219–3232 (2019)
2. Hao, S., Lee, D.H., Zhao, D.: Sequence to sequence learning with attention mechanism for short-term passenger flow prediction in large-scale metro system. Transp. Res. Part C Emerg. Technol. **107**, 287–300 (2019)
3. Jiang, X., et al.: Attention scaling for crowd counting. In: 2020 IEEE/CVF Conference on Computer Vision and Pattern Recognition (CVPR) (2020)
4. Lee, M., Sohn, K.: Inferring the route-use patterns of metro passengers based only on travel-time data within a Bayesian framework using a reversible-jump Markov chain Monte Carlo (MCMC) simulation. Transp. Res. Part B Methodol. **81**, 1–17 (2015)

5. Noursalehi, P., Koutsopoulos, H.N., Zhao, J.: Dynamic origin-destination prediction in urban rail systems: a multi-resolution spatio-temporal deep learning approach. IEEE Trans. Intell. Transp. Syst. **23**(6), 5106–5115 (2021)
6. Wang, Q., Breckon, T.P.: Crowd counting via segmentation guided attention networks and curriculum loss. IEEE Trans. Intell. Transp. Syst. **23**(9), 15233–15243 (2022)
7. Zhao, J., Qu, Q., Zhang, F., Xu, C., Liu, S.: Spatio-temporal analysis of passenger travel patterns in massive smart card data. IEEE Trans. Intell. Transp. Syst. **18**(11), 3135–3146 (2017)
8. Zhao, J., Tian, C., Zhang, F., Xu, C., Feng, S.: Understanding temporal and spatial travel patterns of individual passengers by mining smart card data. In: 2014 IEEE 17th International Conference on Intelligent Transportation Systems (ITSC), pp. 2991–2997. IEEE (2014)
9. Zhao, J., et al.: Estimation of passenger route choice pattern using smart card data for complex metro systems. IEEE Trans. Intell. Transp. Syst. **18**(4), 790–801 (2016)
10. Zhao, J., et al.: GLTC: a metro passenger identification method across AFC data and sparse WiFi data. IEEE Trans. Intell. Transp. Syst. **23**(10), 18337–18351 (2022)
11. Zheng, F., Zhao, J., Ye, J., Gao, X., Ye, K., Xu, C.: Metro OD matrix prediction based on multi-view passenger flow evolution trend modeling. IEEE Trans. Big Data (2022)

Enhanced HMM Map Matching Model Based on Multiple Type Trajectories

Yuchen Song[1,2], Juanjuan Zhao[1(✉)], Xitong Gao[1], Fan Zhang[1], and Kejiang Ye[1]

[1] Shenzhen Institute of Advanced Technology, Chinese Academy of Sciences, Shenzhen, China
jj.zhao@siat.ac.cn
[2] University of Chinese Academy of Sciences, Beijing, China

Abstract. Map matching (MM) aims to align GPS trajectory with the actual roads on a map that vehicles pass through, essential for applications like trajectory search and route planning. The Hidden Markov Model (HMM) is commonly employed for online MM due to its interpretability and suitability for low GPS sampling rates. However, in complex urban areas with notable GPS drift, existing HMM methods face efficiency and accuracy challenges due to the use of a uniform road search radius and imprecise real-time road condition understanding. This paper proposes an improved HMM method using multiple trajectory types based on the following key ideas: Vehicle trajectories can be divided into two types: fixed trajectories (e.g., bus) and free trajectories (taxis, private cars). The relatively accurate information of fixed trajectories can help us more accurately measure the error distribution, as well as accurate road conditions. The novelty of our approach lies in the following aspects: i) Using fixed bus trajectories to estimate region-specific GPS error distribution, optimizing observation probabilities and reducing candidate road search costs. ii) Utilizing real-time fixed trajectories for accurate, real-time road state estimation, enhancing dynamic state transition probabilities in HMM. Empirical analysis, based on real bus and taxi trajectories in Shenzhen over half a year, demonstrate that our method outperforms existing methods in terms of map matching efficiency and accuracy.

Keywords: Map matching · trajectories · spatio-temporal data analysis

1 Introduction

Online map matching technique is a process of associating a sequence of positions collected by vehicle positioning devices with the road network in a digital map. HMM and its variants are commonly used due to their advantages in interpretability, noise tolerance, and robustness in different sampling frequencies. HMM-based map matching treats GPS trajectories as observed states and considers all possible candidate points as hidden states.

D.-N. Yang et al. (Eds.): PAKDD 2024, LNAI 14649, pp. 350–362, 2024.
https://doi.org/10.1007/978-981-97-2262-4_28

The process typically involves three key steps. Firstly, a candidate road selection radius is determined based on GPS error distribution. Within this radius, a set of potential candidate points on the road is obtained to form the hidden state sequence. The observation probability for each hidden state is calculated based on the distance with observed point. Secondly, candidate paths between consecutive hidden state sequences are sought based on digital map, and state transition probabilities are calculated to construct a HMM chain. Finally, decoding the HMM chain yields the optimal path, serving as the final matching sequence. HMM-based methods perform well in scenarios with simple road network structures and relatively stable traffic. However, they face two limitations in handling scenarios with diverse positioning error distribution, dense road networks, and dynamic traffic conditions:

(1) They assume a uniform distribution of trajectory point errors across all urban areas, ignoring the variations caused by environmental factors (e.g., mountains, buildings) in different regions. Consequently, a standardized approach for candidate point search radius and observation probability calculation is employed. If the radius is set too large, unrelated roads may be selected, increasing computational load and impacting matching efficiency. Conversely, a radius set too small might overlook correctly matched roads, and the calculation of observation probability may lack accuracy.
(2) HMM-based map matching method combines the spatiotemporal transit relationship among trajectory points to get the real route traveled. During the process, the real-time road speed information is crucial. Existing methods rely on road speed limits or historical average speed, resulting in to inaccurate calculations of state transition probabilities.

To address the aforementioned limitations, this paper proposes a novel and scalable HMM map matching method based on trajectories of multiple types of vehicles. It is primarily based on the following insights: the trajectory of urban vehicles can be divided into the trajectories with fixed paths and those with free paths. Map matching for vehicles with fixed routes is relatively easier and more reliable, the matching results can aid in estimating the GPS error distribution in different regions and real-time road speeds, thereby addressing the shortcomings of existing HMM methods. In summary, the contributions of this article can be summarized as follows:

(1) It proposes an enhanced HMM-based map matching approach that effectively leverages fixed travel trajectories to obtain precise GPS error distributions and current road conditions.
(2) It improves the configuration of candidate radius, observation probability, state transition probabilities of HMM by employing a regional division based on GPS error homogeneity, an area-specific GPS error fitting using various distribution functions, and the correlation between bus and free vehicle speeds.
(3) We use the real GPS trajectory dataset of Shenzhen buses and taxis over half year, and enhanced our method on the existing two types of HMMs. The results indicate that our method can significantly improve the MM performance of the existing methods.

2 Related Work

The map matching (MM) problem can be classified into two main types based on processing scenarios: offline and online map matching.

In offline MM, complete vehicle trajectory data is available, various methods have demonstrated effective performance. The capability to utilize the entire global path information, including both starting and ending points, contributes to the success of many approaches in this scenario. Examples include the method based on Multiple Hypothesis Testing (MHT) [6] and shortest path [14,15].

Online MM mainly include statistics based methods and learning based methods. The learning-based matching methods [11] and [17], lack interpretability and necessitate an extensive number of training samples. At present, statistical methods still the mainstream approach. HMM is the most widely used statistical online MM methods, and first formally introduced by Newson et al. [9]. That calculated state transition probabilities by measuring the proximity of candidate path distances between points and the straight-line distances between corresponding pairs of trajectory points. However, the method performed poorly when dealing with significant bends in candidate paths. Lou et al. [7] introduced geometric topological constraints and road speed constraints into state transition probability calculation. Another method introduced in [13] not only considered the spatiotemporal constraints of GPS trajectories but also designed a weighted interaction strategy to simulate the influence among trajectory points. However, existing HMM matching methods often overlook the regional differences in calculating GPS error distributions within urban road networks. Additionally, they do not accurately acquire real-time road segment speeds when considering speed constraints. As a result, these methods still have limitations in terms of matching efficiency and accuracy when applied to online matching.

3 Preliminary

3.1 Some Definitions

Definition 1 (Trajectory Point, z): It is recorded by a vehicle's location device at a particular moment, including four attributes: longitude, latitude, current speed, and timestamp, denoted as $z = (lon, la, speed, t)$.

Definition 2 (Trajectory, Z): A time ordered sequence of trajectory points is recorded when a vehicle continuously travels over a certain period, denoted as $Z = (z_1, z_2, \ldots, z_n)$.

Definition 3 (Road Network, G): The vehicular traffic road network within a specific geographical area, denoted as $G = (V, E)$, where V is a set of connecting points between road (segments), and E represents the set of road (segments).

Definition 4 (Candidate Road, c): A candidate road is a road segment within a specified range where a observed trajectory point may potentially be matched.

Definition 5 (Candidate Point, s): The candidate point is a trajectory point on the candidate road that is closest to the observed trajectory point.

Definition 6 (Path, P): A path P is composed of a series of adjacent and connected road segments within the road network, represented as $e_1 \rightarrow e_2 \rightarrow \ldots \rightarrow e_n$. The starting and end points of the path are $e_1.start$ and $e_n.end$.

The map matching problem is defined as follows: Given a GPS trajectory sequence Z, find the actual vehicle path P within the road network G.

3.2 HMM Map Matching

To facilitate introduce our extended HMM matching method, we first give the processing of basic HMM map matching method involving three specific steps:

Step 1: For each trajectory point z_i in the GPS trajectory sequence Z, it first collects all relevant candidate roads $C_i = (c_i^1, c_i^2, \ldots, c_i^m)$ within a specified radius and the corresponding candidate points $S_i = (s_i^1, s_i^2, \ldots, s_i^m)$. The number of candidate roads m, may be vary due to the varying road density in different city areas. In this model, GPS trajectory points are considered as observed states, while candidate points are treated as potential hidden states. Each candidate point, denoted as s_i^j, is initially assigned an observation probability $p_o(z_i|s_i^j)$, representing the likelihood of matching the trajectory point. Previous research [3, 5,7–9] generally use a global Gaussian distribution to measure the likelihood based on the distance between two points:

$$p_o(z_i|s_i^j) = \frac{1}{\sqrt{2\pi\sigma^2}} e^{-\frac{(d_{sz}-\mu)^2}{2\sigma^2}} \tag{1}$$

where d_{sz} represents the Euclidean distance, μ and σ are the mean and standard deviation of the Gaussian distribution.

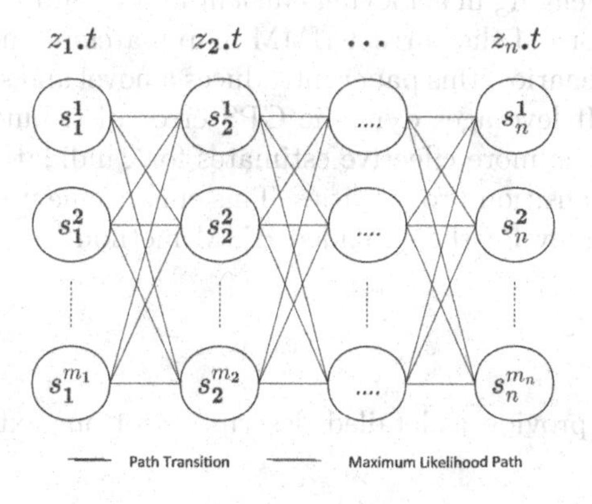

Fig. 1. Construct HMM chain

Step 2: then construct the HMM chain as illustrated in Fig. 1 based on hidden state transition probabilities between adjacent candidate points. The probability from s_i^j to s_{i+1}^k refers to the likelihood of the vehicle actually traveling from s_i^j to s_{i+1}^k given the two consecutive observed points z_i and z_{i+1}. That is usually calculated based on two ways: i) the similarity between the length of the path $e_1 \rightarrow e_2 \rightarrow \ldots \rightarrow e_n$ traversed from s_i^j to s_{i+1}^k (denoted as $w(s_i^j \rightarrow s_{i+1}^k)$) and the Euclidean distance between $z_i \rightarrow z_{i+1}$ (denoted as $d(z_i \rightarrow z_{i+1})$) [9]. ii) The similarity between the average travel speed $\overline{v}_{i\rightarrow i+1}$ along the path $s_i^j \rightarrow s_{i+1}^k$ and the constrained speed of the road segments traversed by the path [1,7,16] or historical average speeds [5,10,12]. The state transition probabilities is calculated by:

$$p_s(s_i^j \rightarrow s_{i+1}^k) = \frac{d(z_i \rightarrow z_{i+1})}{w(s_i^j \rightarrow s_{i+1}^k)} \cdot \frac{\sum_{u=1}^{n}(e_u.v \cdot \overline{v}_{i\rightarrow i+1})}{\sqrt{\sum_{u=1}^{n}(e_u.v)^2} \cdot \sqrt{(\sum_{u=1}^{n}(\overline{v}_{i\rightarrow i+1})^2)}} \tag{2}$$

Step 3: Decoding the HMM chain to find the hidden state sequence with the highest probability of observation and state transitions. Typically, the Viterbi algorithm is considered an effective method [2,3,5,8,9].

The timeliness and accuracy are two crucial aspects that online MM need to address. Firstly, the complexity of HMM depends mostly on the number of potential points in a known road network. The number is determined by the candidate radius, which, as we discussed earlier, is linked to GPS errors. Setting the candidate radius too large or too small can lead to delays or algorithm failure. Secondly, the accuracy is largely determined by the observation probabilities and transition probabilities. Observation probabilities depend on the GPS error distribution, and transition probabilities rely on the availability of roads over time and their speed limits. So only with accurate estimates of GPS error distribution and road speed constraints can the correct matching path be determined.

Using predefined parameters such as fixed GPS error distributions, road speed limits, or historical average speeds to estimate candidate radii or probabilities can be challenging in achieving efficient and accurate map matching. To address the limitations of the current HMM map matching models in such complex urban traffic scenarios, this paper introduces a novel and scalable HMM map matching method. It leverages dynamic GPS error distributions and real-time road speeds to provide more effective estimates for candidate radii, observation probabilities, and transition probabilities. This enhancement results in improved timeliness and accuracy for the extended HMM method.

4 Algorithm

In this section, we provide a detailed description of our extended HMM map matching algorithm.

4.1 Overview

We divide trajectories into two groups: trajectories with fixed routes (public buses) and those with flexible routes (taxis). The map Matching of trajectories with fixed routes is simpler and more dependable since it follows predefined paths. By utilizing the accurate matches, we can obtain GPS errors in various city areas and real-time road speeds with precision, which are the foundation for map matching of trajectories with flexible routes. The overall framework of our proposed algorithm is illustrated in Fig. 2. It can be divided into two stages: offline Analysis and online HMM map matching.

Offline Analysis: it first conducts bus trajectory map matching to estimate GPS error distributions in different city areas through combining the bus lines. Given the operational disparities between public transit buses and free vehicles, bus speeds do not accurately represent the speeds of free vehicles on the same roads. To address this, we perform offline map matching using historical complete free vehicle trajectories. Finally, we establish the connection between bus speeds on roads and free traffic speeds. We employed an offline map algorithm based on the shortest path to accomplish the matching of historical GPS data, a method that has demonstrated favorable performance in previous literature [14,15].

Online HMM map matching: It dynamically adapts the candidate radius and computes observation probabilities according to the GPS error distribution in the current area. Moreover, real-time road speeds are estimated for free trajectories using the bus-to-free traffic speed relationships, which are integrated into the calculation of state transition probabilities.

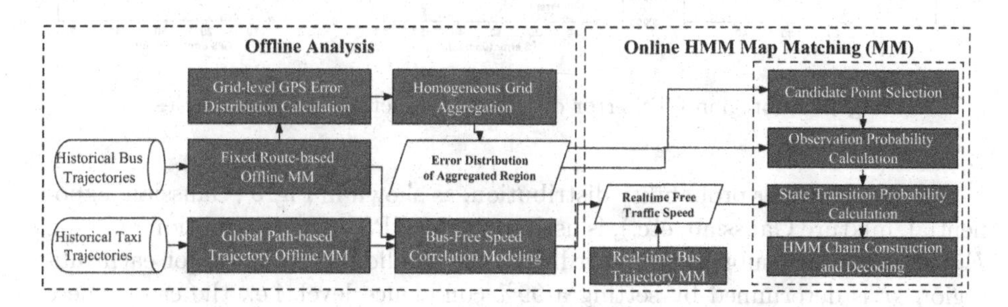

Fig. 2. The architecture of extended HMM-based MM method.

4.2 Offline Analysis

Candidate Radius and Observation Probability Definition. In order to set the candidate road search radius for trajectory points and calculate the observation probabilities more accurately and effectively, we employ precise matching of historical GPS data from fixed-route buses to obtain GPS error distributions in different areas of the city.

Grid-Level GPS Error Distribution Calculation: Firstly, we divide city region R into multiple 100m× 100m grids, denoted as $R = \{grid_1, grid_2, \ldots, grid_n\}$. For historical bus trajectories, we perform map matching and calculate GPS error d_i as the distance of each trajectory point z_i to its matched point s_i^k.

Then the GPS error of each grid is calculated as the average GPS error of the bus trajectory points (total number defined as q_j) within the grid:

$$d_{aver}(grid_j) = \frac{1}{q_j} \sum_{i=1}^{q_j} d_i \tag{3}$$

Homogeneous Grid Aggregation: Due to the incomplete coverage of urban areas by bus trajectories, it can be challenging to estimate GPS errors in all grids. To address this issue, we first estimate the missing values of grids by averaging data from nearby grids. Then, we use the spatial clustering method [4] to cluster adjacent grids and aggregate them based on $d_{aver}(grid_j)$, ultimately dividing the map into multiple sub-regions with similar GPS error distributions. This clustering process divides R into multiple sub-regions $R = \{sr_1, sr_2, \ldots, sr_n\}$.

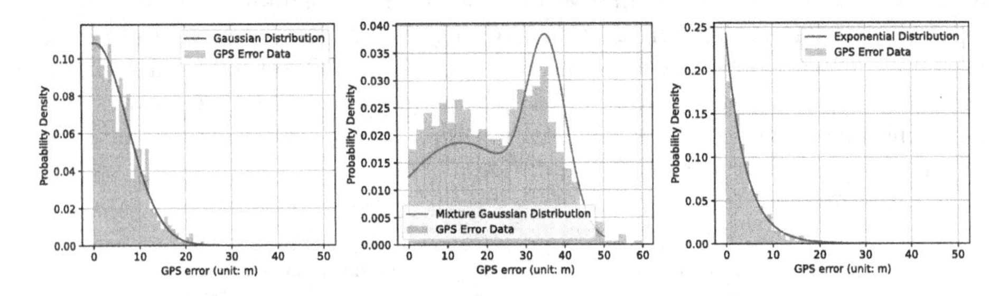

Fig. 3. Subregion GPS error distribution function fitting examples.

Finally, a proper probability distribution, as shown in Fig. 3 (Gaussian, exponential, mixture Gaussian, etc.), is used to fit the GPS error distribution function $F_{sr_i}(d_{sz})$ for each subregion. Accordingly, the candidate radius r_{sr} for each subregion sr is determined by setting a 99% confidence level, i.e., the error where the cumulative distribution function reaches 99%. We use $F_{sr_i}(d_{sz})$ from the subregion to calculate the observation probabilities for candidate points:

$$p_o(z_i|s_i^j) = F_{sr_i}(d_{sz}) \tag{4}$$

State Transition Probability Calculation. We perform offline MM on historical trajectories to estimate the speeds of bus and free vehicles on each road, respectively, then build their relationship, which are finally used to estimate the state transition probability for free trajectories in real-time.

Bus Speed estimation: we first exclude the segments of the bus trajectories near bus stops to eliminate the impact of slowing down and boarding/alighting on calculating the average road speed. Next, we align the filtered trajectory with the bus routes to determine the bus speed on the roads traveled at the corresponding timestamps. We divide a day into 288 time slots (every 5 min) and record the average bus speed on each road for each time slot.

Free Traffic Speed Estimation: We use complete historical free travel trajectories, combining effective routes between the starting and ending pointspath [15], and consider path similarity for offline road matching. This allows us to ultimately compute the average speed of free vehicles traveling on each matched road segment for each time interval.

Speed Relationship Modeling: By analyzing the average speed of buses and free traffic in the same space-time context, we can calculate their trend similarity, with the average speed of buses being lower than that of free traffic. Therefore, we consider modeling the speed relationship using the following formula:

$$v_f = w_1(\lambda v_b) + w_2 v_{hf} \tag{5}$$

Here, we comprehensively consider the average speed of buses v_b during a specific time period and the historical average speed of free traffic v_{hf} with weights $w_1 = 0.8, w_2 = 0.2$, and a speed coefficient $\lambda = 1.2$. Based on the relationships, we can estimate the free traffic speed v_f of vehicles on road segments based on the real-time speed of buses.

Using real-time free traffic speed, we calculate the state transition probabilities by Eq. 2.

4.3 Online HMM Map Matching

In this section, we introduce the online HMM map matching process based on the previously defined candidate radius, observation probabilities, and state transition probabilities. This process consists of four main steps:

Step 1: Real-time Speed Maintenance: Before running the algorithm matching process, we initiate a real-time maintenance process for estimating free traffic speeds. This process calculates the current speed for each road segment using real-time GPS data from buses. It constantly receives the latest GPS trajectory sequences from buses, performs map matching, and calculates the real-time average speed for each road e, the free traffic speed $e.v_f$ is estimated based on Eq. 5. These free traffic speeds are stored in a dynamic hash table, which is continuously updated in real-time:

$$vmap = Map < e.id, e.v_f > \tag{6}$$

Step 2: Candidate Point Preparation. For each trajectory point z awaiting matching, we first get the corresponding candidate point set S based on the subregion related candidate radius r_{sr_i}. Then the Euclidean distance d_{sz} between each candidate point s and trajectory point z is calculated, which is used to calculate the observation probability for each candidate point based on the subregion's error distribution function $F_{sr_i}(d_{sz})$ using the Eq. 4.

Step 3: HMM Chain Construction. A directed graph structure $G_h = (V_h, E_h)$ is used to represent the HMM chain, where V_h consists of the candidate point sets for all trajectory points $\{S_1, S_2, \ldots, S_n\}$ within the a predefined matching window. Each edge in E_h is a reachable path $P_{s_i^j \to s_{i+1}^k} = e_1 \to e_2 \to \ldots \to e_n$ between adjacent candidate points. We retrieve the real-time road speed $e.v_f$ in the path from *smap* for state transition probability calculation. The probability of matching $P_{s_i^j \to s_{i+1}^k}$ in the result path sequence is represented as a combination of observation probability and state transition probability:

$$p_o(z_{i+1}|s_{i+1}^k) \cdot p_s(s_i^j \to s_{i+1}^k) \tag{7}$$

Step 4: Decoding the Matching Result. Utilizing the Viterbi algorithm to identify the candidate point sequence $s_1^i \to s_2^j \to \ldots \to s_n^k$ that maximizes the sequence probability from $G_h = (V_h, E_h)$, with the aim of obtaining the corresponding road match.

5 Experiments

Dataset Description: We collected GPS data for buses and taxis in Shenzhen, China, with a 20-second interval, from January 1 to June 30, 2023.

Baselines: By Using the candidate radius, observation probabilities, and state transition probabilities specified in this paper, we enhanced two widely used HMM-based methods, namely HMMM [9] and ST-Matching [7]. The improved versions are referred to as E-HMMM and E-ST-Matching.

In our experiments, we set the candidate road search radius to commonly used fixed values of 50 m for original HMMM and 100 m for ST-Matching. To estimate observation probabilities, we applied a Gaussian distribution with parameters $\mu = 0$ and $\sigma = 20$ m as specified in the baseline papers [7,9]. The road speed constraint parameters were determined based on the speed limits specified in the OSM map.

Evaluation Methods. To evaluate our algorithm's performance, we used two metrics across various sampling rates: precision (*Prc*) and average matching time (*Mt*). *Prc* is the ratio of correctly matched path total length to the total length of matched paths. *Mt* represents the average time taken to obtain the matching result for each trajectory point.

5.1 Experiment Results

Prc at Different Sampling Rates: As depicted in Fig. 4, our method enhances the precision of the baseline methods. The precision tends to decrease for all methods as the sampling rate drops. This is because a longer sampling interval increases the distance between sample points, leading to more potential candidate paths and higher matching error rates. However, our method excels in accurately evaluating real-time speed constraints on road segments, allowing for more precise evaluation of vehicle likelihood on each candidate path in complex traffic conditions. As illustrated in Fig. 5, due to the proximity of the point z_2 to e_3' and the similarity in the path length between $e_3' \to e_4'$ and $e_3 \to e_4$, the baseline method matches the traveled path to $e_3^{'} \rightarrow e_4^{'}$. However, in reality, z_2 experiences significant drift in GPS positioning in this area due to obstructive structures like tall buildings. Additionally, $e_4^{'}$ has lower traffic flow speed and experiences congestion during the time period. Our method effectively addresses these issues, providing the correct matched path $e_3 \rightarrow e_4$.

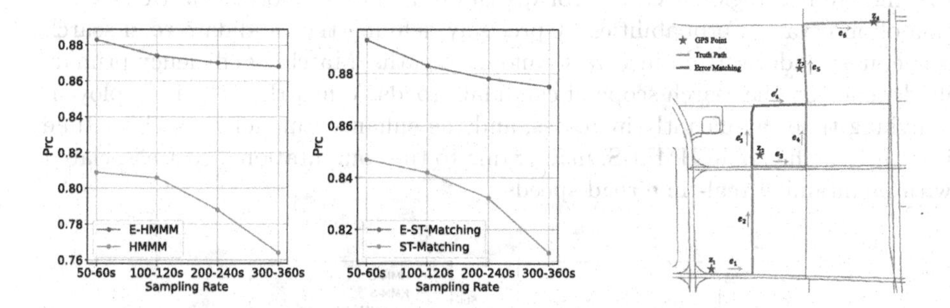

Fig. 4. Comparison of *Prc* metric.

Fig. 5. MM example.

Mt at Different Sampling Rates. As shown in Fig. 6, our method decreases the matching time for all baseline approaches. In contrast to the baseline methods that use a fixed candidate search radius, our approach dynamically adjusts the search range based on error distribution in various regions. This effectively reduces the number of candidate matching points at the start and end points, thus decreasing the number of candidate paths between them, resulting in reduced matching time and enhanced matching efficiency.

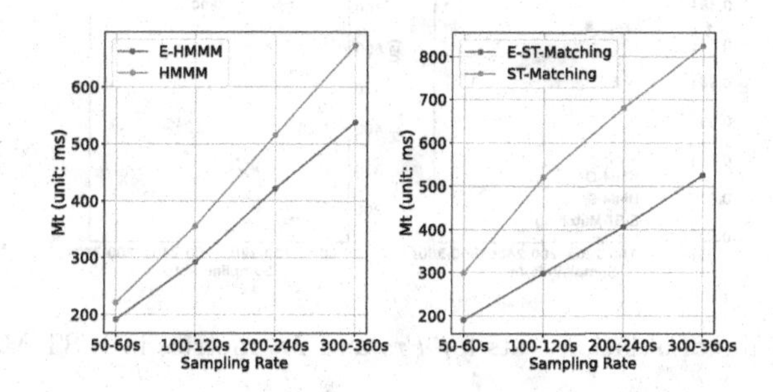

Fig. 6. Comparison of *Mt* metric.

Ablation Test. We conducted an ablation test for the two enhancement methods: E-HMMM and E-ST-Matching. We use HMM-O and HMM-S to represent the ablated methods that eliminate region-dependent candidate radius settings and use the original fixed radius instead, and eliminate the setting of real-time road speed and use fixed road speed limit instead.

The result is shown in Fig. 7 and Fig. 8. It is evident that real-time road speed setting has a significant influence on improving matching precision. This could be attributed to factors like urban traffic congestion, which often prevent vehicles from consistently adhering to speed limits. Consequently, using road speed limits as parameters for estimation might bring some level of distortion. Furthermore, HMM-S also shows some loss in matching precision. This is partly because the fine-grained regional GPS error distribution enables more accurate calculation of observation probabilities by precisely defining the candidate road search scope, it excludes certain incorrect candidate paths. Matching efficiency primarily depends on the search scope of candidate roads. When HMM-O is employed, matching time significantly increases, and the enhanced model's matching time is slightly higher than HMM-S, mainly due to the computational cost associated with maintaining real-time road speeds.

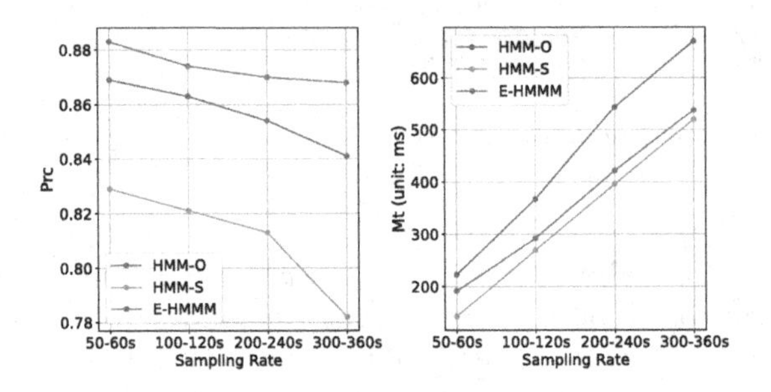

Fig. 7. Results of ablation tests for *Prc* and *Mt* metrics in E-HMMM.

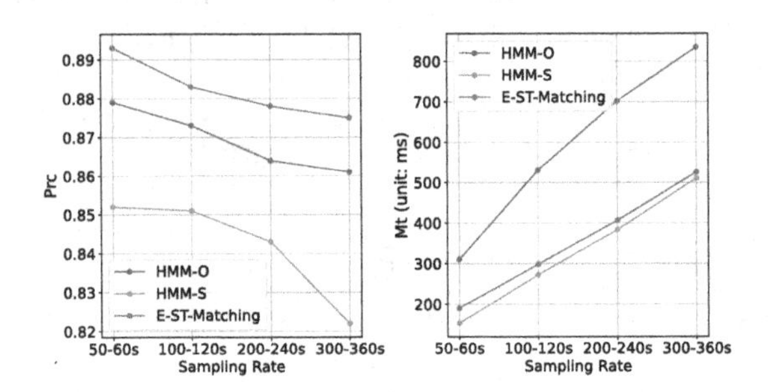

Fig. 8. Results of ablation tests for *Prc* and *Mt* metrics in E-ST-Matching.

6 Conclusion

In this paper, we utilizes fixed-route bus trajectory to enhance any HMM map matching approach. We specifically use the accurate road matching abilities of fixed-route buses to determine GPS error distribution in different urban areas, addressing the limitations of existing HMM methods that rely on fixed error patterns. This advancement improves candidate road selection and observation probability calculations, leading to faster matching and better accuracy. Additionally, we suggest estimating a road's free traffic speed using real-time bus speeds, enabling more precise assessment of real-time speed constraints during the matching process and enhancing state transition probability calculations. We tested our method on real traffic data from Shenzhen, and the experiments showed that our method outperforms the baselines.

Acknowledgement. This study was funded by the National Natural Science Foundation of China (No. 62372443, No. 62376263), Shenzhen Industrial Application Projects of undertaking the National key R & D Program of China (No. CJGJZD20210408091600002).

References

1. An, Q., Feng, Z., Chen, S., Huang, K.: A green self-adaptive approach for online map matching. IEEE Access **6**, 51456–51469 (2018)
2. Chao, P., Xu, Y., Hua, W., Zhou, X.: A survey on map-matching algorithms. In: Borovica-Gajic, R., Qi, J., Wang, W. (eds.) ADC 2020. LNCS, vol. 12008, pp. 121–133. Springer, Cham (2020). https://doi.org/10.1007/978-3-030-39469-1_10
3. Chen, R., Yuan, S., Ma, C., Zhao, H., Feng, Z.: Tailored hidden Markov model: a tailored hidden Markov model optimized for cellular-based map matching. IEEE Trans. Industr. Electron. **69**(12), 13818–13827 (2021)
4. Guo, D., Wang, H.: Automatic region building for spatial analysis. Trans. GIS **15**, 29–45 (2011)
5. Hu, G., Shao, J., Liu, F., Wang, Y., Shen, H.T.: If-matching: towards accurate map-matching with information fusion. IEEE Trans. Knowl. Data Eng. **29**(1), 114–127 (2016)
6. Knapen, L., Bellemans, T., Janssens, D., Wets, G.: Likelihood-based offline map matching of GPS recordings using global trace information. Transp. Res. Part C Emerg. Technol. **93**, 13–35 (2018)
7. Lou, Y., Zhang, C., Zheng, Y., Xie, X., Wang, W., Huang, Y.: Map-matching for low-sampling-rate GPS trajectories. In: Proceedings of the 17th ACM SIGSPATIAL International Conference on Advances in Geographic Information Systems, pp. 352–361 (2009)
8. Mohamed, R., Aly, H., Youssef, M.: Accurate real-time map matching for challenging environments. IEEE Trans. Intell. Transp. Syst. **18**(4), 847–857 (2016)
9. Newson, P., Krumm, J.: Hidden Markov map matching through noise and sparseness. In: Proceedings of the 17th ACM SIGSPATIAL International Conference on Advances in Geographic Information Systems, pp. 336–343 (2009)
10. Ozdemir, E., Topcu, A.E., Ozdemir, M.K.: A hybrid hmm model for travel path inference with sparse GPS samples. Transportation **45**, 233–246 (2018)
11. Taguchi, S., Koide, S., Yoshimura, T.: Online map matching with route prediction. IEEE Trans. Intell. Transp. Syst. **20**(1), 338–347 (2018)
12. Xu, M., Du, Y., Wu, J., Zhou, Y., et al.: Map matching based on conditional random fields and route preference mining for uncertain trajectories. Math. Probl. Eng. **2015** (2015)
13. Yuan, J., Zheng, Y., Zhang, C., Xie, X., Sun, G.Z.: An interactive-voting based map matching algorithm. In: 2010 Eleventh International Conference on Mobile Data Management, pp. 43–52. IEEE (2010)

14. Zhang, D., Dong, Y., Guo, Z.: A turning point-based offline map matching algorithm for urban road networks. Inf. Sci. **565**, 32–45 (2021)
15. Zhang, D., Guo, Z., Guo, F., Dong, Y.: An offline map matching algorithm based on shortest paths. Int. J. Geogr. Inf. Sci. **35**(11), 2238–2261 (2021)
16. Zhang, Y., He, Y.: An advanced interactive-voting based map matching algorithm for low-sampling-rate GPS data. In: 2018 IEEE 15th International Conference on Networking, Sensing and Control (ICNSC), pp. 1–7. IEEE (2018)
17. Zhao, K., et al.: Deepmm: deep learning based map matching with data augmentation. In: Proceedings of the 27th ACM SIGSPATIAL International Conference on Advances in Geographic Information Systems, pp. 452–455 (2019)

A Multimodal and Multitask Approach for Adaptive Geospatial Region Embeddings

Rajjat Dadwal[1,2](✉), Ran Yu[1,2], and Elena Demidova[1,2]

[1] Data Science and Intelligent Systems Group (DSIS), University of Bonn, Bonn, Germany
{dadwal,ran.yu,elena.demidova}@cs.uni-bonn.de
[2] Lamarr Institute for Machine Learning and Artificial Intelligence, Bonn, Germany
https://lamarr-institute.org

Abstract. Geospatial region embeddings are vital in developing predictive models tailored to urban environments. Such models enable critical applications, including crime rate prediction and land usage classification. However, state-of-the-art methods typically generate embeddings based on fixed administrative regions. These regions may not always align with specific tasks or areas of user interest. Creating fine-grained embeddings tailored to specific tasks and regions of user interest is labor-intensive and requires substantial resources. In this paper, we propose *MAGRE* – a novel approach that generates fine-granular adaptive geospatial region embeddings by leveraging multimodal and multitask learning. The embeddings generated by *MAGRE* can be flexibly aggregated to suit various region boundaries, rendering them effective in diverse urban applications. Our experimental results demonstrate that *MAGRE*'s embeddings outperform state-of-the-art embedding baselines, resulting in a 25.73% reduction in root mean squared error for crime rate prediction and a 19.08% reduction for check-in count prediction.

Keywords: Adaptive Geospatial Embeddings · Multitask Learning

1 Introduction

Real-world applications that rely on geographic data often require embeddings of the regions of interest (ROIs) for a particular user and a task. Geospatial region embeddings play an essential role in consolidating information across sources and enable capturing complex spatial relationships within and across regions. Such embeddings have proven beneficial in various applications, including land use classification and crime rate prediction [12,16]. However, embeddings created by existing methods may not align with the regions and tasks of user interest.

The mismatch between the embedding provided by the state-of-the-art methods and the ROIs results from the substantial limitations of the existing geospatial region embedding approaches. First, conventional geospatial embeddings rely on fixed administrative boundaries, such as districts [12,15]. Second,

D.-N. Yang et al. (Eds.): PAKDD 2024, LNAI 14649, pp. 363–375, 2024.
https://doi.org/10.1007/978-981-97-2262-4_29

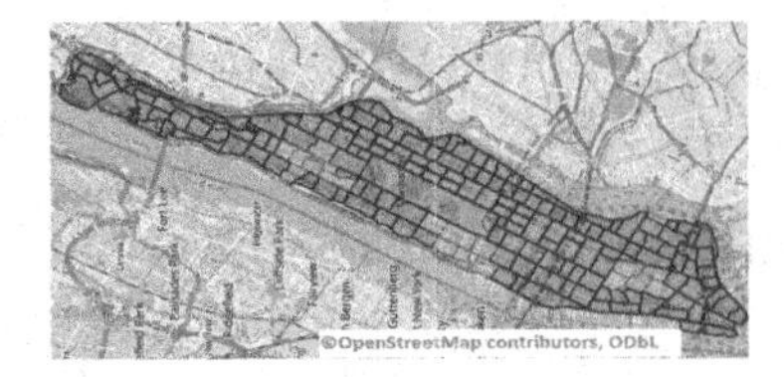

Fig. 1. Manhattan division based on administrative boundaries (left) and based on hexagonal grids (right). The user ROIs are marked in orange. Map data: ©OpenStreetMap contributors, ODbL. (Color figure online)

geospatial embeddings created for smaller geometric-shape regions typically rely on the skip-gram model [11], which is unsuitable for embedding aggregation [1]. Third, existing approaches use satellite imagery [13] as multimodal contextual information. However, satellite imagery has limited accessibility and requires substantial data acquisition and preprocessing effort. Finally, region embeddings are often designed for specific tasks [12], neglecting significant factors of urban dynamics and patterns, and may fail to generalize to unseen tasks.

For example, Fig. 1 illustrates the Manhattan division based on administrative boundaries and smaller hexagonal-shaped grid cells. A user may be interested in assessing crime rates for property purchases for the ROIs (encoded in orange). The spatial misalignment between the user's ROIs and the pre-computed region embeddings based on administrative boundaries may result in an inaccurate crime rate assessment. In contrast, the union of the grid-cell-based embeddings can capture the ROI more precisely.

This paper introduces a novel approach for obtaining geospatial region embeddings efficiently, focusing on adaptable regions of interest (ROIs). Our idea involves generating adaptive region embeddings by embedding smaller geospatial units (grid cells) and dynamically aggregating them into an ROI flexibly on demand. However, such aggregation is challenging. Due to limited data, the representation of individual grid cells might lack context and broad applicability. Furthermore, the semantics of the ROI as a whole may differ from that of the union of its constituent grid cells. We tackle these challenges with a multimodal and multitask approach, incorporating rich visual cues and graph context.

We propose *MAGRE* – a novel multitask and multimodal adaptive geospatial region embedding approach. In contrast to conventional methods, *MAGRE* partitions the geospatial region into smaller hexagonal grid cells, which can be flexibly aggregated to match the specific ROI. In addition to features from various cross-modal sources such as Points of Interest (POIs) and mobility data, *MAGRE* also extracts the image for each grid cell from OpenStreetMap (OSM)[1] to obtain visual information for creating comprehensive embeddings. We generate various graphs utilizing the extracted features. These graphs capture similarities and

[1] OpenStreetMap: https://www.openstreetmap.org/. The OpenStreetMap name is a trademark of the OpenStreetMap Foundation and is used with their permission. We are not endorsed by or affiliated with the OpenStreetMap Foundation.

rich context of grid cells based on various factors, including mobility patterns, locality, and infrastructural attributes. To train *MAGRE*, we propose a multi-task learning approach, which enables the embedding to learn region semantics from different perspectives and reduces overfitting through shared representations. Based on the fine-grained grid embeddings, *MAGRE* can efficiently generate embeddings for any ROI by embedding aggregation. The aggregated region embeddings effectively preserve the semantic information, as demonstrated in our experiments on several downstream tasks. Our contributions are as follows:

- We propose *MAGRE* – an adaptive region embedding approach, which creates representations that accurately capture the spatial properties and relationships between the hexagonal grid cells and can embed regions of flexible shape and size through efficient aggregation.
- *MAGRE* leverages multimodal data from various sources to build region embeddings. To the best of our knowledge, we are the first to incorporate visual cues from map images into region embedding, effectively capturing the context and features of urban regions. Our feature analysis results demonstrate the importance of map images across different tasks.
- To enhance embedding generalizability, *MAGRE* embraces multitask learning, where we train our model on two tasks and test the geospatial embeddings on unseen tasks. Experimental results demonstrate that *MAGRE* outperforms the state-of-the-art methods, leading to a root mean squared error reduction of 25.73% and 19.08% for crime rate prediction and check-in count prediction, respectively.

2 Definitions and Problem Formulation

In this section, we introduce the relevant definitions and formulate the problem of spatial region embeddings.

Definition 1 (Geospatial grid cell). *A geospatial grid cell, denoted as g, is a minimal spatial unit characterized by specific geometric boundaries. A grid cell is associated with features in different categories. Features of a grid cell g_i based on a feature category f are denoted as a vector $\vec{h}^i_f$.*

We adopt hexagonal geospatial grid cells. Feature category f can represent the frequency of different types of POIs or mobility patterns. We represent the relationships between the grid cells according to f as a grid graph.

Definition 2 (Grid graph). *We denote a grid graph as $\mathcal{G}_f = (\mathcal{V}, \mathcal{E}, \mathcal{A}_f)$, where $\mathcal{V} = \{g_1, ..., g_n\}$ represents the set of grid cells, $\mathcal{E}$ is the set of edges. $\mathcal{A}_f$ denotes the weighted adjacency matrix associated with the feature category f. $A^{ij}_f = sim(\vec{h}^i_f, \vec{h}^j_f)$, where $sim(\cdot)$ denotes the similarity function.*

Grid cell similarity can be computed as the cosine similarity between their feature vectors. To enable efficient grid cell representation, we rely on embeddings.

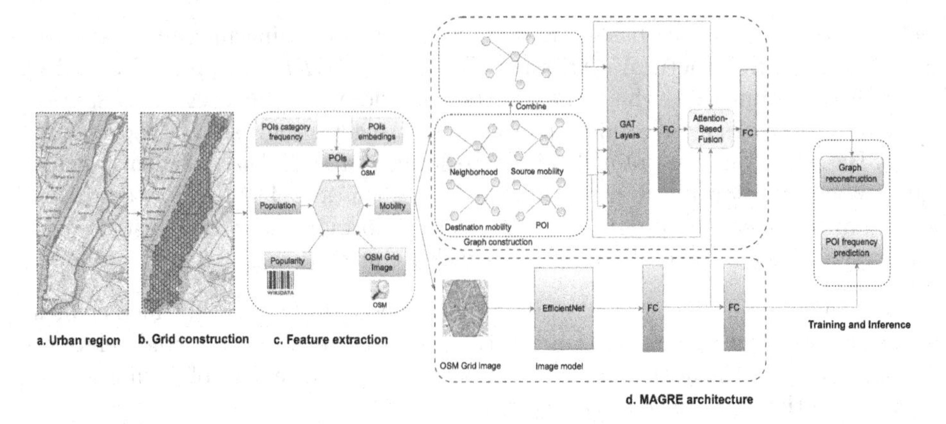

Fig. 2. The overall architecture of the proposed *MAGRE* approach. Map data: ©OpenStreetMap contributors, ODbL.

Definition 3 (Grid cell embedding). *For a grid cell* g_i*, an embedding* $e_i = \phi(g_i)$*,* $e_i \in R^d$ *is a d-dimensional dense vector representation of* g_i*.* $\phi(\cdot)$ *is an embedding function capturing semantic and contextual information of* g_i*.*

In this paper, we address the problem of adaptive spatial region embedding for spatial regions of user interest, e.g., a district or a new business area.

Definition 4 (Spatial region). *A spatial region, denoted as* r*, is a geographic area defined by specific boundaries. A spatial region* r *can be represented by a set of spatial grid cells* $\{g_1, ..., g_n\}$ *it contains or intersects with.*

We aim at generating embeddings tailored to any geospatial region, based on the embeddings of the contained spatial grid cells.

Definition 5 (Spatial region embedding). *For a given spatial region* r*, the geospatial embedding* e_r *is constituted by the aggregation of the grid cell embeddings* $e_r = \gamma(\{\phi(g_i)\})$*, where* $g_i \in r$ *and* $\gamma(\cdot)$ *is an aggregation function, such that* e_r *retains the semantics of* r*.*

3 The *MAGRE* Approach

The architecture of the proposed *MAGRE* approach is illustrated in Fig. 2. In this section, we provide a detailed description of each step.

3.1 Grid Construction and Feature Extraction

In this step, we partition the entire geographic area into hexagonal grid cells and extract features from multimodal data for each grid cell.

Grid Construction. We opt for a hexagonal grid to partition urban regions compared to other geometric shapes, as illustrated in Fig. 2b. The hexagonal grid has several advantages. First, hexagonal grids experience less distortion caused by the earth's curvature compared to the shape of a fishnet grid [3]. Second, due to the consistent length of each side, the centroids of neighboring cells are equidistant [3]. We set the size of each side of the hexagonal grid to 250 m, such that the resulting hexagon area is comparable to [11].

Feature Extraction. We express each grid cell as feature vectors representing POIs count, mobility patterns, OSM images, population statistics, and popularity count, as illustrated in Fig. 2c.

- **POIs.** The type distribution of POIs in a region provides important semantic indicators, such as urban types. To capture such semantics, we extract all the POIs corresponding to each grid cell and map them to OSM categories, resulting in 12 categories. Each grid cell contains a feature vector of length 12 representing the *POIs categories frequency* with categories as amenity, barrier, highway, leisure, man-made, natural, office, power, public transport, railway, shop, and tourism. Additionally, to maintain POI semantics, we aggregate the names of all POI venues within a particular grid cell and utilize the sentence transformer [7] to generate *POIs embeddings*. The concatenated feature vector of *POIs category frequency* along with *POIs embeddings* is denoted $\vec{h}_{poi}$.
- **Mobility patterns.** Human mobility patterns are pivotal in understanding the underlying correlations between regions [10]. Regions with similar incoming or outgoing mobility patterns often have similar functions and are closely connected from the human mobility perspective [14]. The number of trips originating from and ending at a grid cell is concatenated, denoted as $\vec{h}_{mob}$.
- **Population and popularity.** A region's population can reflect socioeconomic indicators. We aggregate population statistics as a grid feature. We extract the popularity count of each grid cell using POI Wikidata links. A higher number of POI Wikidata links in a grid cell acts as a proxy for popularity. We denote the population and popularity frequency as $\vec{h}_{pp}$.
- **Map images.** The visual representation of spatial regions helps to recognize and distinguish various characteristics. For example, OSM distinct colors to represent different objects facilitate visual map interpretation. We partition the map into multiple images, each capturing a specific grid cell. These images can reveal substantial patterns, such as the POI density.

3.2 *MAGRE* Model Architecture

In this section, we present our model architecture in detail. We consider two tasks to design the objective functions: grid graph reconstruction, and POI frequency prediction. To learn the joint multitask representation, we apply an attention-based fusion, followed by training and inference, as illustrated in Fig. 2d.

Grid Graph Reconstruction. We first construct different grid graphs, capturing the semantic and spatial similarity between grids. Each graph $\mathcal{G}_f$ is constructed by computing an adjacency matrix $\mathcal{A}_f$ of all grid cells as described in Definition 2. This results in a grid-graph $\mathcal{G}_{poi}$ based on POIs ($\vec{h}_{poi}$), two grid-graphs based on mobility patterns using source and destination frequency ($\vec{h}_{mob}$), represented as $\mathcal{G}_{src}$ and $\mathcal{G}_{dst}$, respectively. In addition, we create a grid graph $\mathcal{G}_{nbh}$ for the neighborhood information, capturing the relationship between grids based on their geospatial proximity. We also build a grid graph, which is a combination of the average of all the adjacency matrices from different grid graphs, represented as $\mathcal{G}_{cmb} = (\mathcal{V}, \mathcal{A}_{cmb})$, such that $\mathcal{A}_{cmb} = \frac{1}{|f|}\sum_{i=1}^{|f|} \mathcal{A}_i$, where $f \in \{poi, src, dst, nbh\}$, as illustrated in Fig. 2d. The intuition behind averaging adjacency matrices is that grid cells with high average weights in the combined matrix indicate strong and consistent connections across different modalities.

We employ Graph Attention Networks (GAT) [9] to extract meaningful representations from grid graphs. GAT is specially designed for graph-structured data and employs an attention mechanism. This mechanism facilitates the update of grid representations by efficiently propagating information to neighboring grids within each grid graph. We represent grid cell feature as $\vec{h^i}$ where $\vec{h^i} \in \{\vec{h}^i_{poi}||\vec{h}^i_{mob}||\vec{h}^i_{pp}\}$ and $||$ represents the concatenation operator. The GAT layer updates the grid representations through the following steps. First, we incorporate edge weights A^{ij} as an additional feature along with the grid features $\vec{h^i}$ and $\vec{h^j}$ in the learning process, given as $c_{ij} = \exp(\mathrm{ReLU}(\vec{a}^T[W\vec{h^i}||W\vec{h^j}||W_e A^{ij}]))$. The c_{ij} calculation is performed only for grids $j \in N_i$, where N_i denotes the set of the top N neighbors of the grid i, ranked according to adjacency matrix weights for the grid i (including i). To ensure the comparability of coefficients across different grids, we normalize them using the softmax function $\alpha_{ij} = \mathrm{softmax}(c_{ij})$. Next, we compute the updated grid representation $\vec{h^{i'}}$ by applying a weighted sum of the neighboring grid representations as $\vec{h^{i'}} = \sigma(\sum_{j\in N_i} \alpha_{ij} W\vec{h^j})$, where the weights are given by α_{ij} and σ denotes the activation function. To improve model convergence, we implement the skip connection mechanism, wherein certain layers in the neural network are skipped, and the output of one layer is directly fed to the subsequent layers. We concatenate the feature vector $\vec{h^i}$ with $\vec{h^{i'}}$, resulting in $\vec{h^{i''}}$ which denotes the representation of the grid cell i for a grid graph. We utilize a multi-head attention mechanism within each GAT layer to enhance performance, as proposed by [9]. In practice, we apply three GAT layers [9] followed by fully connected (FC) layers on each grid graph, namely $\mathcal{G}_{src}$, $\mathcal{G}_{dst}$, $\mathcal{G}_{poi}$, $\mathcal{G}_{nbh}$ and $\mathcal{G}_{cmb}$. This process yields the hidden representations $E_G = \{\vec{h}''_{src}, \vec{h}''_{dst}, \vec{h}''_{poi}, \vec{h}''_{nbh}, \vec{h}''_{cmb}\}$.

POI Frequency Prediction. In this task, we leverage Convolutional Neural Networks (CNN) to extract meaningful representations of grid images. We aim to train a model based on POI frequency, capable of learning object distribution within a grid image. We develop a regression model ψ incorporating the EfficientNet architecture [8] as its base model. EfficientNet is an image classification model known for its state-of-the-art accuracy, achieved with fewer model parameters. We customized EfficientNet for our regression task, which predicts the number of POIs in a given grid image. Formally, given a grid image dataset I_{img}, where $I_{img} \in \{i_1, .., i_k, .., i_n\}$ such that i_k represents the image for a given grid cell k. We apply the regression model ψ which predicts the POI frequency $\hat{y}$. We extract the intermediate representation of the model, i.e., E_{img}.

Attention-Based Fusion. Finally, a multi-head attention-based fusion is applied to the embeddings from the multiple grid graphs and the grid images. This fusion helps to propagate knowledge across the representations of different modalities, given as $E = MultiHeadAtt(E_G||E_{img})$. To reduce the dimensionality, we apply an FC layer, i.e., $e = FC(E)$, representing the grid cell embeddings.

Training and Inference. We employ the graph reconstruction task to train the graph reconstruction module in an unsupervised way. That is, having obtained the different representations for each grid graph, we reconstruct the original adjacency matrix $\mathcal{A}_f$ with $\hat{\mathcal{A}}_f = sigmoid(e.e^T)$. We employ Mean Square Error (MSE) loss to compute the reconstruction loss, represented as $\mathcal{L}_f^{rec} = ||\mathcal{A}_f - \hat{\mathcal{A}}_f||^2$. The smooth L1 loss [5] is utilized as the loss function for predicting POI frequency in grid images which combines the benefits of both L1 and L2 loss, making it suitable for handling outlier values. For instance, the contrasting frequency of POIs between the Manhattan Central Park grid, which has very few POIs, and other regions illustrates this variability. The formal definition of the smooth L1 loss between the original POIs count (y) and the predicted values ($\hat{y}$) is computed as in [5]:

$$\mathcal{L}_{img}^{smooth} = \begin{cases} \sum_{i=1}^{n} 0.5(\hat{y}_n - y_n)^2/\beta & \text{if } |\hat{y}_n - y_n| < \beta \\ \sum_{i=1}^{n} |\hat{y}_n - y_n| - 0.5 * \beta & \text{otherwise,} \end{cases} \tag{1}$$

where β specifies the threshold for switching between L1 and L2 loss.

We define the loss function for our model as a combination of the loss of the two objective tasks: $\mathcal{L}_{tot} = \sum_{k \in f} \mathcal{L}_k^{rec} + \mathcal{L}_{img}^{smooth}$. During training, model parameters and all the embeddings are learned through backpropagation.

3.3 Embedding Aggregation for Spatial Regions

Once the grid cell embeddings are generated, the next task is to aggregate embeddings for a given region r. We sum the grid cell embeddings for a region r, represented by the set of spatial grid cells it contains or intersects with, and obtain $e_r = \sum_{i=1}^{m} e_i$, where m is the number of grid cells in the region r. Then, we utilize embedding e_r for the downstream tasks. Following [16], for the regression tasks, the aggregated embeddings are passed through a fully connected neural network, followed by the prediction layer resulting in the prediction value $\hat{o_i}$. To optimize the regression tasks, we use the MSE loss function as $\mathcal{L}_{agg} = \frac{1}{p} \sum_{i=1}^{p} (\hat{o_i} - o_i)^2$, where p is the number of spatial regions in the downstream task.

4 Experimental Setup

We assess the generalizability and effectiveness of the spatial region embeddings generated by *MAGRE* on unseen tasks. This section describes the downstream tasks, datasets, baselines, parameter settings, and evaluation metrics.

Downstream Tasks and Datasets. We experiment with three distinct downstream tasks, including two regression tasks – crime rate prediction and check-in count prediction, and one classification task – land use classification. For evaluation, we consider the following datasets containing publicly available statistical and geographical data:

- Crime rate statistics: We use crime statistics, i.e., the count of crimes per region, provided by [16].
- Check-in count statistics: We use the count of check-ins per region, provided by [16].
- Land use classification: We use the district divisions determined by the community boards [2] as the reference, which corresponds to 12 categories [16].

Following past works [12], we select the Manhattan City area. For feature generation, we consider the following data:

- POI data: approx. 48,000 POIs extracted from the OpenStreetMap.
- Taxi data: anonymized data which contains start and end locations of approximately 5 million taxi trips in 2015[2].
- Images: We extract images of each grid cell from OSM.
- Popularity data: We extract the Wikidata tag for each POI from OSM, and compute the number of Wikidata links for each POI with a SPARQL query.
- Population statistics for 3,930 administrative regions, aggregated to our grids (see footnote 2).

We map our hexagonal grids to the existing division of the Manhattan region, consisting of 180 census blocks based on street boundaries. This alignment ensures a fair and accurate comparison with the baselines. Particularly for the land classification task, we further cluster the aggregated embeddings into 12 groups, to align with the number of distinct labels in the ground truth [12].

Baselines. We compare our *MAGRE* method with the baseline methods for region embeddings. HREP [16] captures both intra-region and inter-region correlations by integrating statistical taxi data and POI data. MG-FN [12] is a joint learning approach, utilizing mobility patterns for region representation. MVURE [15] is a multi-view graph representation approach that uses region correlations based on POIs, taxi statistics, and check-in statistics to learn urban region embeddings. Hex2Vec [11] relies only on OSM data incorporating a skip-gram model to create vector representations of hexagonal regions. Similarly, RegionDCL [6] considers only OSM building footprints and employs dual contrastive learning for region embeddings. MV-PN [4] constructs a multi-view POI-POI network using POIs and human mobility data based on autoencoder. Moreover, we explore variations of *MAGRE* where we employ a Graph Autoencoder (GAE) and node2vec instead of GAT layers on the grid graphs, represented as $MAGRE_{n2v}$ and $MAGRE_{gae}$, respectively.

[2] https://opendata.cityofnewyork.us/.

Table 1. Performance of *MAGRE* and baselines on different tasks. %improv. shows the percentage improvement of *MAGRE* over the best baseline result.

Methods	Crime rate prediction			Check-in count prediction			Land use	
	MAE↓	RMSE↓	R^2 ↑	MAE↓	RMSE↓	R^2 ↑	NMI↑	ARI↑
RegionDCL	118.31	156.45	0.18	464.41	732.73	0.12	0.43	0.17
Hex2Vec	109.31	144.02	0.05	400.78	651.47	0.37	0.39	0.13
MV-PN	93.14	125.27	0.28	476.12	783.12	0.07	0.39	0.15
MVURE	65.41	91.63	0.61	297.12	494.36	0.63	**0.75**	0.55
MG-FN	77.34	98.32	0.56	321.44	510.04	0.61	0.74	0.55
HREP	66.66	85.13	0.67	273.27	411.98	0.75	**0.75**	0.53
$MAGRE_{n2v}$	60.46	82.99	0.69	302.35	483.93	0.65	0.66	0.42
$MAGRE_{gae}$	86.45	118.91	0.35	297.84	507.52	0.61	0.20	0.03
MAGRE	**35.47**	**63.22**	**0.82**	**209.39**	**333.34**	**0.83**	**0.75**	**0.57**
% improv	45.77	25.73	22.38	23.37	19.08	10.66	0.0	3.63

Evaluation Metrics. In the regression tasks, we compute all the metrics using 5-fold cross-validation with a train-test split of 80-20%. We utilize Root Mean Square Error (RMSE) and Mean Absolute Error (MAE) to evaluate the performance of the prediction models. Furthermore, to measure how well the regression model fits the observed data, we use the coefficient of determination, denoted by R^2. To assess the quality of the clustering results, we utilize Normalized Mutual Information (NMI) and Adjusted Rand Index (ARI).

Parameter Settings. For the hexagonal grid, the number of neighbors (N_i) for each grid cell is seven, including the grid itself. For the grid embedding, we chose an embedding dimension of 128 and utilized the Adam optimizer with a learning rate of 0.001 [16]. For L1 smooth loss, β is set to its default value 1. *MAGRE* is trained for 2000 epochs. For downstream tasks, we adopt the hyperparameters as in [16].

5 Evaluation Results

In this section, we present evaluation results of the *MAGRE* compared to baselines and analyze the importance of features.

General Performance. First, we discuss the overall performance of our approach on different downstream tasks. In regression tasks, *MAGRE* outperforms the baseline methods in both MAE, RMSE, and R^2, as illustrated in Table 1. HREP, being the best-performing baseline in both regression tasks regarding RMSE, demonstrates its effectiveness with prompt learning, which replaces the direct use of region embedding in the downstream tasks. RegionDCL's poor performance is attributed to its sole reliance on OSM building data for region embeddings, failing to capture sufficient semantics for effective prediction on downstream tasks. Comparing the variations of our model, we observe a similar trend as reported in [16], i.e., $MAGRE_{n2v}$ outperforms $MAGRE_{gae}$. For the land use classification, as can be seen in Table 1, *MAGRE* achieves the highest scores in terms of ARI and surpasses the best-performing baselines by 3.63%. Regarding NMI, *MAGRE* achieves comparable performance with the best-performing baselines MVURE and HREP. The overall evaluation results on all three tasks demonstrate the effectiveness of the *MAGRE*'s embeddings.

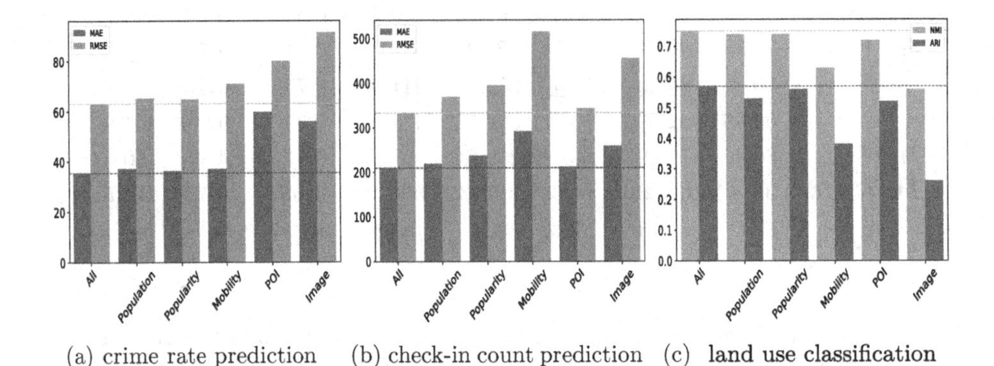

(a) crime rate prediction (b) check-in count prediction (c) land use classification

Fig. 3. Results of removing *MAGRE* features one at a time.

Feature Analysis. To analyze the impact of each feature category on the model performance, we systematically remove one feature category at a time, as shown in Fig. 3. We observe that the best results in all three tasks are obtained by using all features, demonstrating the effectiveness of our model in capturing the region semantics. Removing the OSM image leads to a notable increase in MAE and RMSE in regression tasks (Fig. 3a and 3b) and a decrease in NMI and ARI for land use classification (Fig. 3c). Specifically, this leads to a 44.62% increase in RMSE for crime rate prediction, a 36.67% increase in RMSE for check-in count prediction, and a 54.38% decrease in NMI. This finding emphasizes that OSM images contain useful information for learning region embeddings. The absence of the mobility feature negatively affects check-in count prediction, which is intuitive given the close relationship between check-in statistics and mobility patterns. Furthermore, POI features play a crucial role in crime rate prediction.

6 Case Study: Crime Rate Prediction on ROIs

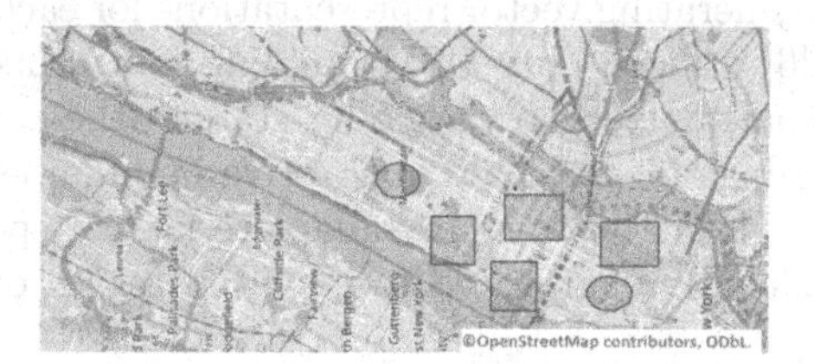

Fig. 4. Example of different ROI shapes.

Table 2. Crime rate prediction on ROIs.

Methods	MAE↓	RMSE↓
HREP	82.81	113.50
MVURE	113.40	166.41
MAGRE	**30.13**	**54.45**

Our case study is designed to showcase the adaptability and flexibility of *MAGRE*. To demonstrate its capabilities, we predict crime rates in ROIs with varying sizes and shapes. We randomly chose 200 locations within the Manhattan boundary. At each of these selected locations, we employ a randomization process to generate one of three distinct shapes for spatial regions: square, rectangle, or circle, as illustrated in Fig. 4. The area of each shape is randomly generated, with the upper bound of the area of the largest administrative region of Manhattan. Subsequently, we aggregate the grid embeddings to create region embeddings for these 200 ROIs, following the steps in Sect. 3.3. For baseline methods, we chose the top-2 best-performing baselines, i.e., HREP and MVURE. We conduct a five-fold cross-validation to obtain prediction results. As baseline methods can only compute predictions for administrative regions, we compute a weighted sum of the administrative region's prediction scores for a fair comparison. The weights are determined by the proportion of the overlap of administrative regions within the ROIs. As shown in Table 2, *MAGRE* outperforms the selected baseline methods, leading to an MAE reduction of 63.61% and a RMSE reduction of 52.02% as compared to the best-performing baseline, i.e., HREP. The gap in crime rate prediction scores between the ROIs and administrative boundaries (Table 1 and Table 2) for the two baseline methods indicates a lack of adaptability in these approaches. These outcomes highlight the adaptability and effectiveness of *MAGRE* in handling varying ROIs.

7 Related Work

This section briefly summarizes the related works in the representation learning of geospatial regions. Some recent works incorporate POIs and mobility data to construct meaningful region embeddings for fixed administrative boundaries. For instance, Zhang et al. [15] introduced a multi-view graph representation approach, which considered POI data and mobility patterns to generate a representation of fixed-size urban regions. Similarly, Zhou et al. [16] utilized a prompt learning method by leveraging both POI and mobility data. Wu et al. [12] focused on leveraging mobility data alone and trained for a specific task, i.e., mobility distribution. Xi et al. [13] integrated the satellite imagery alongside POI data.

Fu et al. [4] built a multi-view POI-POI network utilizing POI data and employed an autoencoder for region embedding. Woźniak et al. [11] relied only on OSM data incorporating a skip-gram model, generating vector representations for each hexagonal region. Similarly, Li et al. [6] utilized only OSM building footprints for region representation with dual contrastive learning. The state-of-the-art approaches either create region embeddings for fixed administrative boundaries or rely on limited data sources. With *MAGRE*, we acquire adaptive latent representations of grid cells that can be flexibly aggregated to a spatial region of any shape and size.

8 Conclusion

We proposed *MAGRE* – a novel approach that leverages multitask learning and multimodal spatial embeddings to create an adaptive representation of urban regions. *MAGRE* leverages fine-grained hexagonal grid cells, enabling a more precise and detailed depiction of their spatial characteristics. *MAGRE* can efficiently generate embeddings of any ROI by embedding aggregation and effectively preserves the semantics, as demonstrated in our experiments. Experimental results on three downstream tasks demonstrate that *MAGRE* exhibits superior performance compared to baseline methods, highlighting the benefits of multitasking and multimodal approach for learning latent representations of urban regions.

Acknowledgements:. This work was partially funded by the Federal Ministry for Economic Affairs and Climate Action (BMWK), Germany ("ATTENTION!", 01MJ22012C).

References

1. Bartunov, S., Kondrashkin, D., Osokin, A., Vetrov, D.P.: Breaking sticks and ambiguities with adaptive skip-gram. In: AISTATS 2016. JMLR.org (2016)
2. Berg, B.F.: New York City Politics: Governing Gotham. Rutgers University Press, New Brunswick (2007)
3. Birch, C.P., Oom, S.P., Beecham, J.A.: Rectangular and hexagonal grids used for observation, experiment and simulation in ecology. Ecol. Model. **206**(3–4), 347–359 (2007)
4. Fu, Y., Wang, P., Du, J., Wu, L., Li, X.: Efficient region embedding with multi-view spatial networks: a perspective of locality-constrained spatial autocorrelations. In: AAAI 2019, pp. 906–913. AAAI Press (2019)
5. Girshick, R.B.: Fast R-CNN. In: IEEE, ICCV 2015, pp. 1440–1448. IEEE Computer Society (2015)
6. Li, Y., Huang, W., Cong, G., Wang, H., Wang, Z.: Urban region representation learning with openstreetmap building footprints. In: ACM SIGKDD. ACM (2023)
7. Reimers, N., Gurevych, I.: Sentence-bert: sentence embeddings using siamese bert-networks. In: EMNLP-IJCNLP 2019. ACL (2019)
8. Tan, M., Le, Q.V.: Efficientnet: rethinking model scaling for convolutional neural networks. In: ICML 2019, pp. 6105–6114. PMLR (2019)

9. Velickovic, P., Cucurull, G., Casanova, A., Romero, A., Liò, P., Bengio, Y.: Graph attention networks. In: ICLR 2018. OpenReview.net (2018)
10. Wang, H., Li, Z.: Region representation learning via mobility flow. In: ACM, CIKM 2017, pp. 237–246. ACM (2017)
11. Wozniak, S., Szymanski, P.: hex2vec: context-aware embedding H3 hexagons with openstreetmap tags. In: GeoAI@SIGSPATIAL 2021, pp. 61–71. ACM (2021)
12. Wu, S., et al.: Multi-graph fusion networks for urban region embedding. In: IJCAI 2022 (2022)
13. Xi, Y., Li, T., Wang, H., Li, Y., Tarkoma, S., Hui, P.: Beyond the first law of geography: learning representations of satellite imagery by leveraging point-of-interests. In: WWW 2022, pp. 3308–3316. ACM (2022)
14. Yao, Z., Fu, Y., Liu, B., Hu, W., Xiong, H.: Representing urban functions through zone embedding with human mobility patterns. In: IJCAI 2018 (2018)
15. Zhang, M., Li, T., Li, Y., Hui, P.: Multi-view joint graph representation learning for urban region embedding. In: IJCAI 2020, pp. 4431–4437. ijcai.org (2020)
16. Zhou, S., He, D., Chen, L., Shang, S., Han, P.: Heterogeneous region embedding with prompt learning. In: AAAI 2023. AAAI Press (2023)

Attention Mechanism Based Multi-task Learning Framework for Transportation Time Prediction

Miaomiao Yang[1], Tao Wu[1], Jiali Mao[1,2](✉), Kaixuan Zhu[1], and Aoying Zhou[1,2]

[1] East China Normal University, Shanghai, China
{51215903097,52195100007,51215903072}@stu.ecnu.edu.cn,
{jlmao,ayzhou}@dase.ecnu.edu.cn

[2] Shanghai Engineering Research Center of Big Data Management, Shanghai, China

Abstract. Transportation time prediction (*TIP*) of a truck is one of key tasks for supporting the services in bulk logistics like route planning. But *TIP* prediction is challenging as it involves travel time prediction and dwell time prediction, which are influenced by various complex factors. Besides, there exists mutually constrained effects between travel time prediction and dwell time prediction. In this paper, we propose an *Attention Mechanism* based *Multi-Task prediction* framework consisting of travel pattern learning, stay pattern learning and transportation time modeling, called *AMP*. In view of that low prediction performance resulted by uncertain dwell time and mutually constrained effects between travel time and dwell time, we put forward a stay pattern learning module based on transformer and multi-factor attention mechanism. Furthermore, we design a multi-task learning based prediction module embedded with a mutual cross-attention mechanism to enhance overall prediction performance. Experimental results on a large-scale logistics data set demonstrate that our proposal can reduce *MAPE* by an average of 9.2%, *MAE* by an average of 19.5%, and *RMSE* by an average of 23.0% as compared to the baselines.

Keywords: Bulk logistics · Attention mechanism · Multi-task learning

1 Introduction

Transportation time prediction (or *TIP* for short) task plays a most significant role in bulk logistics platform, which provides the supports for critical logistics services such as route planning, vehicle-cargo matching and transportation capacity scheduling, etc. Its goal is to estimate the arrival time of transporting cargoes according to given departure time of truck and the route between the origin and destination of transportation. In real applications, most of transporting routes need to travel across several cities, and dwell on some places along the routes for long time due to various staying requirements like refueling, resting and dining, etc. Therefore, the time spent on each transporting route not

D.-N. Yang et al. (Eds.): PAKDD 2024, LNAI 14649, pp. 376–388, 2024.
https://doi.org/10.1007/978-981-97-2262-4_30

only involves the travel time on numerous roadways, but also the dwell time in multiple stay hotpots along the route.

Over the past decades, a lot of researches have been conducted on the travel time prediction issue. But all of them focus on estimating the travel time of the vehicles for given routes in an urban road network, which are usually shorter than the transportation routes in bulk logistics and do not need to dwell on any places for long time. So the existing solutions for travel time prediction issue cannot be directly utilized to tackle *TIP* issue in bulk logistics field. It is necessary for designing an appropriate *TIP* method that not only considers time estimation for the trucks' driving on the roads, but also take into account dwell time prediction for all stay hotspots along the transport route.

However, the problems such as uncertain dwell time and mutually constrained effects between travel time and dwell time, present severe challenges for attaining high-precision *TIP*.

- ***Challenge 1: The factors that affect dwell time of the trucks are various complex.***
  According to the traffic law, each truck shall stay for at least 20 min every 4 h during long-distance transporting trip. The truck drivers may dwell on multiple places sequentially in a transporting trip for distinct requirements. Due to the dwell time spent on various types of stay hotspots and dwell preferences of the truck drivers are distinct, the transportation time may be different even for the transportation route having same origin and destination pair. Also, dwell time prediction is influenced by a series of factors such as departing time of transporting, long dwell time on the previous stay hotspots, busyness levels of stay hotspots and working hour limitation of the cargo recipient company. This further increases uncertainty of transportation time prediction.
- ***Challenge 2: There are mutual restrictions between the travel time prediction and dwell time prediction.***
  Due to the working time limitation of recipient for receiving the goods, the truck driver usually needs to make a trade-off between dwell time at stay hotspots near the destination and travel time of the remained transporting trip. To be specific, if the truck cannot reach the destination during the consignee's working hours, the driver tend to dwell at a temporary rest area nearby the destination for a long time, until the consignee restart to work. On the contrary, if the remaining transporting trip requires long travel time to finish, the truck driver tend to shorten dwell time at the stay hotspots as far as possible to ensure reaching the destination in time.

To tackle the issue of complex diverse influencing factors of dwell time prediction, we first leverage Transformer to extract the effect of dwell time of preceding stay hotspot on that of subsequent ones. Then we introduce attention mechanism to capture differentiated influences of various factors on dwell time of each stay hotspot. In addition, aiming at the issue of reciprocally constrained effects between travel time and dwell time, we extract influence relationship between the travel route and the stay hotspot sequence by introducing cross-attention mechanism. Further, we put forward a parameter-sharing based multi-task learning method to obtain potential route representation shared by travel time prediction and dwell time prediction task to enhance the performance of *TIP* prediction. The contributions of this work are four-fold:

- We address the issue of transportation time prediction in bulk logistics, and then propose an *Attention Mechanism* based *Multi-Task prediction* framework, called *AMP*, which is composed of travel pattern learning, stay pattern learning and transportation time modelling.
- To respectively capture various complex influencing factors of dwell time prediction, we present a stay pattern learning module based on transformer and attention mechanism.
- To proliferate prediction performance, we design a multi-task learning based prediction module embedded with a mutual cross-attention mechanism to fuse mutual constrained effects between dwell time and travel time prediction task.
- We evaluate our proposal based on a large scale of real data set, and observe the improvement of decreasing *MAPE* by an average of 9.2%, *MAE* by an average of 19.5%, and *RMSE* by an average of 23.0% compared with state-of-the-art baselines.

The rest of this paper is organized as follows. Section 2 reviews relevant researches about transportation time prediction. Section 3 provides preliminaries and the problem definition. In Sect. 4, we elaborate the details of *AMP*. In Sect. 5, we evaluate the performance of *AMP* on a large scale real logistics data set. Finally, we conclude the paper in the last Section.

2 Related Work

In recent years, numerous studies have been conducted on the issue of travel time prediction, which can be classified as follows.

Traditional Methods. They mainly include *road segment-based* ones and *path-based* ones. *road segment-based* methods roughly estimated the travel time of entire path by summing up the travel time of all road segments [13]. They may exacerbate prediction deviation when predicting for long distance transportation routes due to cumulative errors. The route-based methods attempted to predict travel time of whole path using the techniques such as nearest neighbors search and regression [5,7,10,14]. Since the aforementioned methods do not consider complex influencing factors related with the transport process, they are not suitable for our proposed *TIP* issue.

Deep Learning Based Methods. Wang et al. proposed a model by combining Wide&Deep model with recurrent model, which can improve prediction accuracy by employing *LSTM* to capture contextual information among the roads [12]. Wang et al. transformed the trajectory to feature maps, and then employed

CNN and $LSTM$ on them to learn spatial and temporal dependencies for travel time prediction [9]. Zhang et al. first represented a trajectory as a sequence of grid cells, and then extracted manually-craft features to train a travel time prediction model using CNN and $LSTM$ [15]. Fu et al. represented a trajectory as a sequence of *generalized images*, and then designed a two-dimensional $CNNs$ to capture complex moving patterns from them for travel time estimation [2]. In view of graphical structure of the road network, Fang et al. proposed a spatial-temporal graph neural network incorporated with graph attention mechanism for travel time prediction [1]. Jin et al. presented a multi-scale spatial-temporal graph convolution network to capture spatial-temporal dynamics for travel time estimation [4]. Nevertheless, all the aforementioned methods cannot be directly used to tackle *TIP* issue because they do not consider long-time dwell behavior during transportation.

Multi-task Learning based Methods. Owing to multi-task learning can leverage useful information contained in multiple tasks to improve the generalization performance of all the tasks, it was introduced into travel time prediction task to improve prediction performance by estimating each road and entire path simultaneously [1,8,9]. Wang et al. proposed a multi-task framework for travel time estimation by simultaneously estimating travel time and transition probabilities among road segments and intersections [11]. Inspired by them, we employ multi-task learning technique for *TIP* prediction by modeling mutual constrained effects between travel patterns and stay patterns.

3 Preliminaries

Definition 1 (Trajectory of Truck). *A trajectory of the truck j refers to a sequence of consecutive positional points, denoted as $T_j = \{p_1, p_2, \cdots, p_n\}$, where n is the number of points in a trajectory, and each point $p_i = (lng_i, lat_i, t_i)$ contains the longitude, latitude, and timestamp.*

Definition 2 (Waybill). *A waybill refers to a transport task assigned to a truck, denoted as $W_j^l = (StaID, DesID, t_s, t_e)$, where l represents the l-th transport task among all tasks assigned to the truck j, $StaID$ and $DesID$ denote transport origin and destination of a waybill respectively, t_s and t_e denote the start timestamp and the arrival timestamp of transporting separately.*

A truck trajectory is split into several *waybill trajectorys* according to the start and arrival timestamps of each transportation task. Then the *waybill trajectories* are mapped to *road network* by map matching to obtain *transport routes.*

Definition 3 (Road Network). *Road network is typically represented as a graph $G = (V, E)$, where the set of vertexes V $(v \in V)$ includes all road segments, and the set of edges E indicates reachable relationship among the road segments.*

Definition 4 (Transport Route). *The transport route of waybill w_j^l is represented as a sequence of consecutive road segments in the road network connecting the pair of transport origin and destination, denoted as $R = \{v_1, v_2, \cdots, v_m\}$, where m represents the number of road segments in a route.*

To model staying behavior during transport, the sequence of zero-speed trajectory points is extracted from *waybill trajectorys*, called *stay point.*

Definition 5 (Stay Point). *Given a stay duration threshold* thr_{dur}*, a contiguous zero-speed trajectory point sequence, denoted as* $\{p_e, p_{e+1}, \cdots, p_f\}(e < f)$*, refers to one section of a trajectory, in which the first point* p_e *is viewed as a stay point if the time gap between* p_e *and* p_f *is beyond* thr_{dur}*, and the timestamp of* p_e *is viewed as visiting time.*

Definition 6 (Stay Hotspot). *A stay hotspot is a uniquely identified place where the truck drivers usually stay for a rest or a meal during transportation, which has the attributes including functional category* h_{type}*, and geographical location* h_l *represented by longitude and latitude coordinates.*

Each stay point in a *waybill trajectory* can be mapped to the nearest *stay hotspot.* Moreover, each *waybill trajectory* can be converted into a sequence of stay hotspots based on the order in which it passed through.

Definition 7 (Problem Definition). *Given a set of trajectories (denoted as* Tr_s*), a set of waybills (denoted as* W_s*), a transport route R and multiple external factors, our goal is to estimate the transportation time along R based upon modeling travel pattern and dwell pattern.*

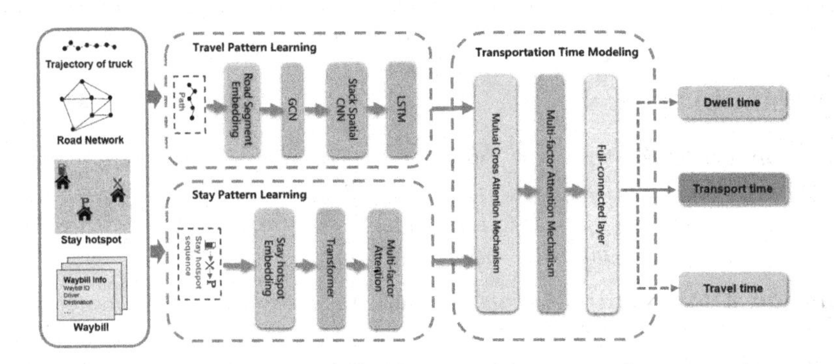

Fig. 1. Overview of *AMP*

4 Solution

We propose an *Attention Mechanism* based *Multi-Task* transportation time prediction framework, called *AMP*. It is composed of *travel pattern learning*, *stay pattern learning* and *transportation time modeling*, as shown in Fig. 1.

4.1 Travel Pattern Learning

Initially, we split the trajectories of the trucks into *waybill trajectories* according to the start and arrival timestamps of different waybills, and project *waybill trajectories* into the transport routes by map-matching [6]. Then we

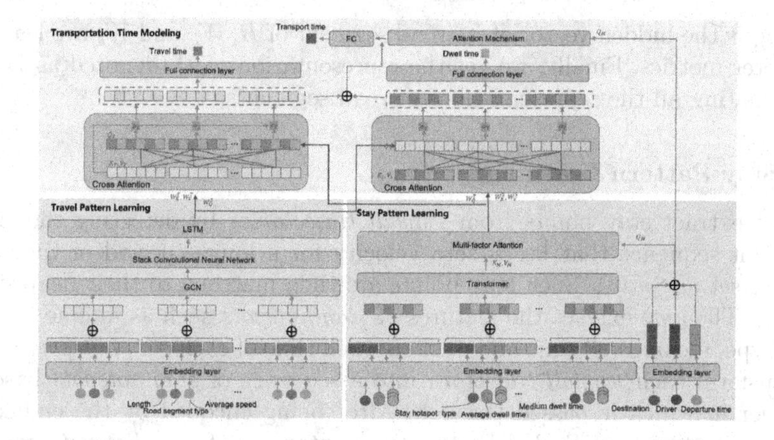

Fig. 2. Details of *AMP*

extract the features related with road segments, including respective *id*, *length*, *type* (e.g., primary, secondary, tertiary, etc.), and the mean as well as the median travelling velocity. Subsequently, we obtain the road segment embedding by converting the categorical features into a dense vector and concatenating it with numerical features. Thereafter, the transport route is represented as $R = \{segment_1, segment_2, \cdots, segment_i, \cdots, segment_m\}$, where $segment_i$ is i_{th} road segment embedding of R. Then, R is put into *Graph Convolution Network* (*GCN*) layer to learn local spatial correlation between neighboring road segments, which is defined as:

$$\hat{segment_i} = \sigma_g(\sum_{j \in N_i} segment_j \star f_{\Theta_{i,j}}) \tag{1}$$

where σ_g is activation function, N_i is the neighbor segment set of $segment_i$ in road network, $\star$ denotes the filter operation, and $f_{\Theta_{i,j}}$ denotes the weight parameter in the filter between $segment_i$ and $segment_j$.

The output of *GCN* is further fed into the *Convolutional Neural Network* (*CNN*) layer to learn cross-segment relationship, which is defined as:

$$subroute_i = \sigma_c(W_f * \hat{segment}_{i:i+k} + b) \tag{2}$$

where b is the bias, $segment_{i:i+k}$ represents consecutive segments from index i to $i + k - 1$ ($k = 3$ in subsequent experiments), W_f is the parameters of the convolution filter, and σ_c is the activation function. In addition, a multi-layer $CNNs$ is constructed to shorten the sequence length of the transport route, which can reduce the computational overhead.

The final layer of this module is *Long Short Term Memory*, which learns temporal dependence among road segments for travel time. It is defined as follows:

$$h_i = \sigma(W_h * h_{i-1} + W_s * subroute_i) \tag{3}$$

where h_i is the hidden vector of i_{th} road segment of R, W_h and W_s are learnable parameter metrics. Finally, we get the representation of R (denoted as H_T) by concatenating all the representations of road segments.

4.2 Stay Pattern Learning

We first extract stay points from *waybill trajectories* by detecting the trajectory point sequence that keeps zero velocity for a longer period of time (here thr_{dur} is set as $8min$). Such stay points are then matched to their nearest *stay hotspots*. Then we extract the features of *stay hotspots* such as unique id, functional type, mean dwell time and median dwell time of each *stay hotspot*. Finally, we transform each *waybill trajectory* into a sequence of *stay hotspots* based on the order in which it passed through. After being encoded by the embedding layer (as illustrated in Sect. 4.1), the representations of *stay hotspot* sequences can be represented as $SR = \{stayhp_1, stayhp_2, ..., stayhp_{|SR|}\}$. To model the effect of the dwell time of preceding *stay hotspots* on that of subsequent ones, we employ *Transformer* encoder for SR, as shown in *Stay pattern learning* of Fig. 2. The reason for using *Transformer* is that it has powerful long-distance sequence modelling capability, and can capture the correlation of *stay hotspots* at longer distances in the sequence.

Transformer. In view of that remaining travel distance of a transport route has an impact on the dwell time of *stay hotspot* to be visited, we take remaining distance into account in modelling. Specifically, the embedding of i_{th} stay hotspot in SR is replaced with $st\hat{a}yhp_i = Concat(stayhp_i, pe_i, rd_i)$, where pe_i denotes the positional embedding, and rd_i is the remaining distance embedding. Subsequently, SR are respectively multiplied by the parameter matrices W_q, W_k, $W_v \in R^{d_{sh} * d_h}$ (here d_{sh} denotes the embedding size of stay hotspot, and d_h denotes the hidden size) to obtain vectors Q_S,K_S and V_S. Then we obtain the output as:

$$H_S(Q_S, K_S, V_S) = ||_{h=1}^{|h|} softmax(\frac{Q_S^h \cdot K_S^{h\,T}}{\sqrt{d}})V_S^h \tag{4}$$

where $||$ represents the concatenation operation, $|h|$ is the number of heads, d is a normalization factor which is consistent with the feature dimension of Q_S^h, and H_S denotes final output of transformer layer.

Multi-factor Attention. To learn different impacts of external factors including driver, destination and departure time, we introduce multi-factor attention mechanism, as defined below:

$$H_S(Q_F, K_F, V_F) = ||_{h=1}^{|h|} softmax(\frac{Q_F^h \cdot K_F^{h\,T}}{\sqrt{d}})V_F^h \tag{5}$$

where K_F and V_F comes from the outputs of transformer layer H_S, Q_F comes from external factors. The updated H_S is the output of the stay pattern learning module.

4.3 Transportation Time Modeling

Mutual Cross Attention. As shown in *transport time modeling* of Fig. 2, we regard H_T as query, and H_S as keys and value. Then we model mutual influence between *transport route* (H_T) and *stay hotspot* sequence (H_S) by using the formulas:

$$\begin{gathered} Q_T^F = W_q^T \cdot H_S, K_T^F = W_k^T \cdot H_T, V_T^F = W_v^T \cdot H_T, \\ \hat{H_T} = softmax(\frac{Q_T^F \cdot K_T^{FT}}{\sqrt{d}}) \cdot V_T^F \end{gathered} \tag{6}$$

In turn, we view H_S as query, and H_T as keys and values. Then we get such cross attention's output $\hat{H_S}$. After that, we fuse two parts by concatenating two hidden vectors, which can be defined as $H_{fusion} = \hat{H_T}||\hat{H_S}$. Besides, considering the influence of external factors, we adopt another multi-factor attention to capture the impact on entire transportation route (the equation is same as Eq. (5)).

Multi-task Learning. Next, we view the transportation time prediction as main task, and travel time and dwell time as auxiliary tasks.

For the travel time estimation task, we feed the representation of transport route ($\hat{H_T}$) into a *MLP* layer, which can be defined as $\hat{y_T} = MLP(Concat(\hat{H_T}, F_e))$, where F_e is the embedding of external features. We adopt average absolute percentage error ($MAPE$) as the loss function, and calculate the loss by $L_T = \frac{1}{|D|}\sum_i^{|D|}\frac{|y_T - \hat{y_T}|}{y_T}$, where $\hat{y_T}$ denotes the prediction of the route's travel time, y_T is the ground truth, and $|D|$ is the size of training dataset.

For the dwell time estimation task, we feed representation of *stay hotspot* sequence ($\hat{H_S}$) into a MLP layer, which can be defined as $\hat{y_S} = MLP(Concat(\hat{H_S}, F_e))$, where F_e is the embedding of external features. In view of that the label of dwell time may be zero value, we use the masked $MAPE$ loss function, and calculate the loss of dwell time by $L_S = \frac{1}{|SD|}\sum_i^{|SD|}\frac{|y_S - \hat{y_S}|}{y_S}$, where y_S is the ground truth for the dwell time, $\hat{y_S}$ denotes the prediction value, and $|SD|$ represents the size of sub-dataset whose ground truth value is not zero.

Based on the above tasks, we combine their outputs to predict the transportation time by $\hat{y_{TP}} = MLP(Concat(H_{fusion}||H_S||H_T, F_e))$. Then we combine three types of losses for joint optimization, as defined below:

$$L = (1 - \alpha - \beta) \cdot \frac{1}{n}\sum_i^n \frac{|y_{TP} - \hat{y_{TP}}|}{y_{TP}} + \alpha \cdot L_T + \beta \cdot L_S \tag{7}$$

where α and β are the combination coefficients that balance the trade-off between L, L_T and L_S. We minimize the weighted loss in train phase and take the transportation time as our final estimation in test phase.

5 Experiments

5.1 Datasets and Settings

We evaluate our model on a real logistics dataset of 4 months (from Nov. 1st, 2020 to Mar. 1st, 2021) from a steel logistics enterprise. The dataset consists of the trajectories and waybills of three transportation routes departing from a steel company in *Shandong* Province, China. We divide them into three datasets according to the cities of transport destinations (hereafter termed *Linyi*, *Rizhao*, and *Qingdao* respectively). Table 1 reports the description and statistics of three datasets. In addition, we obtain the road network of *Shandong* city from *OpenStreetMap* for map matching, which contains 435,905 edges and 644,070 nodes. The stay hotspot data set comes from manual annotation, including gas stations, maintenance stations, restaurants, deserted spaces (for resting), etc.

Table 1. The statistics of logistics datasets.

Dataset	*Linyi*	*Rizhao*	*Qingdao*
Trajectories	28,320	44,513	20,415
Average segments per route	140	157	149
Average distance of route	21,138.93	23,464.72	22,560.05
Average travel time	286.32	318.46	302.79

We split each dataset into a training set, a validation set and a test set with a splitting ratio of 7:2:1. All experiments are conducted on a GPU-CPU platform with Tesla V100. The program and baselines are implemented in Python 3.8.

5.2 Metrics

We adopt *mean absolute error* (*MAE*), *mean absolute percentage error* (*MAPE*) and *root-mean-squared error* (*RMSE*) as the performance metrics.*MAE* intuitively reflects absolute error between the prediction values and the ground truth, which is defined as $MAE = \frac{1}{n}\sum_{i=1}^{n}|\hat{y} - y|$. *RMSE* is more sensitive to larger errors and can reflect the stability of the prediction, which is defined as $RMSE = \sqrt{\frac{1}{n}\sum_{i=1}^{n}(\hat{y} - y)^2}$. *MAPE* measures relative error to avoid the evaluation bias caused by excessive label differences, which is defined as $MAPE = \sum_{i=1}^{n}|\frac{\hat{y}-y}{y}|$, where $\hat{y}$ is the predicted transportation time and y is the ground truth.

5.3 Comparison Approaches

To verify the performance of our proposal, we compared it with the following travel time prediction methods:

- *AVG*: It calculates average travel time of road segments per hour, and then estimates the given route by summing the average travel time of each segment.
- *TEMP* [10]: It is a collective estimation method that estimates the travel time based on historical trajectories with nearby origin and destination.

- *WDR* [12]: It integrates machine learning and neural networks for sequence learning and enriches features through feature intersection.
- *DeepTTE* [9]: It directly captures the spatial and temporal dependency from raw trajectory points and simultaneously predicts the travel time of each sub-path and entire route through multi-task learning.
- *DeepTravel* [15]: It learns temporal dependence of route with a dual interval loss mechanism for auxiliary supervision.
- *HetETA* [3]: It considers turning and transfer relationships based on historical trajectories by constructing a heterogeneous road network graph.
- *STGNN-TTE* [4]: It introduces an multi-scale module to enhance the lower hidden layer information for improving accuracy.

5.4 Overall Evaluation

The comparison results of all methods on three datasets are reported in Table 2. From the results, we can observe that our proposal outperforms other methods at least 7.53%, 19.43%, 23.88% in $MAPE$, MAE, $RMSE$ on three datasets. Among all the methods, AVG works worst due to that it simply considers average historical speed of road segments per hour. $TEMP$ performs better as it considers temporal dynamics and regional differences. But it's prediction performance is still limited without capacity to model complex influence of different factors. WDR and *DeepTTE* achieve better performance owing to they have better modeling capabilities of neural networks. In addition, $HetETA$ and $STGNN-TTE$ attain better prediction accuracy by taking the structural road network into account and learning the transfer relationship between road segments. However, due to the above methods disregard the dwell time for *stay hotspots* along the transport route, all of them behave worse than *AMP* in *TIP* prediction.

Table 2. Results of comparison approaches.

Dataset	Linyi			Rizhao			Qingdao		
Metric	MAPE	MAE(min)	RMSE(min)	MAPE	MAE(min)	RMSE(min)	MAPE	MAE(min)	RMSE(min)
AVG	94.3%	232.29	294.9	96.75%	291.97	382.40	95.19%	287.4	354.61
TEMP	65.25%	187.98	254.64	56.67%	172.16	266.28	58.87%	180.20	224.94
WDR	56.12%	150.88	212.78	58.28%	163.72	240.42	53.12%	161.38	198.80
DeepTTE	45.54%	141.09	187.48	47.77%	159.36	216.84	46.23%	147.62	178.43
DeepTravel	30.87%	89.91	138.06	29.94%	95.60	154.18	33.29%	109.59	154.26
HetETA	29.66%	83.49	101.71	28.39%	87.46	118.37	27.65%	80.32	112.51
STGNN-TTE	24.84%	70.61	92.67	26.63%	80.74	105.41	23.39%	71.59	97.27
AMP	**15.13%**	**46.84**	**68.49**	**16.32%**	**51.20**	**76.75**	**15.86%**	**52.16**	**73.39**

5.5 Ablation

In this section, we evaluate the impact of each module to verify the effectiveness of the framework in our proposal (detailed in Table 3). First, we remove $Transformer$ and put *stay hotspot* sequence into next multi-attention layer. We can observe that $MAPE$ decreased by 1.10%, 1.68%, 0.72% for *Linyi*,*Rizhao*,*Qingdao*. Then we remove multi-factor attention from the stay pattern learning module. We can see that $MAPE$ drops more on all datasets, but MAE and $RMSE$ increase due to the loss of the ability to capture the effects of external factors. Subsequently, we delete mutual cross-attention and just concatenate the output of stay pattern learning module and travel pattern learning module for prediction. It resulted in a 1.97%, 2.29%,1.4% decrease in $MAPE$, a 6.07,13.16,9.42 min decrease in MAE, and a 7.64,13.37,7.04 min decrease in $RMSE$. It shows the significance of modeling mutual effect between *stay hotspot* sequence and transport route. At last, we discard the travel time prediction and dwell time prediction component. Obviously, the experimental results proves the importance of multi-task learning module.

Table 3. Results of Ablation.

Dataset	Linyi			Rizhao			Qingdao		
Metric	MAPE	MAE(min)	RMSE(min)	MAPE	MAE(min)	RMSE(min)	MAPE	MAE(min)	RMSE(min)
w/o transformer	16.23%	50.12	71.72	17.90%	55.16	83.08	16.58%	56.39	76.94
w/o multi-factor attention	19.33%	57.90	80.65	20.67%	63.72	85.90	18.32%	67,78	82.80
w/o mutual cross-attention	18.10%	52.91	74.13	18.61%	64.36	90.12	17.26%	61.62	80.43
w/o multi-task learning	17.92%	53.88	76.31	18.03%	65.23	93.84	19.23%	69.20	80.34
AMP	**15.13%**	**46.84**	**68.49**	**16.32%**	**51.20**	**76.75**	**15.86%**	**52.16**	**73.39**

5.6 Parameter Experiments

In our experiments, the epoch is set as 100. We train the model using Adam algorithm under the learning rate of 0.01. In the road segment embedding and *stay hotspot* embedding component, we embed *weekID* to R^3, *hourID* to R^6, *driverID* to R^9, *road segment type* to R^6, *stay hotspot type* to R^3, and *destination* to R^9. For $LSTM$ layer and $Transformer$ layer, the size of hidden vector is set as 128. In the multi-task learning module, we conduct experiments under different parameter α and β from 0.0 to 0.99 for the multi-task learning module. From the result as show in Fig. 3, we observe that the best performance is attained when both α and β are set as 0.2, and prediction performance gets worse as α and β continue to increase or decrease.

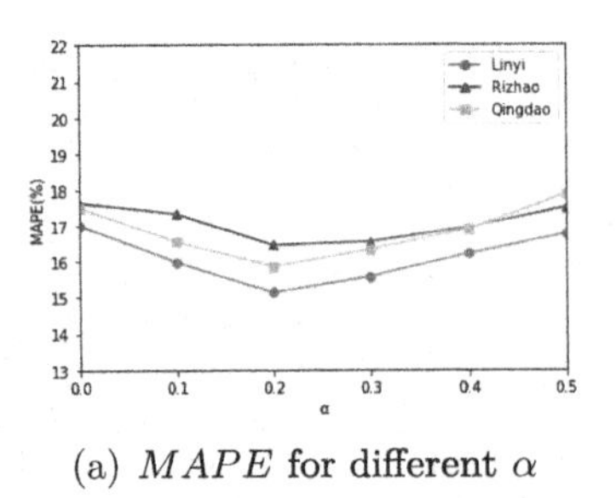

(a) $MAPE$ for different α

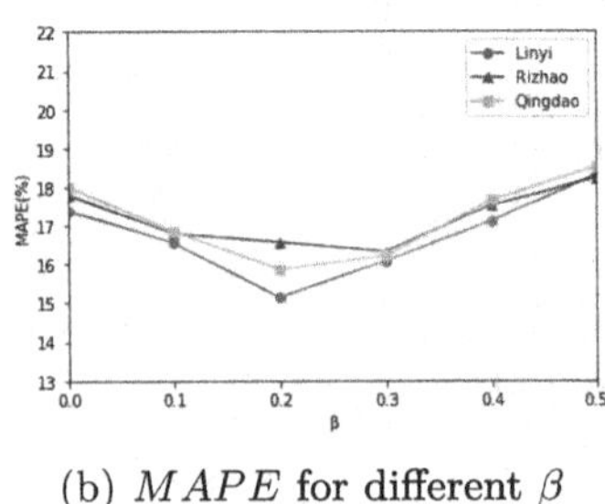

(b) $MAPE$ for different β

Fig. 3. $MAPE$ for different α and β

5.7 Case Study

We present a real-world case to prove the availability and reliability of *AMP*. As shown in Fig. 4, when a driver accepts the transport task and departs at 19:50, *route planning* is invoked to generate the transport route (marked by red lines). Its transportation time is predicted by using *AMP*, and further used for available capacity query (marked by green box). Also, we show actual travelling time and dwell time of the transport route. We observe that *AMP* generates a predicted transportation time of $11h36min$ and a dwell time of $6h3min$, which is close to real transportation time ($13h25min$) and dwell time ($7h18min$). Finally, we deploy *AMP* a steel logistics platform, serving over 800 transport lines and approximately 69,000 trucks.

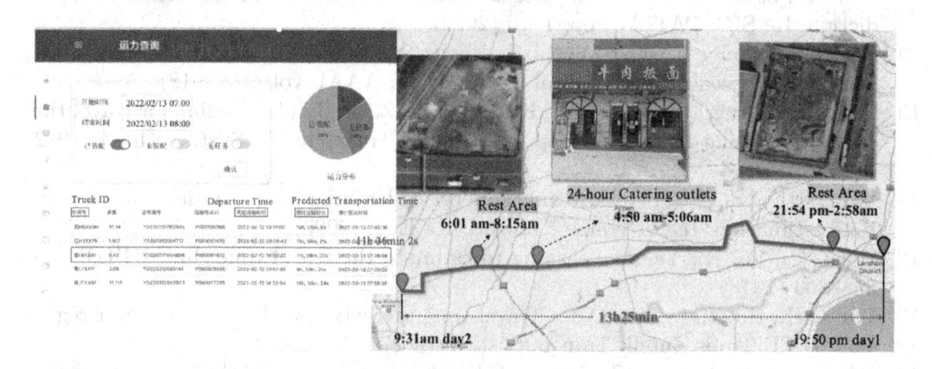

Fig. 4. A case of transportation time prediction

6 Conclusion

In this paper, we focus on the issue of transportation time prediction in bulk logistics and propose an *Attention Mechanism* based *Multi-Task prediction* framework *AMP*. The model captures the effect of the dwell time of preceding stay hotspot on that of subsequent ones by leveraging transformer and introduces multi-factor attention mechanism to capture differentiated influences of various factors on dwell time of each stay hotspot. To address the issue of reciprocally constrained effects between travel time and dwell time, *AMP* learns mutual influence relationship between the travel route and the stay hotspot sequence by introducing mutual cross-attention mechanism. Finally, *AMP* employs multi-task learning method to proliferate overall prediction performance.

References

1. Fang, X., Huang, J., Wang, F., Zeng, L., Liang, H., Wang, H.: Constgat: contextual spatial-temporal graph attention network for travel time estimation at baidu maps. In: SIGKDD, pp. 2697–2705 (2020)

2. Fu, T.Y., Lee, W.C.: Deepist: deep image-based spatio-temporal network for travel time estimation. In: CIKM, pp. 69–78 (2019)
3. Hong, H., et al.: Heteta: heterogeneous information network embedding for estimating time of arrival. In: SIGKDD, pp. 2444–2454 (2020)
4. Jin, G., Wang, M., Zhang, J., Sha, H., Huang, J.: STGNN-TTE: travel time estimation via spatial-temporal graph neural network. Futur. Gener. Comput. Syst. **126**, 70–81 (2022)
5. Luo, W., Tan, H., Chen, L., Ni, L.M.: Finding time period-based most frequent path in big trajectory data. In: SIGMOD, pp. 713–724 (2013)
6. Newson, P., Krumm, J.: Hidden Markov map matching through noise and sparseness. In: Agrawal, D., et al. (eds.) SIGSPATIAL, pp. 336–343. ACM (2009)
7. Tiesyte, D., Jensen, C.S.: Similarity-based prediction of travel times for vehicles traveling on known routes. In: SIGSPATIAL, pp. 1–10 (2008)
8. Wan, F., et al.: Mttpre: a multi-scale spatial-temporal model for travel time prediction. In: SIGSPATIAL, pp. 1–10 (2022)
9. Wang, D., Zhang, J., Cao, W., Li, J., Zheng, Y.: When will you arrive? Estimating travel time based on deep neural networks. In: AAAI, vol. 32 (2018)
10. Wang, H., Tang, X., Kuo, Y.H., Kifer, D., Li, Z.: A simple baseline for travel time estimation using large-scale trip data. ACM Trans. Intell. Syst. Technol. **10**(2), 1–22 (2019)
11. Wang, H., et al.: Multi-task weakly supervised learning for origin-destination travel time estimation. IEEE Trans. Knowl. Data Eng. (2023)
12. Wang, Z., Fu, K., Ye, J.: Learning to estimate the travel time. In: SIGKDD, pp. 858–866 (2018)
13. Wu, C.H., Ho, J.M., Lee, D.T.: Travel-time prediction with support vector regression. IEEE Trans. Intell. Transp. Syst. **5**(4), 276–281 (2004)
14. Yang, B., Dai, J., Guo, C., Jensen, C.S., Hu, J.: PACE: a PAth-CEntric paradigm for stochastic path finding. VLDB **27**, 153–178 (2018)
15. Zhang, H., Wu, H., Sun, W., Zheng, B.: Deeptravel: a neural network based travel time estimation model with auxiliary supervision. In: Lang, J. (ed.) IJCAI, pp. 3655–3661 (2018)

MSTAN: A Multi-view Spatio-Temporal Aggregation Network Learning Irregular Interval User Activities for Fraud Detection

Wenbo Zhang[1], Shuo Zhang[1], Xingbang Hu[1], and Hejiao Huang[1,2](✉)

[1] School of Computer Science and Technology, Harbin Institute of Technology, Shenzhen, China
{21s151165,22s151067}@stu.hit.edu.cn, soco868@sina.com, huanghejiao@hit.edu.cn

[2] Guangdong Provincial Key Laboratory of Novel Security Intelligence Technologies, Shenzhen, China

Abstract. Discovering fraud patterns from numerous user activities is crucial for fraud detection. However, three factors make this task quite challenging: Firstly, previous research usually utilize just one of the two forms of user activity, namely sequential behavior and interaction relationship, leaving much information unused. Additionally, nearly all works merely study on a single view of user activities, but fraud patterns often span across multiple views. Moreover, most existing models can only handle regular time intervals, while in reality, user activities occur with irregular time intervals. To effectively discover fraud patterns from user activities, this paper proposes MSTAN (Multi-view Spatio-Temporal Aggregation Network) for fraud detection. It addresses the above problems through three phases: (1) In short-term aggregation, SIFB (Sequential behavior and Interaction relationship Fusion Block) is employed to integrate sequential behavior and interaction relationship. (2) In view aggregation, 2-dimensional multi-view user activity embedding is obtained for simultaneously mining multiple views. (3) In long-term aggregation CTLSTM (Convolutional Time LSTM) is designed to deal with irregular time intervals. Experiments on two real world datasets demonstrate that our model outperforms the comparison methods.

Keywords: Fraud detection · Spatio-temporal aggregation · Neural networks · Attention mechanism

1 Introduction

Recently, there has been a significant rise in the number of fraudsters engaging in illicit activities on online platforms and communication systems. These

Supported by Shenzhen Science and Technology Program under Grant No. GXWD20220817124827001 and No. JCYJ20210324132406016.

D.-N. Yang et al. (Eds.): PAKDD 2024, LNAI 14649, pp. 389–401, 2024.
https://doi.org/10.1007/978-981-97-2262-4_31

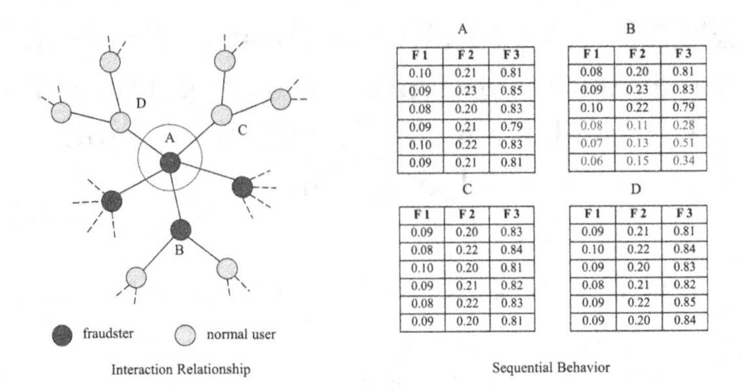

Fig. 1. Sequential Behavior and Interaction Relationship. The parts highlighted in red indicate fraud patterns. (Color figure online)

fraudsters encompass various types, including telecom fraudsters [1], spam publishers [2], webpages malicious editors [3], and online advertising fake visitors [4]. Their malicious actions are aimed at obtaining illegal profits, resulting in serious financial losses and posing a significant threat to social security. Such magnitude losses highlight the urgent need for effective fraud detection measures. However, in reality, fraudsters often disguise themselves as ordinary users and camouflage their illegal activities within a vast sea of legitimate user activities. This deceptive behavior complicates the task of fraud detection [5].

Discovering fraud patterns from numerous user activities is the key to fraud detection. However, due to the inadequate utilization of the following three key inherent properties of user activities, existing methods fail to achieve satisfactory results. **(1) User activities can be presented in two forms: sequential behavior and interaction relationship** [6]**. Both forms offer distinct perspectives for uncovering fraud patterns and should be jointly captured.** Specifically, sequential behavior refers to the temporal sequence of user actions, revealing fraud patterns from temporal perspective. Meanwhile, interaction relationship refers to the connections between different users or between users and other entities (such as products or merchants), illuminating fraud patterns from spatial perspective. As shown in Fig. 1, both A and B are fraudsters. The fraud patterns of A primarily manifest through interaction relationship, while the fraud patterns of B are mainly evident in sequential behavior. As existing methods [7,8] utilize only one form for fraud detection, they perform poorly when fraud patterns are mainly hidden in another form. Jointly modeling both sequential behavior and interaction relationship can provide a more comprehensive understanding of fraud patterns, but there is currently limited research in this area. **(2) In some scenarios, user activities happen in multiple views, which need to be considered together to accurately mine fraud patterns** [9]. For instances, in activities where user purchases products from merchants, there exist two views: user-merchant view and user-product view. Fraud patterns may hide in any view or lay in multiple views. Therefore, fraud can only be effectively detected if multiple views of user activities are considered simultaneously, which has seldomly been involved in previous work [8,17]. **(3) The irregular time intervals between successive user activities greatly complicates the modeling of fraud patterns.** In practice, user activities always happen at irregular time intervals. However, most existing models [10,11] can only handle regular time intervals and therefore struggle to effectively capture fraud patterns from user activities in real-world settings. Thus, in order to accurately model fraud patterns, special structure needs to be designed to deal with irregular time intervals.

To solve above problems, we propose a MSTAN (Multi-view Spatio-Temporal Aggregation Network) for fraud detection. It addresses the above problems through three phases: (1) it integrates the two forms of user activities in short-term aggregation; (2) it mines the multiple views jointly in view aggregation; (3) it deals with the irregular time intervals in long-term aggregation. To be specific, SIFB (Sequential behavior and Interaction relationship Fusion Block) is employed in short-term aggregation to integrate sequential behavior and interaction relationship. Moreover, for simultaneously mining multiple views, 2-dimensional multi-view user activity embedding is obtained through view aggregation. Finally, CTLSTM (Convolutional Time LSTM) is designed to deal with irregular time intervals in long-term aggregation.

Our contributions can be summarized as follows:

- A novel model MSTAN is proposed for fraud detection in online platforms and communication systems. To the best of our knowledge, this is the first model to comprehensively consider all three properties of user activities.
- SIFB is designed to fuse sequential behavior and interaction relationship, enabling the model to leverage both temporal and spatial information.
- CTLSTM is proposed to aggregate spatio-temporal information, in which a specially designed time gate is used to deal with irregular time intervals that most models cannot handle. As far as we know, this is the first model that can handle spatio-temporal sequences with irregular time intervals.

2 Problem Statement

In fraud detection, we are given a set of users and their activities. All the user activities happen on T timestamps $\{t_1, t_2, ..., t_T\}$. And for each user u, his/her activities happen on a series of timestamps $\{t_{a_1}, t_{a_2}, ..., t_{a_i}, ..., t_{a_l}\}$, where $1 \leq a_i \leq T$ and $a_2 - a_1, a_3 - a_2, ..., a_l - a_{l-1}$ are not equal, which means the time intervals between $\{t_{a_1}, t_{a_2}, ..., t_{a_i}, ..., t_{a_l}\}$ are irregular. Meanwhile, each user activity happens in m views $v_1, v_2, ..., v_m$. From activities of each user u, sequential behavior set B_u and interaction relationship set I_u can be created. Specifically, $B_u = \{B_u^1, B_u^2, ..., B_u^m\}$, where m means m views, and for each view v, $B_u^v = \{(B_u^v)_{t_{a_1}}, (B_u^v)_{t_{a_2}} ..., (B_u^v)_{t_{a_l}}\}$. Similarly, $I_u = \{I_u^1, I_u^2, ..., I_u^m\}$, and for each view v, $I_u^v = \{(I_u^v)_{t_{a_1}}, (I_u^v)_{t_{a_2}} ..., (I_u^v)_{t_{a_l}}\}$. Our task is to detect whether a user u is fraudulent by learning B_u and I_u with irregular intervals under m views.

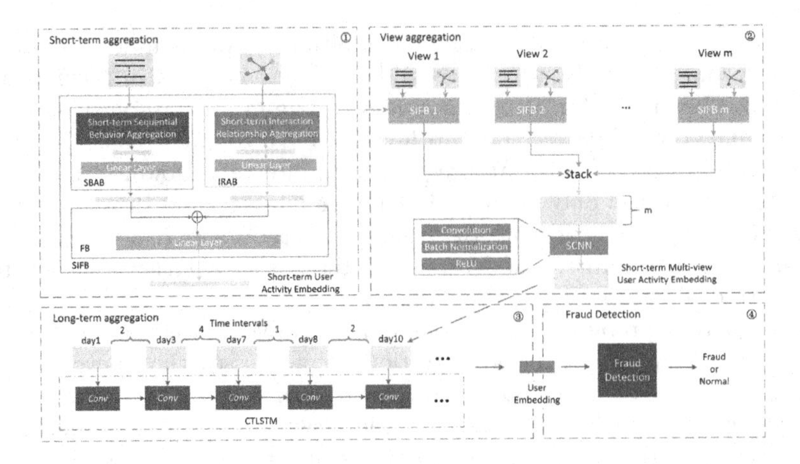

Fig. 2. Overall Framework. The model consists of four phases: (1) Short-term aggregation. (2) View aggregation. (3) Long-term aggregation. (4) Fraud detection. A user embedding is obtained through (1), (2), (3), and then it is used in (4) for detecting whether the user is a fraudster or not.

3 Overall Architecture

3.1 Architecture

As shown in Fig. 2, the model consists of four phases: short-term aggregation, view aggregation, long-term aggregation and fraud detection. Firstly, SIFB is used to fuse sequential behavior and interaction relationship in short-term aggregation. Then, 2-dimensional multi-view user activity embedding is obtained through view aggregation for simultaneously mining multi-view user activity information. Next, CTLSTM is employed in long-term aggregation to deal with irregular time intervals. Finally, the user embedding obtained through the above steps is used for detecting whether the user is a fraudster or not.

3.2 Short-Term Aggregation

The goal of short-term aggregation is to aggregate user activities on each timestamp t_{a_i}. As explained before, the two forms of user activities display fraud patterns from different perspectives and should be captured together. What's more, in reality, there are dynamics in interaction relationship [14]. Dynamics means that the connection between users and entities will disappear or occur over time (A more detailed description is provided in the supplementary material). However, most of existing fraud detection models aggregates the interactions statically, resulting in insufficient exploration of interaction relationship. Therefore, it is crucial to effectively mine the dynamic graph sequence to fully explore fraud patterns from interaction relationship.

SIFB is designed to solve the above problems. It firstly employs SBAB to aggregate short-term sequential behavior and employs IRAB to aggregate short-term interaction relationship. Then it fuses the outputs of the two modules by FB to obtain the short-term user activity embedding. In this way, each short-term embedding integrates the sequential behavior and interaction relationship of the corresponding timestamp. Moreover, since the interactions of each timestamp is included in the corresponding short-term embedding, the dynamics of interactions can be captured when using CTLSTM to aggregate the embedding sequence.

Sequential Behavior Aggregation Block. SBAB is used to aggregate short-term sequential behavior on t_{a_i}. Firstly, the module extracts features from the short-term sequential behavior. Suppose that $\{b_i, b_{i+1}, ..., b_k\}$ is the sequential behavior of user u happened in timestamp t_{a_i}. SBAB extracts short-term sequential behavior features $s = \phi\{b_i, b_{i+1}, ..., b_k\}$. ϕ is the pre-processing transform. On purpose of take full use of the information for each sequential behavior and extract the most expressive features, this paper chooses ϕ as statistical functions [15]. In this way, s is a set of summary statistics of the sequential behavior. Next, s is digitized to obtain the short-term sequential behavior feature vector v_{seq}. In order to discover more complicated behavior patterns from the extracted features, v_{seq} is mapped to the short-term sequential behavior embedding x_{seq} through a fully connected layer. Equation 1 and Eq. 2 shows this process:

$$v_{seq} = Digitized(s) \tag{1}$$

$$x_{seq} = FC(v_{seq}) \tag{2}$$

Interaction Relationship Aggregation Block. IRAB is used to aggregate short-term interactions on t_{a_i}. Firstly, all the user interactions in t_{a_i} are integrated into an interaction relationship graph, in which each user or entity is a node, an edge between two nodes means that they interact at t_{a_i}, and the edge weight denotes the number of interactions. Because in practice, each neighbor is of different importance to the user, IRAB employs a Graph Neural Network with attention mechanism to aggregate users' neighbors.

Suppose that $h = \{h_1, h_2, ..., h_n\}$ is the set of initial feature vectors of each node. At first, the importance of node j to node i, e_{ij}, is calculated by Eq. 3, where W is a parameter matrix and a is the attention mechanism. Considering that the number of interactions reflects the importance of a neighbor to the user, e_{ij} is multiplied by the edge weight ω_{ij} to obtain the normalized attention coefficient α_{ij}. Then, the linear aggregation of the neighbors for each user is calculated by Eq. 5.

$$e_{ij} = a(Wh_i, Wh_j) \tag{3}$$

$$\alpha_{ij} = softmax(\omega_{ij}e_{ij}) = \frac{exp(\omega_{ij}e_{ij})}{\sum_{k\in N(i)} exp(\omega_{ik}e_{ik})} \tag{4}$$

$$v_{in} = \sigma\left(\sum\nolimits_{j\in N(i)\cup\{i\}} \alpha_{ij}Wh_j\right) \tag{5}$$

v_{in} is the aggregated features of user's interaction relationship. Finally, a fully connected layer is added to discover more complex interaction, and short-term interaction embedding x_{in} is obtained.

$$x_{in} = FC(v_{in}) \tag{6}$$

Fusion Block. FB is used to fuse x_{seq} and x_{in}. In different scenarios, sequential behavior and interaction relationship may have different importance to discover fraud patterns. Therefore, we give different weights to x_{seq} and x_{in}, so that they can be complementary to each other sufficiently. The formula of feature aggregation is given as follows:

$$x_{sh} = \alpha x_{seq} + (1 - \alpha)x_{in} \tag{7}$$

where α is a learnable variable. α is set as 0.5 initially, meaning that the sequential behavior and interaction relationship are given equal initial weights. Which form is more representative for the specific scenario is determined by users' habits and fraud patterns.

3.3 View Aggregation

View aggregation is used to aggregate the user activities of all views $v_1, v_2, ..., v_m$. In the multi-view scenes, fraud patterns may hide in any view or lay in multiple views. At the same time, there are complex correlations between different views. Therefore, considering multiple views at the same time and discovering fraud patterns from these correlated views is crucial for effective fraud detection. To solve the problems, view aggregation is introduced.

For each view v_i, SIFB is utilized to obtain the short-term user activity embedding of that view:

$$x_{sh}(i) \leftarrow SIFB_i \tag{8}$$

where $x_{sh}(i)$ is the short-term user activity embedding of view i. In this way, the short-term user activity embedding of each view can be obtained: $x_{sh}(1), x_{sh}(2), ..., x_{sh}(m)$, m is the number of views. In order to consider multiple views at the same time, the short-term user activity embeddings of each view are stacked together to obtain a 2-dimensional short-term multi-view user activity embedding.

$$X_{sh} = Stack(x_{sh}(1), x_{sh}(2), ..., x_{sh}(m)) \tag{9}$$

In order to extract more complex information from the correlated views, short-term multi-view user activity embedding needs to pass through a Shallow CNN (SCNN) layer. The SCNN is composed of three consecutive operations: a 3×3 Convolution (Conv), Batch Normalization (BN) and a Rectified Linear Unit (ReLU).

$$X_{sh} = SCNN(X_{sh}) \tag{10}$$

3.4 Long-Term Aggregation

Long-term aggregation aims at aggregating user activities of all timestamps $\{t_{a_1}, t_{a_2}, ..., t_{a_l}\}$. Through short-term aggregation and view aggregation for each

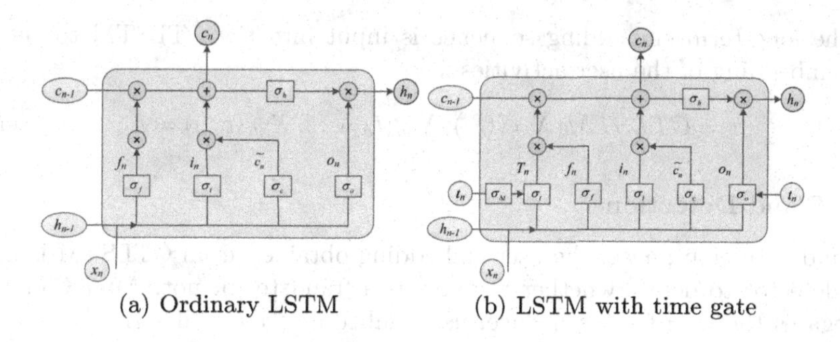

(a) Ordinary LSTM (b) LSTM with time gate

Fig. 3. The recurrent unit structure of ordinary LSTM and LSTM with time gate.

timestamp, a user activity embedding sequence $\{X_{sh}(t_{a_1}), X_{sh}(t_{a_2}), ..., X_{sh}(t_{a_l})\}$ can be obtained. The sequence can be regarded as a time series, where each timestamp corresponds to a 2-dimensional embedding. Since these timestamps are not consecutive, the sequence has irregular time intervals.

Convolutional Time LSTM. CTLSTM is proposed to mine fraud patterns from the long-term embedding sequence. Firstly, a time gate is employed to deal with irregular time intervals. The recurrent unit structure of ordinary LSTM and LSTM with time gate is shown in Fig. 3.

Δt_n is the time interval between the n-th timestamp and the $(n-1)$-th timestamp. T_n is the time gate. As shown in Eq. 15, history information c_{n-1} is filtered by not only the forget gate f_n, but also the time gate T_n, which means that T_n can control the influence of c_{n-1} on current state and therefore capturing the correlation between adjacent user activities. Meanwhile, through the time gate T_n, the time interval information Δt_n is saved in c_n and then transferred to the subsequent states $c_{n+1}, c_{n+2}, ...$. Therefore, T_n can help store long-term time interval information, thus model the irregular time interval of user activities.

In order to deal with the 2-dimensional embedding, convolution is used to replace linear layer in LSTM. The convolution operation can simultaneously capture the dependencies between different feature dimensions and the correlations between different views. So, more in-depth and representative features can be extracted through convolution. Equation 11–Eq. 18 shows the form of CTLSTM, where * denotes the convolution operator and $\circ$ denotes the Hadamard product:

$$i_n = \sigma_i(W_{xi} * X_n + W_{hi} * H_{n-1} + b_i) \tag{11}$$

$$f_n = \sigma_f(W_{xf} * X_n + W_{hf} * H_{n-1} + b_f) \tag{12}$$

$$T_n = \sigma_t(W_{xt} * X_n + \sigma_{\Delta_t}(W_{tt} * \Delta t_n) + b_t) \tag{13}$$

$$\tilde{C}_n = \sigma_c(W_{xc} * X_n + W_{hc} * H_{n-1} + b_c) \tag{14}$$

$$C_n = T_n \circ f_n \circ c_{n-1} + i_n \circ \tilde{C}_n \tag{15}$$

$$o_n = \sigma_o(W_{xo} * X_n + W_{to} * \Delta t_n + W_{ho} * H_{n-1} + b_o) \tag{16}$$

$$H_n = o_n \circ \sigma_h(C_n) \tag{17}$$

$$h_n = flatten(H_n) \tag{18}$$

The long-term embedding sequence is input into the CTLSTM to get the final embedding of the user activities x.

$$x = CTLSTM(X_{sh}(t_{a_1}), X_{sh}(t_{a_2}), ..., X_{sh}(t_{a_l})) = h_{t_l} \tag{19}$$

3.5 Fraud Detection

In fraud detection phase, the user embedding obtained from CTLSTM is input to a detector to detect whether the user is a fraudster or not. After CTLSTM aggregates the long-term sequence, user embedding x encodes the information of the whole sequence. Then, it is input to the detector. This work employs a SoftMax classifier as the detector and the equation is shown as follow:

$$P(\hat{y} = k|x) = \frac{exp(w_k^T x + b_k)}{\sum_{k'=1}^{K} exp(w_{k'} x + b_{k'})} \tag{20}$$

where k is the number of classes. In the fraud detection problem, there are two classes: fraud and normal, so $k = 2$. $\hat{y}$ is the predicted class of the user. w_k and b_k are the learnable parameters, and w_k^T indicates the transpose of w_k.

The loss function of this framework is as follows, where y is the true label of the user, N is the total number of users:

$$L = \frac{-1}{N} \sum_{N} [yln(\hat{y}) + (1 - y)ln(1 - \hat{y})] \tag{21}$$

4 Experiment

4.1 Experimental Setup

Dataset: (1) **Reddit** is an online forum where the users can comment under different reddits. This dataset contains public Reddit comments data from September 2019, including 6000 users and 2943 reddits [16]. Two views are extracted from this dataset: 1) U-R means that user comments under reddit. 2) U-R-U means that two users comment on the same reddit. (2) **Wiki** is an encyclopedia collaborative project which allows users to freely read and edit the content of most pages [7]. This dataset contains records of the user editing the pages between January 2013 and July 2014, including 6,000 users and 52403 pages. Two views are extracted from this dataset: 1) U-P means that user edits Wikipedia page. 2) U-P-U means that two users edit the same page.

User activities in both datasets occur with irregular time intervals. Detailed statistics for the datasets and the information of views extracted from the datasets are shown in Supplementary Material.

Baselines: We compare the performance of our model with six popular fraud detection models:

- GCN (Graph Convolutional Networks) [17]: A neural network which uses convolution to aggregate the neighborhoods on the graphs. The model considers only interaction relationship.

- GAT (Graph Attention Networks) [12]: A neural network which uses the attention mechanism to aggregate the neighborhoods on the graphs. The model considers only interactions.
- EvolveGCN (Evolving Graph Convolutional Networks) [13]: This model adapts the graph convolutional network (GCN) model along the temporal dimension. This model uses only interaction relationship, but it evolves the GCN parameters by RNN to capture the temporal pattern of the graph sequence.
- HetGNN (Heterogeneous graph neural network) [18]: A graph neural network that handles the heterogeneous graphs of multiple types of nodes. It uses only interaction relationship.
- M-LSTM (Multi-source Long-short Term Memory Network) [7]: It uses Multi-source LSTM to model user behavior sequence and encodes each user into a low dimensional user embedding. The model uses only sequential behavior.
- STAGN (Spatial-Temporal Attention-based Graph Network) [19]: It uses a spatial-temporal network to aggregate user activities, and then fed the aggregated result to a 3D convolution network. But it builds spatial relationships based on the similarity of users, without using interactions.

Evaluation Metrics: We choose Precision, Recall, F1-Score, Accuracy and AUC (the area under the ROC curve) to evaluate the fraud detection performance. In fraud detection, we focus more on whether fraud is detected, so Recall is more important than Precision. F1-Score measures the performance of the model on unbalance dataset. Accuracy indicates the percentage of samples that are correctly detected. AUC is used to measure the generalization ability of the model. Among these metrics, F1-score and AUC are the most important for our fraud detection task.

Implementation Details: The hyper-parameters of MSTAN is set empirically as follows: In SIFB, we employ 2 GNN layers in IRAB, 1 linear layer in SBAB, IRAB and FB separately, and the output dimension of the GNN layers and linear layers are all 128. For the SCNN in view aggregation, the convolution kernel is set as 3*3. During long-term aggregation, we utilize 1 layer of CTLSTM, and the number of input channels for each CTLSTM layer is 1, while the hidden channels for each CTLSTM layer is 3. The user embedding dimension is set as 64. The learning rate of Adam is set as 0.01. We test all the assembly of hyper-parameter to ensure the best performance, and the same operation is performed when dealing with other models.

4.2 Experimental Results

Comparison. We repeat the experiment several times for each model and each dataset and then use the best result as the final result. Table 1 shows the fraud detection performance of MSTAN and comparison models. It can be observed that MSTAN outperforms the comparison models on both the datasets in terms of AUC score, Accuracy score, F1-score and Recall. Next, we analyze the performance of these methods in detail.

Table 1. Fraud detection performance of MSTAN and the comparison models.

Model	Reddit					Wiki				
	AUC	ACC	F1	Recall	Precision	AUC	ACC	F1	Recall	Precision
GCN	0.782	0.862	0.604	0.583	0.627	0.818	0.890	0.682	0.657	0.709
MLSTM	0.752	0.836	0.470	0.6452	0.352	0.916	0.884	0.745	0.684	0.819
GAT	0.876	0.882	0.624	0.5914	0.660	0.885	0.890	0.722	0.784	0.669
Evolvegcn	0.872	0.886	0.640	0.6025	0.684	0.939	0.903	0.736	0.798	0.675
Hetgnn	0.886	0.886	0.669	0.693	0.646	0.928	0.912	0.762	0.782	0.742
STAGN	0.922	0.894	0.721	0.717	**0.725**	0.947	0.931	0.802	0.814	0.791
MSTAN	**0.944**	**0.909**	**0.756**	**0.803**	0.7136	**0.969**	**0.953**	**0.862**	**0.868**	**0.857**

Table 2. Fraud detection performance of MSTAN and its variants.

Model	Reddit					Wiki				
	AUC	ACC	F1	Recall	Precision	AUC	ACC	F1	Recall	Precision
MSTAN-ni	0.762	0.864	0.524	0.500	0.551	0.954	0.944	0.830	0.849	0.812
MSTAN-ns	0.902	0.895	0.710	0.698	**0.722**	0.940	0.934	0.805	0.825	0.785
MSTAN-sv	0.912	0.898	0.725	0.731	0.719	0.957	0.946	0.833	0.845	0.822
MSTAN-lstm	0.912	0.891	0.719	0.763	0.681	0.951	0.937	0.826	0.807	0.846
MSTAN	**0.944**	**0.910**	**0.756**	**0.804**	0.713	**0.969**	**0.953**	**0.862**	**0.868**	**0.857**

Both GCN and MLSTM performs bad on the two datasets because they only use one of the two forms of user activities separately, leaving a large amount of information unutilized. Moreover, on the Reddit dataset, GCN performs better than MLSTM, while on the Wiki dataset, GCN performs worse than MLSTM. It is because that on Reddit dataset fraud patterns are mainly hidden in interaction relationship, so GCN that uses interaction relationship performs better than MLSTM that uses sequential behavior. Oppositely, on Wiki dataset fraud patterns are mainly hidden in sequential behavior, so MLSTM performs better than GCN. Because MSTAN uses both sequential behavior and interaction relationship, it performs better than GCN and MLSTM on the two datasets.

GAT, EvolveGCN, and HetGNN are all graph-based methods. GAT employs attention mechanism but ignores the edge weights. EvolveGCN captures temporal pattern but cannot handle irregular time intervals. HetGNN aggregates various type of relations but cannot deal with multiple views simultaneously. Moreover, none of the three considers sequential behavior. So, they perform better than GCN but not as well as MSTAN.

STAGN employs a spatial-temporal network to aggregate both of the spatial information and temporal information of user activities, so it performs well on both datasets. However, due to the limitations of not utilizing interactions and the inability of considering multiple views and handle irregular time intervals, it cannot perform as well as MSTAN.

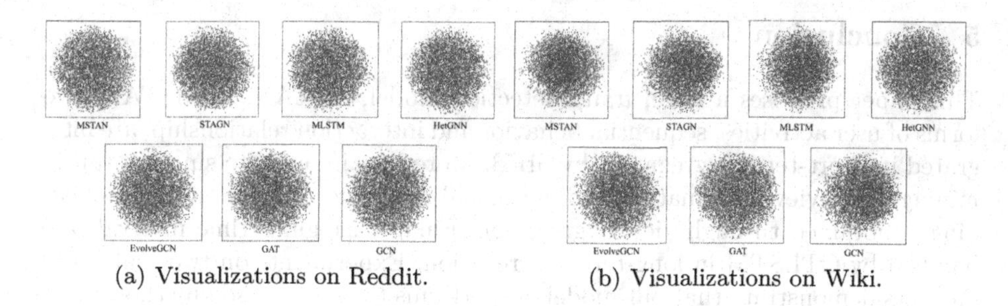

(a) Visualizations on Reddit. (b) Visualizations on Wiki.

Fig. 4. User embedding visualizations on two datasets. Blue points: "Normal Users", red: points: "Fraudsters". (Color figure online)

Ablation. In order to verify the effectiveness of each component employed in our model, ablation experiment is conducted and analyzed in this section. We reconfigure MSTAN to create four variants described as follows:

(1) MSTAN-ns: sequential behavior is not used in SIFB;
(2) MSTAN-ni: interaction relationship is not used in SIFB;
(3) MSTAN-sv: only single view information is considered;
(4) MSTAN-lstm: CTLSTM is replaced with ordinary LSTM.

The experiment results are shown in Table 2. For MSTAN-sv, we chose the result of the best performing view for each dataset. From the results, it is clear that all of SIFB, multi-view information, and CTLSTM help to improve the model's performance.

Visualization. In order to intuitively observe the learning ability of the model, we utilize the t-SNE method [20] to map the learned user embeddings to two-dimensional space and visualize them. Figure 4 shows the visualization of user embeddings given by MSTAN and baseline models. We can observe that the embeddings learned by MSTAN can better separate the blue normal users and red fraudsters.

Firstly, in the visualization of MSTAN, the boundary between blue and red points is the clearest, and the blue and red regions overlaps the least. This suggests that MSTAN could accurately distinguish between fraudulent and normal users. However, in the baselines, a large number of blue and red points are mixed together, indicating that these models could not accurately distinguish fraudsters from normal users. Secondly, the distribution of red points in MSTAN is concentrated. This means that MSTAN can recognize the general characteristics of most samples in the fraud class, and gather them together. However, in baseline models, the distributions of red points are scattered. This shows that they have not learned the major characteristics of the fraud class, so they cannot gather fraudsters together.

5 Conclusion

This paper proposes a novel fraud detection model, MSTAN. In MSTAN, two forms of user activities, sequential behavior and interaction relationship, are integrated in short-term aggregation by SIFB. Moreover, in order to simultaneously mining multi-view information, 2-dimensional multi-view user activities embedding is obtained through view aggregation. Finally, irregular time intervals are handled by CTLSTM in long-term aggregation. Experiments on two real world datasets demonstrate that our model outperforms the comparison methods, and the SIFB, multi-view information, and CTLSTM used in the model contribute to improving fraud detection performance.

References

1. Guo, J., Liu, G., Zuo, Y.: Learning sequential behavior representations for fraud detection. In: ICDM, pp. 127–136 (2018)
2. Li, A., Qin, Z., Liu, R.: Learning sequential behavior representations for fraud detection. In: CIKM, pp. 2703–2711 (2019)
3. Tempelmeier, N., Demidova, E.: Attention-based vandalism detection in OpenStreetMap. In: WWW, pp. 643–651 (2022)
4. Zhu, F., Zhang, C., Zheng, Z.: Click fraud detection of online advertising-LSH based tensor recovery mechanism. In: TITS, pp. 9747–9754 (2021)
5. Hooi, B., Song, H.A., Beutel, A.: Fraudar: bounding graph fraud in the face of camouflage. In: SIGKDD, pp. 895–904 (2016)
6. Hilal, W., Gadsden, S.A., Yawney, J.: Financial fraud: a review of anomaly detection techniques and recent advances. Expert Syst. Appl. **193**, 116429 (2022)
7. Hilal, W., Gadsden, S.A., Yawney, J.: Wikipedia vandal early detection: from user behavior to user embedding. In: ECML PKDD, pp. 832–846 (2017)
8. Liu, Y., Ao, X., Qin, Z.: Pick and choose: a GNN-based imbalanced learning approach for fraud detection. In: WWW, pp. 3168–3177 (2021)
9. Peng, Z., Luo, M., Li, J.: A deep multi-view framework for anomaly detection on attributed networks. IEEE TKDE **34**(6), 2539–2552 (2020)
10. Nyajowi, T., Oyie, N., Ahuna, M.: CNN real-time detection of vandalism using a hybrid-LSTM deep learning neural networks. In: IEEE AFRICON, pp. 1–6 (2021)
11. Benchaji, I., Douzi, S., El Ouahidi, B.: Enhanced credit card fraud detection based on attention mechanism and LSTM deep model. J. Big Data **8**, 1–21 (2021)
12. Velickovic, P., Cucurull, G., Casanova, A.: Graph attention networks. In: ICLR, pp. 1–10 (2017)
13. Pareja, A., Domeniconi, G., Chen, J.: Evolvegcn: evolving graph convolutional networks for dynamic graphs. In: AAAI, pp. 5363–5370 (2020)
14. Skarding, J., Gabrys, B., Musial, K.: Foundations and modeling of dynamic networks using dynamic graph neural networks: a survey. IEEE Access **9**, 79143–79168 (2021)
15. Whitrow, C., Hand, D.J., Juszczak, P.: Transaction aggregation as a strategy for credit card fraud detection. DMKD **18**, 30–55 (2009)
16. Zhao, T., Ni, B., Yu, W.: Action sequence augmentation for early graph-based anomaly detection. In: CIKM, pp. 2668–2678 (2021)
17. Kipf, T.N., Welling, M.: Semi-supervised classification with graph convolutional networks. In: ICLR, pp. 1–10 (2016)

18. Zhang, C., Song, D., Huang, C.: Heterogeneous graph neural network. In: SIGKDD, pp. 793–803 (2019)
19. Cheng, D., Wang, X., Zhang, Y.: Graph neural network for fraud detection via spatial-temporal attention. IEEE TKDE **34**(8), 3800–3813 (2020)
20. Van der Maaten, L., Hinton, G.: Visualizing data using t-SNE. JMLR **9**(11) (2008)

Author Index

D. Yang et al. (Eds.): PAKDD 2024, LNAI 14649, pp. 403–404, 2024.
https://doi.org/10.1007/978-981-97-2262-4